D0991370

THE

INTERNATIONAL SERIES

OF

MONOGRAPHS ON PHYSICS

GENERAL EDITORS

W. MARSHALL D. H. WILKINSON

CRYSTALS WITH THE FLUORITE STRUCTURE

ELECTRONIC, VIBRATIONAL, AND DEFECT PROPERTIES

EDITED BY

W. HAYES

CLARENDON PRESS . OXFORD
1974

Oxford University Press, Ely House, London W.1

GLASGOW NEW YORK TORONTO MELBOURNE WELLINGTON
CAPE TOWN IBADAN NAIROBI DAR ES SALAAM LUSAKA ADDIS ABABA
DELHI BOMBAY CALCUTTA MADRAS KARACHI LAHORE DACCA
KUALA LUMPUR SINGAPORE HONG KONG TOKYO

ISBN 0 19 851277 5

PRINTED IN NORTHERN IRELAND AT
THE UNIVERSITIES PRESS, BELFAST

PREFACE

SOME materials with the fluorite structure provide model systems for the study of a variety of problems in the physics and chemistry of the solid state. Most interest has centred on alkaline earth fluorides since they are stable compounds and large single crystals are easily grown. This monograph is primarily concerned with these materials although others are also dealt with such as strontium chloride, cadmium fluoride, lead fluoride, the oxides of uranium, cerium and thorium and rare-earth hydrides.

Rare-earth ions readily dissolve in alkaline earth fluorides and are sometimes responsible for the colours observed in fluorite (naturally occurring calcium fluoride). Alkaline earth fluorides doped with rare-earth ions figured prominently in the early days of laser research in the solid state. Subsequently much of the interest in these systems centred on the charge-compensation problem arising from replacement of a divalent lattice cation by a trivalent impurity. This complex phenomenon has been more thoroughly studied in fluorites than in other solids.

Alkaline earth halides provide an attractive medium for colour-centre research and rapid strides have been made in our understanding of colour-centre phenomena in these materials in recent years. The work has now advanced to an extent that a general understanding of radiolysis in some fluorites is beginning to emerge. This advance has also contributed significantly to investigations of the phenomenon of photochromism in rare-earth doped alkaline earth fluorides.

Much of this monograph is concerned with electronic and vibrational properties of point defects in alkaline earth fluorides, with rare-earth impurities and colour centres figuring prominently. To provide a general perspective the monograph contains a detailed examination of the thermodynamics of defects and defect interactions in fluorites and also a discussion of crystal-field effects for rare-earth ions in these materials. A full understanding of the properties of point defects must rely eventually on information about the electronic and vibrational properties of the bulk crystals and these aspects are also covered.

The writing of any book benefits from a single author and a uniform presentation. However, this monograph is intended to be of use to people who are actively engaged in research on defects in insulating crystals and it seemed unlikely that a single author with a variety of commitments could properly evaluate the large and growing literature on fluorites in a reasonably short space of time. A team of authors brings rapidity and additional expertise but inevitably there are differences of style. It is part of the work of the editor to reduce the impact of these differences and to try to make the book a cohesive whole. The team of authors who wrote this monograph cooperated fully in this effort.

W. Hayes
Clarendon Laboratory
March 1974

CONTENTS

1

ELECTRONIC PROPERTIES

1.1. Introduction

In this monograph we are considering the properties of some of the chemical compounds with formula RX_2 which crystallize in a lattice similar to that of calcium fluoride (CaF_2 in its naturally-occurring form is known as fluorite). In this chapter we discuss those properties of the pure solid which are dominated by the electrons, in particular the energy levels and the eigenstates of the electrons in the absence of lattice defects or impurities. From these energies derive not only the cohesive energy of the crystal but also its preferred lattice structure. The optical properties of the solid, measured at photon energies above those appropriate to lattice vibrations, are dominated by the transitions of electrons from the occupied core and valence levels to the normally empty conduction states. These latter conduction levels together with the highest valence levels determine the electronic transport properties of the crystal.

Most of the fluorite materials are photo-conducting insulators, but CdF_2 may be converted to a semiconductor by chemical doping. Many of the hydrides of the rare-earth metals are metallic compounds. Some materials, such as uranium oxide and rare-earth hydrides, are paramagnetic at room temperature but ordering of the magnetic moments may be observed at lower temperatures.

The properties of defects in the fluorite compounds must ultimately be understood in terms of the properties of the pure crystal. It will become apparent that our present understanding of the electronic properties of the pure crystals is very inadequate. The extreme difficulty experienced in growing single crystals of high chemical purity and good lattice order has so far prevented the application to any of the fluorite solids of cyclotron resonance or other experimental techniques giving detailed information on their electronic band structure. Some information may, however, be deduced from a study of the structure and the cohesive energy of the fluorite lattice and from the optical and magnetic properties of the various fluorite compounds. Furthermore, useful analogies may be drawn with other compounds of related structure whose electronic energy levels are well understood.

The techniques which are used for the growth of single crystals of the fluorite compounds are summarized in an appendix to the chapter.

1.2. Qualitative properties of the fluorite lattice

The fluorite lattice structure is shown in Fig. 1.1. In a compound RX_2 each ion of species R is surrounded by eight equivalent nearest-neighbour ions of species X forming the corners of a cube of which R is the centre. Each ion of species X is surrounded by a tetrahedron of

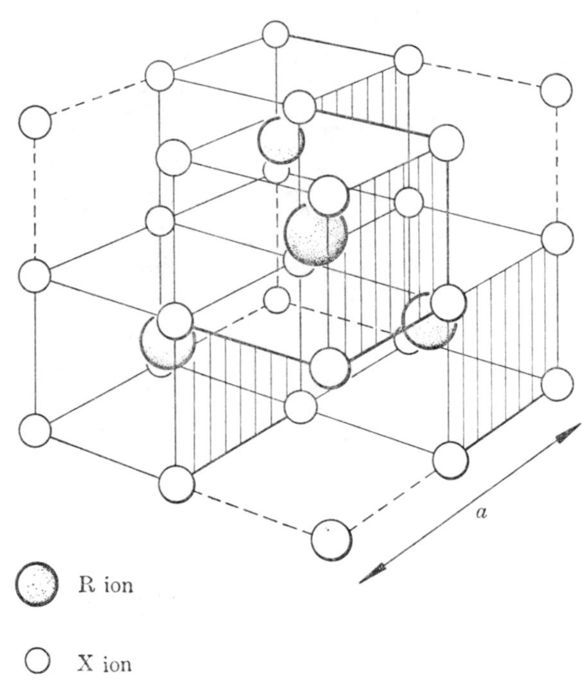

FIG. 1.1. The structure of the fluorite RX_2 lattice. 'a' is the lattice parameter of the conventional cubic unit cell. For clarity, the X ions are drawn smaller than the R ions.

four equivalent R ions. More fundamentally the structure has a face-centred-cubic translational group and a space lattice of symmetry O_h^5. If the structure is interpreted in terms of a primitive cube of side a, as marked in Fig. 1.1, it comprises three inter-penetrating face-centred-cubic lattices. The first is a lattice of species R with its origin at the point (0, 0, 0) and with primitive translational vectors of $(0, \frac{1}{2}a, \frac{1}{2}a)$; $(\frac{1}{2}a, 0, \frac{1}{2}a)$; and $(\frac{1}{2}a, \frac{1}{2}a, 0)$ in the cube of side a. The X species are located on two further lattices with similar translational vectors but with origins at $(\frac{1}{4}a, \frac{1}{4}a, \frac{1}{4}a)$ and at $(\frac{3}{4}a, \frac{3}{4}a, \frac{3}{4}a)$. The site of the R ion has O_h symmetry and the site of the X ion has T_d symmetry. The interstitial site again has O_h symmetry, being at the centre of a cube of eight X ions. The crystal is not piezoelectric.

It is apparent that the fluorite structure provides close contact between the different species of atom or ion. Furthermore if the ions of species R are sufficiently large, close contact between the ions of species X is prevented. If the constituent species are regarded as hard spheres with radii of $r(\mathrm{R})$ and of $r(\mathrm{X})$, contact occurs between the R and X ions to the exclusion of X–X contact and of R–R contact (Wyckoff 1965) when the radii satisfy the condition

$$4{\cdot}45 \geqslant \frac{r(R)}{r(X)} \geqslant 0{\cdot}73.$$

The possibility of close contact between dissimilar ions suggests that the fluorite structure will be favoured by those strongly ionic compounds with formula RX_2 which possess large ions of type R. Study of a

TABLE 1.1

The 'crystal radii' of certain ions, after Zachariasen (1967) (coordination corrections are neglected)

Be^{2+}	0·30 Å	Th^{4+}	0·99 Å
Mg^{2+}	0·65 Å	F^-	1·33 Å
Ca^{2+}	0·94 Å	Cl^-	1·81 Å
Sr^{2+}	1·10 Å	Br^-	1·96 Å
Ba^{2+}	1·29 Å	I^-	2·19 Å
Zr^{4+}	0·77 Å	O^{2-}	1·46 Å
Ce^{4+}	0·92 Å		

self-consistent table of ionic radii, such as that of Zachariasen (1967) summarized in Table 1.1, shows that one is unlikely to find a hypothetical compound in which contact between R ions could occur. This would require that the R ions be exceptionally large, with

$$r(\mathrm{R}) > 4{\cdot}45r(\mathrm{X}).$$

In fact, the R ions are normally relatively small and it is possible to find several series of compounds in which one is able to pass through the other limiting value, enabling contact between X ions to occur. For example, among the halides of cadmium or of barium one finds that the fluorite lattice structure occurs for the smaller halide (X) ions while the iodides possess orthorhombic or sheet-like structures.

The materials found to possess the fluorite structure have been divided by Wyckoff (1965) into four main classes:

(1) The halides of the larger divalent cations (all but two are in fact fluorides)
e.g. CaF_2, SrF_2, BaF_2, CdF_2, PbF_2, $SrCl_2$, $BaCl_2$, and HgF_2.
(2) The oxides of certain large quadrivalent cations
e.g. ZrO_2, CeO_2, ThO_2, and UO_2.
(3) The oxides and other chalcogenides of alkali and other univalent cations
e.g. K_2O, Na_2S, and Li_2Se.
(4) Miscellaneous intermetallic compounds and the oxyfluorides and hydrides of rare-earth elements.

In this book we consider primarily compounds of Class 1 but some attention will also be given to materials in Classes 2 and 4. In the materials of the first class, the absence of $CaCl_2$, of bromides and of iodides may be interpreted in terms of a violation of the radius requirement, the anions being relatively too large. It is apparent by inspection of Table 1.1 that close contact between R and X ions cannot occur for the light cations. Indeed the fluorite structure is stable only at high temperatures in $BaCl_2$ and in ZrO_2. In compounds such as CdI_2 which violate the radius criterion we find sheet-like structures which approach close packing of the iodine ions. Considerable covalent bonding occurs and the smaller cations fill the interstices. A systematic study has yet to be made of the degree of ionicity of the RX_2 compounds, but as Phillips (1970) explains, the large forbidden energy gap between the valence and the conduction bands of the alkaline earth fluorides suggests a strong ionic bond, consistent with the presence of the fluorite structure.

The compounds in Class 1 possess ions of well-defined valence and have a well-defined stoichiometry. In contrast, the valence of the uranium ion is ill defined and uranium oxides UO_x exist for values of x ranging from 2·0 to 3·0. UO_2 possesses the fluorite structure but must be kept in an oxygen-free environment to prevent a gradual oxidation. As the value of x increases from 2 to 2·25 the extra oxygen ions fill randomly the empty interstitial sites in the fluorite lattice. Ultimately when x is 2.25 the added oxygen atoms form an ordered array and U_4O_9 possesses a very complex body-centred cubic structure (Willis, 1963*a*). Annealing at 1500°C in a hydrogen atmosphere may slightly reduce the surface layers of specimens of UO_2, but the bulk of the material is not reduced below $x = 2·0$ (Matzke and Ronchi, 1972).

Distortions of the true fluorite structure may occur with change in temperature. For example, a study of the diffraction of neutrons by UO_2 (Willis, 1963*b*) showed a gradual displacement of the oxygen ions in a $\langle 1, 1, 1 \rangle$ direction towards the empty interstitial site. This can easily be interpreted in terms of anharmonic vibrations of the oxygen ions in their tetrahedral environment. The anharmonicity of anion vibrations is discussed in § 2.4.2.

Considerable study has been made of the hydrides of the rare-earth metals. A review of their properties is given by Libowitz (1965, 1972). All react readily with oxygen and with moisture. The dihydrides EuH_2 and YbH_2 have an orthorhombic structure but the dihydrides of all the other elements from lanthanum to lutecium inclusive, possess the fluorite structure. The stoichiometry of these compounds with the fluorite structure is not well defined. Vacancies may occur in the hydrogen lattice and excess hydrogen may be incorporated in the body-centre interstitial sites of the lattice. It is not clear whether the excess hydrogen is located in a regular array in the available interstitial sites. The dihydrides of La, Ce, Pr, and Nd retain the fluorite structure as additional hydrogen dissolves up to the composition RH_3, but other rare-earth dihydrides change from the fluorite structure to a hexagonal structure as the trihydride composition is approached. The composition ranges of the different structures are given in Table 1.8. Most of the rare-earth ions prefer to exist in a trivalent state but the ions Eu^{2+} and Yb^{2+} are particularly stable, possessing respectively a half full and a completely full 4f shell of electrons. The compounds EuH_x and YbH_x therefore possess a stoichiometry better defined than that of the other rare-earth hydrides. EuH_3 has not been prepared yet. A compound of composition $YbH_{2\cdot55}$ has been prepared, but is stable only under a high pressure of hydrogen (Warf and Hardcastle, 1961).

Certain structural and thermodynamic properties of materials with the fluorite structure will be found in Table 1.2.

1.3. The cohesion of the fluorite lattice

A study of Table 1.2 will show that the compounds with the fluorite structure have considerable thermal stability. The oxides of quadrivalent cations are refractory materials with melting points ranging from 2600 °C to 3050 °C. The alkaline earth fluorides have melting points near 1250°C. The lowest true melting point, that of PbF_2, occurs at 855 °C. It is obvious that the lattice is stable and strongly bound and we wish to discuss in this section the origin of this binding

TABLE 1.2

Certain properties of crystals with the fluorite structure

Crystal	Molecular weight[a]	Specific gravity[a]	Lattice constant[b] Å	Heat of formation[a] ΔH kcals/mole	Approximate melting point[a] °C	Madelung potential at +ve ion[d] Volts	Madelung potential at −ve ion[d] Volts
CaF_2	78·08	3·18	5·4629	286·26	1360	−19·96	+10·74
SrF_2	123·63	4·24	5·7996	288·89	1450	−18·79	+10·10
BaF_2	175·36	4·83	6·2001	287·70	1280	−17·59	+9·45
CdF_2	150·41	6·64	5·388	172·50	1100	−20·20	+10·85
PbF_2	245·21	8·24	5·927	159·40	855	−18·38	+9·89
HgF_2	238·61	8·95	5·54		(645)	−19·68	+10·58
$SrCl_2$	158·54	3·05	6·977	197·85	873	−15·62	+8·39
$BaCl_2$	208·27		7·34	207·55	962	−14·85	+7·99
ZrO_2	123·22	5·6	5·07		2715	−43·0	+23·15
CeO_2	172·13	7·3	5·411	234·95	2600	−40·28	+21·67
ThO_2	264·12	10·03	5·600	330·9	3050	−38·94	+20·94
UO_2	270·07	10·9	5·468	256·63	2805[c]	−39·86	+21·40
CeH_2	142·13		5·590				

[a] From Hodgman (1959)
[b] From Wyckoff (1965), being the length at room temperature of one edge of the conventional cubic unit cell, i.e. twice the separation of nearest X ions in RX_2
[c] From Hausner (1965)
[d] Madelung potentials calculated from the lattice spacing of column 3, assuming point ions with the full electronic charge.

energy. Full quantum mechanical treatments of the binding of a many electron system are not yet practicable and it will be necessary to use a semi-empirical approach in which the crystal is built up of well-defined ions which interact with each other by forces which are weak relative to their internal binding energies. Most of the calculations on the fluorite compounds have been made on CaF_2 but some calculations are available for SrF_2, BaF_2, and UO_2. Our discussion will concentrate on the alkaline earth fluorides.

We have seen in the Introduction that the existence of the fluorite structure suggests that the lattice is built of charged ions held together by Coulomb attraction. Phillips (1970) has emphasized that the presence of a large forbidden band gap, as observed in the alkaline earth fluorides, is a sign of a strong ionic contribution to the binding energy. Strong Reststrahl absorption associated with the transverse optical phonons occurs in the infrared region (see § 2.2.2).

The observed separation in energy of longitudinal optical and transverse optical phonon modes also suggests the presence of charged ions. It will be shown later that the magnetic properties of UO_2 are consistent with those of a lattice containing U^{4+} ions. Nevertheless the strongest evidence that the alkaline earth fluorides comprise an assembly of singly- and doubly-charged ions, with little covalent bonding, comes from the comparison of the observed cohesive energy with that calculated on the assumption of an ionic lattice. The agreement between the observed and the calculated energy is close. Such a calculation assumes an assembly of distinguishable and recognizable ions. This assumption may be tested by a study of the electronic distribution in the unit cell.

1.3.1. *The distribution of electrons in the unit cell*

In principle, the density of electrons at all points in the unit cell may be determined by a study of the diffraction of X-rays by the crystal. Such studies have been made on CaF_2 by Togawa (1964), by Weiss, Witte, and Wölfel (1957) and by Warren (1961). Typical measurements are shown in Fig. 1.2, (from Witte and Wölfel, 1958). Certain general features may be observed. The crystal is built of well-defined atoms or ions with a low density of electrons between them. No evidence can be seen of a strong covalent bonding which would produce bridges of high electron density between the atoms. The available measurements have been analysed by Maslen (1967) who showed that it is very difficult to distinguish from the X-ray data alone between a crystal built of atoms and a crystal built of ions. The scattering form factor is dependent

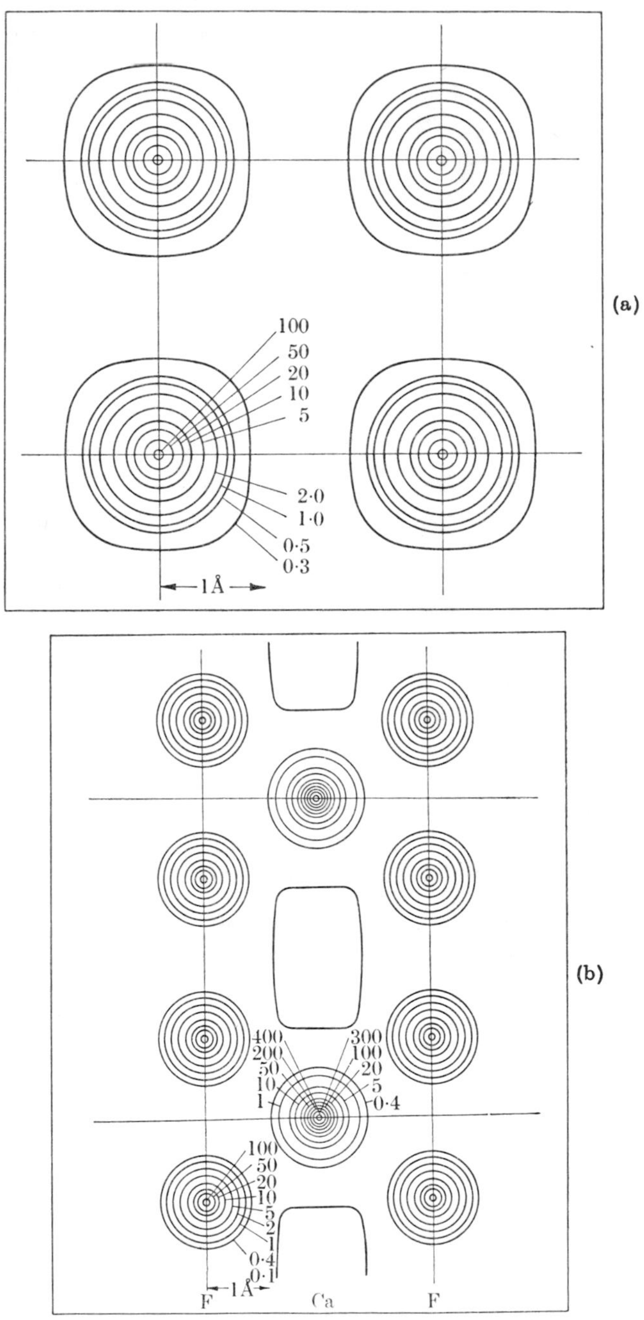

FIG. 1.2. The density of electrons in CaF_2 (after Witte and Wölfel (1958)). a—the density in the $(x, y, \frac{1}{4})$ plane which passes through the fluorine ions. b—the density in the (x, x, z) plane which passes through both the calcium and the fluorine ions.

not only on the number of electrons in the ion but also on their radial distribution within the ion. This distribution is not well known for an ion in a crystal.

Maslen calculated the form factor for the scattering of X-rays using three different models of the CaF_2 crystal. The crystal is built by the superposition, without orthogonalization, of the electronic distributions of the following ionic species:

(a) The 'Free-ion' Model. The Hartree-Fock wavefunctions calculated for Ca^{2+} and for F^- ions existing in free space are used.
(b) The 'Crystal-ion' Model. The wavefunctions of the Ca^{2+} and F^- ions are recalculated, adding the electrostatic Madelung potential present in the fully ionic crystal.
(c) The 'Neutral-atom' Model. An angular average is taken of the Hartree-Fock wavefunctions of the Ca and F atoms calculated in free space.

For each of the three models, the full form factor for the scattering of X-rays was calculated with appropriate corrections for lattice vibrations in the harmonic approximation and for anomalous dispersion. The electronic density, $\rho(r)$, of the neutral atom and of the free ion were almost identical near to the nucleus. Differences occurred only at large values of the radius r. The form factor for forward scattering by the free ions will be different from that for the free atom, being proportional to the number of electrons which are present. However, even if forward scattering could be studied experimentally, the addition of the Ca and the F intensities would eliminate any distinction between an atomic and an ionic crystal. On the other hand when the form factor for large-angle scattering is calculated for the individual atoms or ions the outer parts of the electronic distributions will tend to be unimportant because of phase cancellations. Useful information can therefore be obtained only from scattering of low but non-zero order.

Maslen's wavefunctions for the 'crystal ions' differ considerably from those of the 'free ions'. The Ca^{2+} ion is placed in a negative potential resulting in a more diffuse wavefunction. The electronic density is reduced near the nucleus but increased at a large radius. Maslen was able to show that his 'crystal-ion' wavefunctions were incompatible with the available measurements. The final choice, namely between a crystal made of free atoms and a crystal made of free ions could only be made from the observed intensity of that (2, 0, 0) reflection in which the calcium and fluorine contributions were in antiphase. For no other reflections were the differences between the models large enough to be significant. The model of a crystal built up of 'free ions' was the only one able to fit the measurements. It is clear that the distinction between the various models is not made as sharply as one might wish. It is not obvious why little success was found with the lattice built of 'crystal ions'. Further and more extensive calculations of the wavefunctions in the solid would be valuable. It will be clear, as Maslen (1967) emphasizes, that attempts such as that of Kurki-Suonio and Meisalo (1966) to use X-ray diffraction measurements to deduce information concerning distortions of the ions from a spherical form are highly speculative.

In conclusion we have seen that the available X-ray studies do not give an accurate measure of the charge on the calcium or the fluorine ion. Detailed studies of the infrared absorption by the lattice would not necessarily be of value as the effective oscillating charge responsible for infrared absorption or for polaron phenomena bears no simple relation to the static charge assumed in a calculation of cohesive energies. We will see however in the following section that if the cohesive energy of the lattice is calculated on a simple ionic model good agreement is found with the measured lattice energy. The dominant term is in fact the attractive electrostatic or Madelung energy calculated on a point-ion picture. One is led to the conclusion that the simple model of a solid built of ions whose charges are integral multiples of the electronic charge is a good approximation for the alkaline earth fluorides.

1.3.2. *Experimental determination of the cohesive energy*

We will define the lattice energy U of a solid compound RX_2 as the energy required to form the solid, at a defined temperature, from its constituent ions which are assumed to be infinitely separated from each other. If the solid is stable, U will be a negative quantity and hence be cohesive. The lattice energy may be determined experimentally from the Born–Haber thermodynamic cycle using the equation

$$-\Delta H = U+D+L+I-2E.$$

ΔH is the heat emitted in the formation of the solid compound RX_2 from the respective elements (see Table 1.2). D and L are respectively the heats of atomization of the electropositive and of the electronegative elements. I is the sum of the first and the second ionization potentials of the metal (R). E is the electron affinity of the electropositive element (X).

The knowledge of these parameters enables experimental values of the lattice energy U to be deduced (see Table 1-3).

1.3.3. *Empirical calculation of the cohesive energy*

We may think qualitatively of ions attracted by simple electrostatic and by van der Waals' forces, but prevented from collapsing together by hard, short-range repulsive forces between the ion cores. Two approaches are then used in attempts to calculate the nett binding energy. We concentrate in this section on the use of semi-empirical methods but will first consider the possibility of a fully quantum mechanical treatment. This latter may be regarded as a full calculation of the electronic energy levels performed from first principles. Such a calculation, discussed for instance by Löwdin (1956), attempts to solve the Schrödinger equation for the many-particle problem using self-consistent techniques and estimating the various forms of exchange and correlation energy which occur as the ions begin to overlap. No such calculation has yet been completed for any fluorite compound.

In the second and semi-empirical approach developed by Born, Mayer, and others and reviewed by Tosi (1964) the interactions between the ions are approximated by theoretically reasonable empirical force relations. The necessary parameters such as ionic 'radii' and the repulsive force constants are deduced by comparison with measurements of the lattice constant and of the compressibility as a function of pressure. It is necessary for simplicity to assume that the forces between ions are of a two-body type and also that they are central forces.

In order of descending magnitude the terms entering the empirical calculation are (a) the electrostatic attraction between the charged ions, (b) the short-range repulsive interaction between the ion cores, (c) the Van der Waals attraction between polarizable ions and (d) the vibrational energy of the lattice. These various contributions to the lattice energy are listed in Table 1.3 for certain crystals with the fluorite structure and are discussed below.

(a) *The Madelung energy* (the electrostatic interaction between charged point-ions which we will assume to possess integral electronic charges). In the fluorite lattice the nearest neighbours of any given ion are of the opposite polarity and a strongly attractive potential exists. We will define the Madelung potential as the energy of a unit point charge inserted at a defined place in the crystal. Several ways

TABLE 1.3

The major contributions to the cohesive lattice energy of certain fluorite crystals. All energies are expressed in kcals per mole

Material	CaF_2[a]	CaF_2[b]	SrF_2[b]	BaF_2[b]	ThO_2[c]	UO_2
Electrostatic energy	−710	−711·5	−670·8	−627·4		
Van der Waals' energy	−19	−21·2	−21·6	−22·7		
Repulsive energy	+98	+102·4	+102·3	+94·3		
Vibrational energy	+6	+3·2	+2·4	+1·9		
Calculated lattice energy	−625	−627·1	−587·7	−553·8	−2373	−2360
Experimental lattice energy[d]	−617		−584	−549	−2413	−2461

[a] Reitz, Seitz, and Genberg (1961).
[b] Benson and Dempsey (1962).
[c] Benson, Freeman, and Dempsey (1963).
[d] Harries and Morris (1959).

exist of calculating this potential in terms of the known potentials of other and simpler lattices. One technique is to consider the fluorite structure as built by the superposition of NaCl and of CsCl lattices with appropriate lattice parameters. The regular cubic array of Cl ions in CsCl will provide the fluorine ions of the CaF_2 lattice. Let us therefore take a crystal of CsCl whose Cl to Cl spacing equals the F to F spacing of CaF_2. To convert the CsCl to CaF_2 we must eliminate alternate Cs ions along $\langle 1, 0, 0\rangle$ directions and double the charge of those which remain. This may be achieved by superimposing an NaCl lattice whose Na to Cl spacing equals the Cs to Cs spacing of CsCl. All of the Na and Cl ions are therefore placed on Cs ions, doubling the charge of some and eliminating the charge of others. The Cl ions of the original CsCl are not affected by this process. The Madelung potentials in CaF_2 may then be calculated from the known potentials of the NaCl and the CsCl structures (Tosi, 1964).

Let us define the potential ϕ at a point in the lattice as

$$\phi = \frac{Ae,}{a}$$

where A is the Madelung constant, e is the electronic charge, and a is the side of the conventional cube (being twice the separation of

nearest-neighbour X ions). We then find the values of the Madelung constant given below for an undistorted lattice:
At a Ca^{2+} ion site (in the absence of that ion) $A = -7{\cdot}56582$.
At a F^- ion site (in the absence of that ion) $A = +4{\cdot}07070$.
At an interstitial site (body centred with respect to the X lattice) $A = -0{\cdot}57558$.

It is easy to show that the fluorite lattice may be decomposed in other ways e.g. into the sum of a CsCl lattice and a zinc blende lattice.

(b) *The van der Waals attraction.* The instantaneous dipole or higher-order moments present on one ion will polarize neighbouring ions and an attraction will then exist between the moments. The lowest-order term is that between dipoles, giving an energy varying as the inverse sixth power of separation. Higher-order terms will decrease faster with increase in separation. Only nearest neighbours of R ions and nearest and next-nearest neighbours of X ions need be considered. The validity of such expansions is questionable at the small ionic separations actually present in the solid but the energies involved are only about 3% of the cohesive energy, as shown in Table 1.3. The relevant matrix elements and excitation energies may be deduced from the polarizability of free ions and from the ultraviolet absorption by the crystal. Reitz, Seitz, and Genberg (1961) included only the dipole-dipole interaction while Benson and Dempsey (1962) considered in addition the dipole-quadrupole energy.

(c) *The repulsion between ion cores.* In the Born–Mayer treatment of lattice energies, the repulsive energy U_{ij} between ion i and ion j with 'radii' r_i and r_j, separated by distance R_{ij} is assumed to have the form

$$U_{ij} = b\,\exp\{r_i+r_j-R_{ij})/\rho\}.$$

In the fluorite lattice, it is normally sufficient to consider the interaction of an R ion with its first and second neighbours and of an X ion with three shells of neighbours. Further terms are of negligible magnitude. The ionic 'radii' r_i and r_j are found by a study of the lattice spacings of a series of crystals with a given structure. Several different, but internally self-consistent, sets of 'radii' have been reported. The parameters b and ρ are assumed to be the same for R–R as for R–X and X–X interactions and are deduced by detailed comparison of the calculated and the measured equation of state of the crystal. Full details of such calculations are found in the review by Tosi (1964). Specific calculations on CaF_2 are described by Reitz *et al.* (1961), and calculations on several fluorite crystals are given by Benson and Dempsey (1962) and by Benson, Freeman, and Dempsey (1963).

The relatively small vibrational energy is readily calculated from the phonon spectrum of the solid (see Chapter 2).

The results of such empirical calculations of the cohesive energy of the alkaline earth fluorides and a summary of such calculations in ThO_2 and on UO_2 are shown in Table 1.3. The agreement between the calculated and experimental cohesive energies (Harries and Morris, 1959) is seen to be good and justifies our original assumption of a crystal built of weakly interacting ions whose charges are close to integral multiples of the charge of an electron. Detailed calculations on other materials with the fluorite structure would be valuable. It may be noted that the absolute magnitude of the calculated lattice energy of the alkaline earth fluorides exceeds the experimental value. The sign of the difference is the same as that found in the most ionic alkali halides. An error of the opposite sign, as in UO_2, is found in the less ionic of the alkali halides.

1.4. The electronic energy-band structure of the fluorite compounds

Our discussion above of the cohesive energy of the fluorite lattice included implicitly an average over the energies of all the electrons in the solid. Let us now consider more closely the symmetry and the energy of the eigenstates available for electrons in the crystal. We will consider not only the normally full core and valence states but also the normally empty conduction levels. At the time of writing, no detailed calculations of these energy levels have been completed for any solid with the fluorite structure. Some experimental measurements relevant to the band structure have been made and can be interpreted within the constraints set by the symmetry of the lattice structure and the known properties of the constituent ions.

The absolute energies of electrons present in the solid may be studied by measurements of photo-emission. The solid is illuminated with ultraviolet or X-ray photons of well-defined energy and the distribution in energy of the externally emitted electrons is measured. Detailed studies of the highest filled valence bands and of the lowest conduction bands can be made by study of the electrical transport processes of the solid. All the fluorite materials are strongly polar and allowance must be made for the coupling of an electron or a hole to the polarization of the lattice. If this coupling is relatively weak, as is expected for the conduction electrons, we find the well known 'Large Polaron' phenomena, familiar in the alkali and silver halides (Hodby, 1972). When the coupling to the lattice is strong the carrier may be self-trapped in a potential well created by the motion of nearby ions. Such effects are seen in the 'self-trapped hole' well known in the alkali halides and in the alkaline earth fluorides (§ 4.6.1). Any detailed study of the electronic transport properties of a solid requires a high degree of chemical purity and of crystalline perfection. The high melting points of the fluorite materials have made the achievement of purity and crystal perfection much more difficult than in the alkali halides. Detailed and useful transport measurements relevant to the band structure have been made only on CdF_2. The development of materials whose mobility exceeds 10^4 cm^2 V^{-1} s^{-1} will enable detailed studies of the band extrema to be made by techniques such as cyclotron resonance.

Most of our information on the energy levels of electrons in the fluorite compounds has been obtained from measurements of the dielectric function ϵ as a function of frequency. Measurements have been made by study of the interband optical absorption and reflectivity and also by study of the energy loss spectra of fast electrons traversing thin films of the material. The latter gives a value of $\mathrm{Im}(\epsilon^{-1})$. The forbidden band gap of the fluorite materials is large, being about 12 eV for CaF_2 and above 7.6 eV for CdF_2 (see Table 1.7). The optical measurements are therefore difficult. Furthermore the interpretation of the measurements is not easy. An electron is excited from a filled core or valence state to an empty state in the conduction bands or the continuum levels. The excited electron is often associated in space with the empty state from which it was excited. An exciton of 'core' or 'valence' type is thereby made. The relative oscillator strengths of the excitonic and the simple interband transition depend on several factors. When the electronic dielectric constant is small and the relevant band masses are high, as in the fluorite materials, the exciton will have a small radius. The oscillator strength of the optical transitions will then be concentrated on the generation of excitons to the exclusion of simple interband transitions (Elliott, 1957). Studies of the interband magneto-optical effects of free carriers are then difficult. In principle, information on the effective masses of electrons and of holes may be deduced from the binding energies of the excitons. No experimenters have yet been able to resolve the exciton spectra of a fluorite compound sufficiently well to enable an unambiguous interpretation to be made. The reasons are not clear. The exciton spectra may be broadened by strain or by impurities in the lattice. Dynamic broadening by the zero-point motion of the lattice should not be serious. Even if high resolution of the spectra could be achieved, the interpretation of the measurements would be difficult. The initial state involved in the optical absorption process is well defined. Unfortunately this is not the case for the final state because of the effects of lattice relaxation. Two extreme cases are possible. In the first, the optical transition is assumed to occur in a time too short for any lattice relaxation to occur. The relevant effective mass of the hole and of the electron will then be the band mass and the relevant dielectric constant will not exceed the optical dielectric constant. In the other extreme case, when lattice relaxation is able to occur, the relevant masses will be the heavier polaron masses and the dielectric constant may approach the static value, ϵ_s. A tentative analysis of these effects is at present possible only in the thallium halides and a comparison of the exciton data of Kurita and Kobayashi (1971) with the cyclotron resonance measurements of Hodby, Jenkin, Kobayashi, and Tamura (1972) shows that both the reduced mass of the exciton and the effective dielectric constant are intermediate between their respective limiting values. It is thus difficult to interpret measurements on excitons to obtain information relevant to the band structure of the host lattice.

The dielectric constants of the alkaline earth fluorides are small (see Table 2.2) and the band masses of the electrons are likely to lie between 0·5 and 1 free-electron masses. The orbital frequency of the electron about the hole is much larger than the characteristic phonon frequencies. This would suggest that the screening of the electron-hole interaction by polarization of the lattice will be small. In this case it is natural to use the optical dielectric constant ϵ. The radius of the exciton is found to be close to the interatomic spacing of about 2·5 Å. It then becomes logical to interpret the optical spectra in terms of tightly-bound Frenkel excitons rather than the Wannier model widely applied in the common semiconductors (Rubloff, 1972).

Let us now consider how we might calculate the electronic band structure of a fluorite compound. We will commence with a lattice of ions fixed in space at the appropriate positions. When considering the potential seen by an electron near any one nucleus, the effect of the other ions will be small. A good approximation will be made by adding the relevant Madelung potential. Those electronic eigenstates which form the core levels of the ion and are spatially concentrated near the nucleus are therefore raised or lowered in energy by the Madelung potential with respect to their energies in a free ion. In contrast, the wavefunctions of the outer electrons will overlap with neighbouring ions. For these electrons it is necessary to consider the dependence of their energy on the crystal wave vector **q**. In the consideration of the core and valence levels of a solid containing N electrons, a common potential function may be assumed for all filled levels. This potential will also be appropriate for the study of excitons. When we proceed to the study of the normally empty conduction states a new potential should be used, that appropriate to a solid with $N+1$ electrons, thereby eliminating any interaction between holes and the conduction electrons.

1.4.1. *The core and valence bands*

In Table 1.4 we see the electronic configurations of the ions found in the fluorite compounds considered here. The energies of the electrons in the highest filled states of the positive ions are known from their optical spectra (Moore, 1970). The ionization energies of the negative halide ions have been measured by Bailey (1958) from the equilibrium constant of chemical reactions, and that of the negative oxygen ion has been calculated by Morris and Schmeising (1958) by extrapolating along its isoelectronic sequence.

As a first approximation to the true crystal let us assemble the constituent ions, ignoring the overlap between them. The energies of the core and valence levels will be raised or lowered by the relevant Madelung potential given in Table 1.2. In Fig. 1.3 we display the resulting energies of the highest filled states of the doubly-charged metal ion and of the F^- ion in four fluorite crystals. Let us first consider the results for CaF_2. The outermost filled shell of the Ca^{2+} ion, the $3p$ shell, is raised from -51.2 eV with respect to the vacuum level to -31.25 eV. The outermost fluorine-ion level initially at -3.59 eV drops to -14.3 eV and will form the basis of the highest filled valence band of the crystal. The Ca^{2+} levels are well separated in energy from this F^- ion level and little admixture is expected. We may therefore anticipate a valence band with maximum at the centre of the Brillouin zone, as found in those alkali halide crystals whose valence band is of pure halide-ion character. In contrast to CaF_2 a more complex valence band is probable in CdF_2, PbF_2, and HgF_2. In these materials a near degeneracy exists between the outermost filled states of the positive and negative ions. In AgBr, a crystal which also possesses a face-centred-cubic lattice a similar degeneracy produces a valence band with its

TABLE 1.4

Electronic configurations of ions. The letter e marks those shells which contain the additional electrons found in the atom

Ion	1*s*	2*s*	2*p*	3*s*	3*p*	3*d*	4*s*	4*p*	4*d*	4*f*	5*s*	5*p*	5*d*	5*f*	6*s*	6*p*	6*d*	7*s*
Ca^{2+}	2	2	6	2	6		*e*											
Sr^{2+}	2	2	6	2	6	10	2	6			*e*							
Cd^{2+}	2	2	6	2	6	10	2	6	10		*e*							
Ba^{2+}	2	2	6	2	6	10	2	6	10		2	6			*e*			
Ce^{4+}							2	6	10	*e*	2	6			*e*			
Hg^{2+}							2	6	10	14	2	6	10		*e*			
Pb^{2+}							2	6	10	14	2	6	10		2	*e*		
Th^{4+}											2	6	10		2	6	*e*	*e*
U^{4+}											2	6	10	2	2	6		*e*
H^-	2																	
F^-	2	2	6															
Cl^-	2	2	6	2	6													
O^{2-}	2	2	6															

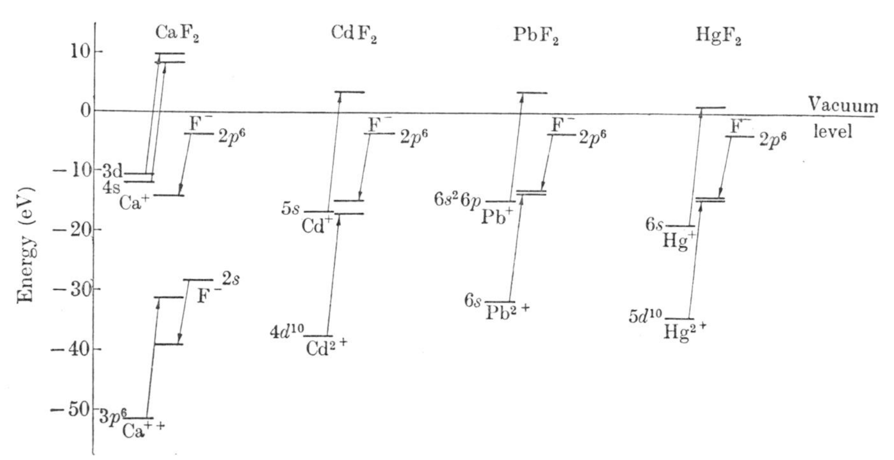

FIG. 1.3. The energy levels of electrons in certain fluorite compounds are approximated by the addition of the Madelung potential (arrowed) to their energy levels in the free ions.

maximum at L, on the [1, 1, 1] face of the Brillouin zone shown in Fig. 1.4. Those optical transitions in which electrons are excited from the valence band must therefore be interpreted with caution in CdF_2, PbF_2, and HgF_2. Lower valence states will involve the 2*s* states of the F^- ion whose energy has been calculated by Hayes, Koch, and Kunz (1971) to be about 24 eV below that of the 2*p* levels.

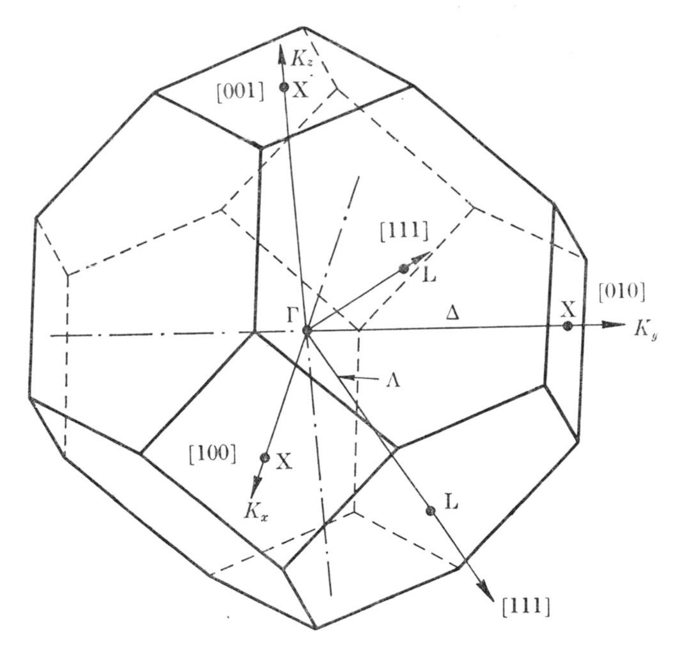

FIG. 1.4. The first Brillouin zone of the face-centred-cubic fluorite lattice.

Measurements of the X-ray stimulated photo-emission of certain fluorite compounds have been reported by several authors (Jorgensen, 1971; Jorgensen, Berthou, and Balsenc, 1971; Jorgensen, 1972; Chadwick and Graham, 1972; and McGuire, 1972). The measured binding energies of electrons in certain cation and fluorine-ion levels are given in

TABLE 1.5

The measured and calculated binding energies of core electrons in certain fluorite compounds

Compound	Level	Measured binding energy (eV)	Free ion binding energy (eV)	Madelung potential V	Calculated binding energy (eV) cf. § 1.4.1
CaF_2	Ca^{2+} $3p$	33·4[b]	50·91[e]	−19·96	30·95
	F^- $2p$	16·15[b]	3·59[d]	−10·74	14·33
	F^- $1s$	692·7[c]		+10·74	
SrF_2	Sr^{2+} $4p$	28·6[b]		−18·79	
	F^- $2p$	17·0[b]	3·59[d]	+10·10	13·69
	F^- $1s$	693·5[c]		+10·10	
BaF_2	Ba^{2+} $5p$	23·6[b]		−17·59	
	F^- $2p$	16·3[b]	3·59[d]	+9·45	13·04
	F^- $1s$	693·2[c]		+9·45	
CdF_2	Cd^{2+} $4d$	19·15[b]	37·48[e]	−20·20	17·28
	F^- $2p$	13·65[b]	3·59[d]	+10·85	14·44
	F^- $1s$	692·1[c]		+10·85	
PbF_2	F^- $1s$	690·1[c]		+9·89	
UO_2	U^{4+} $4f^{5/2}$	391·5[a]		−39·86	
	U^{4+} $4f^{7/2}$	380·7[a]		−39·86	

The measurements by the photo-emission of electrons were made by the following authors:

[a] Chadwick and Graham (1972)

[b] Jorgensen (1971)

[c] Jorgensen, Berthou, and Balsenc (1971)

The binding energies in the free ion are derived from:

[d] Bailey (1958)

[e] Moore (1970).

Table 1.5 and are compared with the binding energies calculated by the addition of the Madelung potential to the binding energies found in the free ions (Moore, 1970). Unfortunately no accurate measurements are available of the energy of the $1s$ state in the free F^- ion. Where comparisons are possible, the agreement between the observed and the estimated energies of the core states is seen to be good. In nearly every case the binding energies in the solid are slightly greater than expected. An exception is provided by the $2p$ level of F^- in CdF_2 which has a lower binding energy than expected. This may be caused by the close proximity of the $4d$ levels of the Cd^{2+} ion. Nevertheless these comparisons must be treated with caution because of the difficulty in defining the vacuum level at the surface of an insulating solid. Electrostatic charging of the surface may produce errors of several volts. As an extreme example, current measurements of the binding energies of the $4f$ levels of U^{4+} in UO_2 differ among themselves by 7 eV.

It should be pointed out that the energy of a particular core state of an ion incorporated in a solid is normally very close to that of the same eigenstate of the free atom from which the ion is derived (Citrin, Shaw, Packer, and Thomas, 1972). For example, in making the cation Ca^{2+} from the atom Ca^0 the classical potential of the two outer electrons is removed. When the ion is incorporated into a solid the charges of the oppositely-charged ions situated nearby will compensate closely for the two missing electrons.

1.4.2. *The conduction bands*

The rather naive tight-binding arguments used above have only a limited utility in a discussion of the levels forming the conduction bands of the solid. We are concerned with the addition of a further electron to the solid and not with the excitation of an electron already present in the solid. Detailed calculations show that the possibility of an F^{2-} ion may be ignored and that a first approximation is achieved by the addition of the electron to a Ca^{2+} ion. We thereby form a Ca^+ ion whose sequence of energy levels is shown in Fig. 1.3. A $4s$ level with ionization energy of 11·9 eV in the free ion is followed by $3d$ and $4p$ levels higher in energy respectively by 1·7 and by 3·1 eV. These states will form the basis of the lowest conduction bands. The naive application of the Madelung potential shown in Fig. 1.3 is invalid for conduction states because of the overlap of wavefunctions onto neighbouring ions. When the detailed effects of the lattice potential have been included, one will expect a band minimum with s-like symmetry at the centre of the Brillouin zone. A separate minimum derived from d-like states will be found near the X point of the zone. A close similarity is expected between the energy bands of conduction electrons in CaF_2 and those in KI (Onodera, Okazaki, and Inui, 1966). K^0 is isoelectronic with Ca^+ and the translation groups of KI and of CaF_2 are identical. The lowest conduction-band minimum of KI, with symmetry Γ_{6+}, is derived primarily from s states of K^0, and the X minimum with symmetry X_{7+} is composed mainly of d states of K^0. It is not clear which minimum will be the lowest in CaF_2.

A schematic band structure for CaF_2 showing the highest valence band and the lowest conduction band states is shown in Fig. 1.5 (Starostin, 1969, and Rubloff, 1972). Double-group notation is used for those levels where spin-orbit splitting is significant. Inclusion of the

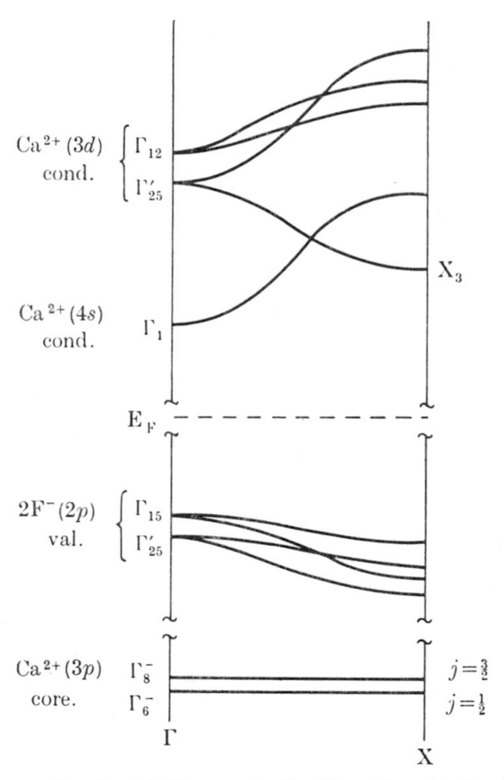

FIG. 1.5. The energy bands of CaF_2 near to the Fermi level (after Rubloff (1972)).

spin-orbit splitting of the F^- ions (0·03 eV) would split the upper level into two with symmetries Γ_8^- and Γ_6^-. The presence of two fluorine ions in the unit cell produces the state with symmetry Γ'_{25} which is absent in KI. Optical dipole transitions from Γ'_{25} to the conduction states with symmetries Γ_1, Γ_{12}, and Γ'_{25} are forbidden by parity conservation. Transitions are however allowed from Γ_{15} to Γ_1 and also from X'_5 and X'_2 to X_3. Further discussion must await the completion of detailed calculations of the energy bands of electrons in the fluorite compounds.

Let us now consider the types of exciton which may be predicted from our knowledge of the band structure of the alkaline earth fluorides. From Fig. 1.5 it is clear that the excitons of lowest energy will probably be associated with the transitions of an electron from the Γ_{15} valence band to the Γ_1 conduction band minimum of CaF_2. In the Frenkel picture this may be visualized as the transfer of an electron from a fluorine ion to a nearest-neighbour Ca^{2+} ion. Excitons of higher energy will be associated with the transfer of electrons from deeper core levels to the conduction band. The first of such transitions will generate holes in the $3p$ band of the Ca^{2+} ions. As the conduction band is primarily of calcium $4s$ character, these first core excitons may be regarded as an internal transition of the Ca^{2+} ion. A series of levels is indeed found to be associated with the $3p^5 3d$ and $3p^5 4s$ configuration of the free Ca^{2+} ion. The initial state in the optical transition is of ^1S character. The final states of most interest in the interpretation of optical phenomena will be of P character. Hayes, Koch, and Kunz (1971) have shown that the lowest states of P character are a ^3P state from the configuration $3p^5 3d$, followed by ^3P and ^1P states from $3p^5 3d$ configuration.

The spin-orbit splittings of the core and the valence levels of the ions may be used to assist in the assignment of the initial states involved in the observed transitions. Certain relevant atomic splittings are given in Table 1.6. The splittings of levels in the ions will be similar to those of the atoms.

TABLE 1.6

The spin-orbit splitting (eV) *of the electronic energy levels of certain free atoms*

Atom \ Level	$2p$	$3p$	$4p$	$5p$	$6p$
F	0·03[a]				
Cl	1·6[b]	0·07[a]			
Br	46·1[b]	7·8[b]	0·31[a]		
O	<0·03[a]				
Ca	3·6[b]	0·4[c]			
Sr	67·2[b]	10·7[b]	1·3[c]		
Ba	376·6[b]	74·5[b]	12·1[b]	2·0[b,c]	
Cd	189·5[b]	34·2[b]	7[b]		
Pb		487·8[b]	119·4[b]	18·8[b]	
U			227·7[b]	64[b]	10[b]

The values are derived from the following works:
[a] Phillips (1964)
[b] Bearden and Burr (1967).
[c] Hayes, Koch, and Kunz (1971).

1.5. The optical properties of the fluorite compounds

The wide band-gap of the fluorite compounds has restricted the measurement of their optical properties. Calcium fluoride is commonly used as a window material able to transmit ultraviolet radiation with photon energies of up to 10 eV. Only recently, with the use of synchrotron radiation, have measurements of the real and the imaginary parts ϵ_1 and ϵ_2 of the dielectric constant been made over a wide range of energy.

Measurements of the reflectance of the alkaline earth fluorides have been reported by Stephan, Le Calvez, Lemonier, and Robin (1969), Tomiki and Miyata (1969), Hayes *et al.* (1971), and Rubloff (1972). Sanchez and Cardona (1972) have studied the piezo-birefringence of the alkaline earth fluorides with photon energies of up to 10 eV, their results suggesting a complex band edge. The band edge of CdF_2 has been studied in detail by Forman, Hosler, and Blunt (1972). They conclude that CdF_2 possesses a direct allowed band edge. A careful study of the reflectance of CdF_2 by Eisenberger and Adlerstein (1970) suggested an exciton binding energy of 1·8 eV. The reduced mass calculated using ϵ_∞ is 0·8 times the mass of the free electron.

Preliminary measurements on films of PbF_2 by Beaumont (1971) show a first absorption peak at 5·7 eV at 77 K which may however be caused by impurities. The band gap exceeds 6·4 eV.

The band gap of $SrCl_2$ has been estimated by Saar and Elango (1972) from measurements of the $L_{2,3}$ absorption spectra of the chlorine ion.

These measurements are summarized in Table 1.7.

TABLE 1.7

The band gap and the first exciton peak at low temperatures of certain fluorite compounds. The energies are expressed in electron-volts

	CaF_2	SrF_2	BaF_2	CdF_2	PbF_2	$SrCl_2$
Band gap	12·2[a,b]	11·44[b]	10·59[b]		⩾6·4[c]	5·8[d]
First exciton peak	11·18[a]	10·6[a]	10·0[a]	7·6[e]	—	—

The values are derived from the following works:
[a] Rubloff (1972)
[b] Tomiki and Miyata (1969)
[c] Beaumont (1971)
[d] Saar and Elango (1972)
[e] Forman, Hosler, and Blunt (1972).

1.5.1. *The alkaline earth fluorides*

In Fig. 1.6 we see measurements by Rubloff (1972) of the optical reflectance spectra of calcium, strontium, and barium fluoride at near

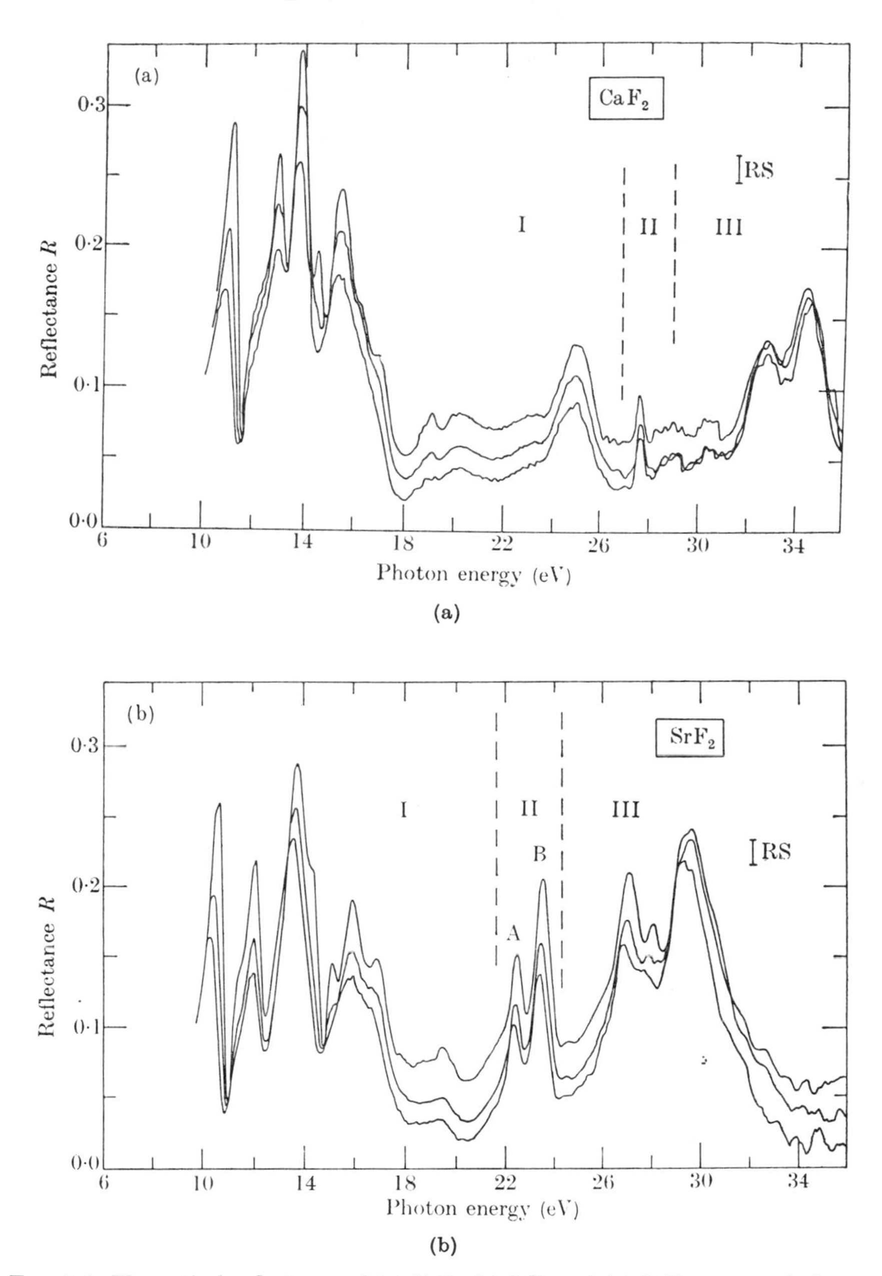

FIG. 1.6. The optical reflectance of (a) CaF_2 (b) SrF_2 and (c) BaF_2 measured at near normal incidence (after Rubloff (1972)). The ordinate scale is appropriate to the central curve which displays measurements at 300 K. Measurements at 90 K and at 400 K are shifted upwards and downwards respectively by the increment, RS. The wavelength resolution is 5 Å. The spectral regions I, II, and III correspond to the formation of different types of excitation and are discussed in § 1.5.1.

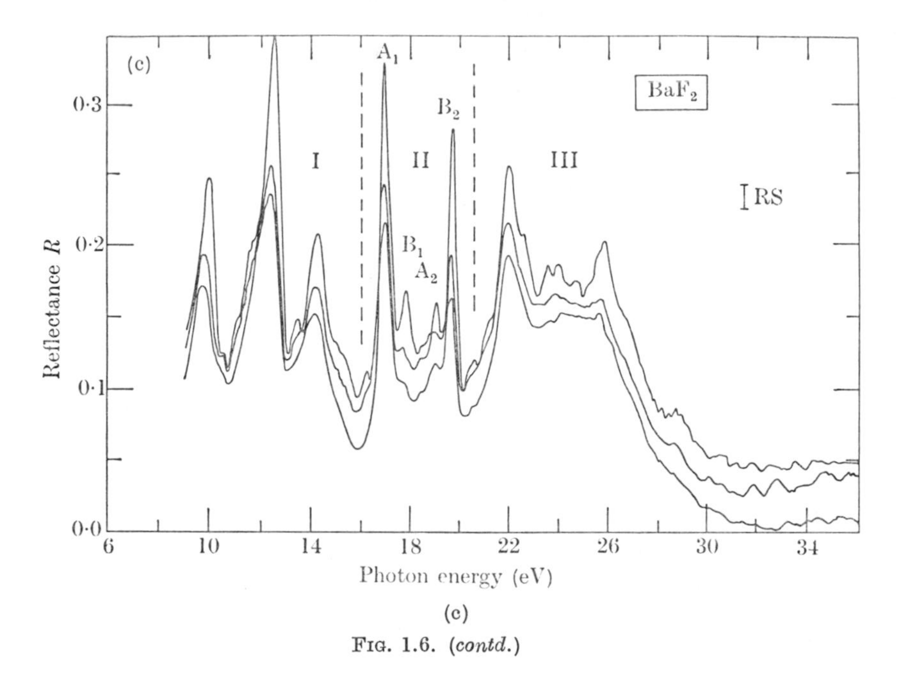

(c)

FIG. 1.6. (*contd.*)

normal incidence. Superimposed are spectra measured at 90 K, 300 K, and 400 K. At lower energies the specimens, when pure, are transparent until the onset at long wavelength of the absorption and dispersion associated with phonon generation (Chapter 2).

In Fig. 1.6 we note a profusion of sharp peaks in the reflectance extending upward in photon energy from the onset of absorption. At a sufficiently high energy broad bands appear, for example above 26 eV in SrF_2.

Measurements of the energy-loss spectra of fast electrons traversing thin films of the alkaline earth fluorides have been made by Frandon, Lahaye, and Pradal (1972) following the pioneering work of Best (1962). The energy-loss spectra, such as that of CaF_2 shown in Fig. 1.7, give $\mathrm{Im}(\epsilon^{-1})$ as a function of energy, ϵ being the dielectric function. By use of the Kramers–Kronig relations the various optical constants of the material may be calculated. A comparison between such a calculated reflectance spectrum of CaF_2 and the direct measurements of Rubloff (1972) is shown in Fig. 1.8. The agreement is seen to be good. The energy-loss spectra are particularly sensitive to the excitation of longitudinal plasmons. Plasmons associated with valence electrons are identified at 17 eV in CaF_2 and 17·2 eV in SrF_2. Plasmons associated with the p electrons of the metal ions are identified at 35·8 eV in CaF_2,

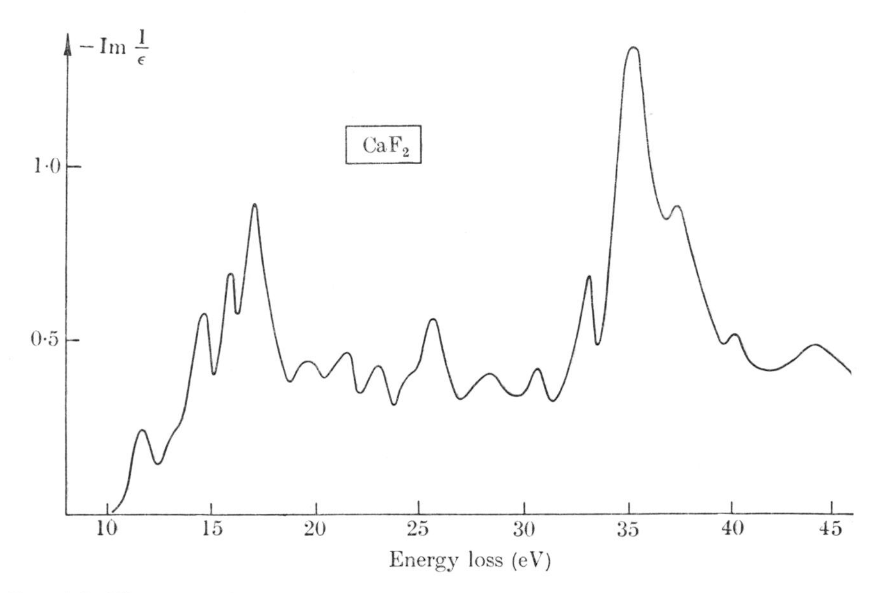

FIG. 1.7. The energy-loss spectrum of fast electrons in CaF_2 at room temperature, after Frandon *et al.* (1972).

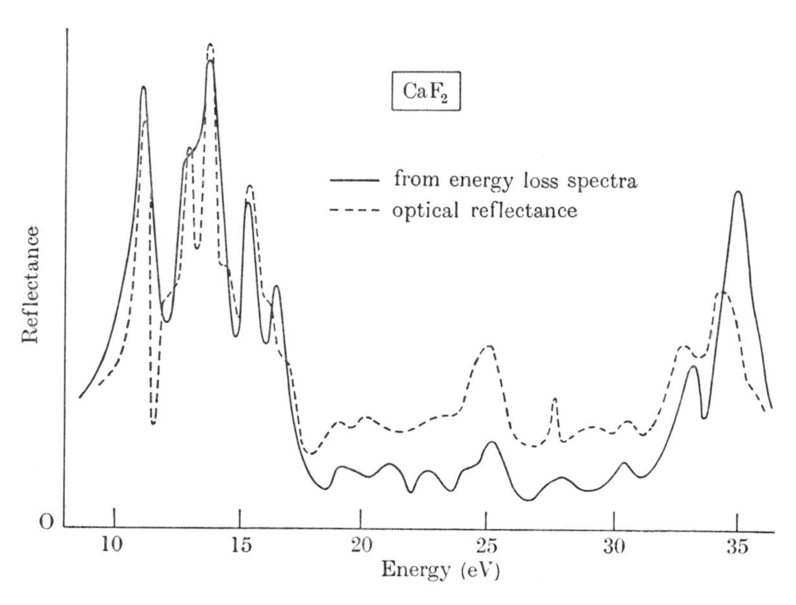

FIG. 1.8. The optical reflectance of CaF_2. The optical measurements at 90 K of Rubloff (1972) are compared with the reflectance calculated from the room-temperature electron energy-loss spectra of Frandon *et al.* (1972)

30·6 eV in SrF_2, and 26·4 eV in BaF_2. These plasmons have no direct optical analogue.

In the absence of detailed calculations of the electronic energy band structure a full interpretation of the optical measurements is not useful, but certain general observations may be made. In a study of the optical spectra of the alkali halides, Rubloff (1972) showed that the spectra could be divided into several distinct parts. By analogy with their spectra Rubloff similarly divided the spectra of the alkaline earth fluoride into Regions I, II, and III as marked in Fig. 1.6.

Region I corresponds to the generation of excitons by the excitation of an electron from the upper valence bands to the lower conduction bands. The lowest peaks are attributed to excitons associated with transitions from Γ_{15} to Γ_1 levels and the peaks at higher energy in Region I to transitions involving *d*-like conduction states. Tomiki and Miyata (1969) observed a fine structure in the exciton peaks of lowest energy in CaF_2, SrF_2, and BaF_2. A plausible interpretation then allowed a 1*s* exciton binding energy of about 1 eV in CaF_2, 0·9 eV in SrF_2, and 0·65 eV in BaF_2. Using the optical dielectric constant the reduced exciton masses deduced are then 0·30, 0·29, and 0·22 times the free-electron mass respectively for CaF_2, SrF_2, and BaF_2. Other workers have since attempted, without success, to reproduce this fine structure. Further studies are needed.

Region II is attributed to excitons associated with transitions in which an electron is excited from the outermost core states of the metal ion e.g. from the 3*p* levels of the Ca^{2+} ion. The increase in the splitting of the lines when moving from CaF_2 to SrF_2 and BaF_2 reflects the greater spin-orbit splittings of the *p* states of the heavier elements (see Table 1.6).

Region III comprises broad peaks attributed by Rubloff to interband and ionizing transitions of core electrons. Frandon *et al.* attribute energy loss peaks near 140 eV in SrF_2 and between 98 and 140 eV in BaF_2 to the excitation of electrons to conductions bands from the 3*d* levels of Sr^{2+} and from the 4*d* levels of Ba^{2+}.

This general assignment of transitions is supported by the calculations by Hayes *et al.* (1971) of the energy levels of the free Ca^{2+} and Sr^{2+} ions. They show that the spectrum of CaF_2 from 25 to 36 eV may be interpreted in terms of transitions from the $3p^6\,{}^1S$ level of the Ca^{2+} ion to the manifold of excited states formed by the configuration $3p^54s$ and $3p^53d$ of Ca^{2+}. A similar assignment is made to explain the structure from 22 to 33 eV in SrF_2.

Particular interest attaches to the strong dependence on temperature of certain peaks in the spectrum, such as the peak in CaF_2 at 14·4 eV and that in BaF_2 at 26 eV. It will be seen that the spectral peaks observed here have widths of about 0.5 eV whereas the thermal energy, kT, associated with a temperature of 400 K is only 0·03 eV. Similar effects in the photo-emission spectrum of AgBr were interpreted by Bauer and Spicer (1970) in terms of fluctuation in the overlap integrals between nearly degenerate eigenstates associated with different ions in the solid. As the temperature rises, an increase in the amplitude of vibration of the ions will increase the variation in the overlap integrals, causing a rapid broadening of the levels. The observation of such anomalous temperature dependences may be very valuable in the detailed comparison between the observed optical spectra and the results of detailed energy-band calculations.

1.6. Electronic transport

1.6.1. *The alkaline earth fluorides*

The electronic mobility of a conduction electron is determined by its scattering both by phonons and by defects in the lattice, the latter being either chemical impurities or lattice imperfections. At a sufficiently low temperature the scattering by phonons becomes negligible but the difficulties experienced in the growth of pure crystals of the alkaline earth fluorides are such that the highest mobility observed so far at low temperature is about 3000 $cm^2\,V^{-1}\,s^{-1}$ (Crowder and Hodby, 1972). Such a low mobility prevents direct measurements by cyclotron resonance of the mass and symmetry of the carrier.

The probability of the scattering of an electron by optical phonons depends on the effective mass of the electron. This mass may be deduced by absolute measurements of the mobility in those temperature ranges where it is limited by its interaction with optical phonons. Such measurements on CaF_2 were made by Seager (1971) who used non-contacting electrodes appropriate for the study of the properties of an insulating solid (Hodby, 1972). Electrons were photoexcited into the conduction band from *F* centres introduced by additive coloration. The mobility was found to fall with increasing temperature more rapidly than predicted by the usual theories of interaction between phonons and polarons (Thornber and Feynman, 1970). It is possible that this deviation is an experimental effect caused by dark current or surface-leakage currents in the specimen. Ignoring this possibility a naive application of polaron theory, assuming a spherical conduction band minimum at the centre of the Brillouin zone, suggests a band mass for the electron of $(0{\cdot}75 \pm 0{\cdot}25)m_e$ and a polaron mass of about $2m_e$, where m_e is the mass of the free electron. Bennett (1968) in a calculation of the position of the '*F*' absorption band in CaF_2 assumed that the polaron mass would be about $0{\cdot}6m_e$, obtaining good agreement with the observed band. A much lower value for the polaron mass in CaF_2 was deduced by Dresner and Heyman (1971), but their use of standard semiconductor techniques is open to gross errors when applied to CaF_2.

Further and more definitive measurements on the alkaline earth fluorides must await the development of specimens of higher chemical purity and lower dislocation content. Both the mobility and the lifetime before recapture of photo-excited carriers would thereby be increased.

1.6.2. *Cadmium fluoride*

Pure CdF_2 is a colourless insulator with band gap greater than 7·6 eV. Kingsley and Prener (1962) showed that CdF_2 could be converted into a low-resistance semiconductor by doping with certain transition metal or rare-earth ions, followed by coloration in cadmium vapour. Extensive investigations followed, reviewed by Feldman (1972). References may also be found in the paper by Adlerstein, Pershan, and Feldman (1971).

Many crystals with the fluorite structure may be doped by the addition of transition metal or rare-earth compounds to the molten salt. Most of the rare-earth elements and certain transition elements, such as yttrium, are incorporated in CaF_2, SrF_2, BaF_2, and CdF_2 as the trivalent ion substituting for a divalent metal ion of the host. In the absence of impurities, such as oxygen, the extra positive charge of the doping ion is normally compensated by the incorporation of an extra fluorine ion in a nearby interstitial site (Chapter 6). Exceptions to this general behaviour are found near the ends or the centre of the series of rare-earth ions. Certain electronic configurations may then possess particularly high stability. An example is provided by the doubly-ionized state of Europium. This possesses a half-filled $4f$ shell. Europium is normally incorporated as the Eu^{2+} ion, substituting directly for a divalent metal ion.

Many of the fluorite materials may be coloured by chemical treatment. Heating in the vapour of the host metal alters the electronic properties of suitably doped crystals of the alkaline earth fluorides and also of cadmium fluoride. When initially undoped samples of the alkaline earth fluorides are heated in the vapour of the metal, additive coloration occurs with the formation of *F* centres and their aggregates (section

4.1). Similar chemical treatment can change the valence state of some trivalent ions incorporated into the alkaline earth fluorides. Divalent ions are often formed by the trapping of an electron at a trivalent ion (Chapters 5 and 7). Rather different results follow the heating in cadmium vapour of pure or doped cadmium fluoride. Such chemical treatment does not produce F centres in pure CdF_2. Some yellow coloration occurs, probably due to precipitation of colloidal cadmium metal. However, when CdF_2 doped with Nd, Sm, Gd, Tb, Dy, Ho, Er, Tm, Yb, Lu, Y, or Sc is heated in cadmium vapour, the crystal becomes a pale blue semiconductor whose detailed transport properties do not depend on the chemical identity of the doping ion. The dependence of the electrical conductivity on the temperature of the specimen suggests a donor centre with a thermal depth of 0·1 eV below the bottom of the polaron conduction band. The optical properties of the impurity ion are consistent with a trivalent state (Weller 1965) although the immediate environment of the ion in the crystal may be disturbed (Prener and Kingsley, 1963). Detailed studies of the optical absorption and of the photoconductivity of semiconducting CdF_2 (Feldman and Pershan, 1972) show internal transitions between the ground and the higher states of a roughly hydrogenic donor beginning at 0·128 eV. Multi-phonon side bands are observed at higher energies. Photo-conductivity begins at 0·15 eV and possesses oscillatory phonon structure at higher energy. Some doped crystals of CdF_2 do not behave in this way, and crystals lightly doped with La, Ce, Pr, Eu, and U remain insulating after heating in cadmium vapour.

The absence of *F* centres after the heating of pure CdF_2 in cadmium vapour is not fully understood (cf. § 4.8.1). Tan and Kramp (1970) and Kessler and Caffyn (1972) have shown that fluorine ion vacancies are mobile in CdF_2 (see section 3.3). When the crystal is heated in cadmium vapour we expect the reaction

$$Cd + 2F^- \rightarrow CdF_2 + 2e^-$$

to occur at the surface. Fluorine ion vacancies and free electrons are liberated at the surface and should be able to migrate into the crystal. The formation of *F* centres would then seem to be possible. The apparent absence of stable *F* centres may be caused by the high electron affinity of the Cd^{2+} ion. In CaF_2 as in the alkali halides, electrons may be trapped stably in the coulomb potential of a negative ion vacancy. However, the binding energy of the outermost electron in the Cd^+ ion is 5 eV greater than that of the outermost electron in the Ca^+ ion. The energies of electrons in the conduction band of CdF_2 are therefore lower. It is possible that this lowering of the conduction band prevents an electron from being held within a negative ion vacancy in CdF_2. The point-ion potentials in the two crystals differ by only 0·1 eV.

Similar arguments may be applied to the coloration of crystals doped with trivalent rare-earth ions. Fluorine ion vacancies and free electrons will be present in the crystal. Trivalent rare-earth ions may thereby have a significant probability of losing their charge-compensating fluorine ions. An electron approaching an uncompensated ion is then trapped by the strong central coulomb field. The conduction band of CdF_2 will lie several electron volts lower in energy than that of CaF_2. In CaF_2 the electron is trapped by the trivalent ion, changing its valence state. In CdF_2 however, the lower conduction band will favour a large semiconductor-like orbital. The formation of such a donor state of large radius explains why the transport properties of semiconducting CdF_2 show no dependence on the identity of the original doping ion. One exception to the normal rule is provided by In^{3+} in CdF_2 whose ground state after coloration is divalent. In^{3+} is isoelectronic with Cd^{2+} and would be expected to possess a high electron affinity. Deep capture of the extra electron is therefore probable.

Many of the early studies of semiconducting CdF_2 were performed on heavily doped specimens. It has since become clear that the resulting overlap of the wavefunctions of the bound electrons made a correct analysis of the measurements very difficult. For instance, a microwave absorption attributed at first to the cyclotron resonance of free carriers (Eisenberger, Pershan, and Bosomworth, 1968) has since been shown to be a transition between the symmetric and antisymmetric states of electrons trapped at close pairs of impurity ions (Adlerstein *et al.*, 1971). Recent measurements on specimens containing a low concentration of impurity ions have provided a consistent picture. The ground-state orbit of the electron is intermediate between the Wannier and the tight-binding models. Its Bohr radius is close to the lattice spacing and its effective mass is about 0·76 times the mass of the free electron. The small spatial extent of the wavefunction will cause splittings of the energy of the centre depending on its detailed symmetry. A transition with energy of 178 cm^{-1}, visible both in infra-red and in Raman spectra, is probably associated with such a splitting.

A detailed interpretation of the many measurements now available is not yet possible as the nature of the lowest conduction band states is not known. The measurements of Forman *et al.* (1972) showed that transitions across the fundamental band gap are probably direct and allowed. It is not yet known whether the gap is at Γ or elsewhere in the Brillouin zone. A fuller understanding awaits the success of measurements of cyclotron resonance in pure samples.

1.6.3. *The hydrides of the rare-earth metals*

The hydrides of the rare-earth metals are alone among the fluorite compounds in possessing a wide range of electrical properties. EuH_2 and YbH_2 are ionic semiconductors. The other dihydrides are metallic. When excess hydrogen is dissolved their resistivities rise and they become semiconductors. Detailed study has been made in CeH_x which possesses a metal-to-semiconductor transition at a composition close to $CeH_{2.8}$. A very small structural distortion of the lattice seems to occur in the high resistance state (Libowitz, Pack, and Binnie, 1972).

Detailed calculations of the band structures by Switendick (1970) have shown that a band gap can indeed appear as the concentration of hydrogen is increased, thus helping to explain the transition from metallic to semiconducting behaviour. Much overlap occurs between those energy levels with rare-earth parentage and those with hydrogen parentage and it is not meaningful to discuss the bonding of the solid in simple ionic terms. This conclusion is confirmed by measurements of the angular correlation of gamma rays emitted in the annihilation of positrons in the dihydrides of Ce, Er, Gd, Ho, and Yb (Chouinard, Gustafson, and Heckman, 1969; Chouinard and Gustafson, 1971). They were able to show in all cases that the structure is metallic, electrons from both the rare-earth element and from the hydrogen entering the conduction band of the dihydride.

1.7. Magnetic order in fluorite compounds

Cooperative magnetic processes are known in several materials with the fluorite structure. The U^{4+} ion in UO_2 and also many of the rare-earth ions in the hydrides RH_x possess unfilled inner shells of electrons. The gound state of the ion may therefore be degenerate. The resulting magnetic moment provides a paramagnetic susceptibility at room temperature. At lower temperatures cooperative transitions to a magnetically ordered state are possible if exchange interactions or other coupling mechanisms are sufficiently powerful.

1.7.1. *Magnetic order in* UO_2

The quadruply charged uranium ion U^{4+} contains two electrons in addition to its various closed shells. These two electrons might be found either in the $5f$ or in the $6d$ shells. The ion is therefore expected to possess a magnetic moment. The resulting magnetic properties of the various oxide species have been widely studied. A useful review of the earlier measurements is given by Frazer, Shirane, Cox, and Olsen (1965). The magnetic susceptibility of the various oxide species UO_x has been studied by Dawson and Lister (1952) and by Leask, Roberts, Walter, and Wolf (1963). Dawson and Lister show that the magnetic moment of the U^{4+} ion is consistent with an electronic configuration of $5f^2$ with a ground multiplet $^3H_4$. Such a configuration in L-S coupling could have a magnetic moment of up to 3·58 Bohr magnetons in the presence of the full orbital moment. The observed magnetic moment in the paramagnetic region was 3·2 Bohr magnetons in UO_2 (Dawson and Lister, 1952, and Nasu, 1966), decreasing below 3·0 Bohr magnetons and approaching the spin-only value of 2·83 Bohr magnetons (Slowinski and Elliott, 1952) when diluted with ThO_2. Above 31 K UO_2 is paramagnetic and its susceptibility shows a dependence on temperature of Curie–Weiss type. At 30·8K a very sharp phase transition, probably of first order, is observed both in susceptibility and in specific heat and the material becomes magnetically ordered. Neutron diffraction measurements (Frazer *et al.*, 1965) show a simple antiferromagnetic superlattice in the ordered state. Those uranium ions which lie in a [1, 0, 0] plane perpendicular to one of the 4-fold axes of the conventional cubic unit cell have parallel moments lying in that [1, 0, 0] plane. The moments in each plane are anti-parallel to those in the plane next above. The actual orientation of the moment within each plane is not available from the neutron diffraction studies. Measurement of the neutron form factor shows that the outer electrons are indeed in a $5f^2$ configuration, most probably $^3H_4$.

The measurements of neutron diffraction do not determine the mechanism responsible for the appearance of the ordered state. A clue is provided by the observation of a large difference between the frequencies of the transverse acoustic phonon modes when measured at temperatures above and below the magnetic transition temperature. A strong phonon-magnon interaction is suggested. Allen (1968) describes a consistent microscopic theory both of the electronic ground state of the ordered system and of the spin-wave excitations. Point-ion calculations show that a magneto-strictive distortion around each U^{4+} ion, caused by the crystal field, produces indirect interactions between the U^{4+} ions. These indirect interactions are biquadratic in the spin and are of the same magnitude as the direct exchange energies. The parameters needed for this magnetostrictive interaction may be deduced from the observed spin-wave spectra. A consistent picture emerges in which the magnetic moments are quantized along ⟨1, 1, 0⟩ axes. A competition between these cooperative Jahn–Teller forces and the direct exchange energy produces a ground state whose spins are not fully polarized.

Similar calculations by Sasaki and Obata (1970) show the importance of dynamic Jahn–Teller interactions in determining the magnetic properties of dilute solutions of UO_2 in ThO_2. Good agreement exists with the detailed susceptibility measurements of Comly (1968) and of Slowinski and Elliott (1952).

1.7.2. *Magnetic order in the rare-earth hydrides*

Most of the rare-earth metals form cooperatively ordered magnetic states at a sufficiently low temperature. The dominant exchange interaction between the $4f$ electrons on different atoms occurs via the sea of conduction electrons in the metal. The magnetic transition temperatures occur typically around 100 K, e.g. 135 K for holmium metal.

At a sufficiently low temperature many of the rare-earth hydrides with composition ReH_x change from the paramagnetic behaviour found at room temperature to a magnetically ordered state. In Table 1.8 are given the composition ranges of the various hydrides in which ordering has been observed or has been shown not to occur. It will be noted that certain compounds, e.g. ErH_x show no magnetic ordering down to 3 K over the composition range $1{\cdot}96 \leqslant x \leqslant 3{\cdot}02$. Other compounds e.g. NdH_x order at a low temperature unless x exceeds about 2·5.

A detailed analysis of these magnetic properties has yet to be made. Most authors, e.g. Kubota and Wallace (1963*b*) assume that the exchange interaction in the hydrides is again mediated by the conduction electrons. The disappearance of ordering processes as the composition approaches the trihydride is then readily interpreted in terms of the disappearance of free conduction electrons in these materials (cf. § 1.6.3). Further evidence of the role of the conduction electrons is the observation by Kopp and Schreiber (1967*b*) of a strong Knight shift of the nuclear magnetic resonance of the hydrogen nuclei as $CeH_{2{\cdot}45}$, $NdH_{2{\cdot}0}$, and $GdH_{2{\cdot}1}$ enter a magnetically ordered state. The hydrogen nuclei are coupled to the conduction electrons by hyperfine interactions and are sensitive to the spin polarization of these electrons.

In the paramagnetic temperature range the magnetic susceptibilities of the hydrides follow a Curie–Weiss law with an effective magnetic

TABLE 1.8

The crystal structures and magnetic properties of the rare-earth hydrides ReH_x

	Crystal structure ranges of x		Magnetic order: Ordering observed			Magnetic order: No order observed		
	Cubic	(g) Hexagonal	Form of order	Phase limits	Transition temperature	Phase limits	Minimum temperature of observation	Reference
La	$2 \rightarrow 3$					$x \leqslant 2.0$		
Ce	$2 \rightarrow 3$		Ferro.	$2.1 \leqslant x \leqslant 2.45$	≈5 K	$2.6 \leqslant x$	4 K	e,f
Pr	$2 \rightarrow 3$					$2.0 \leqslant x \leqslant 2.71$	4 K	b,f
Nd	$2 \rightarrow 3$		Ferro.	$2.0 \leqslant x \leqslant 2.30$	≈6 K	$2.6 \leqslant x \leqslant 2.75$	4 K	b,f
Pm								
Sm	$1.93 \rightarrow 2.55$	$2.59 \rightarrow 3$				$1.99 \leqslant x \leqslant 2.88$	4 K	b
Eu	Orthorhombic		Ferro.	2	25 K			
Gd	$1.82 \rightarrow 2.3$	$2.85 \rightarrow 3$		2.1	30 K			
Tb	$1.90 \rightarrow 2.15$	$2.81 \rightarrow 3$	Antiferro.	2	40 K			d
Dy	$1.94 \rightarrow 2.08$	$2.86 \rightarrow 3$	Antiferro.	$1.97 \leqslant x \leqslant 2.03$	8 K	$2.9 \leqslant x \leqslant 3$	4 K	
Ho	$1.95 \rightarrow 2.24$	$2.64 \rightarrow 3$		$1.6 \leqslant x \leqslant 2.31$	8 K	$2.6 \leqslant x \leqslant 3$	4 K	a,d
Er	$1.95 \rightarrow 2.31$	$2.82 \rightarrow 3$				$1.96 \leqslant x \leqslant 3$	3 K	c
Tm	$1.99 \rightarrow 2.41$	$2.76 \rightarrow 3$				$1.99 \leqslant x \leqslant 2.95$	3 K	c
Yb	Orthorhombic							
Lu	$1.85 \rightarrow 2.23$	$2.78 \rightarrow 3$						

The values are derived from the following works.

[a] Kubota and Wallace (1962)
[b] Kubota and Wallace (1963a)
[c] Kubota and Wallace (1963b)
[d] Cox, Shirane, Takei, and Wallace (1963)
[e] Kopp and Schreiber (1967a)
[f] Kopp and Schreiber (1967b)
[g] Libowitz (1965).

moment close to that of the free tripositive rare-earth ion. Deviations from the Curie–Weiss law are observed at low temperatures and are attributed to splitting of the rare-earth energy levels in the local crystal field (Kubota and Wallace, 1963*b*). The ordering temperatures, 5 to 40 K, are low when compared with those in the rare-earth metals and are interpreted in terms of the greater separation of the rare-earth ions and the lower density of free conduction electrons present in the hydrides.

APPENDIX

1.8. The preparation of single crystals

It is convenient to divide into three classes those materials with the fluorite structure which we are considering in this monograph. The halides whose melting points range up to 1360 °C may be grown by the usual techniques of crystallization from the molten salt. The refractory oxides whose melting points range from 2805 °C for UO_2 to 3050 °C for ThO_2 have been grown by arc melting, from solution and by sublimation. The rare-earth hydrides decompose on heating and are normally grown by direct reaction, the pressure of the hydrogen gas defining the composition of the hydride.

1.8.1. *The halides*

We consider here the growth of the alkaline earth halides CaF_2, SrF_2, BaF_2, $SrCl_2$, and $BaCl_2$ and also of CdF_2, and PbF_2. The widespread use of the alkaline earth fluorides as host crystals in maser and laser applications encouraged attempts to grow single crystals with low impurity content. The difficulties have already been mentioned in § 1.6. CaF_2 occurs naturally in single crystal form but contains about 500 parts per million of Al and Fe as well as rare-earth impurities (O'Connor and Chen, 1963). Synthetic material of higher purity may be made by reacting the purified oxide or chloride with gaseous HF (O'Connor and Chen, 1963, Guggenheim, 1963).

All three alkaline earth fluorides hydrolyse readily when warmed in moist air. Precipitates of $Ca(OH)_2$ or CaO form in the crystal. Heating in dry oxygen can also introduce O^{2-} ions into the lattice. Kaiser and Keck (1962) showed that removal of O^{2-} and OH^- ions reduced the scattering of light in laser host materials. Three methods have been used to remove these ions from fluoride salts. Guggenheim (1963) sintered the powdered salt in pure HF gas at temperatures just below the melting point, the crystal being grown under an atmosphere of HF gas. Alternatively lead fluoride, PbF_2, or ammonium bifluoride, NH_4HF_2, are mixed with the powdered salt before melting under vacuum. In the first case lead oxide evaporates preferentially from the melt followed by any excess of lead fluoride. Ammonium bifluoride acts similarly, any excess evaporating from the melt. The purification of CaF_2 by horizontal zone refining in high-purity graphite boats was studied by Duke (1964) who showed that effective segregation of O, Mg, Si, P, and Fe was achieved. Little segregation of Sr, Ba, K, or Cl was found.

Single crystals of the alkaline earth fluorides can be grown by all the standard techniques if adequate precautions are taken to eliminate oxygen and water from the atmosphere of the growing process. All three alkaline earth fluorides are available commercially from several sources, all using the Stockbarger-Bridgman technique of growth. CaF_2, SrF_2, and BaF_2 may be pulled from the melt although care must be taken to reduce thermal gradients in the solid crystal (Guggenheim, 1961). It is generally found to be difficult to grow single crystals of CaF_2 from melts which are low in O^{2-}, and OH^- ions. Such chemically pure crystals will normally possess curved cleavage faces. Single crystals are more easily grown from pure BaF_2. Nassau (1961) showed that the segregation coefficients of many rare-earth ions incorporated in molten CaF_2 at about 1 per cent concentration were close to unity. Doping is therefore easy. Strontium chloride single crystals have been grown by the Stockbarger-Bridgman technique from material zone refined in HCl gas (Hukin and Ritchie, 1971). The preparation of pure CdF_2 single crystals is described by Trautweiler, Moser, and Khosla (1968). Pure Cd metal was dissolved in nitric acid. CdF_2 was precipitated with NH_4F, zone refined and then grown by the Bridgman technique in a graphite crucible in an atmosphere of dry helium gas.

CaF_2, SrF_2, and BaF_2 cleave readily along [1, 1, 1] planes which meet to form $\langle 1, 1, 0 \rangle$ cleavage edges, these directions being defined with respect to the conventional unit cube. The cleavage is of value in the alignment of specimens. $SrCl_2$ and CdF_2 may be cleaved but with increasing difficulty as the purity of the specimen is increased.

The epitaxy of films of CaF_2 deposited by evaporation in vacuum onto [1, 0, 0] cleavage faces of NaCl was studied by Bujor and Vook (1969). When the substrate temperature was between 150 °C and 250 °C during evaporation the film was complex. Good epitaxial monocrystals were however produced when the NaCl substrate was held at 350 °C during the evaporation. The [1, 0, 0] face of the CaF_2 film lay parallel to the [1, 0, 0] face of the NaCl substrate. The ⟨1, 1, 0⟩ direction in the film lay parallel to the ⟨1, 0, 0⟩ direction in the NaCl surface. The fluorine ions of CaF_2 are thereby enabled to lie in the potential minima of the NaCl face.

1.8.2. *The refractory oxides*

The fluorite structure is stable in pure ZrO_2 only at temperatures above 1400 °C. This structure may however be stabilized at lower temperatures by the addition of small quantities of CaO. In the growth of the uranium oxides the uranium-to-oxygen ratio is controlled by the ambient pressure of oxygen, UO_2 being produced by growth in an inert or slightly reducing atmosphere. The oxides have been grown from the molten salt, by sublimation and from solution.

(a) *Growth from the molten salt.* This is the most common method of producing single crystals of the refractory oxides. An arc is struck between carbon electrodes immersed in a loosely compacted mass of the powdered oxide. The centre of the mass fuses while the outer parts provide thermal insulation. Single crystals up to 1·5 cm on side are found on cooling. The impurity content of such crystals can be less than 0·1 per cent. Inclusions of uranium metal or of polycrystalline UO_2 are common in single crystals of UO_2 grown in this way (Brite and Anderson, 1962). UO_2 has been grown by a modification of the vertical floating zone technique (Chapman and Clark, 1965). UO_2 powder is compacted by pressure to form a rod about $1\frac{1}{2}$ cm in diameter. The core of the rod is melted by a radio-frequency loop after preheating. The surface of the rod remains solid because of radiant loss of heat. Single crystals up to 5 cm long and 1 cm diameter may be produced with oxygen-to-uranium ratio of 2·000 to 2·003 when grown in a reducing atmosphere.

(b) *Vapour growth.* These techniques were applied to UO_2 by Van Lierde, Strumane, Smets, and Amelinckx (1962). A hollow tube of compacted UO_2 powder was held axially in vacuum between two heavy water-cooled electrodes which covered the end of the tube. After preheating, a current passed between the electrodes heated the tube of UO_2. Material sublimed from the inner wall of the tube onto the cooled electrodes, growing there as single crystals. Oxygen to uranium ratios of 2·00014 were achieved.

(c) *Solution growth.* This relies on the discovery of a suitable solvent or flux whose melting point is lower than that of the crystal to be grown. Single crystals of ThO_2 were grown by Willis (1963*a*) from solution in lead metaborate on cooling the solution from 1500 °C to 800 °C. Similarly UO_2 may be deposited on a platinum cathode by electrolysis of UO_2Cl_2 dissolved in the eutectic mixture of NaCl and KCl (Robins 1961).

Willis (1963*a*) studied the mosaic spread in crystals of UO_2 grown by these different techniques. Typical mosaic spreads in a small crystal were 5 min of arc when grown from the fused salt or by vapour growth. Solution growth gave much better crystals with mosaic spread of less than 20 sec of arc.

1.8.3. *The rare-earth hydrides*

The rare-earth hydrides are normally grown by direct reaction of the metal with hydrogen. The pressure of the hydrogen gas determines the composition of the final hydride ReH_x (Hardcastle and Warf, 1966). A typical procedure is that used by Pebler and Wallace (1962). The purified rare-earth metal is heated under high vacuum to 500 °C. Hydrogen gas is admitted to the required pressure and the reaction is allowed to proceed. The specimen is then annealed at between 200 °C and 300 °C for 4 h before cooling to room temperature. The specimen is normally kept under oil to prevent oxidation.

References

Adlerstein, M. G., Pershan, P. S., and Feldman, B. J. (1971). *Phys. Rev.* **B4,** 3402.

Allen, S. J. (1968). *Ibid.* **166,** 530.

Bailey, T. (1958). *J. chem. Phys.* **28,** 792.

Bauer, R. S. and Spicer, W. E. (1970). *Phys. Rev. Lett.* **25,** 1283.

Bearden, J. A. and Burr, A. F. (1967). *Rev. mod. Phys.* **39,** 125.

Beaumont, J. H. (1971). Private communication, Clarendon Laboratory.

Bennett, H. S. (1968). *Phys. Rev.* **169,** 729.

Benson, G. C. and Dempsey, E. (1962). *Proc. R. Soc.* **A266,** 344.

——, Freeman, P. I., and Dempsey, E. (1963). *J. Am. Ceram. Soc.* **46,** 43.

Best, P. E. (1962). *Proc. phys. Soc. Lond.* **80,** 1308.

Brite, D. W. and Anderson, H. J. (1962). *U.S.A.E.C. report TID-7637,* 408.

Bujor, M. and Vook, R. W. (1969). *J. appl. Phys.* **40,** 5373.

Chadwick, D. and Graham, J. (1972). *Nature, Physical Science* **237,** 127.

Chapman, A. T. and Clark, G. W. (1965). *J. Am. Ceram. Soc.* **48,** 494.

Chouinard, M. P., Gustafson, D. R., and Heckman, R. C. (1969). *J. chem. Phys.* **51,** 3554.

—— —— (1971). *Ibid.* **54,** 5082.

Citrin, P. H., Shaw, R. W. Jnr., Packer, A., and Thomas, T.D. (1972). In *Electron Spectroscopy* (ed. D. A. Shirley), p. 691. North-Holland Publishing Company, Amsterdam.

Comly, J. B. (1968). *J. appl. Phys.* **39,** 716.

Cox, D. E., Shirane, G., Takei, W. J., and Wallace W. E. (1963). *Ibid.* **34,** 1352.
Crowder, J. G. and Hodby, J. W. (1972). Unpublished measurements, Clarendon Laboratory, University of Oxford.
Dawson, J. K. and Lister, M. W. (1952). *J. chem. Soc.* 1952, 5041.
Dresner, J. and Heyman, P. M. (1971). *Phys. Rev.* **B3,** 2869.
Duke, J. F. (1964). National Physical Laboratory, Teddington, *U.K. Metallurgy Report* 29.
Eisenberger, P. and Adlerstein, M. G. (1970), *Phys. Rev.* **B1,** 1787.
——, Pershan, P. S., and Bosomworth, D. R. (1968). *Phys. Rev. Lett.* **21,** 543.
Elliott, R. J. (1957). *Phys. Rev.* **108,** 1384.
Feldman, B. J. (1972). Thesis, Harvard University.
—— and Pershan, P. S. (1972). *Solid State Commun.* **11,** 1131.
Forman, R. A., Hosler, W. R., and Blunt, R. F. (1972). *Ibid.* **10,** 19.
Frandon, J., Lahaye, B., and Pradal, F. (1972). *Phys. Stat. Sol.* (b) **53,** 565.
Frazer, B. C., Shirane, G., Cox, D. E., and Olsen, C. E. (1965). *Phys. Rev.* **A140,** 1448.
Guggenheim, H. (1961). *J. appl. Phys.* **32,** 1337.
—— (1963). *Ibid.* **34,** 2482.
Hardcastle, K. and Warf, J. C. (1966). *Inorg. Chem.* **5,** 1728.
Harries, H. J. and Morris, D. F. C. (1959). *Acta Crystallog.* **12,** 657.
Hausner, H. (1965). *J. nucl. Mater.* **15,** 179.
Hayes, W., Koch, E. E., and Kunz, A. B. (1971). *J. Phys.* **C4,** L199.
Hodby, J. W. (1972). In *Polarons in Ionic Crystals and Polar Semiconductors* (ed. J. Devreese), pp. 389–459. North-Holland Publishing Company, Amsterdam.
——, Jenkin, G. T., Kobayashi, K., and Tamura, H. (1972). *Solid State Commun.* **10,** 1017.
Hodgman, C. D. (1959). *Handbook of Chemistry and Physics* (41st edition). Chemical Rubber Publishing Company, Cleveland, Ohio.
Hukin, D. and Ritchie, I. (1971). Private communication, Clarendon Laboratory, University of Oxford.
Jorgensen, C. K. (1971). *Chimia* **25,** 213.
—— (1972). *Theoret. Chim. Acta* (Berlin) **24,** 241.
——, Berthou, H., and Balsenc, L. (1971). *J. Fluorine Chem.* **1,** 327.
Kaiser, W. and Keck, M. J. (1962). *J. appl. Phys.* **33,** 762.
Kessler, A. and Caffyn, J. E. (1972). *J. Phys.* **C5,** 1134.
Kingsley, J. D. and Prener, J. S. (1962). *Phys. Rev. Lett.* **8,** 315.
Kopp, J. P. and Schreiber, D. G. (1967*a*). *J. appl. Phys.* **38,** 1373.
—— —— (1967*b*). *Phys. Letters* **A24,** 323.
Kubota, Y. and Wallace, W. E. (1962). *J. Appl. Phys.* **33S,** 1348.
—— —— (1963*a*). *Ibid.* **34,** 1348.
—— —— (1963*b*). *J. chem. Phys.* **39,** 1285.
Kurita, S. and Kobayashi, K. (1971). *J. Phys. Soc. Japan* **30,** 1645.
Kurki-Suonio, K. and Meisalo, V. (1966). *Ibid.* **21,** 122.
Leask, M. J. M., Roberts, L. E. J., Walter, A. J., and Wolf, W. P. (1963). *J. chem. Soc.* **1963,** 4788.
Libowitz, G. G. (1965). *The Solid State Chemistry of Binary Metal Hydrides.* W. A. Benjamin, New York.
—— (1972). In *Proceedings of the International Meeting on Hydrogen in Metals*—Jülich, Germany, 1972.
——, Pack, J. G., and Binnie, W. P. (1972). *Phys. Rev.* **B6,** 4540.

Löwdin, P. O. (1956). *Adv. Phys.* **5,** 1.
Maslen, V. W. (1967). *Proc. phys. Soc. Lond.* **91,** 466.
Matzke, H. and Ronchi, C. (1972). *Phil. Mag.* **26,** 1395.
McGuire, G. E. (1972). D. Phil. thesis, University of Tennessee. ORNL-TM-3820.
Moore, C. E. (1970). National Bureau of Standards. Publication N.B.S.—34.
Morris, D. F. C. and Schmeising, H. N. (1958). *Nature, Lond.* **181,** 469.
Nassau, K. (1961). *J. appl. Phys.* **32,** 1820.
Nasu, S. (1966). *Japan J. appl. Phys.* **5,** 1001.
O'Connor, J. R. and Chen, J. H. (1963). *Phys. Rev.* **130,** 1790.
Onodera, Y., Okazaki, M., and Inui, T. (1968). *J. phys. Soc. Japan* **21,** 2229.
Pebler, A. and Wallace, W. E. (1962). *J. phys. Chem.* **66,** 148.
Phillips, J. C. (1964). *Phys. Rev.* **136,** A1705.
—— (1970). *Rev. mod. Phys.* **42,** 317.
Prener, J. S. and Kingsley, J. D. (1963). *J. chem. Phys.* **38,** 667.
Reitz, J. R., Seitz, R. N., and Genberg, R. W. (1961). *J. Phys. Chem. Solids* **19,** 73.
Robins, E. G. (1961). *J. nucl. Mater.* **3,** 294.
Rubloff, G. W. (1972). *Phys. Rev.* **B5,** 662.
Saar, A. M. E. and Elango, M. A. (1972). *Sov. Phys. Sol. State* **13,** 2985.
Sanchez, C. and Cardona, M. (1972). *Phys. Stat. Sol.* **50,** 293.
Sasaki, K. and Obata, Y. (1970). *J. Phys. Soc. Japan* **28,** 1157.
Seager, C. H. (1971). *Phys. Rev.* **B3,** 3479.
Slowinski, E. and Elliott, N. (1952). *Acta Cryst.* **5,** 768.
Starostin, N. V. (1969). Sov. Phys.—Solid State **11,** 1317.
Stephan, E., Le Calvez, Y., Lemonier, J. C., and Robin, S. (1969). *J. Phys. Chem. Solids* **30,** 601.
Switendick, A. C. (1970). *Solid State Commun.* **8,** 1463.
Tan, Y. T. and Kramp, D. (1970). *J. chem. Phys.* **53,** 3691.
Thornber, K. K. and Feynman, R. P. (1970). *Phys. Rev.* **B1,** 4099.
Togawa, S. (1964). *J. Phys. Soc. Japan* **19,** 1696.
Tomiki, T. and Miyata, T. (1969). *Ibid.* **27,** 658.
Tosi, M. P. (1964). *Solid State Physics* (eds. Seitz and Turnbull), Vol. 16. Academic Press.
Trautweiler, F., Moser, F., and Khosla, R. P. (1968). *J. Phys. Chem. Solids* **29,** 1869.
Van Lierde, W., Strumane, R., Smets, E., and Amelinckx, S. (1962). *J. nucl. Mater.* **5,** 250.
Warf, J. C. and Hardcastle, K. (1961). *J. Am. Chem. Soc.* **83,** 2206.
Warren, B. E. (1961). *Acta Cryst.* **14,** 1095.
Weiss, A., Witte, H., and Wölfel, E. (1957). *Z. phys. Chem.* **10,** 98.
Weller, P. F. (1965). *Inorg. Chem.* **4,** 1545.
Willis, B. T. M. (1963*a*). *Proc. R. Soc.* **A274,** 122.
—— (1963*b*). *Ibid.* **A274,** 134.
Witte, H. and Wölfel, E. (1958). *Rev. mod. Phys.* **30,** 51.
Wyckoff, R. W. G. (1965). *Crystal Structures* (2nd edition) Vol. 1, p. 239. Interscience Publishers.
Zachariasen, W. H. (1967). Quoted by C. Kittel in *Introduction to Solid State Physics* (3rd edition) p. 105. J. Wiley.

2

LATTICE DYNAMICS

2.1. Introduction

In Chapter 1 it was shown that the cohesive properties of some fluorites were well described by a simple model of point charges and short-range repulsive interactions. It will appear in this chapter that the lattice-dynamic and dielectric properties are reasonably well described by a modest generalization of this model.

The lattice-dynamic properties of the fluorite structure are important in understanding the nature of the pure crystal at an atomic level and the behaviour of defects. The analysis of perfect crystal properties given here follows the methods of Born and Huang (1954) or the various extensions reviewed by Cochran (1971). After a brief survey of the content of these methods, we shall discuss more detailed models and their relation to observations of harmonic and anharmonic properties of both perfect and imperfect crystals.

Conventional lattice dynamics is based on three major approximations. These are:

(I) An adiabatic approximation, which is used to separate the electronic and nuclear motion. One result of this approximation is that a unique potential can be defined for each configuration of the nuclei.

(II) The harmonic approximation, which is used to relate the vibrational states of the lattice to those of an exactly-soluble model, an ensemble of harmonic oscillators. This approximation is accurate when the ions do not move far from their perfect lattice sites. Anharmonic effects are frequently observed, e.g. thermal expansion, and will be treated separately in § 2.4.

(III) The dipole approximation, that we can ignore higher multipoles than dipoles on any ion, is introduced to simplify both microscopic models and the fitting of parameters to experiment. These higher multipole terms appear to be small, but there is little evidence of their precise importance.

The equations of motion of the nuclei in the harmonic approximation can be written down classically, or from the Heisenberg equations of motion. If the ionic displacements $\boldsymbol{\xi}(\mathbf{l})$ have a time dependence $\exp(i\omega t)$, then the general form is:

$$-M_l \xi_\alpha(\mathbf{l})\omega^2 + \sum_{\beta \mathbf{l}'} A_{\alpha\beta}(\mathbf{l}, \mathbf{l}')\xi_\beta(\mathbf{l}') = 0. \tag{2.1}$$

Here $\mathbf{l}$ labels the sites, α and β are cartesian coordinates, M_l is the nuclear mass, and $\mathbf{A}$ the force-constant matrix. There are contributions to $\mathbf{A}$ from both the Coulomb interactions of the point charges and from forces of shorter range. The lattice periodicity means that $\mathbf{A}(\mathbf{l}, \mathbf{l}')$ depends only on the relative separation of the nuclei involved, $\mathbf{R}_l - \mathbf{R}_{l'}$, and this can be exploited to give a major simplification. We define the dynamical matrix $\mathbf{D}(\mathbf{q})$ by

$$A_{\alpha\beta}(\mathbf{l}, \mathbf{l}') = \frac{1}{N}\sum_{\mathbf{l}-\mathbf{l}'} D_{\alpha\beta}(\mathbf{q})\exp[-i\mathbf{q}(\mathbf{R}_\mathbf{l} - \mathbf{R}_{\mathbf{l}'})]. \tag{2.2}$$

The vibrational states of the lattice are then simply the same as those of a set of harmonic oscillators whose frequencies are obtained from the secular equation:

$$\det \| \mathbf{D}(\mathbf{q}) - \omega^2 \mathbf{M} \| = 0. \tag{2.3}$$

The normal modes are plane-wave combinations of the displacements ξ, weighted by the eigenvectors of $\mathbf{D}(\mathbf{q})$, and the frequencies $\omega(\mathbf{q})$ are obtained from the eigenvalues. The fluorite structure has three atoms per unit cell, so $\mathbf{D}$ is a (9×9) matrix, and $\mathbf{M}$ is a diagonal (9×9) matrix with three components equal to the cation mass and six equal to the anion mass. For each value of $\mathbf{q}$ there are nine values $\omega(\mathbf{q})$, giving a phonon spectrum with nine branches. Properties of the phonon spectra are analysed in § 2.2, which discusses long-wavelength phonons and related properties, and in § 2.3, which treats phonon dispersion and various lattice-dynamic models.

So far we have considered a perfect lattice, and lattice periodicity has been exploited. If there is a local imperfection then equations similar to those above can be solved (see e.g. Dawber and Elliott, 1963a). The main change is that defect terms appear on the right-hand side of (2.1). The solution is conveniently obtained in terms of the Green function, $G_{\alpha\beta}(\mathbf{l},\mathbf{l}')$, which is defined by:

$$-\omega^2 G_{\alpha\gamma}(\mathbf{l}, \mathbf{l}'') + \sum_{\beta,\mathbf{l}'} \tilde{A}_{\alpha\beta}(l, l')G_{\beta\gamma}(\mathbf{l}'\mathbf{l}'') = \delta_{\alpha\beta}\,\delta(l, l''), \tag{2.4}$$

where $\tilde{A}_{\alpha\beta}(l, l') \equiv A_{\alpha\beta}(l, l')/\sqrt{(M_l M_{l'})}$ is the mass-reduced force-constant matrix. At zero frequency $\mathbf{G}$ reduces to $\tilde{\mathbf{A}}^{-1}$, the reciprocal of the reduced force-constant matrix. If the changes in mass or force constant of the defect are well localized, then the defect problem can be solved without difficulty. The simplest case is the isotopic impurity, in which a host atom of mass M is replaced by an impurity of mass M' without changes in force constants.

The introduction of point defects into ionic crystals may activate three types of infrared absorption and Raman scattering.

(a) Localized modes arising from vibrations of light impurities. This type of vibration occurs at frequencies higher than the band-mode vibrations and gives rise to sharp lines. Examples are the local modes of hydrogen in fluorites treated in § 2.5. The amplitude of a localized vibrational mode is large near the defect and dies away rapidly with distance from the defect.

(b) Resonant modes which may be activated by all impurities. This type of vibration gives rise in general to broader peaks within the band modes of the host crystal. The amplitude of resonance modes is enhanced near the defect but these modes may be transmitted through the lattice and closely resemble unperturbed modes at large distances from the defect. Heavy metal impurities lead to resonant modes in fluorites, and are discussed in § 2.6.

(c) Gap modes in crystals with an energy gap in the band-mode region (the alkaline earth fluorides are not in this category; see Figs. 2.2, 2.3, and 2.4). This type of vibration may be activated by all defects and gives rise to sharp lines and localized excitations.

Reviews of work on defect vibrational modes in crystals have been given by Maradudin (1966*a*, *b*), Elliott (1966), and Newman (1969).

2.2. Long wavelength properties

We begin by considering the long-wavelength (small **q**) properties in the harmonic approximation. These include the elastic, dielectric and some infrared properties of the perfect crystal.

Three of the nine branches in the phonon spectrum are acoustic branches, whose frequency tends to zero at long wavelengths. The other six modes are optic modes, whose frequencies are finite for all wavelengths. The general form of the phonon spectrum for small wave vectors **q** is shown schematically in Fig. 2.1. The classification into longitudinal and transverse polarizations is only exact for certain high-symmetry directions of **q**, but the division of the optic modes into Raman and infrared-active is precise. In the Raman mode the cation remains fixed whilst the two anions move in opposition; in the infrared mode the anions move together in opposition to the cation.

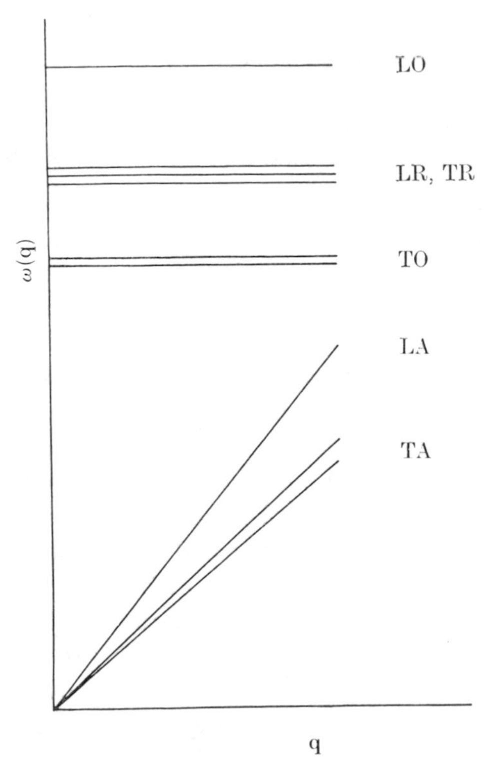

FIG. 2.1. Phonon dispersion relations at long wavelength for the fluorite structure. L = longitudinal, T = transverse, A = acoustic, O = optic and R = Raman.

2.2.1. *Elastic properties*

The long-wavelength acoustic modes determine the elastic properties of the crystal. The values of the elastic constants depend to some extent on the method of measurement. For example, ultrasonic measurements give adiabatic constants and static measurements give isothermal constants. The ultrasonic data are more accurate than the static results at the present time. Second-order elastic constants c_{ij} are given in Table 2.1, together with Debye temperatures. There are several definitions of the Debye temperature, depending on whether it is derived from specific heat, elastic constant or other vibrational data.

Two possible relations among the elastic constants are the condition of elastic isotropy ($c_{11} = c_{12}+2c_{44}$) and the Cauchy relation ($c_{12} = c_{44}$). The Cauchy relations only hold for a crystal in equilibrium under two-body central forces and where each atom is at a site with inversion symmetry. In the fluorite structure, the anion sites lack inversion

TABLE 2.1
Elastic constants and Debye temperatures

		CaF_2	SrF_2	BaF_2	CdF_2	EuF_2	$SrCl_2$	UO_2
Temperature (K)				Elastic constants				
	0	17·124[f] 17·4[g]	12·87[d]	9·81[e]		12·40[j]	7·55[j]	
c_{11}	77	17·096[f]	12.82[d]			12.16[j]		
	300	16·494[f] 16·4[g]	12·35[d]	8·915[e]	18·270[b] 18·6[i]	10·74[j]	6·80[j]	3.95[k]
	0	4·675[f] 5·6[g]	4·748[d]	4·481[e]			1·72[j]	
c_{12}	77	4·655[f]	4·695[d]			5·01[j]		
	300	4·462[f] 5·3[g]	4·305[d]	4·002[e]	6·674[b] 6·8[i]	4·34[j]	1·60[j]	1·21[k]
	0	3·624[f] 3·593[g]	3·308[d]	2·544[e]			1·03[j]	
c_{44}	77	3·6085[f]	3·291[d]			3·17[j]		
	300	3·380[f] 3·593[g]	3·128[d]	2·535[e]	2·175[b] 2·17[i]	2·99[j]	0·945[j]	0·64[k]
Debye Temperatures (K) obtained from								
(a) Specific heat data		508[g,f]		169[e]	328·5[b]	329[j]	378[j]	377[l]
(b) Elastic data		519·4[f] 513·6[g]	380[d]	282[e]	337[b]			

The c_{ij} are in units 10^{11} dynes/cm². All measurements were made by ultrasonic methods.
Sources are [b] Alterovitz and Gerlich 1970b, [d] Gerlich 1974a, [e] Gerlich 1964b, [f] Ho and Ruoff 1967, [g] Huffman and Norwood 1960, [i] Hart 1970, [j] Lauer, Solberg, Kühner, and Bron 1971, [k] Wachtman, Wheat, Anderson, and Bates 1965, [l] Willis 1963. Debye temperatures from various moments of the phonon spectrum are given by Bailey and Yates (1967).
For UO_2 a better description gives the U sublattice a 242 K Debye temperature and the O sublattice an 815 K Einstein temperature (Willis, 1963).
Values quoted at OK are extrapolations.

symmetry, so one does not expect c_{12} and c_{44} to be equal. Axe (1965) has shown that the differences $(c_{12}-c_{44})$ for CaF_2, SrF_2, and BaF_2 can be explained in a shell model without the need to introduce three-body forces of the type necessary in alkali halides (Cochran 1971), although these contributions may also be present. The isotropy condition will only hold fortuitously. Only BaF_2 (and to a lesser extent, EuF_2) approach elastic isotropy, and even for BaF_2 some anharmonic properties are strongly anisotropic (§ 2.4.4).

The elastic constants can be related to the microscopic forces between crystal ions. The most important contributions are from the Coulomb forces and from short-range forces. The general forms (e.g. Srinivasan

1968*b*) are:

$$
\begin{aligned}
c_{11}/(e^2/a_0^4) &= 1{\cdot}5256+\eta_1+\zeta_1, \\
c_{12}/(e^2/a_0^4) &= -2{\cdot}7022+\eta_2-\zeta_2, \\
c_{44}/(e^2/a_0^4) &= -0{\cdot}7628+\eta_1+\zeta_2+I_1.
\end{aligned}
\tag{2.5}
$$

The unit (e^2/a_0^4) can be written as $(1{\cdot}298\times 0{\cdot}10^{13}$ dyne/cm^2)/(nearest neighbour distance in Å)^4. In each case the first term comes from Coulomb forces, the second (η_1 or η_2) from nearest-neighbour anion-cation forces and the third (ζ_1 or ζ_2) from anion-anion interactions. The final term in c_{44} is more complicated, and involves all the previous contributions together with explicit shell-model parameters. The structure of these equations is important in parameterizing models of the lattice dynamics (cf. § 2.3.2). Most of the lattice-dynamic calculations described later give explicit forms for these elastic constants in their own notation.

2.2.2. *Dielectric, infrared, and Raman properties*

The infrared and Raman spectra are determined by the optic modes at long wavelengths. Fluorite crystals have one infrared-active transverse optic mode with symmetry Γ_{4u}. They also have a Raman-active vibration with symmetry Γ_{5g} (Fig. 2.1). Values of the various optic-mode energies at $\mathbf{q} = 0$ are given in Table 2.2. Room-temperature Raman energies are also available for the inter-metallic compounds $AuAl_2$ (267 cm^{-1}, Feldman, Parker, and Ashkin 1968; Brya, 1971), $AuGe_2$ (149 cm^{-1}, Brya), and $AuIn_2$ (124 cm^{-1}, Brya). The zone-centre longitudinal and transverse optic modes are normally split because of the electric field associated with the relative motion of the anions and cations. No such effect occurs for the Raman modes, and ω_{LR} and ω_{TR} are equal at long wavelengths. In the inter-metallic compounds like $AuAl_2$, the conduction electrons are expected to screen the internal electric field so that the zone-centre transverse and longitudinal optic-mode frequencies will be equal (Brya).

The dielectric constants depend on both the long-wavelength optic modes and on the electronic structure of the crystal ions. Values of the dielectric constants are given in Table 2.3. The optic dielectric constant is determined by the electronic polarization of the constituent ions. This is best parameterized within a shell model, although there are several estimates of 'ionic polarizabilities'. The earliest estimates (Fajans and Joos, 1924) were based on rare-gas refractivities and

TABLE 2.2
Zone-centre phonon frequencies obtained from optical data.

	CaF_2	SrF_2	BaF_2	CdF_2	PbF_2	EuF_2	$SrCl_2$
Temperature (K)			Transverse optic modes				
80	267[c]	225[c]	189[c]	224[c]		203[a]	155[y]
100	272[f]	224[f]	193[f]				
300	257[cj]	217[c]	184[cj]	209[f]	106[f]	194[a]	148[y]
	266[f]	219[f]	189[f]	206[c]	102[b]		
	261[w]	223[w]	187·5[w]				
			186[i]				
			Longitudinal optic modes				
5	484[w]	397[w]	346[w]			358[a]	
80	472[c]	384[c]	330[c]	403[c]			
100	474[f]	384[f]	333[f]	410[f]	340[f]		
200	468[c]						
300	463[c]	381[c]	331[c]	380[c]	338[f]	347[z]	(245)[z]
	482[w]	395[w]	344[w]	404[f]			
	475[f]	382[f]	353[i]				
	478[e]	388[x]	326[j]				
			Raman active modes				
80	330[z]	290[z]	249[z]				188[z]
100	327·5[f]	290·2[f]	248[f]		256·4[f]	292[z]	
300	321·5[f]	285·5[f]	241[f]	317·1[f]	256·3[f]	287[z]	183[z]
	322[jn]	284[z]	243[i]		256[z]		
	322[z]		244[m]				
			242[z]				

Units are cm^{-1}.
The sources are [a] Axe and Pettit (1966), [b] Axe, Gaglianello and Scardefield, 1965, [c] Bosomworth 1967, [e] Cribier *et al.* 1962, [f] Denham *et al.* 1970, [i] Hurrell and Minkiewicz 1970, [j] Kaiser *et al.* 1962, [m] Warrier and Krishnan 1964, [n] Russell 1965, [w] Lowndes 1971, [x] Berreman 1963, [y] Hisano, Ohama, and Matumura (1965), [z] Srivastava, Lauer, Chase, and Bron (1971).

properties of ions in solution; subsequent ones (Born and Heisenberg 1924, Pauling, 1927) were based on oscillator strengths and absorption frequencies. These estimates are of free-ion polarizabilities, and agree reasonably well with each other. Thus the Pauling values are 1·04 Å^3 (F^-), 0·47 Å^3 (Ca^{2+}), 0·86 Å^3 (Sr^{2+}) and 1·55 Å^3 (Ba^{2+}). On the other hand, Tessmann, Kahn, and Shockley (1953) obtained estimates for ions in crystals using dielectric data. Their results are very different,

TABLE 2.3
Dielectric constants

Temperature (K)	CaF_2	SrF_2	BaF_2	CdF_2	EuF_2	$SrCl_2$	UO_2
			Static dielectric constant				
4	6·47[f]	6·15[f]	6·96[f]		7·52[c]	6·94[c]	
80	6·38[b]	6·04[b]	6·56[b]	7·78[b]			
	6·4[g]	6·19[f]	7·01[f]				
	6·51[f]						
200	6·53[b]	6·30[f]	7·16[f]				
	6·66[f]						
300	6·63[b]	6·20[b]	6·94[b]	8·49[b]		9·2[c]	24[h]
	6·78[g]	6·48[g]	7·28[g]	8·33[h]			
	6·35[d]		7·02[d]				
	6·7[eH]	6·6[eH]	7·2[eH]				
	6·65[eO]	6·14[eO]	6·73[eO]				
	6·7984[aH]	6·4647[aH]	7·3590[aH]				
	6·8074[aO]	6·4700[aO]	7·3606[aO]				
	6·81[f]	6·50[f]	7·32[f]				
			Optic dielectric constant				
4	2·05[f]	2·08[f]	2·18[f]				
80	2·047[b]	2·07[b]	2·157[b]	2·40[b]	2·42[c]	2·7[c]	
200	2·044[b]						
300	2·040[b]	2·07[b]	2·15[b]	2·40[b]		2·72[c]	5·3[k]
	2·04[f]	2·07[f]	2·16[e]				
			2·17[f]				

Results are due to [a] Andeen *et al.* 1971, [b] Bosomworth 1967, [c] Kühner, Lauer and Bron 1972, [d] Jones 1967, [e] Kaiser *et al.* 1962, [f] Lowndes 1969, [g] Rao and Smakula 1966, [h] Briggs 1964, [k] Ackerman, Thorn, and Winslow 1959. Symbols H and O refer to Harshaw and Optovac crystals.

giving 0·64 Å^3 (F^-), 1·1 Å^3 (Ca^{2+}), 1·6 Å^3 (Sr^{2+}) and 2·5 Å^3 (Ba^{2+}). Neither set can be recommended strongly for crystal calculations. In particular, the divalent cation results of Tessmann *et al.* seem too large, especially when compared with their results for monovalent anions with the same electronic structure.

The infrared properties can be understood in terms of the classical dispersion theory of Born and Huang (1954). This theory gives relations between the real and imaginary parts of the dielectric constant (ϵ', ϵ''), or refractive index (n, κ), and the optic dielectric constant ϵ_∞, the oscillatory strength for one-phonon transitions ρ_1, the transverse optic frequency ω_{TO} and a damping function $\Gamma(\omega)$. In terms of reduced frequencies $x \equiv \omega/\omega_{TO}$ and a dimensionless damping function $\gamma \equiv \Gamma/\omega_{TO}$, the relations are:

$$\epsilon' \equiv n^2-\kappa^2 = \epsilon_\infty+(x^2-1)\Phi, \tag{2.6}$$

$$\epsilon'' = 2n\kappa = x\gamma\Phi, \tag{2.7}$$

where

$$\Phi = 4\pi\rho_1\{(x^2-1)^2+\gamma^2\}^{-1}. \tag{2.8}$$

Other observables, like the absorption coefficient, conductivity and reflectivity, are readily expressed in terms of n and κ. The damping function γ is a function of frequency in general, and is discussed later.

When there is no damping, the dielectric constants are related to the frequencies of the infrared-active modes at long wavelengths by the Lyddane–Sachs–Teller relation:

$$\epsilon_0\omega_{TO}^2 = \epsilon_\infty\omega_{LO}^2. \tag{2.9}$$

Lowndes (1971) has verified that this relation holds with reasonable accuracy for CaF_2, SrF_2, and BaF_2, and discusses some of the effects of damping.

2.3. Phonon dispersion

The most important measurements of phonon frequencies for all wavelengths are those obtained from inelastic neutron scattering. Results are available for CaF_2, SrF_2, BaF_2, and UO_2. Other measurements give more limited information about phonon frequencies, for example, the long-wavelength data of § 2.2. Further instances include observations of two-phonon infrared bands (Fray, Johnson, and Quarrington, 1964), the second-order Raman effect (Kam and Cohen, 1971) and studies of vibronic spectra (Kiel and Scott, 1970; Kam and Cohen; Kühner, Lauer, and Bron, 1972; Hurrell, Kam, and Cohen, 1972). These data are much harder to interpret than the neutron data, and so are less satisfactory as a direct source of information. They do, however, provide detailed information about singularities in the density of states, and can give more precise energies for phonons identified by neutron measurements.

We shall be concerned here with models which allow us to understand the lattice-dynamic and dielectric properties. It will appear that, as for the alkali halides, reasonably simple and plausible models can be devised which fit the harmonic data quite well and even fit some anharmonic results. However, it is not possible to obtain a unique model of the interatomic forces, and various choices are possible.

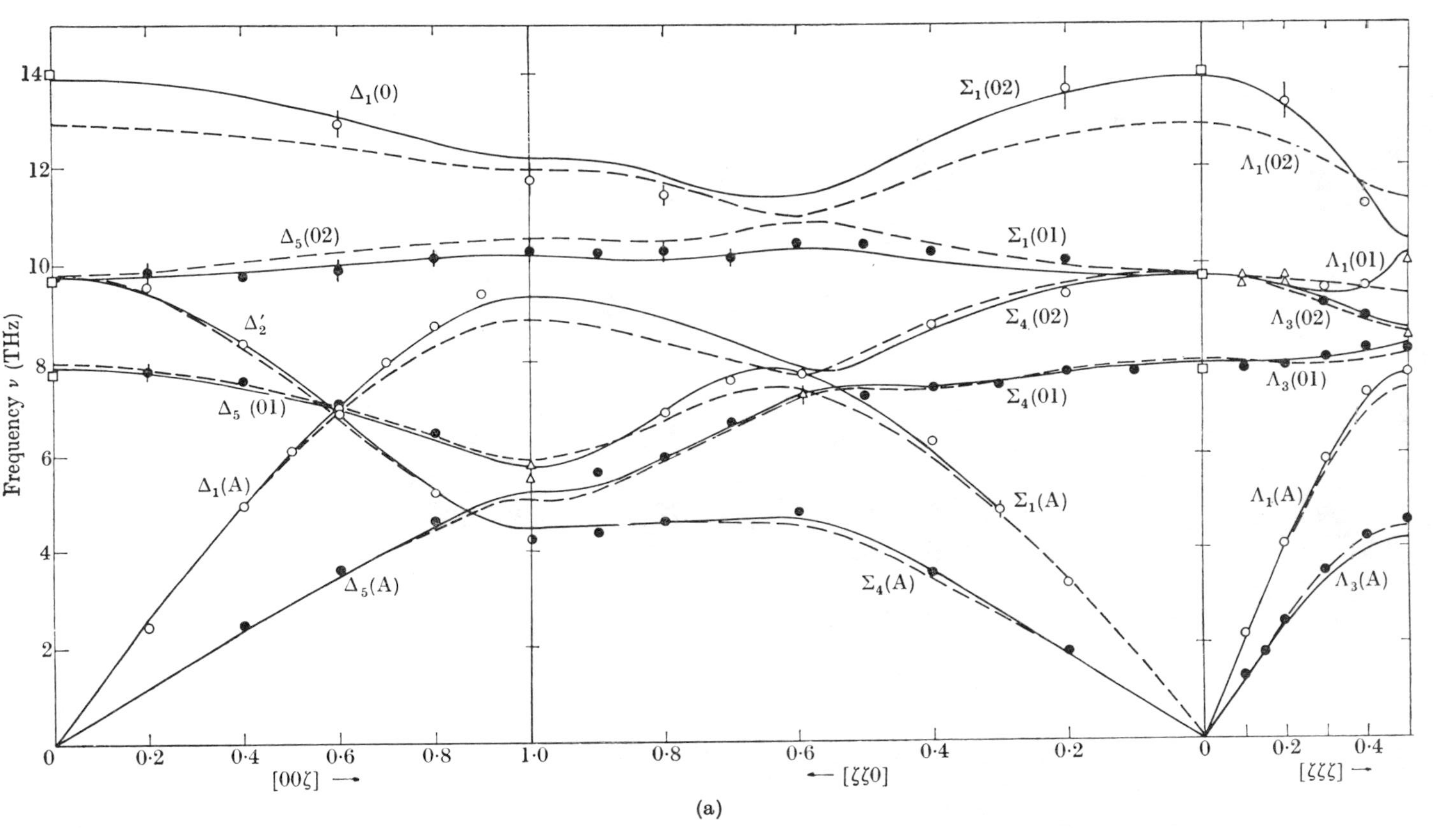

Fig. 2.2. (a) Phonon dispersion relations for CaF_2 at room temperature in the [001], [110], and [111] directions. The dashed and continuous curves represent different shell model fits to the experimental data; (b) single phonon density of states calculated using the shell model parameters giving the continuous curve in (a) (after Elcombe and Pryor, 1970).

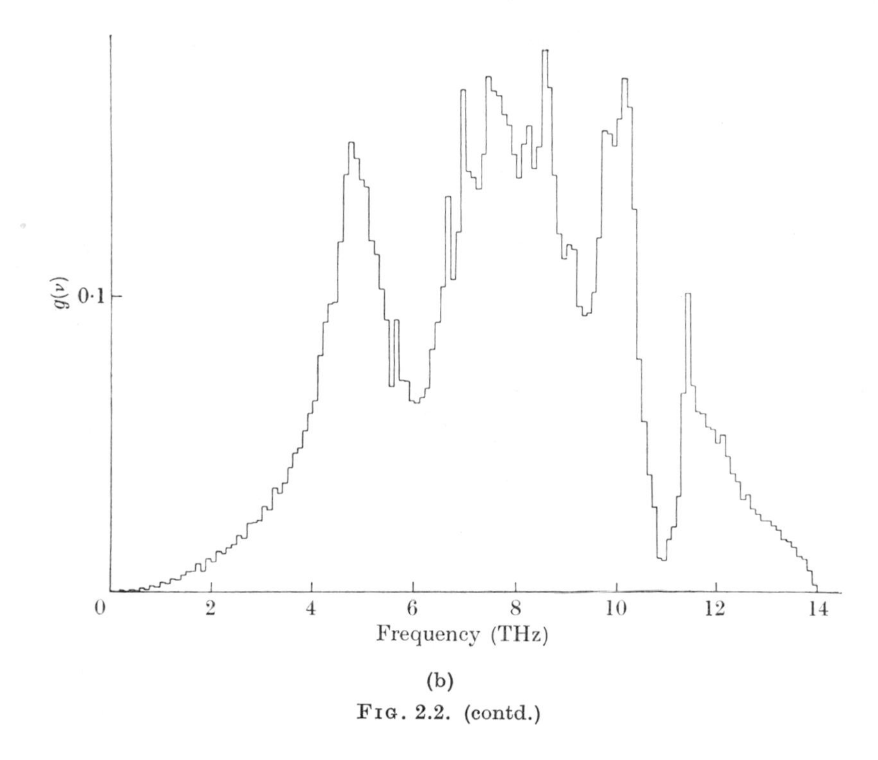

(b)

FIG. 2.2. (contd.)

Usually a model is adopted which provides a good fit to the neutron dispersion curves; examples are shown in Figs. 2.2–2.5. However, even if exact dispersion curves were available, eigenvector information would be needed to remove any ambiguities and no adequate data are available at present (see e.g. Iizumi, 1972).

2.3.1. *General outline of lattice-dynamic models*

The microscopic contributions to the force-constant matrix **A** are dominated by the Coulomb interactions between the ions and by the short-range forces. The different models can be classified according to the form of dipole approximation adopted. Three forms are common. The 'rigid-ion' model ignores electronic polarization; the only contribution to the dipole moment from a given ion is the displacement dipole associated with the rigid motion of the ion. Polarizable 'point-ion' models include both the displacement dipoles and the electronic dipoles which appear because the ions are polarizable. Finally, there are shell models and equivalent treatments. They include displacement dipoles, electronic dipoles, and deformation dipoles. The deformation dipoles appear because the displacement and electronic dipoles are not strictly

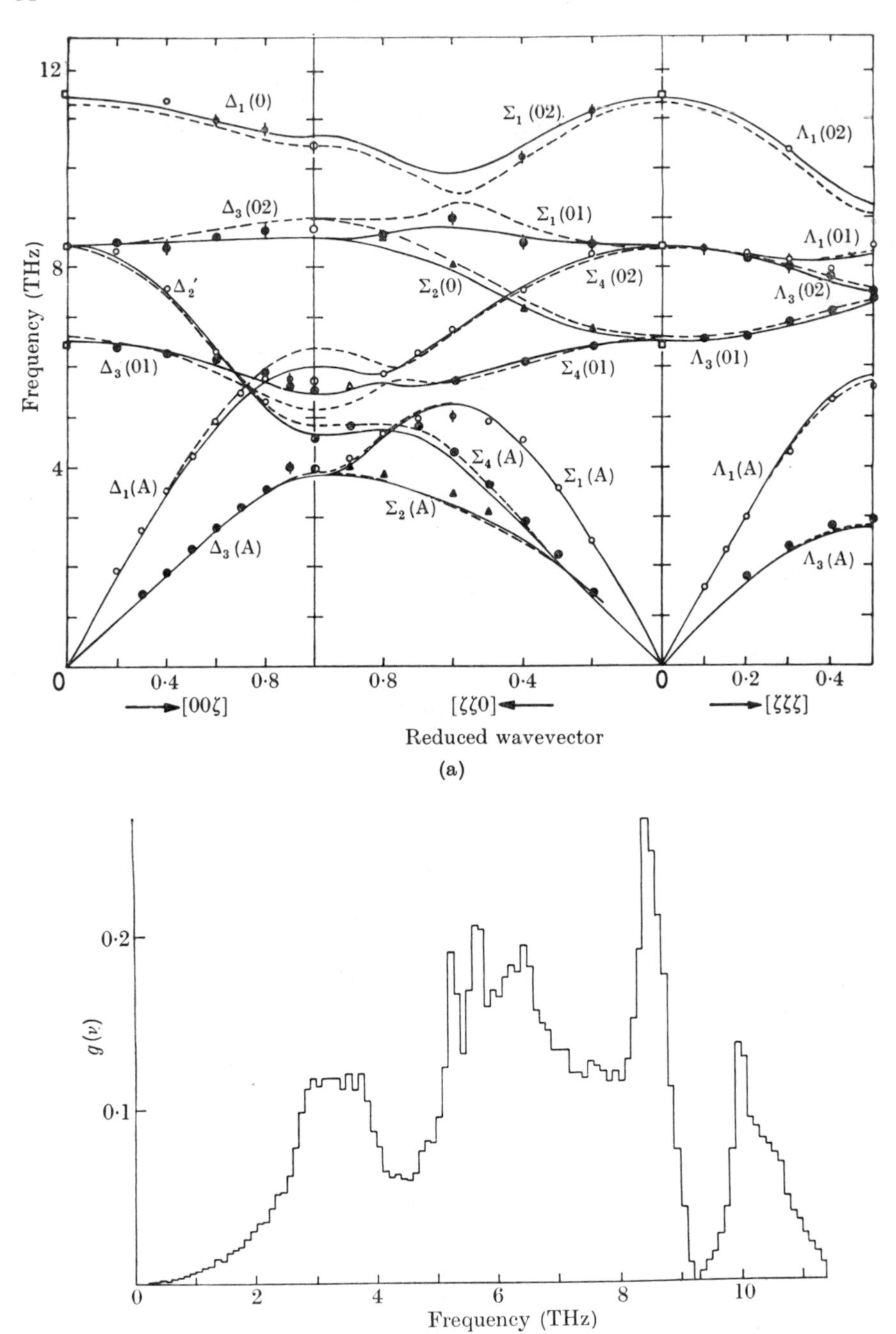

FIG. 2.3. (a) Phonon dispersion relations and (b) single phonon density of states for SrF_2, as in Fig. 2.2 (after Elcombe, 1972).

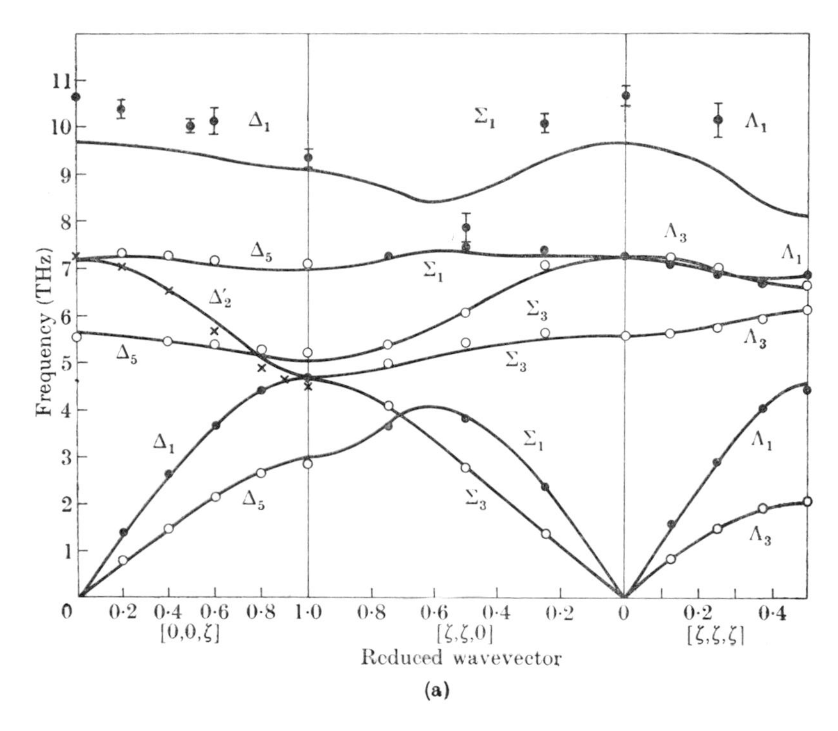

(a)

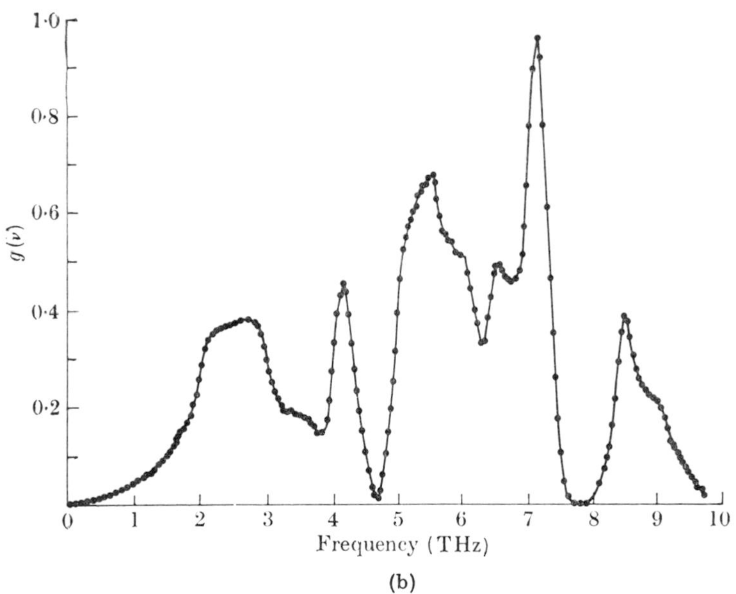

(b)

FIG. 2.4. (a) Measured and calculated phonon dispersion relations for BaF_2 at room temperature; (b) calculated single-phonon density of states (after Hurrell and Minkiewicz, 1970).

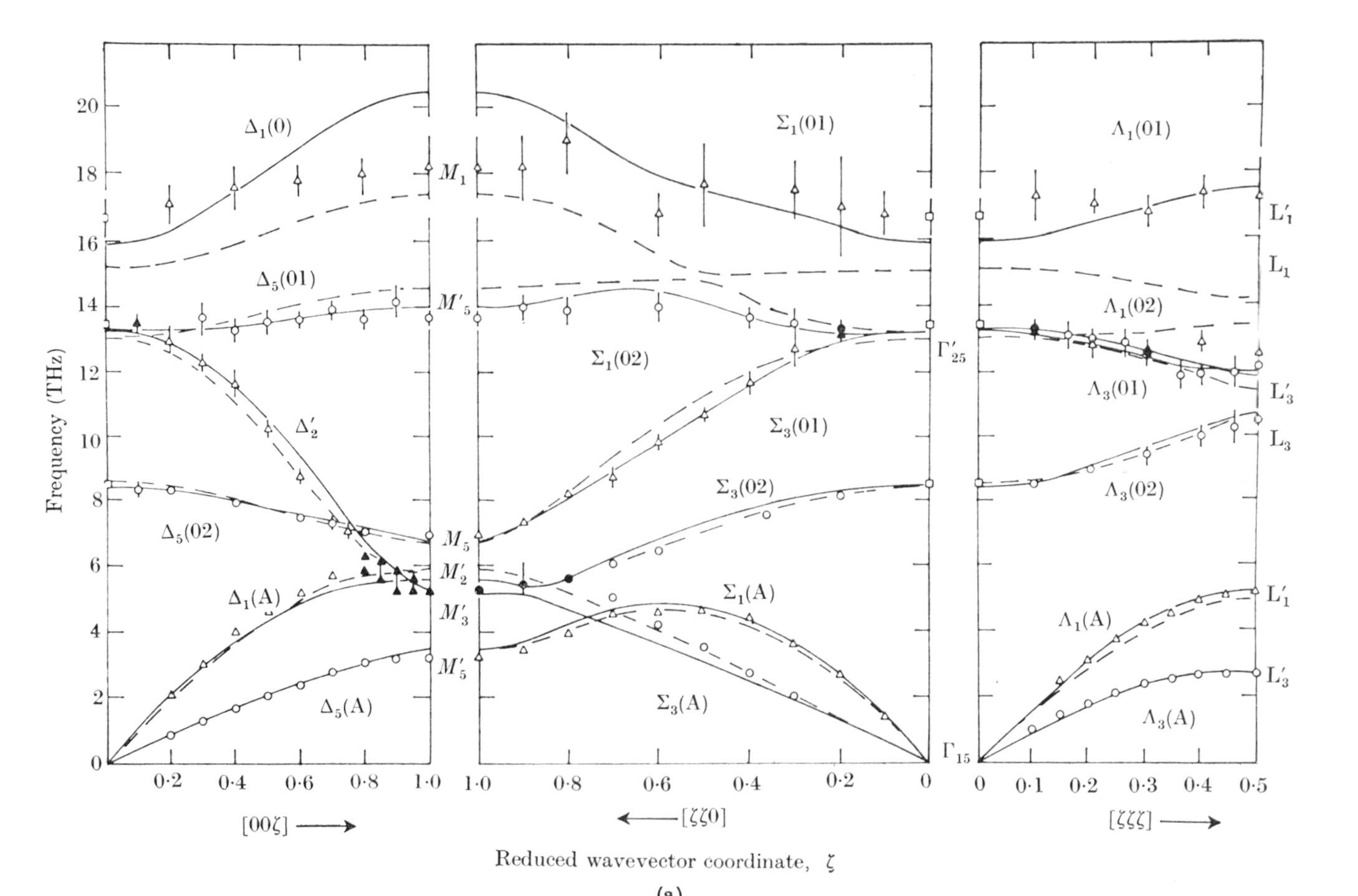

FIG. 2.5. (a) Measured and calculated phonon dispersion relations for UO_2 at room temperature;

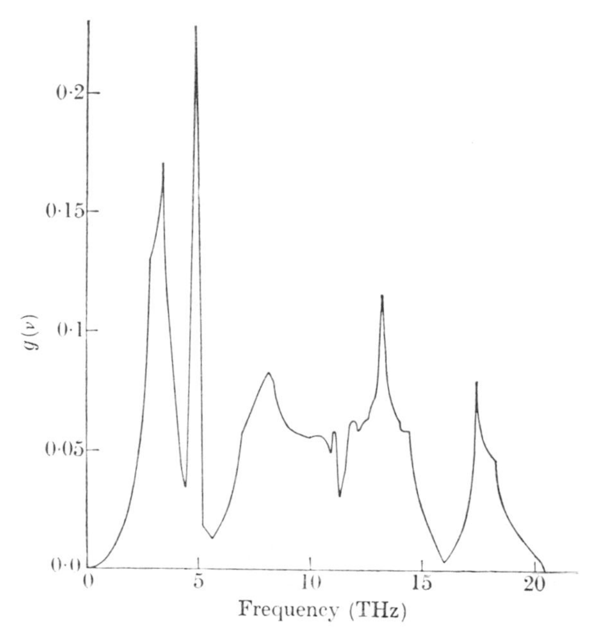

(b) calculated single-phonon density of states (after Dolling *et al.* 1965).

FIG. 2.5 (contd.)

additive; in essence, the electronic polarizability is affected by the relative motion of the neighbouring ions. The various approaches have been critically reviewed by Cochran (1971).

The physical model on which the shell model is based (Dick and Overhauser, 1958) regards each ion as a core with an effective ionic charge Ze and a shell of charge Ye. Loosely, the core corresponds to the nucleus and inner electrons, and the shell includes the outer electrons most strongly affected by the neighbouring ions. The shell and core on a given ion are harmonically coupled, k_- and k_+ being the associated force constants for the anions and cations. The electronic dipole moment comes from the relative motion of shell and core, regarded as point charges. The formalism is established by writing the potential energy as a general expression quadratic in the core displacements $\boldsymbol{\xi}$ and electronic dipoles $\mathbf{p}$:

$$V = \tfrac{1}{2}[\boldsymbol{\xi}\cdot\mathbf{A}^{\mathrm{R}}\cdot\boldsymbol{\xi}+\boldsymbol{\xi}\cdot\mathbf{A}^{\mathrm{T}}\cdot\mathbf{p}+\mathbf{p}\cdot\mathbf{A}^{\mathrm{D}}\cdot\mathbf{p}]. \tag{2.10}$$

The last term can be written as the sum of the dipole self-energies and $\mathbf{p}\cdot\mathbf{A}^{\mathrm{S}}\cdot\mathbf{p}$. The matrices $\mathbf{A}^{\mathrm{R}}$, $\mathbf{A}^{\mathrm{S}}$, and $\mathbf{A}^{\mathrm{T}}$ correspond to the $\mathbf{R}$, $\mathbf{S}$, $\mathbf{T}$ matrices conventionally used (Cochran, 1971). Equations of motion, analogous to (2.1) can now be constructed.

For perfect crystal properties, it is convenient to use the adiabatic approximation to eliminate the dipole moments. If the electronic dipoles adjust to follow the nuclear motion, $\partial V/\partial \mathbf{p} = 0$. The effective harmonic potential is then just:

$$V_{\text{adiabatic}} = \tfrac{1}{2}\boldsymbol{\xi}\,.\,[\mathbf{A}^{\mathrm{R}}-\mathbf{A}^{\mathrm{T}}\,.\,(\mathbf{A}^{\mathrm{D}})^{-1}\,.\,\mathbf{A}^{\mathrm{T}}].\boldsymbol{\xi}. \tag{2.11}$$

However, this adiabatic potential should not be used in static calculations of the distortion near defects, since its second term corresponds to long-range forces of an inconvenient form. Instead, as in Chapter 3, one works explicitly with both shell and core displacements.

We now look in more detail at some of the models proposed for fluorite structures.

2.3.2. *Lattice-dynamic models for the perfect crystal*

There are now many calculations of the lattice dynamics of fluorites. The earliest (Srinivasan, 1958; Reitz, Seitz, and Genberg, 1961; Rajagopal, 1962) used a rigid-ion model with central two-body forces. These models do not account for the difference between c_{12} and c_{44}, and will not be discussed further. Ganesan and Srinivasan (1962) used a similar model in extensive calculations of both lattice dynamics and derived quantities. However, this has been superseded by other calculations, including Srinivasan's own work (Srinivasan, 1968*b*) on the lattice theory of an elastic dielectric. We shall concentrate here on a number of recent models. Some of their important parameters are given in Tables 2.4 and 2.5, including ionic charges, shell and core charges, and the shell-core force constants.

TABLE 2.4

Effective ionic charges Z in the rigid-ion model

	CaF_2	SrF_2	BaF_2	CdF_2	PbF_2
a	0·80	0·84	0·85	0·84	0·94
b	0·78	0·82	0·81	0·83	0·93
c	0·67				
d	0·79	0·84	0·88		

Values are chosen so that the cation has charge $2Z\,|e|$ and the anion $-Z\,|e|$.

The results are [a] Denham *et al.*, 100 K; [b] Denham *et al.*, 300 K; [c] Elcombe and Pryor Model I, 295 K; [d] Verleur and Barker, 'local' ionic charges derived from 100 K and 300 K mixed-crystal data.

TABLE 2.5

Shell charges, Y, shell-core force constants, k, and core charges, Z.

		Y_+	k_+	Y_-	k_-	Z
CaF_2	a	−8·7*	6610	−2·35	215	0·833
	b	+5·23	665	−3·245	978	0·97
	c	7·96	1523	−4·35	1680	1·0
	e	5·24	547·6	−2·38	141·8	1·0
	f	4·11	168·1	−4·04	929·5	1·0
SrF_2	a	−9·9*	5620	−2·35″	258	0·875
	e	7·53	889	−3·70	423·1	1·0
	f	6·65	443·8	−4·93	1491·4	1·0
BaF_2	a	−11·3*	4830	−2·35″	314	0·896
	d	−17·47	11706	−1·92	213	0·94
	e	−16·99	3495·2	−1·59	88·9	1·0
	f	−4·93	580·4	−2·17	97·5	1·0

The units are e for the charges and e^2/(cell volume) for force constants. Sources are [a] Axe, [b] Elcombe and Pryor Model II, [c] Elcombe and Pryor Model III, [d] Hurrell and Minciewicz, [e] Catlow and Norgett Model CNA, [f] Catlow and Norgett Model CNB.
Values marked * were taken from Dick and Overhauser, and values ″ were chosen to agree with the corresponding CaF_2 values.

The first model, due to Denham, Field, Morse, and Wilkinson (1971) is the only rigid-ion model considered in detail here. We discuss it because it shows the strengths and weaknesses of rigid-ion treatments, and because results are available for a wide range of crystals (CaF_2, SrF_2, BaF_2, CdF_2, and PbF_2). Since the electronic polarization is ignored, $\epsilon_\infty = 1$ automatically and this is a serious disadvantage. Denham *et al.* assume short-range interactions between nearest neighbours and next-nearest-neighbours, although two terms (non-axial fluorine-fluorine interactions and one cation-cation term) are dropped, following Ganesan and Srinivasan. A dynamic charge, Z, is chosen to ensure ω_{LO} is given correctly at $\mathbf{q} = 0$; the dynamic cation and anion charges are then $2Ze$ and $-Ze$ respectively. These charges (Table 2.4) are quite distinct from those used to give the correct cohesive energy (§ 1.3.3) and are given by

$$Z^2e^2 = (\omega_{\text{LO}}^2 - \omega_{\text{TO}}^2)\pi a^3 M_+ M_- / (M_+ + 2M_-) \qquad (2.12)$$

(Cochran, 1971). Relations similar to this hold for the more general models discussed later. However, the effective dynamic charge in these later models is altered to $Z\{3\epsilon_\infty/(\epsilon_\infty + 2)\}$ by the electronic polarization. The eight free parameters of the Denham *et al.* model are chosen to fit long-wavelength data, and the results are used to calculate the phonon dispersion curves and densities of states.

Axe (1965) and Catlow and Norgett (1973) have based their shell models on long-wavelength data. Their analyses discussed in detail below, involve two stages. In the first the parameters in some assumed short-range potential are fitted to c_{11}, c_{12}, and the equilibrium condition. In the second, the shell parameters are fitted to other data. The short-range potentials in both models consist of a nearest-neighbour repulsion and central interactions between next-nearest neighbour F^- ions; cation-cation repulsion does not seem important. The F^-–F^- short-range interaction appears to be repulsive in CaF_2 and SrF_2, but attractive in BaF_2.

Axe took different forms of second-neighbour interactions in the different fluorites, each involving one free parameter. This parameter, together with two others from the nearest-neighbour interaction, were fitted to the three equations for c_{11}, c_{12}, and the equilibrium of each crystal. Thus three parameters were fitted to three conditions. Given these values, the shell parameters were obtained from ω_{TO}, ϵ_0, ϵ_∞, and the Pauling ionic polarizabilities. As supplementary conditions, Axe assumed that the cation shell charges were those given by Dick and Overhauser from data for the isoelectronic rare gases. A consistent picture of the long-wavelength properties emerges, with sensible values for all parameters. The difference between c_{12} and c_{44} is satisfactorily accounted for in terms of internal strain. Results are available for CaF_2, SrF_2, and BaF_2. Kühner, Lauer, and Bron have used the same model for EuF_2 and $SrCl_2$; these authors also use a model with deformable shells, but find no real improvement in the fit. We note in passing that a conflict between Axe and Ganesan and Srinivasan is resolved on p. 1056 of Srinivasan (1968b).

Catlow and Norgett use more general forms of second-neighbour interaction involving three parameters. The most successful combined an exponential repulsive term with an attractive r^{-6} term; Morse-like potentials proved less satisfactory. These authors imposed three subsidiary conditions. Firstly, the F^-–F^- interaction is assumed the same in all crystals. Secondly, the short-range part of the F^-–F^- interaction is fitted to results of Hartree-Fock calculations for free ions. Finally, in the nearest-neighbour interactions, proportional to $\exp(-r/\rho)$ the hardness parameters ρ are taken to be the same in all crystals. In consequence there are five free parameters satisfying nine equations for the set of crystals CaF_2, SrF_2, and BaF_2.

Parameters can be found which fit c_{11} and c_{12} to better than 2%, which is very satisfactory. The attractive r^{-6} term proves to be about an order of magnitude larger than one would expect from the van der Waals interaction; its origin is uncertain, and it is best regarded as phenomenological. Given the repulsive parameters, Catlow and Norgett fit the remaining shell parameters in two ways. One model (CNA in Table 2.5) fits ω_{TO}, ϵ_0, and ϵ_∞ and the Tessman-Kahn-Shockley polarizabilities; c_{44} and ω_{R} are not used. The other, CNB, fits ω_{TO}, ω , ϵ_0, ϵ_∞, and c_{44}. The CNB set are closest to the parameters from neutron data discussed below. However, they implicitly assume that all the difference in $(c_{12}-c_{44})$ comes from internal strain, and the deduced polarizabilities do not vary simply from crystal to crystal. For defect calculations (Chapter 3), both sets give comparable results. Both sets also give better third-order elastic constants than the Axe model (see § 2.4.4).

The remaining models have been devised to fit neutron dispersion data. The analyses are those for CaF_2 (Elcombe and Pryor, 1970), SrF_2 (Elcombe, 1972), BaF_2 (Hurrell and Minkiewicz, 1970), and UO_2 (Dolling, Cowley, and Woods, 1965, as corrected by Cochran, 1971). The standard assumptions are of axial short-range forces between the shells of nearest and next-nearest neighbours. Elcombe and Pryor and Elcombe discuss a rigid-ion model (Model I; Table 2.4) and a shell model with the cation charge constrained to $2e$ (Elcombe and Pryor, Model III; Elcombe Models II and III). The other treatments (Elcombe and Pryor Model II; Hurrell and Minkiewicz; Dolling *et al.*) allow variable ionic charges. The shell models improve agreement with experiment considerably over the rigid-ion models, but the various shell models for CaF_2 have a comparable degree of accuracy. Qualitative differences are found from CaF_2 to BaF_2. In BaF_2 the nearest-neighbour interaction dominates to a far greater extent, consistent with the larger F^-–F^- separation, and all the shell charges are negative. The Ca^{2+} shell charge is positive in Elcombe and Pryor's models, contrary to what one would expect of the charge of an electron cloud. This emphasizes once more that the shell model is primarily a scheme for parameterization. One should not expect all results to have the simple significance suggested by the model, and one should be cautious about any prediction sensitive to the detailed values of the parameters.

In summary, the lattice dynamics and related properties are described by a simple extension of the Born model of Chapter 1. The main extension has been to allow ions to polarize. Even here, one is struck by the success of the predictions given by the naive approximations.

2.3.3. *Lattice-dynamic models for mixed crystals*

Infrared and Raman measurements on the mixed crystals $Ca_{1-x}Sr_xF_2$ and $Ba_ySr_{1-y}F_2$ and their analyses have been reported (Chang, Lacina, and Pershan, 1966; Verleur and Barker, 1967; Lacina and Pershan, 1970). The principal results are these. Firstly, the Raman frequency

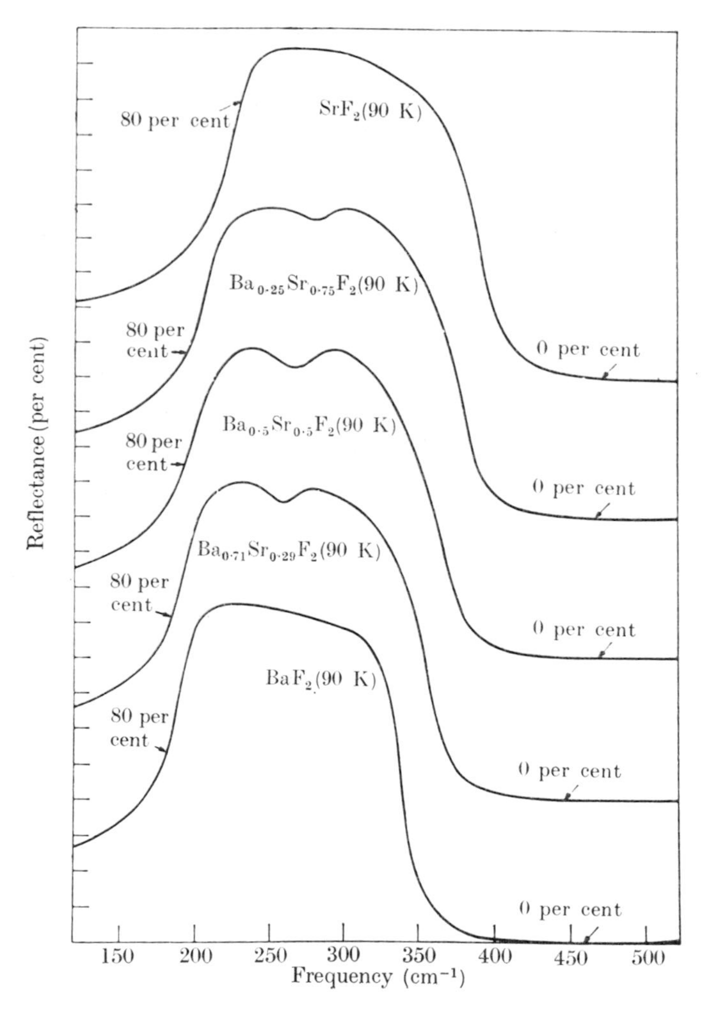

FIG. 2.6. Comparison of experimental reflectance spectra at 90 K of the pure crystals BaF_2 and SrF_2 with three mixed crystals of $Ba_ySr_{1-y}F_2$ (after Verleur and Barker, 1967).

varies roughly linearly with concentration. This is to be expected, for the Raman modes do not involve motion of the cations. Thus one expects a smooth change in value with lattice parameter. Secondly, the infrared modes do not shift linearly with concentration, primarily because the cations are directly involved in these modes. The third feature is that the reststrahlen modes are split in the $Ba_ySr_{1-y}F_2$ case.

Verleur and Barker compare their results with two theoretical models, the 'virtual-crystal' model and the 'mixed-crystal' model. In the virtual-crystal model the mixed system is treated as a perfect array of 'average' ions. Masses, force constants, and effective charges are assumed to scale linearly with concentration. Lacina and Pershan use a very similar model for $Ca_{1-x}Sr_xF_2$, calculating an average Green function to first order in x; their approach starts from the rigid-ion model of Ganesan and Srinivasan and allows for changes in mass and nearest-neighbour force constant. The mixed-crystal model of Verleur and Barker is a cluster model; it emphasizes the short-range order by tabulating the possible configurations of a fluorine ion and its closest cations. Different force constants are adopted for each configuration, and a parameter is introduced to allow for clustering or ordering of ions of a particular species. It can be regarded as a generalized Einstein model and, as such, gives a more general picture of the long-wavelength optic modes.

Both approaches work well for the Raman modes. However, the infrared modes are more complicated because of the relative cation-anion motion. Verleur and Barker show that consistent and reasonable choices of parameters can be made which describe the behaviour well, including the novel splitting of the reststrahlen modes in $Ba_ySr_{1-y}F_2$ shown in Fig. 2.6. They conclude that the splitting is a consequence of the sizable variations in frequency from one configuration to another and of the various probabilities of the different configurations.

2.4. Anharmonic properties

Most lattice-dynamic properties can be understood by assuming that the potential energy of the lattice is a quadratic function of the displacements, as in (2.10). However, the corrections to this harmonic approximation are observed by a number of methods which we now survey. There is no detailed theory of anharmonicity which gives a unified theory of these effects, so we shall concentrate on the experimental results. Most of these results are expressed in terms of a Grüneisen constant γ (not to be confused with the damping function $\gamma(\omega)$

of eqn (2.7)) defined by

$$\gamma_i \equiv -(\partial \ln \omega_i/\partial \ln V)_T. \tag{2.13}$$

The Grüneisen approximation asserts that the frequencies ω_i only depend on the crystal volume V. Naturally, the Grüneisen constant γ_i differs from mode to mode.

2.4.1. *Damping of the infrared absorption and Raman scattering*

Anharmonic interactions determine the reduced damping function $\gamma(\omega)$ of (2.7) and (2.8). The conventional damped-oscillator description of the frequency dependence of ϵ' and ϵ'' was given in (2.6) to (2.8) in terms of $\gamma(\omega)$, an oscillator strength ρ_1, and a frequency ω_{TO}. These equations can be inverted to give

$$\gamma(\omega_{TO}) = \tfrac{1}{2}\frac{\epsilon''(\epsilon_0-\epsilon_\infty)}{(\epsilon'-\epsilon_\infty)^2+(\epsilon'')^2}. \tag{2.14}$$

For real crystals the damping function depends on frequency, although this is sometimes ignored. In using (2.14) it is customary to analyse the behaviour of ϵ' and ϵ'' in terms of a sum of contributions from different oscillators. Usually two oscillators suffice, but as many as five have been used by Verleur and Barker. Table 2.6 lists some of the reduced damping

TABLE 2.6

Reduced damping coefficients $\gamma(\omega_0)$

	CaF_2	SrF_2	BaF_2	CdF_2	PbF_2
a	0·013	0·010	0·010	0·04	0·06
b	0·026	0·025	0·029	0·065	0·10
c	0·018	0·017	0·020		
Raman	0·023				

Results are given for the dominant oscillator at the transverse-optic frequency. The second oscillator has a strength an order of magnitude lower.

Sources are [a] Denham *et al.*, 100 K; [b] Denham *et al.*, 300 K; [c] Kaiser *et al.*, 300 K. Room temperature Raman results of Mirlin and Reshina are also shown.

coefficients. The damping increases with increasing temperature. There is no reliable way of estimating the damping function from first principles; as Zernik (1967) has observed, the alternatives are phenomenology or the use of perturbation theory in terms of unknown anharmonic coefficients. The corresponding damping parameter has also been measured for Raman modes (Mirlin and Reshina, 1971), and is included in Table 2.6. It varies with temperature roughly as $T^{1.6}$.

2.4.2. *Neutron and X-ray diffraction data*

As the temperature increases, certain Bragg intensities in the fluorite structure depart progressively from the predictions of the harmonic approximation. Willis (1965) suggested that anharmonicity was the dominant cause. The nature of the anharmonicity can be seen from a simple Einstein model in which the ions move in a potential parameterized in the form:

$$V = V_0 + \tfrac{1}{2}K(x^2+y^2+z^2) + Bxyz + D(\mathbf{r}). \qquad (2.15)$$

The higher-order terms $D(x, y, z)$ prove negligible, and are omitted in the fit to experiment. The term $Bxyz$ is non-zero for the anion sites, which have tetrahedral symmetry. B is zero by symmetry at the cation site, so only anion anharmonicity is retained. Parameters of (2.15) have been obtained by Cooper (1970) and Strock and Batterman (1972) from X-ray data, and by Lambe and Willis (1963), Dawson, Hurley, and Maslen (1967), Cooper, Rouse, and Willis (1968), Willis (1969), and Cooper and Rouse (1970) from neutron data; some results are given in Table 2.7. These papers also quote values for many related properties

TABLE 2.7
Anharmonic parameters from neutron and X-ray diffraction

	CaF_2	BaF_2	BaF_2(900 K)	UO_2
Cation K	4·1	2·9	2·6	7·5
Anion K	3·1	1·9	1·6	4·4
Anion B:				
neutron	−4·0	−1·9	−1·6	−5·9
X-ray	−2·8			

Data from neutron diffraction are those of Dawson, Hurley, and Maslen for CaF_2 and UO_2, and from Cooper, Rouse, and Willis for BaF_2. X-ray data for CaF_2 are from Strock and Batterman. The harmonic force-constants K are in eV/Å^2 and the anharmonic terms B in eV/Å^3. Results are for room temperature, unless otherwise stated.

of the crystals, including thermal parameters, scattering lengths, and extinction factors. Values of thermal parameters for CaF_2, SrF_2, and BaF_2 deduced from specific heat and thermal expansion data are given by Brada and Bates (1969) and compared with neutron data; further results for CaF_2 are given by Hewat (1972).

Comparison of the measured parameters of (2.15) with those estimated from model potentials based on the usual Born–Mayer short-range forces (Dawson *et al.*) showed disappointing agreement. More recent work by Catlow (private communication, 1972) has resolved some of the differences using the shell model of Catlow and Norgett. Like Dawson *et al.*, Catlow finds the potential parameters too large when the neighbours are kept fixed. But quite good agreement is found if the lattice displacements and electronic polarization of the neighbours follow the anion motion adiabatically. The need for this motion is not too surprising, although one might have expected the behaviour to be between these extremes.

Cooper, Rouse, and Willis have also measured the temperature dependence of K and B in (2.15). Both parameters have the same dependence for BaF_2:

$$\Delta \equiv -\frac{\mathrm{d}\ln K}{\mathrm{d}T} \simeq -\frac{\mathrm{d}\ln B}{\mathrm{d}T} \simeq 3{\cdot}7\times10^{-4}\,\mathrm{K}^{-1}. \tag{2.16}$$

In the simplest model of anharmonicity (Cooper, Rouse, and Willis; Willis, 1969), when there is a unique Grüneisen constant, one would expect Δ to be related to the Grüneisen constant, and to χ, the coefficient of thermal expansion. Specifically, the Grüneisen constant (2.13) is $\Delta/2\chi$ and is found to be 2·1 for BaF_2. Note that the diffraction data provide one of the rare instances where one cannot represent anharmonic effects in perfect crystals merely by volume- and temperature-dependent harmonic constants.

2.4.3. *Thermal expansion*

Interferometric measurements of the linear thermal-expansion coefficients between 20 K and 273 K have been reported by Bailey and Yates (1967), who also present results of earlier investigations by other authors. There is general agreement between the various sets of measurements, apart from some high-temperature results which do not appear to be smooth continuations of data obtained at lower temperatures. The linear expansion coefficients χ given by Bailey and Yates (Table 2.8) increase from zero at the lowest temperatures to room-temperature values of about 1·8 10^{-5}/K. Values for $SrCl_2$ are given by Khan and Deshpande (1968) by a fit

$$\chi = [2{\cdot}08+1{\cdot}129x+4{\cdot}660x^2]\times10^{-5}/\mathrm{K}$$

where x is $(T-273)/1000$.

Bailey and Yates use their data in two ways. Firstly, they calculate a Grüneisen constant γ_A in terms of thermal and elastic properties and the coefficient of linear expansion. Their values are listed in Table 2.9.

TABLE 2.8

Linear coefficients of thermal expansion in units 10^{-5}/K
(Bailey and Yates)

Temperature (K)	CaF_2	SrF_2	BaF_2
20	(0·005	(0·005)	(0·01)
30	0·03	0·04	0·07
40	0·08	0·11	0·16
50	0·16	0·21	0·29
60	0·26	0·33	0·45
80	0·51	0·60	0·74
100	0·76	0·86	0·99
120	0·99	1·07	1·18
140	1·17	1·23	1·33
160	1·33	1·37	1·45
180	1·45	1·48	1·54
200	1·56	1·56	1·61
220	1·65	1·63	1·67
240	1·73	1·68	1·72
260	1·79	1·73	1·76
270	1·82	1·75	1·79

Srinivasan (1968*a*) has calculated effective Grüneisen parameters using Axe's shell model, and finds acceptable agreement with their values. In addition, the data were used to obtain effective Debye frequencies which fit measured moments of the phonon spectrum, and to predict effective Grüneisen constants $\gamma(n)$ for the nth moments. One general feature is that $\gamma(n)$ is an increasing function of n since the modes of higher frequency tend to have stronger anharmonic interactions.

TABLE 2.9

Grüneisen constants

	γ_L	γ_T	γ_R	$\gamma_A(0K)$	γ_A
CaF_2	1·1	3·2 3·2*	1·8	1·0	1·9
SrF_2	(0·8)	3·1 (2.4*)	2·0	0·3	1·62
BaF_2	1·1	2·4 2·9*	1·4	0·3	1·57

Different values are appropriate for different modes, as indicated by the subscripts. These refer to longitudinal (L) and transverse infrared-active (T) modes (Lowndes, 1969, 1971), Raman (R) modes (Kessler and Nicol, 1972 and Acoustic (A) modes (Bailey and Yates 1967). Results * for γ_T are from dielectric constant data; results are for room temperature except for values of γ_A extrapolated to 0 K.

2.4.4. *Third-order elastic constants*

Third-order elastic constants measure the strain dependence of the usual (second-order) elastic constants. Precise definitions are given by Wallace (1970). Gerlich (1968) and Alterovitz and Gerlich (1969, 1970*a*,*b*) have measured the third-order elastic constants for CaF_2, SrF_2, BaF_2, and CdF_2 by examining the effects of external stress on the velocity of sound. Their results are shown in Table 2.10, where they are

TABLE 2.10
Third-order elastic constants

			c_{111}	c_{112}	c_{144}	c_{166}	c_{123}	c_{456}
CaF_2	Experiment		124·6	40·0	12·4	21·4	25·4	7·5
	Rigid Ion	S	95·5	44·2	17·0	29·7	27·9	8·8
	Shell Model	S	95·5	44·2	11·4	25·3	27·9	9·1
	Shell Model	CNA	107·8	33·8	9·3	23·2	17·5	7·8
	Shell Model	CNB	107·8	33·8	10·4	24·2	17·5	7·9
SrF_2	Experiment		82·1	30·9	9·5	17·5	18·1	4·2
	Rigid Ion	S	68·6	36·5	12·9	22·2	23·7	5·3
	Shell Model	S	68·6	36·5	8·8	19·2	23·7	6·0
	Shell Model	CNA	76·7	31·4	6·8	16·5	18·6	4·5
	Shell Model	CNB	76·7	31·4	8·7	18·0	18·6	4·4
BaF_2	Experiment		58·4	29·9	12·1	8·89	20·6	2·71
	Rigid Ion	S	47·0	32·9	8·5	14·2	23·1	0·7
	Shell Model	S	47·0	32·9	5·3	12·9	23·1	1·7
	Shell Model	CNA	54·6	28·4	8·2	13·2	18·5	1·1
	Shell Model	CNB	54·6	28·4	5·4	11·8	18·5	1·8
			$c_{111}+2c_{112}$		$c_{123}+2c_{112}$		$c_{144}+2c_{166}$	
CdF_2	Experiment		263·0		143·1		73·24	
	Rigid Ion		255·4		168·3		89·4	
	Shell Model	A	255·4		168·3		81·7	
	Shell Model	B	255·4		168·3		72·1	

The results are in units of 10^{11} dynes/cm^2, and a minus sign has been omitted throughout. Experimental results are due to Gerlich (1968) for BaF_2, and Alterovitz and Gerlich (1969, 1970*a*, 1970*b*) for CaF_2, SrF_2, and CdF_2. The theoretical values S for CaF_2, SrF_2, and BaF_2 are due to Srinivasan (1968*b*) and the values CNA and CNB are due to Catlow and Norgett (1973). The theoretical values for CdF_2 are due to Alterovitz and Gerlich (1970*b*).

compared with predictions due to Srinivasan (1968*b*) and Catlow and Norgett (1973). It can be seen that the results are in satisfactory agreement. It is interesting to note that whilst the second-order elastic constants of BaF_2 suggest it is almost isotropic (§ 2.2.1), the third-order constants exhibit a pronounced anisotropy.

The third-order elastic constants can be used to predict ultrasonic attenuation. Tor, Tandon, and Rai (1972) find good agreement with experiment for CaF_2 and BaF_2. They observe that the Akhieser (1939) mechanism dominates, involving the relaxation of thermal energy between different phonon branches heated to different extents by the ultrasonic waves.

In passing, it should be mentioned that the generation of acoustic harmonics in alkaline earth fluorides has been observed by Shakin and Lemanov (1972) using Bragg light scattering.

2.4.5. *Temperature and pressure dependence of the infrared and Raman modes*

Whereas third-order elastic constants measure the anharmonicity of long-wavelength acoustic modes, the response of the dielectric constant measures the anharmonicity of long-wavelength optic modes. Lowndes (1969) has used arguments of Szigeti (1961) to derive effective Grüneisen constants γ_T and γ_L for the $\mathbf{q} \simeq 0$ infrared-active transverse and longitudinal optic modes respectively, from dielectric constant data. They should not depend strongly on temperature. Lowndes (1971) has also derived γ_T for transverse modes from infrared data. The results are given in Table 2.9. Those for SrF_2 are bracketed because temperature and pressure derivatives of ϵ_∞ were not available and were assumed to be zero.

Kessler and Nicol (1972) have observed Raman spectra under pressure. The Raman-mode Grüneisen constants, γ_R, which they derive are also given in Table 2.9.

2.4.6. *Photoelastic constants*

The photoelastic constants (for definitions see Nye, 1957) give the change in refractive index per unit applied stress (the $\mathbf{q}$ constants) or strain (the $\mathbf{p}$ constants). The dielectric-constant tensor changes with strain $\mathbf{e}$ as follows:

$$d(\epsilon_\infty)_{ij}/de_{kl} = -\epsilon_\infty^2 p_{ijkl}. \tag{2.17}$$

In quoting experimental results abbreviated suffixes ($p_{mn} = p_{ijkl}$; see Nye) are generally used. Recent work by Rao and Narasimhamurty (1970) completes results for CaF_2 and BaF_2 at 589·3 nm. These authors list previous published and unpublished values. Shakin, Bryzhina, and Lemanov (1972) give values for CaF_2, SrF_2, and BaF_2 at 632·8 nm.

Both sets are listed in Table 2.11. Errors cited are in the 10% range, smaller than some of the differences between the values. It is not clear if there is a genuine wavelength dependence, or merely optimism in the estimates of the errors.

Young and Frederikse (1969) have measured the temperature and pressure dependence of the real part of the dielectric constant in CdF_2.

TABLE 2.11
Photoelastic constants

		CaF_2	SrF_2	BaF_2
p_{11}	RN	0·0258		0·131
	SBL	0·0433	0·080	0·110
p_{12}	RN	0·202		0·277
	SBL	0·276	0·269	0·257
p_{44}	RN	0·0239		0·0264
	SBL	0·0287	0·0185	0·0142

The values are those of Rao and Narasimhamurty (RN) at 589·3 nm and of Shakin, Bryzhina, and Lemanov (SBL) at 632·8 nm.

2.4.7. *Pressure-induced phase changes*

An extreme non-harmonic effect is the phase change of BaF_2 (Samara, 1970, Minomura and Drickamer, 1961, Dandekar and Jamieson, 1969) which occurs under pressure. This first-order transition is from the cubic fluorite structure to the orthorhombic a-$PbCl_2$ structure. The transition pressure $P(T)$ is temperature dependent; it is about 18 kbar at 295 K and 12·5 kbar at 423 K. There is some hysteresis, and also the temperature dependence is not quite linear: $dP(T)/dT$ is about $-2{\cdot}6\times10^{-2}$ kbar/K below 373 K and $-2{\cdot}2\times10^{-2}$ kbar/K between 373 K and 573 K. There is a large volume change, about 11% at room temperature, from which one can deduce an entropy change $\Delta S \sim 2{\cdot}4$ cal/gm mole K. Kessler and Nicol (1972) report the phase change for BaF_2 to be at 23 kbar at room temperature. They find a similar transition for PbF_2 at 4 kbar (see also Schmidt and Vedam, 1966).

Samara compares the BaF_2 transition with the better-known example of RbI. He notes that in the first-order transition the frequencies of a number of modes usually become small, although they need not vanish. By analogy with RbI, he suggests that it is the [110] transverse acoustic modes whose frequencies are reduced by pressure, and that the modes near the zone boundary are affected particularly strongly. Ramachandran and Srinivasan (1972) have suggested that TA modes are also responsible for the PbF_2 phase instability.

2.5. Vibrations of hydrogen impurity ions

Hydrogen in the form of a negatively-charged ion substitutes for halogens in some simple ionic solids. In the alkali halides the resulting defect (H_s^-) is known as a *U* centre. The localized vibrations of *U* centres were first studied by Schaefer (1960) and later by Fritz (1962, 1965). Hydrogen dissolves readily in alkaline earth fluoride crystals in fluorine sites providing an ideal system for the study of vibrations of light impurities in solids (Elliott, Hayes, Jones, Macdonald, and Sennett, 1965).

A technique for dissolving hydrogen in alkaline earth fluorides in concentrations of the order of 10^{19} cm^{-3} was discovered by Hall and Schumacher (1962). This involves firstly heating crystals for several hours under high vacuum in a quartz tube at 500 C in the presence of aluminium. Hydrogen or deuterium gas is then introduced into the tube at pressures of up to two-thirds of an atmosphere, the tube is sealed off and the temperature raised to 800 C for several hours. Concentrations of H_s^- of 10^{20} cm^{-3} were achieved by Shamu, Hartmann, and Yasaitis (1968) using a similar method but with hydrogen pressures of about 50 atm. Concentrations of H_s^- in alkaline earth fluorides of about 6×10^{20} cm^{-3} were produced by Harrington, Harley, and Walker (1970, 1971) by additively colouring crystals in the cation vapour (see § 4.1 for further discussion) and subsequently heating the coloured crystal in hydrogen gas. The vibrations of H_s^- ions will be discussed in §§ 2.5.1, 2.5.2, and 2.5.3.

Exposure of crystals containing H_s^- centres to X-rays at room temperature results in production of neutral interstitial hydrogen atoms (H_i^0 centres; § 4.7.1 contains a description of the paramagnetic properties of these centres). Diffusion of hydrogen into crystals containing trivalent rare-earth ions introduces interstitial negative ions (H_i^- centres). The vibrations of H_i^0 and H_i^- centres will be discussed in § 2.5.4. A description of the vibrations of H_s^-, H_i^-, and H_i^0 ions paired with rare-earth ions is given in § 2.5.5.

2.5.1. *Infrared absorption and Raman scattering of* H_s^- *centres*

In the case of hydrogen impurities the mass defect is so light that to a good approximation the localized modes may be considered as

TABLE 2.12

Frequencies (cm^{-1}) *and line widths* (cm^{-1}) *of vibrations of* H_s^- *and* D_s^- *ions in* CaF_2

Crystal	Line		290 K Peak position	290 K Width	77 K Peak position	77 K Width	20 K Peak position	20 K Width
$CaF_2:H_s^-$	fundamental	Γ_5	$957{\cdot}8 \pm 0{\cdot}6$	8·7	$965{\cdot}1 \pm 0{\cdot}3$	1·0	$965{\cdot}6 \pm 0{\cdot}5$	0·7†
	2nd harmonic	Γ_1	—	—	—	—	1895 ± 3‡	4†
		Γ_5	1903 ± 1	21·2	$1919{\cdot}0 \pm 0{\cdot}3$	1·5	$1919{\cdot}8 \pm 0{\cdot}3$	0·9
	3rd harmonic	$\Gamma_5^{(1)}$	2888 ± 2	35	$2910{\cdot}8 \pm 0{\cdot}6$	4·5	$2912{\cdot}2 \pm 0{\cdot}5$	3·0
		$\Gamma_5^{(2)}$	—	—	$2825{\cdot}4 \pm 0{\cdot}6$	2·3	$2826{\cdot}6 \pm 0{\cdot}8$	1·3
$CaF_2:D_s^-$	fundamental	Γ_5	$689{\cdot}0 \pm 1$	6·5	$694{\cdot}0 \pm 0{\cdot}5$	2·2	$694{\cdot}3 \pm 0{\cdot}5$	2·2
	2nd harmonic	Γ_5	1374 ± 1	12·3	$1383 \pm 0{\cdot}5$	4·0	$1384{\cdot}5 \pm 0{\cdot}4$	3·6
	3rd harmonic	$\Gamma_5^{(1)}$	2075 ± 2	22	—	—	2093 ± 1	5·4
		$\Gamma_5^{(2)}$	—	—	—	—	2047 ± 1	7

† Instrument limited width

‡ Obtained from Raman measurements at 4 K by Harrington *et al.* (1970). All other measurements from Elliott *et al.* (1965).

oscillations of the hydrogen alone in the potential well set up by the static lattice. The hydrogen then moves in a potential well with T_d symmetry which may be expanded as

$$V = V_0+\tfrac{1}{2}\Omega^2(x^2+y^2+z^2)+Bxyz+C_1(x^4+y^4+z^4)+ \\ +C_2(x^2y^2+y^2z^2+z^2x^2) = V_0+V_2+V_3+V_4^{(1)}+V_4^{(2)} \quad (2.18)$$

to fourth order in the displacement of H_s^- from its equilibrium position; Ω, B, C_1, and C_2 are parameters to be determined from experiment. This expression is similar to eqn (2.15) used earlier to describe the anharmonicity of the lattice anions. The anharmonic terms in (2.18) are small and perturbation theory taken to second order in the cubic term and first order in the fourth-order terms gives the positions of the energy levels with precision adequate for comparison with experiment. The energy levels of (2.18) are represented in Fig. 2.7 up to $n = 3$. The corresponding wave functions without anharmonic admixtures are also given in Fig. 2.7. In the occupation number representation the wave function $|1, 1, 1\rangle$, for example, represents unit excitation of independent harmonic oscillators along the x, y, and z axes.

The four low-lying electric-dipole allowed transitions have been observed for H_s^- centres in alkaline earth fluorides using infrared methods. The measurements for H_s^- and D_s^- in CaF_2 are given in Table 2.12 (see also Fig. 2.7). The ratio of Ω for H_s and D_s^- in CaF_2 is 1·39. The small departure from $\sqrt{2}$ confirms that the lattice ions do not play a significant role in this highly-localized mode. These four allowed transitions give enough information to determine the parameters of (2.18) and they are given in Table 2.13 for CaF_2 and SrF_2 at 20 K and for BaF_2 at 4 K (there is negligible change in the parameters between

TABLE 2.13

Parameters† *of* (2.18) *for* H_s^- *ions in* CaF_2, SrF_2 *and* BaF_2

	CaF_2‡ (20 K)	SrF_2§ (20 K)	BaF_2‖ (4 K)
Ω (cm^{-1})	981·1	907·4	817·0
$\|B\|$ (eV/Å^3)	4·92	3·87	2·10
C_1 (eV/Å^4)	−0·145	−0·112	−0·844
C_2 (eV/Å^4)	−0·063	−0·137	+0·191

† Values have been given previously in units of erg cm^{-3} (for $|B|$) and ergs cm^{-4} (for C_1 and C_2). To make the conversion note that 1 eV $= 1{\cdot}602\times10^{-12}$ erg.

‡ Elliott *et al.* (1965).

§ Hayes and Macdonald (1967). ‖ Harrington *et al.* (1970).

20 and 4 K). The positions of the excited states of H_s^- in CaF_2 other than Γ_5 were calculated using the parameters in Table 2.13 (Elliott *et al.*) and these are given in Fig. 2.7.

The $\Gamma_1 \rightarrow \Gamma_5$ ($n = 2$) transition has observable intensity only because of the cubic anharmonicity in (2.18) and this is a consequence of the lack of inversion symmetry associated with the point group T_d.

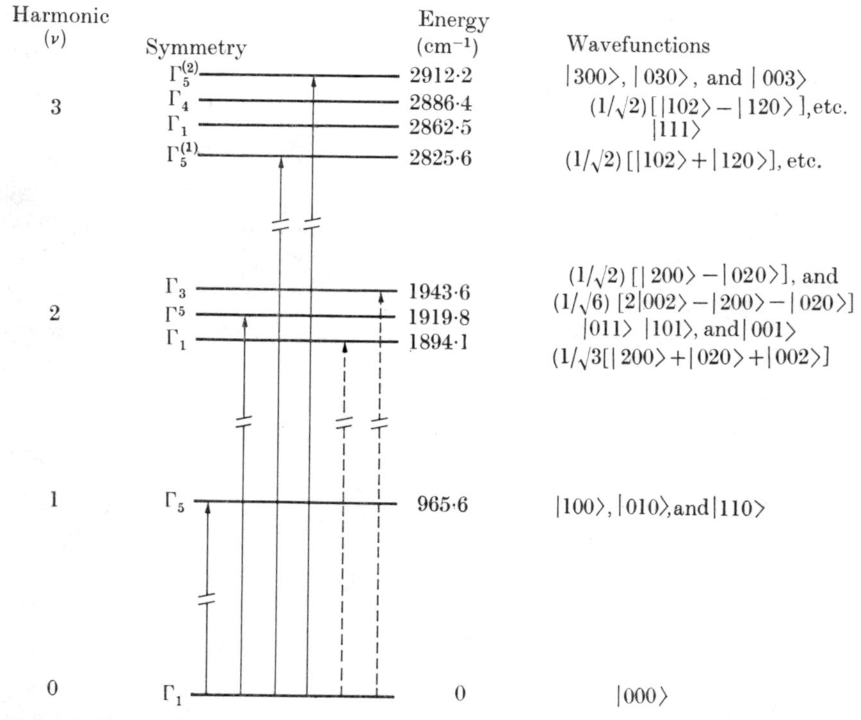

FIG. 2.7. Energy level diagram and wavefunctions for H_s^- in CaF_2. Observed electric dipole transitions in unperturbed crystals are indicated by full vertical lines; the energies given for the other levels were calculated using eqn. 2.18. Additional electric dipole transitions observed in stressed crystals are indicated by dashed vertical lines (after Hayes and MacDonald 1967).

In the alkali halides H_s^- ions have the symmetry O_h, and since only even terms occur in the potential expansion (2.18) the second harmonic is not observed (Fritz, 1968).

The excited states of H_s^- in alkaline earth fluorides were also measured by Harrington, Harley, and Walker (1970, 1971) using Raman scattering methods. The Raman tensor has components of Γ_1, Γ_3, and Γ_5 symmetry for the T_d point group and symmetry-allowed Raman transitions occur to Γ_1 and Γ_3 levels in addition to Γ_5 levels. Harrington *et al.* (1971) measured the position of the Γ_5 ($n = 1$) level for H_s^- and D_s^-

in CaF_2, SrF_2, and BaF_2 and also the positions of the Γ_5 ($n = 2$) and Γ_1 ($n = 2$) levels (Table 2.12).

If we restrict our considerations to (2.18) the intensity of the second harmonic is determined by admixture of the $|1, 1, 1\rangle$ state by the third-order anharmonicity V_3. Taken to second order in perturbation theory this gives

$$\langle 000|\, x\, |011\rangle = \frac{2B}{3\hbar\Omega}\left(\frac{\hbar}{2M'\Omega}\right)^2. \tag{2.19}$$

The intensity of the fundamental is determined by

$$\langle 000|\, x\, |100\rangle = \left(\frac{\hbar}{2M'\Omega}\right)^{\frac{1}{2}}. \tag{2.20}$$

Since the oscillator strength of a line is proportional to $\omega\, |\langle 1|\, x\, |f\rangle|^2$ it follows from (2.19) and (2.20) that

$$\frac{f_\omega}{f_{2\omega}} = \frac{9M'^3\Omega^5}{B^2\hbar}. \tag{2.21}$$

The value of B given in Table 2.13 for H_s^- in CaF_2 gives a calculated ratio (2.21) of 136 whereas the measured ratio is 23. This discrepancy is due to neglect in eqn (2.18) of coupling between the impurity and the lattice.

Assuming that the lattice modes are plane waves the important terms in the coupling between the local and lattice oscillators have the form (see Elliott *et al.* for complete expressions)

$$\mathscr{H}_3 \propto \mathscr{B}Q_h Q_1^2 \qquad \mathscr{H}_4 \propto \mathscr{C}Q_h Q'_h Q_1^2 \tag{2.22}$$

for terms quadratic in the displacement Q_1 of the local oscillator; the Q_h are displacements of the band phonons. The parameters $\mathscr{B}$ and $\mathscr{C}$ have the same dimensions as B and C_1 and C_2 in (2.18) and should have comparable magnitudes. The terms (2.22) give rise to temperature-dependent widths and shifts of the local-mode lines because of their dependence on the excitation of lattice phonons.

The alkaline earth fluorides have an intense reststrahl absorption and the admixture of reststrahl states into the local-mode states gives a contribution to the second-harmonic intensity determined by a series of terms of the form

$$\frac{\langle n_R, 0|\, x\, |n_R+1, 0\rangle\langle n_R+1, 0|\, \mathscr{H}_3\, |n_R, 2\rangle}{\hbar(\omega_R - 2\Omega)}, \tag{2.23}$$

where the states $|n_R, n_L\rangle$ represent excitation of reststrahl states of frequency ω_R and local-mode states. However, the coupling coefficient

$\mathscr{B}$ in $\mathscr{H}_3$ required to explain the observed second-harmonic intensity in CaF_2 through this mechanism is an order of magnitude greater than B in Table 2.13 and this seems unlikely. Elliott *et al.* (1965) concluded that the second-order dipole-moment mechanism (Lax and Burstein 1955) was largely responsible for the intensity of the second harmonic. This mechanism, which contributes to the intensity of two lattice-phonon absorption in solids, gives an admixture of exciton states into local mode states through third-order anharmonic coupling of electron and local-mode excitations; the admixture of exciton states gives a contribution to the intensity.

Cubic anharmonicity also gives rise to forbidden transitions involving a local mode and a band mode (see § 2.5.2 for further discussion and also Fig. 2.9). Absorption due to these transitions may be observed on the low energy side of the local mode (Stokes component) and on the high energy side (anti-Stokes component). Excitation of anti-Stokes components is a summation process requiring the existence of lattice phonons and is strongly temperature dependent. Excitation of Stokes components is a difference process and is less strongly temperature dependent since the lattice will always accept a phonon. These components have been studied for the fundamental of H_s^- in alkaline earth fluorides by Elliott *et al.* (1965). They showed that the integrated intensity of the sideband structure relative to the fundamental due to $\mathscr{H}_3$ (2.22) is

$$\frac{27\mathscr{B}^2\hbar}{4MM'^2\Omega^2\omega_L^3} \tag{2.24}$$

at $T \to 0$ for a Debye spectrum. Experimentally they found this ratio to be about 0·1 for CaF_2 giving

$$|\mathscr{B}| = 4{\cdot}3 \text{ eV/Å}^3 \tag{2.25}$$

and this is comparable in magnitude with B for CaF_2 in Table 2.13.

Because the coupling $\mathscr{H}_3$ of (2.22) is temperature-dependent the transfer of intensity between the local and lattice modes gives rise to a temperature-dependent intensity of the local-mode fundamental. This temperature dependence for the fundamental of H_s^- in CaF_2 was measured by Elliott *et al.* (1965) and it was shown by Mitra and Singh (1966) that the measurements could be fitted to an expression of the form

$$\ln(I_T/I_L) = S\{1+\tfrac{2}{3}\pi^2 T/\Theta\}^2 \tag{2.26}$$

where I_L is the integrated intensity of the fundamental, I_T is the total integrated intensity of the local-mode spectrum including the sideband structure and Θ is the Debye temperature. This expression is derived from the theory of the Mössbauer effect in the Debye approximation and its applicability to localized vibrational modes is justifiable to the extent that $T \ll \Theta$ and $\Omega \gg \omega(\mathbf{q})$ (Hughes, 1968). These conditions hold reasonably well for H_s^- in CaF_2 at low temperatures ($\Omega = 2{\cdot}07\omega_L$ and $\Theta \simeq 500$ K (Table 2.1)). For weak coupling the factor S in (2.26), generally described as the Huang–Rhys factor, is given by (2.24).

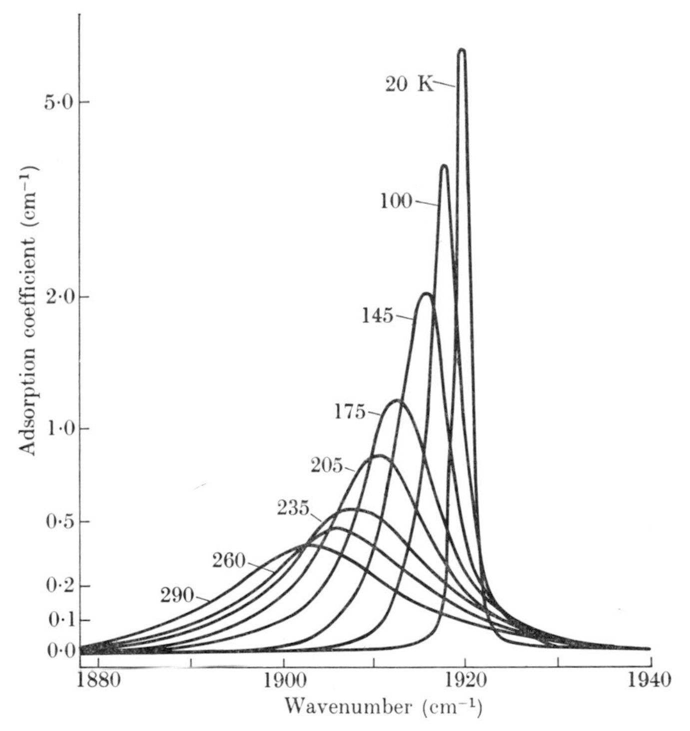

FIG. 2.8. Variation with temperature of the peak position and width of the second harmonic vibrational line of H_s^- in CaF_2 (after Elliott *et al.* 1965).

S is a measure of the strength of the coupling between the local oscillator and the band phonons and for H_s^- in CaF_2 has the relatively low value of about 0·1. The expression (2.26) may be also applied to zero-phonon lines of colour centres and in alkaline earth fluorides some colour centres are found to have considerably larger values of S (§§ 4.3 and 4.4).

Local-mode line widths and peak positions generally show a pronounced temperature dependence (Fig. 2.8). The temperature dependence of the local-mode frequency of H_s^- in alkali halides may be complicated; in the case of KCl the local-mode frequency increases with temperature, reaches a maximum at 70 K and then decreases (Fritz, 1968). In the case of H_s^- in the alkaline earth fluorides the local-mode frequency decreases with increasing temperature and varies linearly with temperature in the region 150–300 K (Elliott *et al.*, 1965). Anharmonic terms such as $\mathscr{H}_4$ (2.22) will give a temperature-dependent contribution to the local-mode frequency. Although $\mathscr{H}_4$ averaged over the band modes gives a peak position varying linearly with T, at high T the magnitude of the predicted effect appears to be too small to account for the experimental results in alkaline earth fluorides. The expansion of the crystal with increasing temperature may give rise to a reduction of force constants and hence of Ω in (2.18) and this effect would contribute to the measured line shift. Other anharmonic terms making positive and negative contributions to Ω have been considered by Bilz, Fritz, and Strauch (1966) but an assessment of the relative contributions of the various effects is difficult.

The contributions to residual line widths of localized vibrational modes at $T \to 0$ were reviewed by Dawber and Elliott (1963*b*). In the case of H_s^- in alkaline earth fluorides residual broadening due to random strain and to random distribution of isotopes in the host lattice is negligible. The main contribution to the residual line width arises from spontaneous decay of excited local modes into several band phonons through anharmonic interactions of the form $Q_h Q_h' Q_1$ with a corresponding anharmonic parameter $\mathscr{B}'$ (see eqn 2.22). For D_s^- in CaF_2 decay into two band-phonons is possible through third-order anharmonicity since $\Omega_D < 2\omega_L$. However, for H_s^- in CaF_2 three phonon decay is required since $2\omega_L < \Omega_H < 3\omega_L$. Elliott *et al.* (1965) found that the cubic anharmonic parameter $|\mathscr{B}'|$ required to account for the residual 2 cm^{-1} line width of D_s^- in CaF_2 is

$$|\mathscr{B}'| \simeq 1{\cdot}0 \text{ eV/Å}^3. \tag{2.27}$$

Although adequate to account for residual line widths the mechanism of spontaneous decay does not explain the magnitude of the observed line widths between 100 and 300 K of H_s^- in alkaline earth fluorides. The dominant line width process at high T in these materials appears to be the elastic scattering of band phonons off the light defect,

$$|n, n(q), n(q')\rangle \to |n, n(q)\pm 1, n(q')\mp 1\rangle \tag{2.28}$$

with conservation of energy ($\omega(q) = \omega(q')$). Since the absorption by the local mode is

$$|0, n(q), n(q')\rangle \to |n, n(q), n(q')\rangle \tag{2.29}$$

the scattering (2.28) removes the crystal from the excited or ground state (2.29) and causes a finite lifetime. The mechanism (2.28) gives a T^2 dependence for the line width at high T in agreement with experiment. It should be added that band phonons may cause transitions between the split energy levels (Fig. 2.7) for the localized vibrational levels with $n > 1$ and hence affect the width of the higher harmonics.

Lee and Faust (1971) investigated the saturation of the $n = 1$ level of H_s^- in CaF_2 using a Q-switched CO_2 laser as a power source and a second CO_2 laser as a monitor. These lasers are tunable over many molecular lines in the 900–1100 cm^{-1} region. They found a decay time for the $n = 1$ level of about 7 ps which implies a line width in order of magnitude agreement with the observed residual width. In addition, they observed the three transitions from the optically pumped $n = 1$ level to the three $n = 2$ levels (Fig. 2.7) and they found that the sum of the measured Γ_1 $(n = 0) \rightarrow \Gamma_5$ $(n = 1)$ and Γ_5 $(n = 1) \rightarrow \Gamma_5$ $(n = 2)$ energies was equal to the energy of the Γ_1 $(n = 0) \rightarrow \Gamma_5$ $(n = 2)$ transition. This latter result is not surprising since observable effects of lattice relaxation are not expected following excitation to the $n = 1$ level of this weakly-coupled system.

2.5.2. *Calculation of the* H_s^- *vibrational spectrum*

It was pointed out in § 2.1 that the vibrational spectrum of impurities can in principle be calculated if the lattice Green function is known. The pure-crystal Green function defined by eqn (2.4) may be calculated by the methods of Dawber and Elliott (1963*a*,*b*). This was done by Hayes and Wiltshire (1973) for CaF_2, SrF_2, and BaF_2 using the shell-model fits of Elcombe and Pryor, Elcombe, and Hurrell and Minkiewicz for neutron diffraction data (see § 2.3.2). Neglecting force constant changes (mass-defect approximation) Hayes and Wiltshire calculated the local-mode frequencies for H_s^- and D_s^- ions from the relation

$$G_{\alpha\alpha}(0, 0) = \frac{M}{(M - M')\omega^2}. \tag{2.30}$$

The calculated frequencies are compared with the measured frequencies in Table 2.14 and it is apparent that the mass-defect approximation gives frequencies from about 18 to 30% too high. In the cases of some alkali halides Fieschi, Nardelli, and Terzi (1965) found that the mass-defect approximation gave local-mode frequencies for H_s^- impurity ions which were about 40 to 60% higher than the measured values.

TABLE 2.14

Comparison of the observed fundamental vibration frequency (cm^{-1}) of the local mode of H_s^- *and* D_s^- *ions in alkaline earth fluorides with values calculated by Hayes & Wiltshire* (1973) *using a mass-defect approximation. A change of nearest-neighbour force constant which will bring* ν_{obs} *and* ν_{calc} *into agreement is also given.*

Crystal	Impurity	ν_{obs}	ν_{calc}	% change of force constant
CaF_2	H_s^-	966	1290	−32
	D_s^-	694	906	−36
SrF_2	H_s^-	893	1158	−27
	D_s^-	641	806	−26
BaF_2	H_s^-	804	976	−16
	D_s^-	578	682	−13

Hayes and Wiltshire expanded the defect space to include the four nearest cations and calculated the reduction of nearest-neighbour central force constant to give agreement with the measured frequencies (Table 2.14). It is of course possible to expand the size of the defect space even further and to include non-central as well as central force constants. However, the complexity of the calculation increases rapidly with the size of the defect space and the problem becomes one of parameterization.

In the previous section we pointed out that side-band structure due to simultaneous excitation of a local phonon and a lattice phonon was associated with the local-mode peaks. These forbidden transitions are due primarily to anharmonicity. There are no symmetry restrictions on the band phonons in the infrared side-band structure for H_s^- ions in alkaline earth fluorides (Loudon, 1964). In an investigation of the relative importance of the various symmetry modes contributing to the side-band structure Hayes and Wiltshire calculated the Γ_5-type motion of the hydrogen alone in the band-mode region. This involves calculating a projected density of states using the mass-defect approximation. This density of states modified by the factor $1/\omega$ (this factor is discussed by Hayes and Wiltshire) is compared with the measured side-band structure associated with the fundamental of H_s^- in BaF_2 in Fig. 2.9. The most pronounced feature in the calculated curve is a sharp resonance at 270 cm^{-1} and it corresponds to an observed sharp resonance at 285 cm^{-1} in the side-band structure associated with the fundamental of H_s^-. It appears from this comparison that the hydrogen

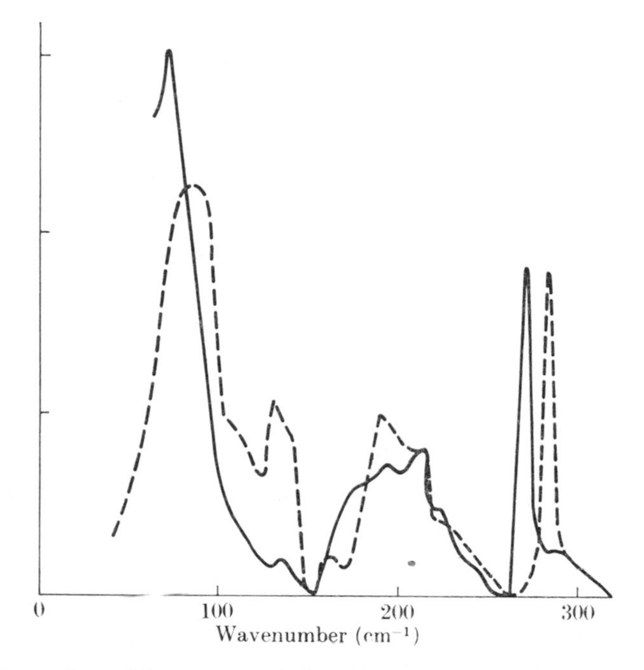

FIG. 2.9. Comparison of the measured Stokes side-band associated with the fundamental local vibration of H_s^- in BaF_2 (dashed curve, given by Harrington *et al.* 1970) with G^{im}/ω (full curve, see text) (after Hayes and Wiltshire (1973)).

motion alone is largely responsible for the side-band structure. However, motion of the near neighbours with a variety of symmetries contributes in the region 100–120 cm^{-1}.

The resonance mode of H_s^- at 285 cm^{-1} in BaF_2 was also observed in Raman scattering by Harrington *et al.* (1971). In addition, they observed weak Raman side-band structure of Γ_1 symmetry in the region 0–200 cm^{-1}. For the mass-defect case the Γ_1 mode is not affected by the impurity and the Raman scattering in this approximation is determined by the pure-crystal Green function of Γ_1 symmetry. The energy distribution of Γ_1 symmetry given by $\mathscr{G}_0(\Gamma_1)$ agrees quite well with observation (Hayes and Wiltshire) although the spectral shape is improved by using the 16% reduction of the central force constant obtained from fitting the local-mode frequency (Table 2.14).

2.5.3. *Effects of uniaxial stress on* H_s^- *centres*

Effects of uniaxial stress on the localized vibrational modes of H_s^- and D_s^- ions in CaF_2 were described by Hayes and Macdonald (1967). Application of stress to the crystal raises the threefold degeneracy of

Γ_5 states and in the case of the $n = 2$ levels causes a sufficient admixture of Γ_5 to Γ_1 and Γ_3 to enable these latter levels to be observed by infrared methods. In the case of coupling to Γ_5 states the only terms allowed in the stress Hamiltonian transform like the representations contained in the symmetric product

$$(\Gamma_5 \times \Gamma_5)_s = \Gamma_1 + \Gamma_3 + \Gamma_5 \tag{2.31}$$

and the Hamiltonian for uniaxial stress takes the form

$$\begin{aligned}\mathscr{H}_p = P\{B_1(S_{11}+2S_{12})r^2+B_3(S_{11}-S_{12}) \times \\ \times[(2n^2-l^2-m^2)(2z^2-x^2-y^2)+3(l^2-m^2)(x^2-y^2)]+ \\ +B_5S_{44}(lmxy+mnyz+nlzx)\}\end{aligned} \tag{2.32}$$

for an arbitrary direction l, m, n of P relative to the crystal cubic axes. The parameters B_1, B_3, and B_5 include the distances from H_s^- of the lattice ions and the strains associated with these parameters transform like the irreducible representations Γ_1, Γ_3, and Γ_5. The term in B_1 is a uniform dilation which shifts the centre of gravity of the Γ_5 states and also affects the position of the Γ_1 ground state. The degeneracy of the Γ_5 states is raised by the B_3 terms which have tetragonal $(2z^2-x^2-y^2)$ and orthorhombic (x^2-y^2) symmetry and by the term in B_5 which represents a shear. The compliance factors S_{ij} are appropriate to the impurity environment and only terms linear in the displacement of the lattice ions and quadratic in the displacement of the light ion are included in $\mathscr{H}_p$ (2.32).

The effect of stress on the fundamental and second harmonic was investigated by Hayes and Macdonald using linearly polarized infrared light, for P along $\langle 100\rangle$, $\langle 110\rangle$, and $\langle 111\rangle$ directions and they list the corresponding eigenvalues and eigenvectors of $\mathscr{H}_p$. Using compliance factors S_{ij} for the perfect lattice as an approximation for the lattice-defect system (a preliminary account of a theoretical investigation of strain at an impurity site in a crystal has been given by Elliott, Krumhansl, and Merrett (1968)) they obtained values

$$\begin{aligned}B_1 &= 2{\cdot}7\pm0{\cdot}1\times10^{20}\ \mathrm{cm^{-1}/cm^2}, \\ B_3 &= 0{\cdot}3\pm0{\cdot}1\times10^{20}\ \mathrm{cm^{-1}/cm^2}, \\ B_5 &= 2{\cdot}3\pm0{\cdot}1\times10^{20}\ \mathrm{cm^{-1}/cm^2},\end{aligned} \tag{2.33}$$

for the parameters of $\mathscr{H}_p$. Using the approximation that the stress splitting arises from motion of the nearest-neighbour lattice ions only one may write $B_n = \frac{1}{2}a\mathscr{B}_n$ where a is the fluorine–fluorine spacing in

CaF_2 (2·73 Å) and $n = 1, 3, 5$. The parameters (2.33) then give

$$\begin{aligned} \mathscr{B}_1 &= 2{\cdot}37 \text{ eV/Å}^3, \\ \mathscr{B}_3 &= 0{\cdot}27 \text{ eV/Å}^3, \\ \mathscr{B}_5 &= 2{\cdot}87 \text{ eV/Å}^3. \end{aligned} \tag{2.34}$$

It is apparent that the magnitude of the cubic anharmonic parameter (2.27) which accounts for the residual width of the D_s^- fundamental line in CaF_2 through a term linear in the displacement of the light ion and quadratic in the displacement of the lattice ions is comparable with (2.34).

Since $\Gamma_1 \times \Gamma_5 = \Gamma_5$ and $\Gamma_3 \times \Gamma_5 = \Gamma_4 + \Gamma_5$ it follows that a perturbation of Γ_5 symmetry will admix Γ_5 to Γ_1 and Γ_3. This admixture will be produced by $\mathscr{H}_p$ (2.32) for $\langle 110 \rangle$ and $\langle 111 \rangle$ directions but not for $\langle 100 \rangle$ directions for which the term in B_5 vanishes. The Γ_1 and Γ_3 states with $n = 2$ are most suitable for investigation, for intensity reasons. The intensity of electric-dipole transitions in a stressed crystal from the ground state to the Γ_1 ($n = 2$) state is determined by the matrix element

$$\frac{\langle \Gamma_1(n=0)|\, \epsilon . P \,|\Gamma_5(n=2)\rangle \langle \Gamma_5(n=2)|\, \mathscr{H}_p \,|\Gamma_1(n=2)\rangle}{E[\Gamma_1(n=2)] - E[\Gamma_5(n=2)]} \tag{2.35}$$

where Γ_5 ($n = 2$) contains anharmonic admixtures of Γ_5 ($n = 1$). A corresponding expression may be written for the Γ_3 ($n = 2$) state. The second-harmonic Γ_1 and Γ_3 levels of H_s^- in CaF_2 were found by Hayes and Macdonald at $1894{\cdot}0 \pm 2{\cdot}0$ and $1945{\cdot}0 \pm 2{\cdot}0$ cm^{-1} at 77 K in stressed crystals (compare Fig. 2.7) and the observed intensities were in agreement with values calculated using (2.33 and 2.35).

It is apparent from inspection of the parameters (2.33) that coupling to the dilating (Γ_1) and shearing (Γ_5) distortions are comparable in size and an order of magnitude larger than the coupling to the distortions of symmetry Γ_3. In the case of H_s^- in KCl Fritz, Gerlach, and Gross (1968) found that the strength of coupling decreases on going from Γ_1 to Γ_3 to Γ_5 distortions. Confining consideration to a nearest-neighbour model one finds two types of shear (Γ_5) distortion of the tetrahedron of Ca^{2+} ions surrounding the hydrogen, one of which involves motion of the hydrogen (see Herzberg, 1945, p. 100, for a pictorial representation). Coupling to both these distortions is allowed in CaF_2 because of lack of inversion symmetry but the two couplings cannot be experimentally distinguished and are included in B_5 (2.32). Coupling to the shear involving motion of the light impurity ion cannot be observed in alkali halides since H_s^- is in a site with inversion symmetry and this may partly explain the relatively small value of shear coupling in these crystals.

Hayes and Macdonald also investigated effects of an applied electric field on the localized vibrations of H_s^- in CaF_2. These effects are smaller than the effects of stress and are more difficult to investigate experimentally. A general discussion of the effects of external perturbations on localized vibrational modes has been given by Maradudin, Ganesan, and Burstein (1967).

2.5.4. *Vibrations of* H_i^- *and* H_i^0 *centres*

Jones, Peled, Rosenwaks, and Yatsiv (1969) found that if CaF_2 crystals containing trivalent rare-earth ions (RE^{3+}) are rapidly quenched after introduction of hydrogen a single vibrational line occurs at 1296 cm^{-1} at room temperature with a line width of 30 cm^{-1}; it occurs at 1310 cm^{-1} at 77 K with a line width of 15 cm^{-1}. The position of the line and its width are only slightly dependent on the rare-earth ion. Jones *et al.* (1969) assigned the line to H_i^- centres remote from rare-earth ions. The crystals also contain H_s^- ions and RE^{3+}–H_i^- complexes (§2.5.5). The H_i^- ions substitute for fluorine interstitials (F_i^-) which provide charge compensation for RE^{3+} ions (see Chapter 6) and are found in an isolated state only in crystals containing rare-earth ions at the middle or latter part of the series. The degree of association of F_i^- ions with RE^{3+} ions follows a similar pattern. The site symmetry of isolated H_i^- ions is O_h and no second harmonic was observed.

The fundamental of D_i^- ions occurs at 932 cm^{-1} at room temperature and the ratio of H_i^- to D_i^- frequencies is 1·39. As in the case of H_s^- ions this ratio is close to $\sqrt{2}$ and again indicates a highly-localized vibration (see § 2.5.1).

Shamu *et al.* (1968) measured the infrared absorption of CaF_2 containing hydrogen after X-irradiation at room temperature. This treatment converts H_s^- centres into H_i^0 centres and the fundamental vibration of H_i^0 centres occurs at 640 cm^{-1} at 100 K. The H_i^0 centres are paramagnetic and by using epr methods to estimate the concentrations of H_i^0 Shamu *et al.* (1968) found the effective oscillating charge of the centres to be about 0·171 $|e|$.

Predictions of the local-mode frequency of H_i^0 in CaF_2 have been given by Shamu *et al.* (1968) and Hartmann, Gilbert, Kaiser, and Wahl (1970). There are two main elements in the theory. The first estimates the interaction of the H_i^0 with the neighbouring fluorine ions. The second is the prediction of the dynamics of the interstitial in terms of this interaction.

The H_i^0–F^- interaction was calculated in several ways. All of these assume that the interactions are additive, so that the potential is a sum of terms, one from each neighbour. Likewise, all assume that the interactions are central and of the form

$$V(R) = V_0 - \alpha/R^4 + \gamma \exp(-R/\rho) + \delta/R^6. \tag{2.36}$$

The term α/R^4 is a monopole-induced dipole interaction, and the δ/R^6 term (usually ignored) contains dipole–dipole and monopole–quadrupole interactions. V_0 does not affect the lattice dynamics. Values of the parameters of eqn (2.36) are listed in Table 2.15; they were estimated

TABLE 2.15
Parameters in the H_i^0-F^- *interaction* (2.36).

Method	α	γ	ρ
Scaling plus free-ion polarizability (HGKW)	2·25	4·3	0·67
Hartree-Fock (SHY)	1·76025	3·47146	0·60898
6 point fit (HGKW)	1·931	3·291	0·624
7 point fit (HGKW)	1·806	4·013	0·596

All units are Hartree atomic units. Results are due to Shamu *et al.* (SHY) and Hartmann *et al.* (HGKW).

from molecular Hartree-Fock calculations and also by using various scaling arguments for the repulsive parameters.

The lattice-dynamic problem was treated in three levels of detail. At the lowest level, the H_i^0 was assumed to move in the harmonic part of the potential of the rigid cage of surrounding F^- ions. At the higher levels, some allowance was made for the anharmonic terms in the potential and for the fluorine motion. In the rigid-cage approximation, the local mode frequency is given by

$$M_H\Omega^2 = \sum_{i=1}^{8} \delta^2 V(R)_i/\delta x^2 = C, \tag{2.37}$$

where the summation covers the interaction with the eight nearest fluorines. The derivatives are readily expressed in terms of

$$T = (\mathrm{d}V/\mathrm{d}R)/a \quad \text{and} \quad b = (\mathrm{d}^2V/\mathrm{d}R^2 - T)T, \tag{2.38}$$

where a is the nearest-neighbour distance. Equation (2.37) may then be expressed in the form

$$C = \tfrac{8}{3}T(b+3). \tag{2.39}$$

There are corrections to C if the fluorine neighbours are allowed to move. These corrections are the source of the deviations from the simple $\sqrt{2}$ isotope mass-dependence mentioned earlier. These corrections may be written in terms of b, T, and the fluorine mass M_F. An Einstein model is used for the fluorine with an effective frequency ω_F, taken to be the longitudinal optic-mode frequency. Essentially one has a coupled oscillator problem. When solved exactly, the change in C is

$$\Delta C = \tfrac{8}{3}T.\frac{gT}{1-gT}\left\{b+3+\frac{b(1+b)}{1-gT(1+b)}\right\}, \tag{2.40}$$

in which g is $[M_F(\Omega^2-\omega_F^2)]^{-1}$. A much simpler derivation assumes the amplitude of the fluorine motion is much less than that of the hydrogen. In this case $gT \ll 1$ and the result,

$$\Delta C = \tfrac{8}{3}T.gT[b+3+b(1+b)] \tag{2.41}$$

is almost as accurate.

The anharmonic terms were treated by perturbation theory. Since the calculation is intended to predict an infrared absorption band, the corrections to the energies of the initial and final states must both be included. The correction to $\hbar\Omega$ is then

$$\frac{\hbar^2}{3(M_H\Omega)^2}\left\{\frac{d^4V}{dR^4}+\frac{4}{a}\frac{d^3V}{dR^3}\right\}. \tag{2.42}$$

The results in the various approximations are summarized in Table 2.16.

TABLE 2.16

Local-mode frequencies (cm^{-1}) *for* H_i^0†

Host lattice	Harmonic + rigid cage	Harmonic + Harmonic F^- motion	Anharmonic + rigid cage	Anharmonic + Harmonic F^- motion	Experiment
CaF_2	578	612	602	633	640
SrF_2	446	476	471	498	
BaF_2	366	401	391	422	

† Calculated by Hartmann *et al.*

The other prediction which can be made concerns the effective charge. Experimentally, this is derived from the integrated absorption using an estimate of the defect concentration and some assumption about the local-field correction. Theoretically, it is estimated from the first moment of the charge distribution as the H_i^0 is displaced relative to the

F^- ions. The Hartree-Fock calculations give

$$\begin{aligned} Z &\simeq -0{\cdot}1254e \qquad \text{(7 point fit)} \\ &\simeq -0{\cdot}1210e \qquad \text{(6 point fit)} \end{aligned} \tag{2.43}$$

compared with the experimental value $|Z| \simeq 0{\cdot}071|e|$.

2.5.5. *Vibrations of* H_s^- *and* H_i^- *paired with rare-earth ions*

Elliott *et al.*, (1965) observed weak satellite lines near the fundamental of the localized vibrational mode of H_s^- in heavily doped crystals and they suggested that the satellite lines were due to H_s^- ions associated with impurities. The satellites were subsequently investigated by Yatsiv, Peled, Rosenwaks, and Jones (1967), Jones, Yatsiv, Peled, and Rosenwaks (1968), and Newman and Chambers (1968) and a review of these investigations was given by Newman (1969). If association with an impurity reduces the H^- site symmetry to trigonal or tetragonal the $n = 1$ level splits into a doublet and a singlet whereas if the site symmetry is orthorhombic or lower it splits into three singlets. These considerations suggest that in the former case the fundamental should split into two components with intensities in the ratio 2:1 whereas in the latter case three equally intense lines should occur.

Yatsiv *et al.* (1967), Jones *et al.* (1968), and Jones *et al.* (1969) investigated the behaviour of H^- ions in crystals of CaF_2 containing rare-earth ions in molar concentrations of the order of 0·1%. In crystals rapidly quenched after the introduction of hydrogen they found H_s^- and H_i^- in cubic sites. In addition, they observed pairs of local-mode lines which they assigned to the fundamental vibrations of hydrogen in RE^{3+}–H_i^- complexes with C_{4v} symmetry (Table 2.17). The H_i^- ions occur in interstitial sites nearest the trivalent rare-earth ion and they replace charge-compensating fluorine interstitials (F_i^-) (a discussion of charge-compensatory mechanisms for trivalent rare-earth ions will be given in Chapter 6). The pairs Eu^{3+}–H_i^- and Yb^{3+}–H_i^- were not reported by Jones *et al.* (1968, 1969) and this is not surprising in view of the fact that Eu and Yb have a tendency to occur in CaF_2 in the divalent state (see Chapter 7). It was found, however, by Chambers and Newman (1969) that these pairs could be produced by X-irradiating hydrogenated crystals containing Eu or Yb at room temperature and they suggested a possible mechanism for production of the pairs. The vibrations of associated hydrogen were also observed as side-bands of electronic excitations in the cases of Gd (Jones *et al.*, 1969) and of Ce (Jacobs, Jones, Ždánský, and Satten, 1971). Yttrium is a common

TABLE 2.17†

Frequencies of infrared absorption lines due to RE^{3+}–H_i^- *pairs with* C_{4v} *symmetry in* CaF_2 (*Jones et al.* 1968).

Rare earth	Low frequency line (cm^{-1})	High frequency line (cm^{-1})
La	976·9	1121·7
Ce	988·9	1130·1
Pr	994·7	1115·0
Nd	1001·5	1119·4
Eu‡	1009·9	1092·9
Sm	1011·6	1117·3
Gd	1017·0	1104·2
Tb	1029·3	1112·8
Dy	1033·1	1103·3
Ho	1035·9	1094·6
Er	1036·2	1086·6
Er	1037·6	1081·2
Tm	1043·5	1087·3
Yb‡	1043·4	1074·6
Tb§	1049·8	1100·7
Lu	1042·5	1072·0
Y	1028·9	1074·8

† Taken from Newman (1969).
‡ Results of Chambers and Newman (1969).
§ These lines decrease in intensity when the temperature is reduced to 4·2 K (see text).

trivalent impurity in CaF_2 and it also forms C_{4v} pairs with H_i^- (Table 2.17).

From results of intensity measurements Jones *et al.* (1969) concluded that the high-energy component of the pair was due to a transition to the doublet. However, subsequent investigations showed considerable departures from an intensity ratio of 2:1 for some pairs (Newman, 1969). Detailed investigation of the Ce^{3+}–H_i^- complex by Jones *et al.* (1969) showed a splitting of 1·3 cm^{-1} for the low energy component of the pair indicating that the low energy component in this case was due to a transition to the doublet and suggesting that intensity measurements were not a reliable guide to transition assignment. It appears that the effective charges associated with the two components of the fundamental vibration are not predictable. A lower energy doublet means that the force constants associated with vibrations of H_i^- in the plane perpendicular to the RE^{3+}–H_i^- axis are smaller than those associated with vibrations along the axis.

Inspection of Table 2.17 shows that the separation of lines in a pair decreases with increasing atomic number of the rare-earth ion involved and this is not inconsistent with the fact that the radius of the RE^{3+}

ions also decreases along the series and that the RE^{3+}–F_i^- complexes are more stable at the beginning of the series (Weber and Bierig, 1964). Table 2.17 also shows that the centre of gravity of the pairs, at about 1070 cm^{-1}, is appreciably different from the fundamental frequency of 1312 cm^{-1} for isolated H_i^- (§ 2.5.4) and it was suggested by Jones *et al.* (1969) that this difference arises from charge transfer from H_i^- toward RE^{3+}.

Jones *et al.* (1969) found that Er^{3+}–H_i^- complexes gave rise to two sets of pairs (Table 2.17) one of which decreased in intensity with decreasing temperature. Chambers and Newman (1969) found a similar effect for Yb^{3+}–H_i^- and they suggested that the doubling was due to a temperature-dependent population of a low lying electronic state of the ion. However, this temperature dependence has not yet been satisfactorily correlated with the electronic level-structure of the rare-earth ions. The splitting of the doublet level in the pair in Ce^{3+}–H_i^- is temperature independent and has been assigned by Jacobs *et al.* (1971) to electron-phonon interaction effects between the cerium $4f$ electronic states and the local-mode phonon of the hydride ion.

Jones *et al.* (1968) found that when crystals of CaF_2 containing Gd^{3+}–H_i^- complexes were irradiated at 77 K with ultraviolet light of wavelength shorter than 2500 Å, or with X-rays, a new vibronic line appeared in the electronic spectrum of gadolinium. They assigned this line which occurred at 767 cm^{-1} from the parent electronic line to one of the fundamental components of vibration of hydrogen in Gd^{3+}–H_i^0 complexes. The corresponding deuterium line occurred at 565 cm^{-1}. The new line was not directly observable in the infrared because of low intensity. The Gd^{3+}–H_i^0 complexes disappear on warming the crystals to room temperature and the intensity of the Gd^{3+}–H_i^- complexes is restored.

The elements Eu, Sm, and Yb may occur in the divalent state in freshly grown crystals of CaF_2 (see Chapter 7). Vibrations of H_s^- ions associated with these divalent ions have been observed in two configurations by Chambers and Newman (1971). The H_s^- centres occur in nearest fluorine sites with site symmetry C_{3v}, giving rise to two fundamental vibrations (Table 2.18). The H_s^- centres also occur in next nearest sites, with site symmetry C_s, giving rise to three fundamental vibrations with smaller overall splitting (Table 2.18). The stronger line of each pair of the C_{3v} centres has the lower energy and has been assigned to the doubly degenerate mode. Corresponding pairs are observed in crystals doped with Nd, Dy, Ho, Er, and Tm (Table 2.18)

TABLE 2.18

Peak positions and weighted means (CG) *of local-mode satellite lines from* Re^{2+}–H_s^- *centres* (*from Chambers and Newman* 1971).

RE	Axial Centre Energy (cm^{-1})		CG (cm^{-1})	Low-symmetry centre Energy (cm^{-1})			CG (cm^{-1})
Nd	1018·7±0·3	924·4±0·3	955·8	981·2±0·3	977·6±0·3	940·5±0·3	966
Sm	1013·2	934·8	960·9	978·4	975·2	944·0	966
Eu	1014·3	937·0	962·8	977·6	974·8	945·9	966
Dy	{1009·6	938·5	961·8	973·4	971·6	952·2	966
	{1007·5						
Ho	1008·0	936·8	960·5	970·8		953·9	
Er	999·6	938·9	959·1			955·6	
Tm	997·8	944·7	962·4			957·3	
Yb	996·2	944·7	961·9	969·3	968·2	958·1	965

but only after X- or γ-irradiation. These ions are readily converted from the trivalent to the divalent state in CaF_2 by ionizing radiations (Chapters 5 and 7). However, the hydrogen-doped crystals do not show RE^{3+}–H_s^- complexes before irradiation and the mechanism by which the irradiation produces RE^{2+}–H_s^- complexes has not been established. It should be mentioned that no trigonal RE^{2+}–H_s^- complexes were produced by irradiation of crystals doped with Ce, Pr, Gd or Tb. These rare earths are the most strongly resistant to reduction to the divalent state in CaF_2 and form photochromic complexes rather than divalent ions under the reducing conditions of additive colouration (Chapter 7). Vibrational spectra of Sr^{2+}–H_s^- and Ba^{2+}–H_s^- complexes in CaF_2 similar to those of RE^{2+}–H_s^- complexes have been observed by Chambers (1971).

As in the case of RE^{3+}–H_i^- complexes the splitting of the fundamental vibration in RE^{2+}–H_s^- complexes decreases along the rare-earth series. However, the centre of gravity of the C_s complexes is close to the fundamental (965 cm^{-1}) of isolated H_s^- ions and it converges on this value along the rare-earth series for the C_{3v} complexes (Table 2.18). The descriptions of RE–H^- complexes given here are consistent with results of epr and endor investigations given in Chapter 6.

Trivalent rare-earth ions occur in alkaline earth fluorides in a variety of environments because of variety in the charge-compensating mechanisms (Chapter 6). Oxygen, which dissolves in anion sites as O^{2-} ions (see §§ 4.7.3 and 6) may provide charge compensation and the presence of RE ions associated with oxygen increases the range and complexity of RE–H^- complexes which may be formed (see Chambers and Newman, 1971, 1972, for a discussion and also Chapter 6).

2.6. Vibrations of heavy metal impurities

An investigation of the far infrared absorption of CaF_2 doped with heavy metal impurities has been reported by Hayes, Wiltshire, Berman, and Hudson (1973). The pure-crystal Green function may be defined by the relations

$$\mathbf{L}_0\mathbf{G}_0 = \mathbf{I}, \qquad \mathbf{L}_0 = \mathbf{A}-\omega^2\mathbf{I}, \tag{2.44}$$

where $\mathbf{A}$ is the force-constant matrix. Similarly, a perturbed-crystal Green function $\mathbf{G}$ may be defined for a crystal containing impurities by the relations

$$\mathbf{L}\mathbf{G} = \mathbf{I}, \qquad \mathbf{L} = \mathbf{L}_0+\mathbf{\Gamma}(\omega), \tag{2.45}$$

where $\mathbf{\Gamma}(\omega)$ contains both the mass change and force-constant changes in the space affected by the impurity. The matrix $\mathbf{G}$ is related to $\mathbf{G}_0$ by

$$\mathbf{G} = \mathbf{G}_0(1+\mathbf{\Gamma}\mathbf{G}_0)^{-1} \tag{2.46}$$

Defining a matrix $\mathbf{T}$ by the relation

$$\mathbf{T} = \mathbf{\Gamma}(1+\mathbf{G}_0\mathbf{\Gamma})^{-1} \tag{2.47}$$

the vibrational absorption coefficient of the perturbed crystal may be written for frequencies well away from ω_{TO} (Klein, 1968; this review contains a full discussion of the theory used here and in § 2.5.2).

$$\alpha(\omega) = \frac{(n_\infty^2+2)^2}{9n(\omega)}\frac{4\pi N}{c}\frac{e^{*2}}{\mu\Omega}\frac{\omega}{(\omega_{\mathrm{TO}}^2-\omega^2)^2}\times \mathrm{Im}\langle 0, \mathrm{TO}_z|\,\mathrm{T}\,|\mathrm{TO}_z, 0\rangle, \tag{2.48}$$

where n_∞ and $n(\omega)$ are the refractive indices of the pure crystal at high frequency and at frequency ω, N is the number of defects in unit volume, Ω is the volume and μ is the reduced mass of a unit cell, e^* is an effective charge and $|\mathrm{TO}_z, 0\rangle$ is the eigenvector of the TO phonon (of frequency ω_{TO}) at $\mathbf{q} = 0$, polarized in the z direction.

Hayes, Wiltshire, Berman and Hudson (1973) studied CaF_2 doped with Co^{2+}, Y^{3+}, La^{3+}, Sm^{2+}, Gd^{3+}, and Tm^{3+} ions. These impurities dissolve in calcium sites and the trivalent ions may be associated with charge-compensating interstitial fluorine ions (Chapter 6). Hayes, Wiltshire, Berman and Hudson (1973) neglected effects of fluorine interstitials for reasons given later in this section and used a model in which the T matrix contains a change of mass at the impurity site and a change of central force constant between the impurity and the eight nearest fluorine neighbours. The site symmetry of Ca^{2+} is O_h and the infrared radiation couples to modes of Γ_4^- symmetry. Transforming the calculation from real space to the space spanned by the irreducible representations of O_h reduces the problem to a calculation in the Γ_4^- subspace. The elements of the matrix **T** are calculated in this subspace using (2.47) and the absorption coefficient is then obtained from (2.48).

It was found that the far infrared absorption induced by La^{3+}, Sm^{2+}, Gd^{3+}, and Tm^{3+} in CaF_2 could be accounted for quite well using a change of the nearest-neighbour central force constant; this varied from an increase of 60% for Sm^{2+} (Fig. 2.10) to a reduction of 5% for

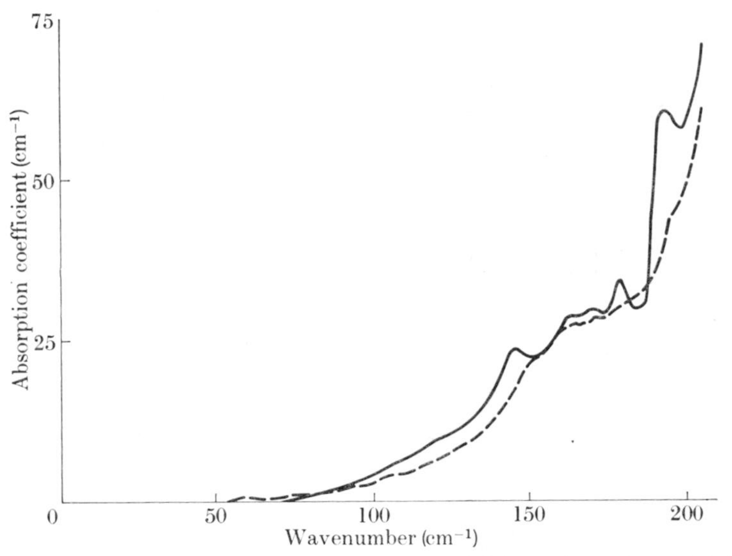

FIG. 2.10. Measured (dashed line) and calculated (full line) absorption at 1·8 K of CaF_2 doped with Sm^{2+} (after Hayes, Wiltshire, Berman, and Hudson 1973).

Tm^{3+} and was clearly correlated with the size of the ion relative to that of Ca^{2+} (Table 2.19). In the case of the small ions Y^{3+} and Co^{2+} the position of observed resonances at 80 and 74 cm^{-1} respectively could be explained by a reduction of the nearest-neighbour central force constant but the calculated resonance intensity was too large. It was clear that for small ions a more elaborate model would be required to explain intensities, possibly including relaxation of the lattice ions around the impurity. Problems of a similar sort were encountered in a more dramatic fashion with the small ion Li^+ in NaCl (Macdonald, Klein, and Martin, 1969). In all cases the values of e^* used in eqn (2.48) was chosen to optimize agreement with experiment.

The increased force constant at the site of the large rare-earth ions affects the motion of the surrounding fluorines also. Hayes, Wiltshire, Manthey, and McClure (1973) reported a phonon side-band with an

TABLE 2.19

Radii (Å) (Pauling 1960), masses of metal ions and percentage change of nearest-neighbour central force constant to give agreement with observed resonances.

Ion	Radius	Mass	% change of force constant
Sm^{2+}	1·18†	150·35	+60
La^{3+}	1·15	138·91	+50
Ce^{3+}	1·11	140	+45
Gd^{3+}	1·02	157·25	+20
Ca^{2+}	0·99	40·08	—
Tm^{3+}	0·95	168·93	−5
Y^{3+}	0·93	88·91	−25
Co^{2+}	0·78	58·93	−30

The results on Ce^{3+} are due to Hayes, Wiltshire, Manthey and McClure (1973); the remaining results are due to Hayes, Wiltshire, Berman and Hudson (1973).

† Taken from Harrington & Walker.

energy of about 490 cm^{-1} in the vibronic spectrum of Ce^{3+} in CaF_2 which they assigned to the symmetric (Γ_{1g}) breathing mode of the surrounding eight fluorines. Different types of charge compensation did not affect this frequency significantly and it appears that interstitial fluorines, for example, do not influence the interaction between the rare-earth ion and its nearest neighbours appreciably. The ionic radius of Ce^{3+} is 1·11 Å (Pauling, 1960, p. 518) and far infrared measurements suggested that the nearest-neighbour central force constant for an ion of this size increased by about 45% (Table 2.19). Using the simple lattice-dynamical model applied to the far infrared data Hayes, Wiltshire, Manthey, and McClure found that a 45% increase in nearest-neighbour central force constant predicted a frequency for the breathing mode of the surrounding fluorines of 490 cm^{-1}, in good agreement with experiment. The fact that this frequency is greater than the highest lattice frequency of CaF_2 ($\sim$465 cm^{-1}) is of interest because of the unusual result that a local mode is induced by a heavy impurity ion.

2.7. Thermal conductivity

In concluding this chapter on lattice dynamics we describe results of some measurements of the thermal conductivity of CaF_2, SrF_2, and BaF_2. Such measurements may be used to study phonon interactions in pure crystals and in crystals containing impurities. Harrington and Walker (1970) found that the undoped crystals were moderately good conductors of heat at low temperatures (Fig. 2.11) (see also Parfan'iva and Smirnov, 1971). Concentrations of the order of 10^{18} cm^{-3} of Li^+ or Na^+ impurities in all three hosts were found to have

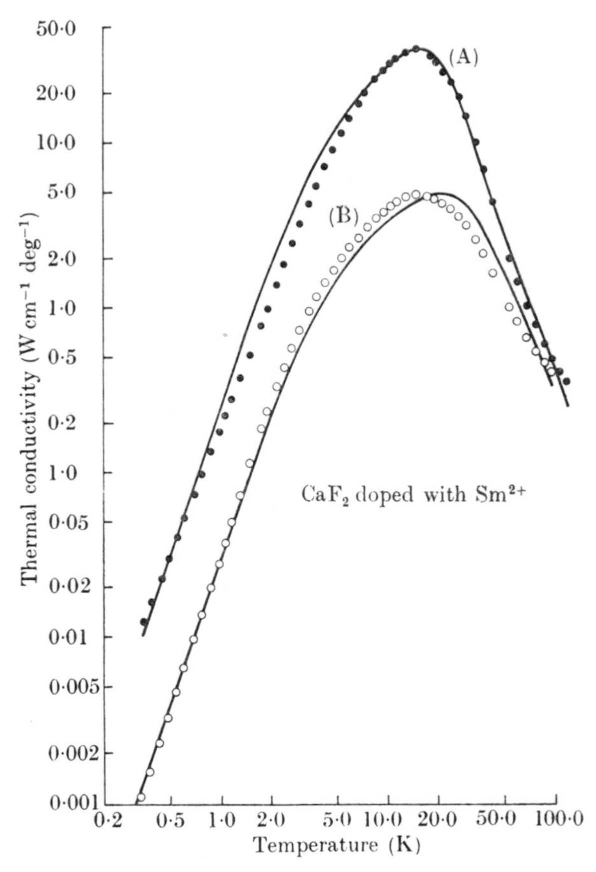

Fig. 2.11. Thermal conductivity as a function of temperature for (A) undoped CaF_2 and (B) CaF_2 containing 3×10^{19} cm^{-3} Sm^{2+} ions. The solid lines are theoretical curves obtained using eqn. (2.49) (after Harrington and Walker 1970).

little effect on the thermal conductivity suggesting that phonon scattering due to point defects in these crystals is weak. In addition, Sm and Eu impurities in the 10^{19} cm^{-3} range were studied in CaF_2 and only point defect scattering of the Rayleigh type was found (Fig. 2.11). Harrington and Walker used a single-term Debye-like integral for the conductivity:

$$\kappa(T) = \frac{\hbar^2}{2\pi^2 vkT^2}\int_0^{\omega_D} \frac{\tau(\omega, T)\exp(\hbar\omega/kT)}{[\exp(\hbar\omega/kT-1)]^2}\,\mathrm{d}w \qquad (2.49)$$

where v is the velocity of sound ($3{\cdot}51\times10^2$ m/s^{-1} in CaF_2 at low temperature), ω_D is the Debye maximum frequency, $\hbar$ and k are Planck's and Boltzmann's constants and $\tau(\omega, T)$ is the phonon relaxation time.

Subsequently Hayes, Wiltshire, Berman, and Hudson measured the thermal conductivity of CaF_2 containing 1% of Y^{3+} ions and found a

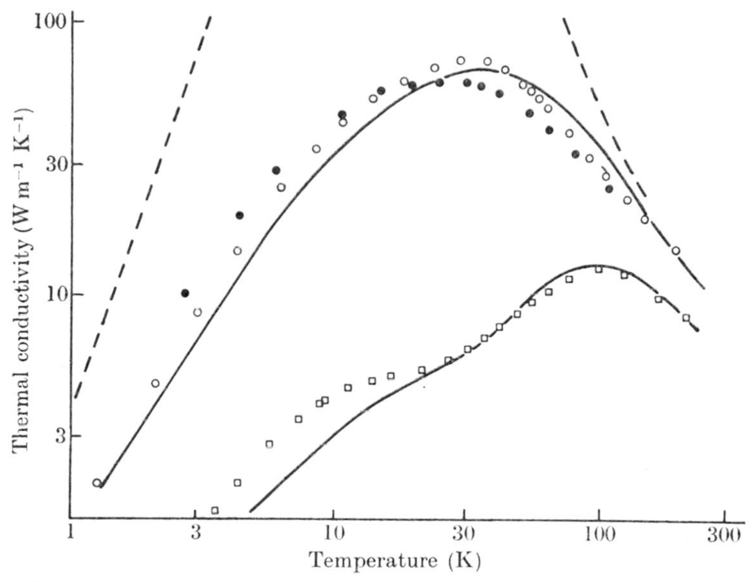

FIG. 2.12. Thermal conductivity of CaF_2 containing impurities with the following concentrations:

○ $2{\cdot}65\times10^{19}$ cm^{-3} of Y^{3+}
□ $2{\cdot}65\times10^{20}$ cm^{-3} of Y^{3+}
● $2{\cdot}65\times10^{19}$ cm^{-3} of Tm^{3+}

--- is pure CaF_2. ——— curves computed according to eqns. (2.49) and (2.50) (after Hayes, Wiltshire, Berman, and Hudson 1973).

resonance dip at about 30 K (Fig. 2.12); they correlated this dip with a strong resonance in the infrared absorption at about 80 cm^{-1} (§ 2.6). They used eqn (2.49) to fit the measured conductivity with the following expression for the total phonon relaxation rate:

$$\tau(\omega, T) = (A+ND)\omega^4 + B\omega^2T^2e^{-C/T} + \frac{NE\,\omega^2}{(\omega_0^2-\omega^2)^2} + F \qquad (2.50)$$

where N is the concentration of added impurity. The first term represents Rayleigh scattering by background defects and by the added impurity and is the term used by Harrington and Walker to explain their data (Fig. 2.11). The second term represents scattering by umklapp processes, the third is the resonance term and the fourth represents boundary scattering. The values of the constants A to F in eqn (2.50) used for CaF_2:Y were as follows:

$A = 1{\cdot}3\times10^{-44}\,s^{-3}$, as used by Harrington and Walker, implying Rayleigh scattering by imperfections in undoped CaF_2 about double that due to calcium isotopes.

$B = 3\times10^{-20} sk^{-2}$.

$C = 69\,k\,(\Theta/7)$ as used by Harrington and Walker.

$D = 5{\cdot}4\times10^{-64}\,s^3\,cm^3$ which is about half the isotope-like scattering to be expected from the mass difference between Y and Ca.

$E = 8\times10^{16}\,s^{-3}$.

$F = 2\times10^{6}\,s^{-1}$ (this factor is determined by the shape of the specimen).

$\omega_0 = 1{\cdot}7\times10^{13}\,s^{-1}$ is close to the measured infrared resonance frequency of $1{\cdot}5\times10^{13}\,s^{-1}$ (§ 2.6).

N was taken to be $2{\cdot}65\times10^{20}\,cm^{-3}$ for 1% Y and $2{\cdot}65\times10^{19}\,cm^{-3}$ for 0·1% Y.

The term $ND\omega^4$ for the specimen with 0·1% Y makes a small contribution to the relaxation rate over most of the temperature range so that the similarity of the thermal conductivity curves for 0·1% Y and 0·1% Tm (Fig. 2.13) suggests that the resonant scattering for these two impurities is similar in magnitude and has about the same value of ω_0. The infrared measurements are consistent with this conclusion (Hayes, Wiltshire, Berman, and Hudson).

It should be emphasized that for 0·1% Sm, Y, and Tm in CaF_2 resonance effects are not obvious and 1% of Y was required to observe the characteristic dip in the thermal conductivity curve. This is in marked contrast with alkali halides where Baumann and Pohl (1967) found pronounced resonance dips for 0·05% Ag and 0·02% Tl. As yet there is no clear understanding of this difference in behaviour.

References

Ackerman, R. J., Thorn, R. J., and Winslow, G. H. (1959). *J. opt. Soc. Am.* **49**, 1107.

Akhieser, A. (1939). *J. Phys. USSR.* **1**, 277.

ALTEROVITZ, S. and GERLICH, D. (1969). *Phys. Rev.* **184,** 999.
—— —— (1970*a*). *Ibid.* **B1,** 2718.
—— —— (1970*b*). *Ibid.* **B1,** 4136.
ANDEEN, C., FONTANELLA, J., and SCHUELE, D. (1971). *J. appl. Phys.* **42,** 2216.
AXE, J. D. (1965). *Phys. Rev.*, **139,** A1215.
—— GAGLIANELLO, J. W., and SCARDEFIELD, J. E. (1965). *Ibid.* **139,** A1211.
—— and PETTIT, G. D. (1966). *J. Phys. Chem. Solids*, **27,** 621.
BAILEY, A. C. and YATES, B. (1967). *Proc. phys. Soc. Lond.* **91,** 390.
BAUMANN, F. C. and POHL, R. O. (1967). *Phys. Rev.* **163,** 843.
BERREMAN, D. W. (1963) *Phys. Rev.*, *130*, 2193.
BILZ, H., FRITZ, B., and STRAUCH, D. (1966). *J. Phys. Paris*, Supp. 27, C2–3.
BORN, M. and HEISENBERG, W. (1924). *Z. Phys.* **24,** 338.
BORN, M. and HUANG, K. (1954). *Dynamical theory of crystal lattices*, Oxford.
BOSOMWORTH, D. R. (1967). *Phys. Rev.* **157,** 709.
BRADA, R. M. and BATES, Y. (1969). *Phys. Stat. Sol.* **36,** 551.
BRIGGS, A. (1964). Report to a panel of the IAEA, Vienna.
BRYA, W. J. (1971). *Sol. St. Comm.* **9,** 2271.
CATLOW, C. R. A., and NORGETT, M. J. (1973). *J. Phys. C*, (*Sol. St. Phys.*), **6,** 1325.
CHAMBERS, D. N. (1971). *Ibid.* **4,** 1977.
CHAMBERS, D. N. and NEWMAN, R. C. (1969), *Phys. Stat. Sol.* **35,** 685.
—— —— (1971). *J. Phys. C.* (*Sol. St. Phys.*) **4,** 3015.
—— —— (1972). *Ibid.* **5,** 997.
CHANG, R. K., LACINA, B., and PERSHAN, P. S. (1966). *Phys. Rev. Lett.* **17,** 755.
COCHRAN, W. (1971). *Crit. Rev. Sol. St. Sci.* **2,** 1.
COOPER, M. J. (1970). *Acta Crystallog.* **A26,** 208.
—— and ROUSE, K. D. (1970). *Ibid.* 214 and "to be published".
—— —— and WILLIS, B. T. M. (1968). *Ibid.* **A24,** 484.
CRIBIER, D. and FARNOUX, B., and JACROT, B. (1962). *Inelastic scatter of neutrons in solids and liquids* **2,** 225 (IAEA, Vienna).
DANDEKAR, D. P. and JAMIESON, J. C. (1969). *Trans. Am. Cryst. Assoc.* **5,** 19.
DAWBER, P. G. and ELLIOTT, R. J. (1963*a*). *Proc. phys. Soc. Lond.* **81,** 453.
—— —— (1963*b*). *Proc. R. Soc.* **A273,** 222.
DAWSON, B., HURLEY, A. C., and MASLEN, V. W. (1967). *Ibid.* **A298,** 289.
DENHAM, P., FIELD, G. R., MORSE, P. L. R., and WILKINSON, G. R. (1970). *Proc. R. Soc.* **A317,** 55.
DICK, B. G. and OVERHAUSER, A. L. (1958). *Phys. Rev.* **112,** 90.
DOLLING, G., COWLEY, R. A., and WOODS, A. D. B. (1965). *Can. J. Phys.* **43,** 1397.
ELCOMBE, M. M. (1972). *J. Phys. C.* (*Sol. St. Phys.*), **5,** 2702.
—— and PRYOR, A. W. (1970). *Ibid.* **3,** 492.
ELLIOTT, R. J. (1966) *Phonons* (ed. R. W. H. Stevenson), p. 377, Oliver & Boyd, Edinburgh & London.
——, HAYES, W., JONES, G. D., MACDONALD, H. F., and SENNETT, C. T. (1965). *Proc. R. Soc.* **A289,** 1.
——, KRUMHANSL, J. A., and MERRETT, T. H. (1968). *Localised excitations in solids* (ed. R. F. Wallis), p. 709, Plenum Press, New York.
FAJANS, K. and JOOS, G. (1924). *Z. Phys.* **23,** 1.
FELDMAN, D. W., PARKER, J. H., and ASHKIN, M. (1968). *Phys. Rev. Lett.* **21,** 607.
FIESCHI, R., NARDELLI, G. F., and TERZI, N. (1965). *Phys. Rev.* **138,** A203.
FRAY, S. J., JOHNSON, F. A., and QUARRINGTON, J. E. (1964). *Lattice dynamics* (ed. R. F. Wallis), p. 377, Pergamon Press.

FRITZ, B. (1962). *J. Phys. Chem. Solids.* **23,** 375.
—— (1965). *Lattice dynamics* (ed. R. F. Wallis), p. 485, Pergamon Press.
—— (1968). *Localised excitations in solids* (ed. R. F. Wallis), p. 480, Plenum Press, New York.
FRITZ, B., GERLACH, J., and GROSS, U. (1968). *Localised excitations in solids* (ed. R. F. Wallis), p. 504, Plenum Press, New York.
GANESAN, S. and SRINIVASAN, R. (1962). *Can. J. Phys.* **40,** 74.
GERLICH, D. (1964*a*). *Phys. Rev.* **135,** A1331.
—— (1964*b*). *Ibid.* **136,** A1336.
—— (1968). *Phys. Rev.*, **168,** 747.
HALL, J. L. and SCHUMACHER, R. T. (1962). *Ibid.* **127,** 1892.
HARRINGTON, J. A. and WALKER, C. T. (1970). *Ibid.* **B1,** 882.
—— HARLEY, R. T. and WALKER, C. T. (1970). *Solid State Commun.* **8,** 407.
——, ——, —— (1971). *Ibid.* **9,** 683.
HART, S. (1970). *Phys. Stat. Sol* (*a*), **3,** K187.
HARTMANN, W. M., GILBERT, T. L., KAISER, K. A., and WAHL, A. C. (1970), *Phys. Rev.* **B2,** 1140.
HAYES, W. and MACDONALD, H. F. (1967). *Proc. R. Soc.* **A297,** 503.
—— and WILTSHIRE, M. C. K. (1973). *J. Phys. C* (*Sol. St. Phys.*), **6,** 1149.
——, ——, BERMAN, R., and HUDSON, P. R. W. (1973). *Ibid.* **6,** 1157.
——, ——, MANTHEY, W. J., and MCCLURE, D. S. (1973). *Ibid.* **6,** L273.
HERZBERG, G. (1945). *Infrared and Raman Spectra*, p. 100, D. van Nostrand Co. Inc., New York.
HEWAT, A. W. (1972). *J. Phys. C* (*Sol. St. Phys.*) **5,** 1309.
HISANO, K., OHAMA, N., and MATUMURA, O. (1965). *J. Phys. Soc. Japan* **20,** 2294.
HO, P. L. and RUOFF, A. L. (1967). *Phys. Rev.* **161,** 864.
HUFFMAN, D. R. and NORWOOD, M. H. (1960). *Ibid.* **117,** 709.
HUGHES, A. E. (1968). *Ibid.* **173,** 860.
HURRELL, J. P., KAM, Z., and COHEN, E. (1972). *Ibid.* **B6,** 1999.
—— and MINKIEWICZ, V. J. (1970). *Solid State Commun.* **8,** 463.
IIZUMI, M. (1972). *J. Phys. Soc. Japan* **33,** 647.
JACOBS, I. T., JONES, G. D., ŽDÁNSKÝ, K., and SATTEN, R. A. (1971). *Phys. Rev.* **B3,** 2888.
JONES, B. W., (1967). *Phil. Mag.* **16,** 1085.
JONES, G. D., PELED, S., ROSENWAKS, S., and YATSIV, S. (1969). *Phys. Rev.* **183,** 353.
—— YATSIV, S., PELED, S., and ROSENWAKS, S. (1968). *Localised excitations in solids* (ed. R. F. Wallis), p. 512, Plenum Press, New York.
KAISER, W., SPITZER, W. G., KAISER, R. H., and HOWARTH, L. E. (1962). *Phys. Rev.* **127,** 1950.
KAM, Z. and COHEN, E. (1971). *Light Scattering in Solids* (ed. M. Balkanski). p. 253, Flammarion Sciences.
KESSLER, J. R. and NICOL, M. (1972). *Bull. Am. Phys. Soc.* **17,** 123.
KHAN, A. A. and DESHPANDE, V. T. (1968). *Acta Crystallogr.* **A24,** 402.
KIEL, A. and SCOTT, J. F. (1970). *Phys. Rev.* **B2,** 2033.
KLEIN, M. V. (1968). *Physics of color centers* (ed. W. Beall Fowler), p. 430, Academic Press, New York & London.
KÜHNER, D. H., LAUER, H. V., and BRON, W. E. (1972). *Phys. Rev.* **B5,** 4112.
LACINA, W. B. and PERSHAN, P. S. (1970). *Ibid.* **B1,** 1765.
LAMBE, K. A. D. and WILLIS, B. T. M. (1963). *A.E.R.E. Report* 4401.
LAUER, H. V., SOLBERG, K. A., KÜHNER, D. H., and BRON, W. E. (1971). *Phys. Lett.* **35,** 219.

LAX, M. and BURSTEIN, E. B. (1965). *Phys. Rev.* **97**, 39.
LEE, L. C. and FAUST, W. L. (1971). *Phys. Rev. Lett.* **26**, 648.
LOWNDES, R. P. (1969). *J. Phys. C (Sol. St. Phys.)*, **2**, 1595.
—— (1971). *Ibid.* **4**, 3083.
LOUDON, R. (1964). *Proc. phys. Soc. Lond.* **84**, 379.
MACDONALD, H. F., KLEIN, M. V., and MARTIN, T. P. (1969). *Phys. Rev.* **177**, 1292.
MARADUDIN, A. A. (1966*a*). *Solid State Phys.* **18**, 273.
—— (1966*b*). *Ibid.* **19**, 1.
——, GANESAN, S. and BURSTEIN, E. (1967). *Phys. Rev.* **163**, 882.
MINOMURA, S. and DRICKAMER, H. G. (1961). *J. chem. Phys.* **34**, 670.
MIRLIN, D. N. and RESHINA, I. I. (1971). *Sov. Phys. Sol. State.* **13**, 2639.
MITRA, S. S. and SINGH, R. S. (1966). *Phys. Rev. Lett.* **16**, 694.
NEWMAN, R. C. (1969). *Adv. Phys.*, **18**, 545.
—— and CHAMBERS, D. N. (1968). *Localised excitations in solids* (ed. R. F. Wallis). p. 520, Plenum, New York.
NYE, J. F. (1957). *Physical Properties of Crystals*, Clarendon Press, Oxford, p. 249.
PARFAN'IVA, L. S. and SMIRNOV, A. I. (1971). *Sov. Phys.-Sol. State.* **13**, 1267.
PAULING, L. (1927). *Proc. R. Soc.* **A114**, 191.
—— (1960). *Nature of the Chemical Bond*, Cornell University Press, p. 518.
RAJAGOPAL, A. K. (1962). *J. Phys. Chem. Solids.* **23**, 317.
RAMACHANDRAN, V. and SRINIVASAN, R. (1972). *Solid State Commun.* **11**, 973.
RAO, K. V. and NARASIMHAMURTY, T. (1970). *J. Phys. Chem. Solids.* **31**, 876.
—— and SMAKULA, A. (1966). *J. appl. Phys.* **37**, 319.
REITZ, J. R., SEITZ, R. N., and GENBERG, R. W. (1961). *J. Phys. Chem. Solids.* **19**, 73.
RUSSELL, J. P. (1965) *App. Phys. Lett.*, **6**, 223.
SAMARA, G. A. (1970). *Phys. Rev.* **B2**, 4194.
SCHAEFER, G. (1960). *J. Phys. Chem. Solids.* **12**, 233.
SCHMIDT, E. D. D. and VEDAM, K. (1966). *Ibid.* **27**, 1563.
SHAKIN, O. V., BRYZHINA, M. F., and LEMANOV, V. V. (1972). *Sov. Phys. Sol. State.* **13**, 3141.
—— and LEMANOV, V. V. (1972). *Sov. Phys. Sol. St.*, **14**, 1189.
SHAMU, R. E., HARTMANN, W. H., and YASAITIS, E. L. (1968). *Phys. Rev.* **170**, 822.
SRINIVASAN, R. (1958). *Proc. phys. Soc. Lond.* **72**, 566.
—— (1968*a*). *J. Phys. C (Sol. St. Phys.)*, **1**, 1138.
—— (1968*b*). *Phys. Rev.* **165**, 1041, 1054.
SRIVASTAVA, R., LAUER, H. V., CHASE, L. L., and BRON, W. E. (1971). *Phys. Lett.* **36A**, 333.
STROCK, H. B. and BATTERMAN, B. W. (1972). *Phys. Rev.* **B5**, 2337.
SZIGETI, B. (1961). *Proc. R. Soc.* **A261**, 274.
TESSMANN, J. R., KAHN, A. H., and SHOCKLEY, W. (1953). *Phys. Rev.* **92**, 890.
TOR, S. K., TANDON, U. S., and RAI, G. (1972). *Ibid* **B5**, 4143.
VERLEUR, H. W. and BARKER, A. S. (1967). *Ibid* **164**, 1169.
WALLACE, D. C. (1970). *Solid State Phys.* **25**, 301.
WACHTMAN, J. B., WHEAT, M. L., ANDERSON, H. J., and BATES, J. L. (1965). *J. nuc. Mater.* **16**. 39.
WARRIER, A. V. R. and KRISHNAN, R. S. (1964). *Naturwissenschaften* **51**, 8.
WEBER, M. J. and BIERIG, R. W. (1964). *Phys. Rev.* **1**, 882.
WILLIS, B. T. M. (1963). *Proc. R. Soc.* **A274**, 134.
WILLIS, B. T. M. (1965). *Acta Cryst.* **18**, 75.
—— (1969). *Ibid* **A25**, 277.

YATSIV, S., PELED, S., ROSENWAKS, S., and JONES, G. D. (1967). *Optical properties of ions in crystals* (ed. H. M. Crosswhite and H. W. Moos), p. 409, Interscience Publishers.

YOUNG, K. F. and FREDERIKSE, H. P. R. (1969). *J. appl. Phys.* **40,** 3115.

ZERNIK, W. (1967). *Rev. mod. Phys.* **39,** 432.

3

THERMODYNAMICS AND KINETICS OF POINT DEFECTS

3.1. Introduction

THE previous two chapters have reviewed the basic model of perfect fluorite lattices, their cohesion, lattice vibrations, etc. In this chapter we shall consider present knowledge of the basic ionic defects in these structures, including vacancies, interstitials, foreign atoms and ions and their aggregates. More specifically we shall be mainly concerned with those defects responsible for physical properties such as diffusion, ionic transport and certain relaxation processes, i.e. with defects in crystals in or near thermodynamic equilibrium.† Later chapters consider such defects in ionized states, e.g. colour centres. Since the book as a whole is rather little concerned with the mechanical properties of these crystals, we shall make only passing reference to dislocation properties, even though space charges formed from an excess of one type of ionic defect exist generally around edge dislocations (as well as at external surfaces) in ionic crystals. In fact, while this feature has been studied to a considerable extent in the alkali and silver halides, rather little has yet been done on the fluorite compounds.

The basic general principles which the interpretations of atomic migration and relaxation measurements rest upon are the same as in the more extensively studied alkali halides. They are three:

(i) the ionic model;

(ii) Frenkel and Schottky defect equilibria and their perturbation by other defects, the most important of which are impurities having a different ionic charge from the corresponding host ion ('aliovalent' impurities);

(iii) the responsibility of these defects, as a result of their thermally activated migration, for mass transport through these crystals.

The first of these is supported by a variety of evidence, e.g. cohesive energies and electron density maps (Chapter 1), lattice vibrations (Chapter 2), the electronic structure of colour centres such as the *F*-centre (Chapter 4), etc. The second, the occurrence of Frenkel and

† For general background to the material of this Chapter see, e.g. Adda and Philibert (1966) for a comprehensive treatment of point defects and atomic migration in crystalline solids or, for a briefer account, see a recent review by Lidiard (1971).

Schottky defects, then follows from rather general statistical thermodynamic arguments. The third, the relation of these defects to observable mass transport effects, has been extensively worked out in the alkali and silver halides and other strongly ionic compounds.

Although we can thus be confident about the general principles governing the theory of ionic transport in these materials, there are some special features sufficiently striking to modify their application here, even though all the ways in which they will do so are not yet clear. Firstly, there are anomalies in the high-temperature specific heats of

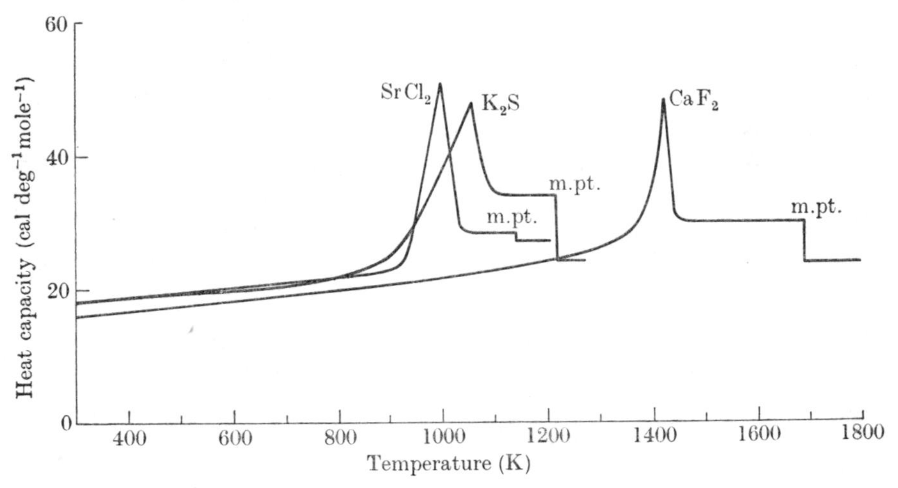

FIG. 3.1. The heat capacity, C_P, of three substances with the fluorite structure as a function of absolute temperature. In each case there is a λ-type anomaly before the melting point (m. pt.) is reached. (After Dworkin and Bredig, 1968).

CaF_2, $SrCl_2$, and K_2S, which are like those associated with order-disorder phase transitions elsewhere, and which have been interpreted as due to a transition from the normal fluorite structure RX_2 to one in which the X ions become randomly distributed over both the tetrahedrally co-ordinated sites (the normal X sites) and the octahedrally coordinated sites (interstitial X sites) associated with a fixed face-centred-cubic lattice of R ions (Dworkin and Bredig, 1968). Examples are shown in Fig. 3.1 and temperatures of the specific heat maxima are listed in Table 3.1. The entropies of transition are about $1-2k$ per molecule and are consistent with such a model. In the case of $SrCl_2$ early X-ray diffraction measurements (Croatto and Bruno, 1946) also showed the occurrence of a high degree of disorder at high temperatures. The increasing anharmonicity and $\langle 111 \rangle$ displacements of the anions in

TABLE 3.1

Approximate temperatures of specific heat maximum and melting point (°C) *of some solids with the fluorite structure* (*after Dworkin and Bredig*, 1968).

Substance	Specific heat maximum	Melting point
CaF_2	1150	1350
$SrCl_2$	720	873
UO_2	2400†	2805
K_2S	780	948

† Estimated.

CaF_2, ThO_2, and UO_2 seen at high temperatures in neutron diffraction experiments by Willis (1963; 1965) may be the preamble to these transitions. If correct, this interpretation means that above these transition temperatures, the fluorite compounds become analogous to some other ionic compounds which disorder in one sub-lattice (e.g. α-AgI, Li_2SO_4). Ionic mobility consequently becomes very high in that sub-lattice and Frenkel and Schottky defect models do not apply (see e.g. Kvist, 1972). However the ionic conductivity of $SrCl_2$ has been measured to 780 °C, i.e. well beyond the temperature of the specific heat maximum, without any obvious irregularity in its temperature dependence although it does rise rapidly to rather high values with a high activation energy (3·2 eV; Hood and Morrison, 1967).

A second special feature of the fluorite compounds, and one which may be closely related to the first, is the ease with which they accept large concentrations of interstitial defects. Thus it has been known for a considerable time that the alkaline earth fluorides will dissolve large proportions of trivalent metal fluorides of the order of 30–50 mole per cent at the melting point, (Ippolitov, Garashina, and Maklasklov, 1967). The foreign cations are incorporated substitutionally and the excess anions interstitially, as is shown by the comparison of lattice parameter and macroscopic density. An example of such results is shown in Fig. 3.2. It is interesting to observe that in CaF_2, the lattice parameter generally increases while in BaF_2 it decreases; in SrF_2 it sometimes increases and sometimes decreases, depending on the particular ion incorporated. An analogous situation arises with UO_{2+x} which has a wide non-stoichiometric field, especially at high temperatures. The excess oxygen again is incorporated interstitially. An interesting system which contrasts with these interstitial systems is

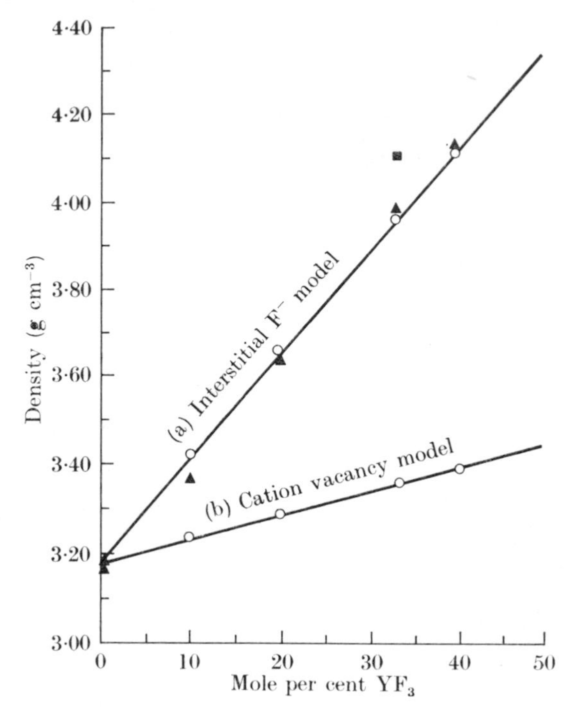

FIG. 3.2. The macroscopic density of CaF_2/YF_3 solid solutions as a function of the mole fraction of YF_3. The experimental pycnometer measurements are shown as ▲ and ■ while the open circles represent theoretical densities calculated from lattice parameters determined by X-ray measurements on the assumption that incorporation of YF_3 leads to the formation (a) of F^- interstitials (in proportion of one to each YF_3 molecule dissolved) (b) of Ca^{2+} vacancies (in proportion of one for each *pair* of YF_3 molecules dissolved). It is seen that such a comparison provides strong evidence for the generation of interstitials. (After Short and Roy, 1963).

ZrO_2 which can be stabilized in a fluorite structure by solution of percentages (12–22 mole per cent) of CaO to give a system containing large numbers of O^{2-} vacancies (Tien and Subbarao, 1963; Etsell and Flengas 1970; Steele 1972). The non-stoichiometric oxides PrO_{2-x}, CeO_{2-x}, and TbO_{2-x} at high temperatures are stable in the fluorite structure up to values of x corresponding to more than 10 per cent of oxygen vacancies while they can also dissolve several mole per cent of lower-valent cations such as Ca^{2+} and Y^{3+}, again with the formation of oxygen vacancies. The fluorite lattice is thus often able to accept unusually high concentrations of defects.

Although the comparison of X-ray lattice parameter and macroscopic density measurements tells us whether there is on average an excess or deficiency of ions per unit cell, i.e. whether we have interstitials or vacancies present, these measurements cannot tell us anything about the *structure* of these defects. The most detailed information about defect structures is to be obtained from endor measurements although, of course, the occurrence of a magnetic moment on the defect is necessary for this to be possible. Experiments of this kind have been carried out on Ce^{3+}- and Yb^{3+}-doped CaF_2 (Baker, Davies, and Hurrell,

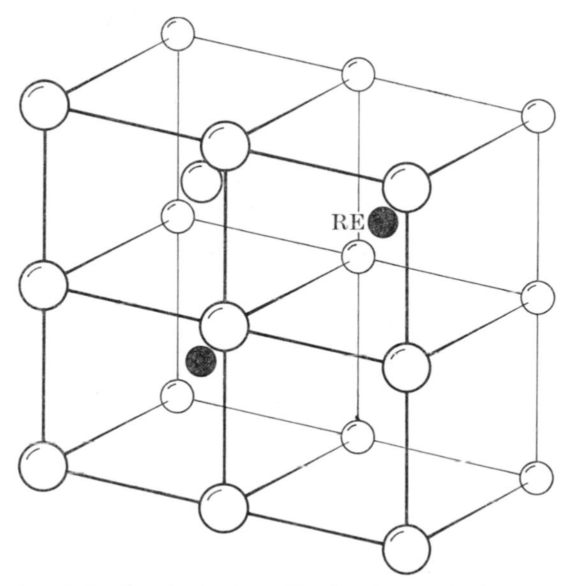

FIG. 3.3. A section of the fluorite lattice RX_2 showing a substitutional trivalent rare-earth ion (RE) with a neighbouring interstitial anion. The anions X are shown as open circles, the cations R as filled circles. The presence of the interstitial anion X next to the rare-earth ion gives a centre with tetragonal symmetry.

1968) which show that when the charge-compensating F^- ions are neighbours of the rare-earth ion, they occupy cube-centre interstitial positions as shown in Fig. 3.3. A large body of epr spectra has also been interpreted on the basis of this model (see Chapter 6). It is thus reasonable to assume that the isolated F^- interstitial also occupies the cube-centre position, an assumption confirmed by detailed lattice calculations (§ 3.5).

However, it is also possible to obtain direct structural information about the more highly defective systems (say >5 per cent defects) by neutron diffraction measurements. Such measurements have been made by Willis (1964) for UO_{2+x} and by Cheetham, Fender, and Cooper, (1971) for CaF_2 doped with between 6 and 32 mole per cent YF_3, at elevated temperatures as well as at ambient temperature. The measurements yield average atomic positions and occupancies. In both systems they showed a normal cation sub-lattice, but a defective anion sub-lattice, the defects being vacant anion lattice sites and anions in *two* abnormal positions, one of which corresponds to a $\langle 111 \rangle$ displacement from a normal anion lattice site and the other to a $\langle 110 \rangle$ displacement from the cube-centre interstitial site. The concentrations of these three defects are the same order of magnitude but the exact ratios depend upon impurity concentration. Although Bragg diffraction measurements do not give direct information about correlations in site occupancy, it is almost certain that the vacant sites and the occupied abnormal or interstitial sites occur close to one another and that the whole rather complex structure is a consequence of defect clustering and not an intrinsic aspect of isolated anion interstitials nor of interstitial-impurity pairs. Thus there is no reason theoretically why such relaxation should occur around an isolated interstitial, and, as we have already remarked, the endor and epr evidence is overwhelmingly in favour of the cube-centre configuration for the interstitial anion. We may thus assume with Cheetham *et al.* (1971) that these complex configurations result only when the Y^{3+} ions and their associated F^- interstitials are clustered to some extent. Indeed Catlow (1973) has recently shown theoretically that the ionic model leads to just such complex relaxations when only two impurity-interstitial pairs come together. We describe these results more fully later (§ 3.5).

In addition to these striking differences between fluorite compounds and other more extensively studied ionic crystals, some aspects of their physical chemistry are also immediately relevant to any discussion of their defect properties. Thus the alkaline earth fluorides take up oxygen rather readily and, of course, the incorporation of a divalent anion, O^{2-}, in place of the fluoride ion immediately alters the defect population. Also, while the solution of YF_3 leads to the introduction of F^- interstitial ions, it is not clear that solution of NaF always gives F^- vacancies. Indeed this is only one of several possibilities (Franklin, 1967; Barsis and Taylor, 1968*b*).

These various features caution us, therefore, against assuming that the interpretation of data on the fluorite compounds will automatically be similar in detail to that for the alkali and silver halides.

The plan of this Chapter is as follows. We deal first with the concentration of defects, the effects of doping with aliovalent ions (i.e. solute ions of different valency from the corresponding solvent ion) and other influences upon the number and distribution of defects (§ 3.2). Then we describe diffusion and other mass transport phenomena involving defect or atomic migration over large numbers of lattice sites (§ 3.3), followed by a consideration of defect relaxation processes (dielectric and mechanical) where observable effects relate to ionic motion over only one or two lattice sites in the immediate vicinity of foreign ions (§ 3.4). Lastly, in § 3.5 we review the theory of defect parameters, e.g. energies of formation and migration. It seems generally desirable that the theoretical structure of the Chapter should begin with the more general and conclude with the specific. In practice, of course, advances are made by various combinations of general and specific arguments and models and we shall not, therefore, hesitate when appropriate to refer forward to aspects not yet covered in full.

3.2. Concentration of defects in thermal equilibrium

As we have already indicated, substances with the fluorite structure are generally strongly ionic and, in constructing models of defect-determined properties of stoichiometric crystals, it is natural to begin from the ideal ionic model. That is, we suppose that the solid is made up of anions and cations bearing the full ionic charge and interacting with one another via electrostatic, van der Waals and overlap repulsion forces. This model then naturally focuses our attention on *ionic* defects, i.e. interstitial ions, ionic vacancies and foreign ions.

In considering the statistical thermodynamics of these systems it is usual to begin with models which omit the interactions between defects. These models provide limiting descriptions valid for very low defect concentrations. As the concentrations increase however defect interactions give rise to quantitiative and qualitative effects; strong attractions, e.g. of a substitutional trivalent rare-earth ion for a fluoride interstitial in CaF_2, lead to the formation of defect pairs (or larger clusters) which stay bound together for relatively long times and these are then treated as distinct defects in the statistical averaging process. Coulomb interactions at long range resulting from the net charges on defects may be described by appealing to Debye-Hückel theory. This approach is widely described in a number of elementary and comprehensive texts.† More rigorous analyses confirm its essential correctness in particular cases (Allnatt and Loftus, 1973). These, and careful comparisons with experiment, show that this approach should be valid,

† See e.g. Kittel (1971) for an introduction or Kröger (1964) for a comprehensive treatment. We also recommend the recent review by Franklin (1972).

generally speaking, up to ~1 mole % of defects. We shall therefore follow it here. However, it should be noted that at higher defect concentrations and when long range order or phase separation sets in, further elaborations of the theory are necessary. Such situations occur most notably in non-stoichiometric oxides (e.g. UO_{2+x}, CeO_{2-x}) and related compounds. Developments of the theory into this region have mostly been made by applying the methods of regular solution theory notably the "mean field" approximation (see e.g. Anderson, 1946, 1970). Such calculations predict certain qualitative features correctly, e.g. the existence of ordered phases, but we shall not report them in detail because they are still at a relatively primitive stage. An accurate theory of the high-concentration region requires a knowledge of the detailed interactions among the defects of the kind now being provided by Catlow's (1973) calculations. The final condition can also depend on the interactions between large clusters of defects which may themselves form an ordered array (Stoneham and Durham, 1973).

In this section we shall therefore start with the description of a system of non-interacting defects and then consider the modifications introduced by long-range Coulomb interactions. We shall not however go through these calculations in full detail since in one form or another they are already widely treated in the literature. Instead we shall sketch the structure which underlies all particular calculations of this class. Our object in doing so is to bring out the generality of certain well known results. To begin with we must obviously specify the defects we expect to be important, e.g. the ionic defects we have already listed. Then by statistical mechanical analysis we establish the Gibbs function $G(P, T; \ldots n_i \ldots)$ for given numbers n_i of the various defects, i. In the simplest case where we suppose the defect concentrations to be low enough that their interactions can be ignored, this Gibbs function for a uniform solid containing m distinct types of defect will have the form

$$G = G_0(P, T) + \sum_{i=1}^{m} n_i g_i - kT \ln \Omega(\ldots n_i \ldots) \tag{3.1}$$

where $G_0(P, T)$ is the Gibbs function of the perfect solid containing the same total numbers of atoms while g_i is the free energy of formation of one particular defect of type i and Ω is the combinatorial function or number of ways in which the set of $\{\ldots n_i \ldots\}$ defects can be distributed over the available sites (and, where appropriate, energy levels). Thermodynamic equilibrium values of these defect numbers, n_i, are obtained as usual by requiring that G be a minimum, i.e.

$$\delta G = \sum_{i=1}^{m} \left(\frac{\partial G}{\partial n_i}\right)_{T,P;\ n_j \neq n_i} \delta n_i = 0 \tag{3.2}$$

subject to any constraints which apply. In general, there are three constraints, (1) structure (2) mass conservation and (3) electroneutrality, all of which are linear in the $\{n_i\}$. The first of these is always important and requires that the crystal structure be maintained, i.e. for fluorite compounds, RX_2, the number of X *sites* must be twice the number of R *sites*. The second requirement is generally taken care of through the notation used but is included for completeness. The third requirement is an approximation resulting from the strength and long range of electrostatic forces. It is relaxed in the vicinity of dislocations, surfaces and other sources and sinks of defects where space charges arise and extend out to distances from these sources of the order of the Debye screening length (κ^{-1}; see eqn (3.23)). We enlarge on this situation below and assume local electroneutrality for the time being. These three constraints, being linear, can be written in matrix notation as

$$\mathbf{n} \cdot \mathbf{A} = \text{a column matrix of constants, } \{a_1, a_2, a_3\} \tag{3.3}$$

where $\mathbf{A}$ is an $(m \times 3)$ matrix of coefficients and $\mathbf{n}$ denotes the row vector of the numbers of defects, n_i, $(n_1 \ldots n_m)$. By the method of Lagrange multipliers the constrained minimization of G (eqn (3.2)) has the solution

$$\boldsymbol{\mu} + \mathbf{A} . \boldsymbol{\xi} = 0 \tag{3.4}$$

where $\boldsymbol{\mu}$ is the column matrix $\{\mu_1 \ldots \mu_m\}$ of derivatives

$$\mu_i = \left(\frac{\partial G}{\partial n_i}\right)_{T,P;\ n_j \neq n_i} \tag{3.5}$$

and $\boldsymbol{\xi}$ is the column matrix $\{\xi_1, \xi_2, \xi_3\}$ of the Lagrange multipliers for the three corresponding constraining equations. In any given calculation these multipliers are found by substitution of the solution (3.4) back into (3.3). The quantities μ_i, by analogy with the usual definition for chemical species, will be called defect chemical potentials. In the dilute limit where (3.1) applies they will always have the form

$$\mu_i = g_i(P, T) + kT \ln([i]/z_i) \tag{3.6}$$

where $[i]$ is the molar fraction of defect i and z_i is its combinatory and internal partition function, i.e. sum over states of the single isolated defect. This quantity z_i must be found for each defect model separately but the form (3.6) follows generally in the non-interacting limit. However the general solution (3.4) is still valid when defect interactions are important and (3.1) and (3.6) are no longer adequate.

Lastly we note that everything we have described is analogous to the theory of chemical systems. In particular the equilibrium equations (3.4) after elimination of the Lagrange multipliers ξ_1, ξ_2, and ξ_3 can all be described as the equilibrium equations for various quasi-chemical defect reactions. Furthermore these equations by (3.6) are just the corresponding mass-action-law equations, so that many results of importance can be written down fairly directly.† This approach is extensively developed in the book by Kröger (1964) and we shall also introduce it here.

3.2.1. *Application to Frenkel and Schottky disorder*

Consider first anion Frenkel defects, i.e. anion vacancies and anion interstitials. These are believed to be the dominant intrinsic defects in alkaline earth fluorides and related compounds generally. We denote these defects by X_v^- and X_i^- (X^- denotes the anion). Then the constraining equations for structure and electroneutrality both give

$$\delta n(X_v^-) - \delta n(X_i^-) + \ldots = 0, \tag{3.7}$$

where ... signifies variations, δn, in the numbers of other defects. The mass conservation equation is automatically satisfied so that the equilibrium condition gives

$$\left.\begin{aligned} (\mu X_v^-) + \xi &= 0 \\ \mu(X_i^-) - \xi &= 0 \end{aligned}\right\}, \tag{3.8}$$

apart from equations involving other defects. Thus

$$\mu(X_v^-) + \mu(X_i^-) = 0. \tag{3.9}$$

We note that this can be regarded as the equilibrium condition for the quasi-chemical reaction

$$X_v^- + X_i^- \rightleftharpoons \text{perfect lattice}.$$

From (3.6) and (3.9) it follows that

$$[X_v^-][X_i^-] = z(X_v^-)z(X_i^-)\exp(-g_F/kT) \tag{3.10}$$

where

$$g_F = g(X_v^-) + g(X_i^-). \tag{3.11}$$

† In particular cases it is however often necessary to go beyond simple quasi-chemical analogy to get the activity coefficients correctly.

All the indications are that the anion vacancy is a simple structure retaining the T_d point-symmetry while the interstitial occupies a cube-centre site of O_h symmetry (equivalent in number and coordination to the Ca^{2+} sites). Since the number of interstitial anion sites is equal to the number of molecules while the number of normal anion sites is twice that number it follows from (3.1) and (3.5) that $z(X_v^-) = 2$ and $z(X_i^-) = 1$; thus (3.10) becomes

$$[X_v^-][X_i^-] = 2\exp(-g_F^-/kT). \tag{3.12}$$

TABLE 3.2

Experimentally determined values (eV) *of the enthalpy of formation of anion Frenkel-defects, h_F, in some fluorite compounds*†.

Substance	h_F	Reference
CaF_2	2·2–2·8	a, b, c, d, e
SrF_2	1·7–2·3	b, c, e, f
$SrCl_2$	1·6–1·8	g
BaF_2	1·9	h, i
CdF_2	2·1–2·8	j, k

† There is a considerable spread in the published values and an objective choice between them is difficult. However, in our view, the higher values for CaF_2 and SrF_2 are more likely to be correct.

[a] Ure (1957); [b] Barsis and Taylor (1966); [c] Bollman *et al.* (1970); [d] Matzke (1970); [e] Lallemand (1971); [f] Knowles and Mahendroo (1970); [g] Barsis and Taylor (1966), but see also Hood and Morrison (1967); [h] Barsis and Taylor (1968*b*); [i] Miller and Mahendroo (1968); [j] Kessler and Caffyn (1972), but see also Tan and Kramp (1970); [k] Süptitz, Brink and Becker (1972).

As usual the Gibbs free energy of defect formation, g_F, can be separated into enthalpy and entropy terms; thus

$$\begin{aligned} g_F &= h_F - Ts_F \\ &= u_F + Pv_F - Ts_F. \end{aligned} \tag{3.13}$$

Some values of the enthalpies of formation, h_F, of these anion Frenkel defects are given in Table 3.2. It will be noticed that there is a sizable spread in the values of different workers. The corresponding entropies are even less well known, although they would not be expected to be greater than 10k. The volume of formation, v_F, has been determined in the alkaline earth fluorides by Lallemand (1971). As in other systems, v_F is of the same order of magnitude as the molecular volume, v_m, so that the Pv_F term in the free energy will not become appreciable until P reaches a few kilobars (i.e. $\sim 10^8$ N/m^2).

We can follow through the analogous argument for Schottky defects, i.e. anion and cation vacancies. For them the corresponding relations are

$$\mu(M_V^{2+}) + 2\mu(X_V^-) = 0 \tag{3.14}$$

or

$$[M_V^{2+}][X_V^-]^2 = 4\exp(-g_S/kT) \tag{3.15}$$

where

$$g_S = g(M_V^{2+}) + 2g(X_V^-) \tag{3.16}$$

TABLE 3.3

Energies of formation of Frenkel and Schottky defects in the alkaline earth fluorides as calculated by Catlow and Norgett (1973). *Details of calculations of this type are given later* (§ 3.5).

Substance	Defects	Energy u(eV)
CaF_2	Anion Frenkel	2·6–2·7
	Schottky	7·0–8·6
	Cation Frenkel	8·5–9·2
SrF_2	Anion Frenkel	2·2–2·4
	Schottky	6·9–8·1
	Cation Frenkel	8·2–8·6
BaF_2	Anion Frenkel	1·6–1·9
	Schottky	6·3–6·9
	Cation Frenkel	7·8–8·0

Lastly, a relation similar to (3.10) holds for cation Frenkel defects. All three relations hold simultaneously. In general the energies of formation are probably in the order $\frac{1}{2}g_F$ (anions) $<\frac{1}{3}g_S<\frac{1}{2}g_F$ (cations) so that anion vacancies and interstitials are the dominant thermally produced defect. In this situation, it is difficult to obtain direct experimental information about the minority defects but some calculated energies (u) of formation are given in Table (3.3).

3.2.2. *Doping with foreign ions*

There have been numerous experiments on fluorite crystals doped with aliovalent ions. Common examples in the alkaline earth fluorides are Na^+, Y^{3+}, trivalent rare-earth ions, and O^{2-}. The presence of such ions is, of course, reflected in the structural and electrical neutrality equations and alters the defect concentration from its intrinsic value. Thus, if we suppose that g_F (anions) is sufficiently small compared to

g_S and g_F (cations) that only the anion Frenkel defects need be considered, the molar fractions $[X_v^-]$ and $[X_i^-]$ must be equal. But the addition, say, of YF_3 to CaF_2 necessarily adds one extra F^- ion for every substitutional Y^{3+} ion so that this equality is disturbed and an excess of F^- interstitials may arise. In this situation (3.12) must be solved in conjunction with the structure equation

$$[Y_s^{3+}]+[F_v^-] = [F_i^-] + 2[Ca_v^{2+}], \qquad (3.17)$$

where $[Y_s^{3+}]$ is the molar fraction of (substitutional) Y^{3+} ions. If we can neglect $[Ca_v^{2+}]$ the solutions are

$$\left.\begin{aligned}[F_i^-] &= \tfrac{1}{2}[Y_s^{3+}]+\tfrac{1}{2}\{[Y_s^{3+}]^2+4\exp(-g_F/kT)\}^{\frac{1}{2}}\\ [F_v^-] &= -\tfrac{1}{2}[Y_s^{3+}]+\tfrac{1}{2}\{[Y_s^{3+}]^2+4\exp(-g_F/kT)\}^{\frac{1}{2}}\end{aligned}\right\}. \qquad (3.18)$$

The effect of the doping with trivalent cations is thus not only to increase the concentration of interstitials but also to decrease the concentration of F^- vacancies to such an extent that, for

$$[Y_s^{3+}] \gg \exp(-g_F/2kT),$$

$[F_v^-]$ is $\simeq \exp(-g_F/kT)/[Y_s^{3+}]$. However, by (3.15) we then observe that the concentration of Ca^{2+} vacancies rises as

$$[Ca_v^{2+}] \simeq [Y_s^{3+}]^2 \exp\{-(g_S-2g_F)/kT\}. \qquad (3.19)$$

Since this increases as $[Y_s^{3+}]^2$ whereas the interstitial concentration increases only proportionally to $[Y_s^{3+}]$ it is evident that the Ca^{2+} vacancy concentration could in principle become greater than that of the F^- interstitials. By (3.12), (3.15) and (3.17) this occurs when

$$[Y_s^{3+}] \geqslant 3\exp\{(g_S-2g_F)/kT\}. \qquad (3.20)$$

If $g_S > 2g_F$ this condition can never be satisfied since the solute fraction $[Y_s^{3+}]$ is necessarily $\ll 1$, but if this is not so, the condition may be satisfied at low enough temperatures. From Table 3.3 we observe that the most recent calculations do indicate $g_S > 2g_F$. Nevertheless, such a changeover in the mechanism of charge compensation in other systems should be kept in mind.

The same principles, of course, apply when monovalent cations, e.g. Na^+, or divalent anions, e.g. O^{2-}, are incorporated substitutionally. In these examples there is a deficiency of anions and F^- vacancies are added; eqns (3.18) still apply except that $[Y_s^{3+}]$ is replaced by $-[Na_s^+]$, $-[O_s^{2-}]$ etc. The generalization of these equations to other examples, e.g. oxides such as CeO_2 containing aliovalent ions, is equally trivial.

At this point we should draw attention to the *assumption* that the foreign ions are present substitutionally in the host lattice. Firstly, it is always possible that solution of foreign ions takes place interstitially; perhaps the best studied examples of this are Cu^+-doped AgCl and AgBr but it is possible that this also occurs when the fluorites are doped with alkali ions. We return to this possibility later (§§ 3.3.2 and 3.5.2). Secondly, there exist definite solubility limits; ions present in precipitates are not, of course, effective in altering defect concentrations in the bulk. If the limit of solubility is exceeded, e.g. by cooling a doped crystal from a high temperature where the solubility is greater, then the excess ions will precipitate from solution, giving rise to time-dependent effects. Although such solubilities and precipitation effects have been studied in detail in some alkali halide systems, rather less is known for the fluorite compounds. In general, however, the solubilities of YF_3 and trivalent rare-earth fluorides in the alkaline earth fluorides are high, being in the range 30–50 mole per cent at the melting points (Ippolitov *et al.*, 1967). The solubilities of alkali ions, and of oxygen ions, appear to be very much lower and thus more 'normal' for aliovalent ions. Of course the above simple equations only apply at very low defect concentrations, certainly less than a fraction of one per cent.

A further important assumption which we have made is that the charge state of these foreign ions is fixed. While this is true for many, there are some ions which can be incorporated in more than one state, with different consequences for the equilibrium defect population. An example is Sm in CaF_2 where an oxidizing atmosphere induces Sm^{3+} ions with F^- interstitials while a reducing atmosphere leads to Sm^{2+} ions, which require no charge compensation. The principles governing such thermal equilibrium valence changes are well established and have been presented in detail for fluorite lattices by Fong (1966), but as quantitative experimental tests on these systems are lacking, we shall not go into detail. The effect of atmosphere upon the charge state of rare-earth ions is, however, qualitatively important (see Chapter 5).

Much more work has been carried out on the variations of stoichiometry of oxides, e.g. UO_{2+x}. Here excess oxygen is believed to be incorporated by the formation of O^{2-} interstitials and electron holes in the U^{4+} sub-lattice, i.e. by U^{5+} ions. The defect equations are then analogous to those for CaF_2:YF_3, except for the replacement of $[Y_s^{3+}]$ by $[U_s^{5+}] = 2x$. The equilibrium value of x is, of course, determined by the oxygen partial pressure, p_{O_2} in the surrounding atmosphere. For the above model of non-interacting defects and for x much greater than the fractions of aliovalent impurities and intrinsic defects it is given by

$$x^6 = \tfrac{1}{16}\exp(\mu_{O_2} - \Delta G)/kT \tag{3.21}$$

where μ_{O_2} is the chemical potential of an oxygen molecule in the gas phase (containing a term proportional to $\ln p_{O_2}$) while ΔG is the increase of Gibbs free energy on incorporating one O_2 molecule, initially in its lowest quantum state at rest, into perfect stoichiometric UO_2. Eqn (3.21) can be obtained most directly by considering the quasi-chemical reaction

$$O_{2,\text{gas}} \rightleftharpoons 2O_i^{2-} + 4U_s^{5+}$$

for which the equilibrium equation is

$$\mu(O_{2,\text{gas}}) = 2\mu(O_i^{2-}) + 4\mu(U_s^{5+}).$$

Eqn (3.21) then follows from (3.6) when we observe that $[O_i^{2-}] = x$ and $[U_s^{5+}] = 2x$. However the experimental indications are of a dependence of x on p in the range $10^{-4} < x < 10^{-2}$ more like $p_{O_2}^{1/4}$ than the $p_{O_2}^{1/6}$ which (3.21) predicts (see e.g. Kröger, 1966). This implies that interactions among the defects are significant, a question to which we return below. Non-stoichiometric fluorides include SmF_{2+x} and EuF_{2+x} (Catalano, Bedford, Silveira and Wickman, 1969) but much less is known of their thermodynamics than is the case for UO_{2+x}.

Before turning to this question of defect interactions, it is appropriate to conclude this section by referring to the hydrolysis of alkaline earth fluorides in water vapour or moist air, since this represents another important way in which defect populations may be modified (sometimes unintentionally). Such hydrolysis is appreciable even at moderate temperatures ($\sim$100 °C) and it increases rapidly as the temperature is raised (see, e.g. Stockbarger, 1949, and Bontinck, 1958, and for an example of the way it can occur under rather unexpected conditions see Ranon and Low 1963). It has been suspected for some years that it leads to the incorporation of O^{2-} and OH^- ions in the lattice (Sierro 1961) but, unfortunately, there is still very little information on the kinetics and detailed nature of the reactions involved. Some important identifications of the end-products formed by the hydrolysis of CaF_2 doped with trivalent rare-earth ions have however been made. Thus it has been known for some years that two trigonal epr spectra (called Tr_1 and Tr_2) are characteristic of $CaF_2{:}RE^{3+}$ crystals which have been subjected to hydrolysis (see Table 6.9) and recently Reddy *et al.* (1971) have made endor studies of $CaF_2{:}Yb^{3+}$ which have enabled them to determine the structure of the centres responsible for these Tr_1- and

Tr_2-spectra, at least for this one impurity but probably more generally. The Tr_1-centre consists of the Yb^{3+} ion whose first neighbours are four O^{2-} ions at the corners of a tetrahedron and one F^- ion (in place of the usual eight F^- ions) while in the simpler Tr_2-centre the Yb^{3+} has just one F^- neighbour replaced by an O^{2-} ion. The ease with which hydrolysis can occur and lead to the incorporation of these centres into fluorite crystals means that a dry atmosphere is demanded in all experiments requiring controlled defect concentrations. If an inert atmosphere, e.g. He, is used care must be taken to ensure that oxygen and water vapour are not released from the glassware or other materials of the enclosure. Unfortunately, this has not always been done.

3.2.3. *Pairing, clustering and interaction of defects*

The above discussion has been based on the assumption that interactions among the defects could be neglected. However, as the interstitial ions, vacancies and aliovalent impurity ions all carry effective electrical charges, it is evident that they must interact and that at large enough separations these interactions are Coulombic as $e_i e_j/\epsilon_0 r_{ij}$, where ϵ_0 is the static dielectric constant. The system is thus analogous to an electrolyte solution. Furthermore a large number of paramagnetic resonance and other experiments (esr and endor) carried out on crystals doped, particularly, with trivalent rare-earth ions provide clear evidence for the formation of close pairs of these trivalent ions with interstitial anions at low and intermediate temperatures (Chapter 6). Interactions must therefore certainly be considered. The approach generally employed in practice is an application of the Bjerrum-Fuoss theory of liquid electrolytes first discussed by Teltow (1949) and Lidiard (1954). This theory regards pairs of defects at the (several) closest separation(s) as distinct, complex defects and handles the pair distribution at larger separations by means of Debye-Hückel theory. This provides not only an equilibrium statistical thermodynamic theory but also a description of transport phenomena, e.g. ionic conductivity and diffusion, and of relaxation phenomena, e.g. dielectric and mechanical relaxation. It is, of course, an approximate theory but more rigorous treatments are beset with considerable formal difficulties, (see however, Allnatt, 1965, 1967 and Allnatt and Loftus, 1973). In addition, it should be said that defect pairs, and under appropriate conditions also higher complexes, have sufficiently long lifetimes that they are physically important concepts; one can meaningfully and selectively study their properties experimentally.

Let us summarize some of the main features of the model. Firstly Debye-Hückel theory shows us that the existence of Coulomb interactions leads to an average distribution of charged defects about any other charged defect which just screens its Coulomb field. In first-order approximation the electrical potential around the given defect (of net charge qe) is simply proportional to

$$(qe/\epsilon_0 r)\exp(-\kappa r), \tag{3.22}$$

where the screening parameter κ is given by

$$\kappa^2 = \frac{4\pi \sum n_i q_i^2 e^2}{\epsilon_0 kT}, \tag{3.23}$$

where n_i is the concentration of defects of type i bearing net charge $q_i e$. This screening of the defects changes the chemical potentials (3.6) by the insertion of additional activity coefficients γ_i, thus

$$\mu_i = g_i(P, T) + kT \ln([i]\gamma_i/z_i). \tag{3.24}$$

In the dilute limit these are given by

$$\ln \gamma_i = -\frac{q_i^2 e^2 \kappa}{2\epsilon_0 kT}, \tag{3.25}$$

but at higher concentrations where the screening distance κ^{-1} is no longer very much greater than the 'distance of closest approach', R, at which the defects are still described as unpaired, it is more accurate to use

$$\ln \gamma_i = -\frac{q_i^2 e^2 \kappa}{2\epsilon_0 kT} \frac{1}{1+\kappa R}. \tag{3.26}$$

As is evident immediately from these formulae, the equations describing the defects bearing no net charge, e.g. Y^{3+}–F_i^- pairs, are unchanged (i.e. $\gamma = 1$). Equations such as those for Frenkel and Schottky defects, however are modified by the appearance of γ^{-1} factors before the factor $\exp(-g/kT)$. The solution of these equations for the defect concentrations must then be done numerically owing to the dependence of κ upon those concentrations. The qualitative effect, however, is to lower the defect formation energies by small amounts ($\sim$0·1 eV) and to lead to higher defect concentrations at high temperatures.

We must also describe the extent of the defect pairing which takes place. To first order these equations are again most simply derived by appealing to the mass-action equations for the corresponding quasi-chemical reaction. Thus for the formation of nearest-neighbour pairs between substitutional Y^{3+} ions and F^- interstitials we have

$$Y_s^{3+}+F_i^- \rightleftharpoons (Y^{3+}-F_i^-) \text{ pairs}$$

and thus

$$\frac{[Y_s^{3+}-F_i^-]}{[Y_s^{3+}][F_i^-]} = 6\gamma(Y_s^{3+})\gamma(F_i^-)\exp(\Delta g/kT), \tag{3.27}$$

where the factor 6 results from the 6 equivalent orientations of the pair and Δg is the free energy gained by bringing the Y_s^{3+} and the F_i^- together ('from infinity'), i.e. the energy of association. (It may be noted that chemists often use the opposite sign convention and refer to $-\Delta g$ as the energy of association). Larger clusters can be described by similar equations.

Although the limits of accuracy of the above equations are not completely clear (Allnatt and Loftus, 1973) one may reasonably expect them to hold to about 1 per cent of defects. Hitherto data on the fluorites has generally not been analysed in the Debye-Hückel approximation but an exception is the work of Barsis and Taylor (1968*a*) on BaF_2.

Franklin and Marzullo (1970*a*,*b*) have attempted to verify (3.27) for Gd^{3+} ions in CaF_2 by monitoring the concentrations of Gd_s^{3+} and $(Gd_s^{3+}-F_i^-)$ pairs by electron spin resonance and dielectric loss measurements made on crystals quenched from various high anneal temperatures. These crystals contained from 1 to 45×10^{-4} mole fraction of Gd^{3+} ions and thus should be well within the range of this theory. Although successful measures of the isolated and paired defects were obtained, these were not in agreement with (3.27) if it is also assumed that the high-temperature equilibrium distributions were retained during the quench. Presumably it is this second assumption which is in error. Thus as we shall see later from the known mobility of F^- interstitials in CaF_2, it probably takes only $\sim10^{-3}$ sec for each interstitial to make 10^4 jumps at 1000 K; the distribution of interstitials relative to Gd^{3+} ions in even the most dilute quenched sample must therefore be characteristic of a considerably lower temperature than the anneal temperature (800–1400 K), but in any case one which depends on the defect concentration. Furthermore as we shall see later (§ 3.5.2) the interactions between pairs of $Gd^{3+}-F_i^-$ pairs are also likely to be large, leading to appreciable formation of low symmetry clusters even at quite high temperatures.

We should probably expect the equations of this pairing model to break down at impurity fractions of several per cent and greater, since large clusters and possibly co-operative effects will become important. We have already discussed the neutron diffraction observations of average atomic positions in CaF_2 containing YF_3 and in UO_{2+x}, which show displacements of anions from normal lattice positions much larger and quite different from those characterizing isolated defects. The occupancy of these abnormal anion positions strongly suggested the existence of clusters. Recent computer simulations in fact also show that pairing of Y^{3+} ions can lead to low symmetry relaxations among the associated interstitials and normal ions which are similar to those observed (Catlow, 1973). The statistical mechanics of these systems containing high concentrations is not very developed, mainly because of ignorance of defect interactions at short distances and their probable complexity. Only rather crude Bragg–Williams–Anderson models have thus been formulated (Thorn and Winslow, 1966 for UO_{2+x}). We shall not go further into these calculations here, partly because their success is not very apparent and partly because they are not directly applicable to CaF_2/YF_3 systems (since they explicitly omit defects in the cation sub-lattice.

3.2.4. *Trapping of rare-gas atoms*

The rare gases are an important class of impurities in many solids at the present time, owing to their production by nuclear transmutations induced by irradiation with fast neutrons, as in a reactor. Their solubility is so low that no true equilibrium thermodynamic situation can be studied, yet in practice at high temperatures one can effectively obtain thermodynamic equilibrium in all components except the rare-gas concentration. A quasi-equilibrium then obtains and the methods of statistical thermodynamics can be used to describe the partition of rare-gas atoms between interstitial sites, vacancy defects, etc. Indeed, even at lower temperatures where radiation-induced vacancy defects are still present, so that there is not thermal equilibrium in defect concentrations, one may have quasi-equilibrium in the distribution of rare-gas atoms between these vacancy defects and interstitial positions. These distributions are evidently important in relation to the average mobility or diffusion coefficient of the gas atoms, since when trapped in vacancies they are immobile while as interstitials they are able to migrate through the lattice. The diffusion coefficients of the rare-gas atoms, which can be measured by studying their rates of escape from the crystal, are thus inversely correlated with the density of vacancy traps whether introduced thermally, by irradiation or in other ways.

This model has provided a useful basis for the interpretation of many experiments on the migration of rare gases in ionic solids and such studies thus provide an additional way in which to obtain further information about defect concentrations and other features. The partition of gas atoms between interstitial sites and the various possible vacancy traps (anion and cation vacancies, vacancy pairs and higher aggregates) is given by elementary statistical mechanics. Thus the fraction of gas atoms which are in interstitial sites, $[G_i]/[G]$ is

$$\frac{[G_i]}{[G]} = \frac{\gamma_i}{\gamma_i + \sum_t [t]\gamma_t \exp(B_t/kT)} \tag{3.28}$$

TABLE 3.4

Calculated (internal) energies (in eV*) of trapping of interstitial rare-gas atoms into vacancies in alkaline earth fluorides.*

System	Trap	
	†Anion vacancy	†Cation vacancy
CaF_2:Ar	1·1–1·3	2·2–2·6
SrF_2:Kr	1·2–1·2	3·2–3·3
BaF_2:Xe	0·7–1·0	3·3–4·1

† The two values given correspond to different choices of gas-ion interaction potentials. (After Norgett, 1971*a*,*b*)

where the sum is over the distinct kinds of trap t and the γ's are the appropriate activity factors for the interstitial sites (i) and traps (t); they are easily worked out for the various possibilities. The molar fraction of trap t is denoted by [t]. In general the energies of binding, B_t, into vacancy defects are ~1–3 eV (see the calculated values in Table (3.4)) so that quite a strong tendency towards trapping exists.

Two limiting cases should be considered. Firstly [t] is constant, corresponding broadly to low temperatures where radiation-induced vacancies are present. In this case $[G_i]/[G] \sim [t]^{-1} \exp(-B_t/kT)$. In the opposite limit of high temperatures the vacancies are present in thermal equilibrium concentrations so that $[t] \sim \exp(-g_{ft}/kT)$ where g_{ft} is an appropriate formation energy. For intrinsic material, g_{ft} is $\frac{1}{2}g_F$ for anion vacancies (by 3.12) and $(g_S - g_F)$ for cation vacancies (by 3.12 and 3.15); other situations, e.g. doped material, can be worked out similarly. The temperature dependence of $[G_i]/[G]$ is then generally slower than for the damaged state, being governed by $\exp\{-(B_t - g_{ft})/kT\}$. This is the converse of a property such as ionic conductivity which contains contributions *directly* proportional to vacancy concentration. It is this feature which makes the rare-gas impurity interesting for defect studies. Experimental results on rare-gas diffusion are considered later in § 3.3.3.

3.2.5. *Surfaces and dislocations*

We have so far concentrated on defect populations in the bulk of uniform crystals. In particular, we have relied upon the condition of electroneutrality in making predictions for doped crystals. However, in the vicinity of defect sources and sinks, this condition is relaxed. Thus if, for example, it requires less energy to form a F^- vacancy than a F^- interstitial, more vacancies than interstitials will tend to enter the lattice from the source. This, of course, would result in a separation of charge between the bulk and the source. However, owing to the long range of Coulomb forces, the electrostatic self-energy of a space charge rapidly becomes very large if these charges extend over macroscopic distances. In thermodynamic equilibrium the net result therefore is the existence of a thin layer of space charge extending a distance $\sim\kappa^{-1}$ into the bulk crystal from the source and leading to an electrical potential difference Φ between the body of the crystal and the source. In the case of an intrinsic alkaline earth fluoride, this potential difference is given by

$$e\Phi = \tfrac{1}{2}\{g(F_v^-)-g(F_i^-)\}. \tag{3.29}$$

The existence of this space charge thus equalizes the separate formation energies, $g_{v,i}\pm e\Phi$, of the complementary defects in the bulk, so that their concentrations become equal there. In the space-charge region there is an excess of the defect with the lower formation energy. Of course, the above balances are altered if the crystal is doped with aliovalent ions; the impurity concentration then also enters into the equation for Φ. The presence of aliovalent impurities, in fact, can change the sign of the space-charge layer.

These effects will be present at external surfaces and at edge dislocations and precipitates. The mathematical theory of them can be developed by extending (3.1) to non-uniform defect distributions with inclusion of Coulomb terms and use of the free-energy minimal principle to determine the equilibrium conditions. The theory again has strong analogies with electrolyte theory, especially that for the equilibrium between electrolytes and colloids in solution. It was first presented in its application to solids by Frenkel (1946) and subsequently developed by a number of others (see e.g. Lehovec, 1953, Eshelby *et al.*, 1958, Brown, 1961, Koehler *et al.*, 1962, Kliewer and Koehler, 1965, Kliewer, 1965,

1966). Very briefly, the conditions of equilibrium which one obtains are the constancy of the electrochemical potentials $\mu_i + q_i\phi$ of the various defects where the μ_i are given by eqn (3.24). The constant values assumed by the electrochemical potentials of the different defects are related to one another in such a way that the defect product-relations (e.g. 3.12 and 3.15) are valid locally. By combining the equations so obtained with the Poisson equation, one obtains the Poisson–Boltzmann differential equation for $\phi(r)$, familiar from electrolyte and colloid theory. As the experimental side is as yet rather undeveloped in the fluorite compounds, we shall not go further into the details of the theory here. The necessary detailed developments are, however, straightforward extensions of the treatments given in the above references.

The physical consequences of these space charges around dislocations can be seen in associated electrical effects as, for example, when the dislocation is moved away from the charge cloud surrounding its core. Changes in the space-charge potential with temperature may also affect the internal friction caused by dislocation motion since this is greatest when the magnitude of the space charge is least (e.g. at the isoelectric temperature where Φ changes sign and no space charge exists). Both these consequences have been clearly demonstrated in the Ag-halides (see Slifkin *et al.*, 1967) and in the alkali halides (see e.g. Nabarro, 1967, especially Chapter IX). The scattering of light from these space-charge regions around dislocations has also been demonstrated in the alkali halides (Nabarro, 1967, especially Chapter XII). So far, however, rather little comparable work has been carried out on the fluorite structures. The core structure of edge dislocations in fluorite has been examined in detail in recent years (Evans and Pratt, 1969; Ashbee and Frank, 1970) but remaining uncertainties here do not invalidate the space-charge model. It is interesting that quantitative measurements of the mobility (under applied stress) of edge dislocations in CaF_2 by Keig and Coble (1968) between 25 and 100 °C showed this to be thermally activated with an activation energy of 0·54 eV. This figure is close to that for motion of free F^- vacancies (section 3.4) and suggests the existence of a vacancy atmosphere around the dislocation line as envisaged in the space-charge theory. However, other explanations involving F_v^-–O_s^{2-} dipoles are also possible (Keig and Coble, 1968).

Space charges at surfaces manifest themselves through the thermopower of ionic conducting crystals (see e.g. the review by Howard and Lidiard, 1964, especially § 5.4). In particular, because they determine the contact potential between the crystal and the electrode they determine the inhomogeneous term in the total thermopower which comes from the temperature derivative of the contact potential. This term thus varies in a predictable way with defect concentration, doping level, etc. There have been a number of precise experiments on alkali, silver and thallium halides which verify this theory (see, e.g. Corish and Jacobs, 1972 for a recent example) but few measurements on fluorite systems. We should also observe, however, that in UO_2 where electrical conduction is by hopping of electron holes (U^{5+} states), the thermopower contains analogous defect-dependent terms since the hole concentration is tied to the defect population (Aronson *et al.*, 1961; De Coninck and Devreese, 1969).

In addition to the above physical aspects it is important also to record that in the alkaline earth fluorides there are significant chemical effects of oxygen to be observed at surfaces and at dislocations. Thus Franklin *et al.* (1967) among others, have shown that heating CaF_2 to a high temperature (>700 °C) in air, leads to the formation of highly-conducting surface regions. Likewise it is known that dislocations in both natural and synthetic CaF_2 can be 'decorated' by heating the crystals in moist air as well as by the more usual techniques of heating in a metal vapour (Bontinck, 1958). The presence of even minute amounts of oxygen in these crystals also reduces the mobility of edge dislocations very considerably (e.g. $\sim 4 \times 10^{-5}$ oxygen reduces it by almost 10 times, Keig and Coble (1968); see also Sashital and Vedam (1972))

3.3. Migration of defects

Under the influence of thermal fluctuations, the vacancy and interstitial defects jump from site to site through the crystal lattice. The frequency of these jumps determines the rate at which the thermodynamic equilibria described in section 3.2 are attained when conditions are changed (e.g. by dropping the temperature, creating excess defects by irradiation, etc.), as well as steady-state responses to external fields (e.g. electrolytic conductivity). It is important to recognize at the outset, however, that to talk of *defect* migration is really to speak of particular modes of migration of the *ions of the lattice*. This is obvious for the vacancy, but is also true for the interstitial ions; thus the calculations of Chakravorty (1971) and of Catlow and Norgett (1973) show that the dominant mode of F^- interstitial displacement in CaF_2 is the replacement or interstitialcy mechanism shown in Fig. 3.4. Direct jumps

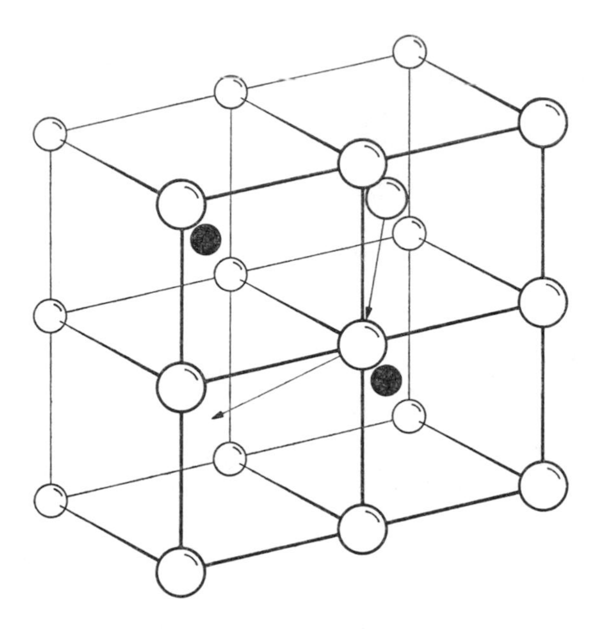

Fig. 3.4. A section of the fluorite lattice, RX_2 showing the interstitialcy mechanism of movement of the interstitial anion, X. The anions X are shown as open circles, the cations as filled circles. The interstitial initially at the centre of the top right cube of anions moves on to a neighbouring anion site and in doing so pushes the anion there into a neighbouring interstitial position.

of the interstitial from one cube-centre to another are also possible but with a very much smaller probability. In other words, a defect is a characteristic of a region of the lattice as a whole, and is not to be identified with any particular ion. This means that we can relate rates of defect migration as inferred, for example, from studies of defect reactions (such as pairing of rare-earth ions with interstitials as followed by epr) to rates of macroscopic diffusion and ion transport. Indeed it should be said explicitly that in otherwise perfect lattices, these defects provide the only seriously considered mechanism of diffusion; the activation energies for direct exchange of two or more like ions are presumed to be much higher than those characterizing defect mechanisms, even though no detailed calculations to confirm this for these materials have been made. Even so, the jumps of ionic defects, ion vacancies or ion interstitials, can be biased by applying an electric field, so giving electrolytic conductivity, whereas, for example, direct exchange involving pairs or groups of like ions would not. Thus the relation of diffusion coefficients to electrolytic conductivity is important;

it is given by a generalized Nernst–Einstein equation. This generalization of the classic Nernst–Einstein relation to defect-determined processes further shows that the precise relation of ionic mobility to diffusion coefficient can involve basic details, geometric and mechanical, of the defect motion.

Even in imperfect regions of these crystals, e.g. in the vicinity of dislocations, grain boundaries and surfaces, the basic mechanisms of atomic migration are probably still determined by point defects. Thus as we have noted in § 3.2.5, space charges consisting of an excess of one kind of defect exist in the vicinity of such sources and sinks of defects. These will lead to changes in ionic mobility in these regions, as a result of the changed concentrations of defects. In a dislocation core it is also possible that the intrinsic mobility of the defects themselves is altered as a result of the distorted ionic arrangements. At present, however, there is little experimental information on fluorite compounds, even though some striking dislocation effects have been seen in other solids (see, e.g. Nabarro, 1967, especially § 6.3).

Lastly, before going into details, it should be pointed out that if the interpretations of the high-temperature specific heat anomalies are correct, then, at temperatures above the λ-points (§ 3.1), it will not be sensible to use simple defect models, at least in the anion sub-lattice.

3.3.1. *Thermally-activated motion of defects*

If we examine, for example, the energy of a CaF_2 lattice as an ion is moved along the line joining it to an adjacent vacancy, it has the double-well form shown schematically in Fig. 3.5. Since potential energy ΔE must be supplied to cause the system to transfer from one minimum to the other, it is physically obvious that the frequency of such transitions in a system in thermodynamic equilibrium, i.e. in this case the frequency of vacancy jumps, will be the frequency of sufficiently large local fluctuations in energy. That is to say, it will be governed by a Boltzmann factor $\exp(-\Delta E/kT)$. Of course, not all energy fluctuations greater than ΔE lead to a transition or vacancy jump; the ions must also be moving in the right directions with the right velocities. Nevertheless, when classical statistical mechanics holds, i.e. generally speaking at temperatures greater than the Debye temperature, we expect the frequency of vacancy jumps to be dominated by a Boltzmann factor, $\exp(-\Delta E_v/kT)$. The same applies to interstitials, though with a different activation energy, ΔE_i.

The standard formulation of these basic ideas in terms of classical statistical mechanics is due to Vineyard (1957), following earlier less general treatments of the same ideas by others. The method is to use classical equilibrium statistical mechanics to give the fraction of systems in transition from one configuration to another per unit time and to equate this to the frequency of vacancy jumps by assuming that a system becomes "deactivated" after passing the saddle point of potential energy (e.g. the point P in Fig. 3.5); that is it does not return

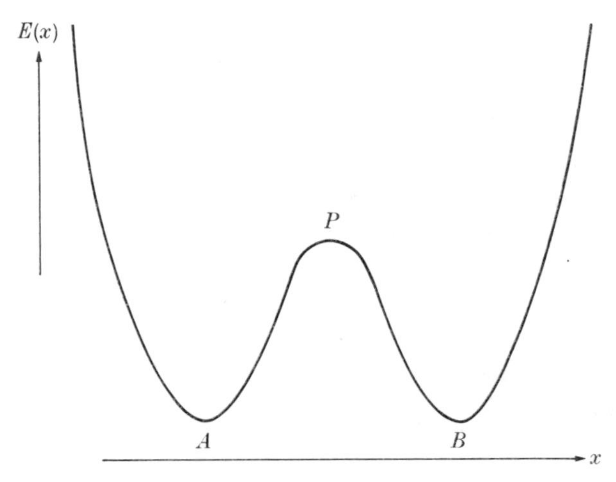

Fig. 3.5. Schematic diagram showing the variation of the potential energy $E(x)$ of a crystal when a normal lattice-ion neighbouring a vacancy is moved along the axis joining it to the vacant site (x-axis). The difference in energy between P and A is termed ΔE_{m} in the text. It is the least energy which must be supplied for the transition A to B to be possible classically.

to its original configuration more rapidly than random or, in other words, that there are no *dynamical* correlations between one jump and the next. For this to be right, it would seem necessary on physical grounds that the average energy of thermal motion of the atoms ($3kT$) be much less than the height of the barrier $E(P)-E(A)$. This is generally well satisfied for defect migration.

The most explicit results of the Vineyard analysis are obtained when harmonic approximations to the crystal potential functions about A and P can be made. Normal mode analysis of the system about the configuration A leads to a set of lattice frequencies ν_j^A. Likewise another set ν_j^P is obtained for motion about P but, since P is a saddle point, one of these frequencies is imaginary, namely that for the decomposition mode corresponding to the system 'rolling off P'. We denote this imaginary frequency by $i\nu_d^p$. The transition frequency is then found to be

$$w = \tilde{\nu} \exp(-\Delta E_m/kT), \tag{3.30}$$

where

$$\Delta E_m = E(P) - E(A) \tag{3.31}$$

and

$$\tilde{\nu} = \nu_d^p(\det|\beta^A|/\det|\beta^P|)^{\frac{1}{2}}, \tag{3.32}$$

in which β^A and β^P are the force-constant matrices for small displacements about A and P respectively. Eqns (3.31) and (3.32) provide the basis for the calculation of activation energies, isotope effects, etc. (see § 3.5). Empirically $\tilde{\nu}$ is found to be about 10^{13} sec^{-1}.

A more general expression for w, not limited by harmonic approximation, but necessarily less explicit than (3.30), is also provided by Vineyard's analysis. It is

$$w = \tilde{\nu} \exp(-\Delta G/kT), \tag{3.33}$$

where ΔG is the free energy of activation. It is generally assumed that the corresponding enthalpy of activation is obtained from

$$\Delta H_m = \partial(\ln w)/\partial(1/kT), \tag{3.34}$$

but this would require $\tilde{\nu}$ to be independent of temperature, whereas this is not generally true outside the harmonic approximation. In practice, the temperature variations of ΔH_m, ΔS_m, $\tilde{\nu}$, etc. may be of little significance.

As we have already mentioned, the above results are obtained from *classical* statistical mechanics and should therefore be valid at temperatures greater than the Debye temperatures. In practice we may be concerned with defect movements at lower temperatures than this and these often appear to follow the same law (eqn 3.30). Nevertheless, quantum effects are to be expected and the theory of them is beginning to be worked out (see e.g. Flynn and Stoneham, 1970). As would be expected, they depend on the nature of the system and there are as yet few useful results for intrinsic defects. These represent the most difficult problem of atoms strongly coupled to the motions of the rest of the lattice.

Experimental information on these defect jump frequencies is obtained in several ways. Properties such as the ionic conductivity and the self-diffusion coefficients (determined for example by radiotracer measurements or by nuclear magnetic resonance) depend directly on the products of defect concentration and defect jump frequency. By varying the defect concentration by doping with foreign ions (as in § 3.2). the properties of the individual defects can often be extracted.

Complications result from the occurrence of interactions between defects and foreign ions and from clustering and precipitation of foreign ions. Considerable experimental and analytical care may then be necessary, as is well shown by some of the recent discussions of these same effects in the alkali halides (see e.g. Jacobs and Pantelis, 1971). In general, the reliability and consistency of the studies in fluorite compounds is less good than that of the more completely studied alkali halides, although for the three alkaline earth fluorides, a reasonably clear picture is emerging.

If now the defect makes Γ jumps per unit time ($=w\times$no. of equivalent steps) each corresponding to a scalar displacement, s, then the assumption of no (dynamic) correlations between successive jumps leads to a root-mean-square displacement

$$\overline{R^2} = \Gamma t s^2, \tag{3.35}$$

at time t later. This allows us to infer a defect diffusion coefficient given by the usual Einstein formula

$$D = \tfrac{1}{6}\Gamma s^2. \tag{3.36}$$

Likewise if we examine the net flow of these defects (of charge qe) in an applied electric field (allowing for the change in activation energy resulting from the field) we obtain an electrical mobility

$$\lambda = \frac{qe\Gamma s^2}{6kT}. \tag{3.37}$$

We assume here that there is only one independent mode of movement, i.e. one jump frequency Γ, and one jump distance s. If there is more than one, then λ is a sum of terms like (3.37). We see, therefore, that *for the defects* λ and D are related by

$$\frac{\lambda}{D} = \frac{qe}{kT}, \tag{3.38}$$

the Nernst-Einstein equation. Such a simple relation does not, however, hold, in general, between the macroscopic ion mobilities and tracer diffusion coefficients.

3.3.2. *Ionic conductivity*

The contribution of defects to the electrical conductivity, σ, is given by a sum of terms

$$\sigma = \sum_i n_i q_i e \lambda_i \tag{3.39}$$

taken over all the defect species, i. If we substitute from eqn (3.37) in this equation we obtain

$$\sigma = \frac{e^2}{6kT} \sum_i n_i q_i^2 \Gamma_i s_i^2. \tag{3.40}$$

The alkaline earth fluorides as normally obtained are, for all practical purposes, exclusively ionic conductors. Furthermore, they and a number of related halides ($SrCl_2$, $BaCl_2$, $BaBr_2$, PbF_2, $PbCl_2$, and $PbBr_2$) conduct current almost entirely via the anions, a fact that has been known since the classic transport number measurements of Tubandt (1932), and confirmed more recently by tracer-diffusion measurements. This in itself is strongly suggestive of anion Frenkel disorder.† CdF_2 also shows ionic conductivity (Kessler and Caffyn, 1972) although when doped with trivalent rare-earth ions it can be converted into an electronic semiconductor by heating in Cd vapour (Kingsley and Prener, 1962; see also § 1.6.2). Stabilized ZrO_2 is well known as an ionic conductor while oxides such as CeO_2 and ThO_2 are predominantly ionic conductors when stoichiometric or when doped with lower-valent cations (Etsell and Flengas, 1970; Steele 1972). Non-stoichiometric oxides such as CeO_{2-x} and UO_{2+x}, by contrast, are semiconductors of n- and p-type respectively. A particularly complete study of the dependence of this electronic conductivity of CeO_{2-x} upon composition, temperature, and oxygen partial pressure has been made by Tuller (1974).

There is insufficient space here to go into the practical aspects of ionic conductivity measurements (e.g. electrode polarization, etc.) for which we refer to earlier articles (Lidiard 1957, Süptitz and Teltow, 1967). We merely indicate the general behaviour of the bulk conductivity expected on the basis of (3.39). The effects of surfaces and dislocations, although demonstrated in some other systems, have not been much studied in the present ones. However, they are not expected to contribute visibly to σ except at the lower temperatures and even then only in very pure material which nevertheless has high dislocation densities or high specific surface area.

The general behaviour of σ with temperature when plotted as $\ln \sigma T$ vs. T^{-1}, an Arrhenius plot, is as shown schematically in Fig. 3.6. The

† The properties of $PbBr_2$ and $PbCl_2$ (which have layer structures and not the fluorite structure) have however been interpreted in terms of Schottky defects with very immobile cation vacancies (see Simkovich, 1963, Verwey and Schoonman, 1967, de Vries, 1965). It is argued that the anions are too large for the interstitial spaces in these structures. The matter cannot yet be regarded as settled, however.

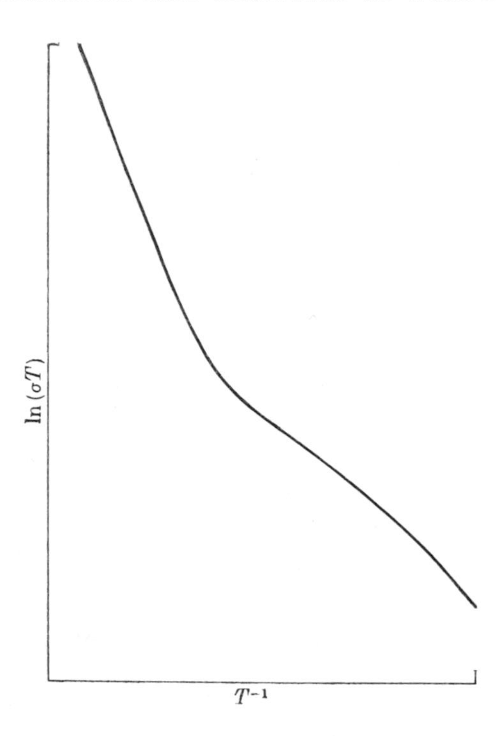

FIG. 3.6. Schematic diagram showing typical behaviour of the ionic conductivity, σ, as a function of absolute temperature (ln σT vs. T^{-1}). The high-temperature region is intrinsic to the material while the behaviour at lower temperatures is sensitive to the presence and state of dispersion of impurities, particularly aliovalent impurities.

two principal regions with different slopes or activation energies arise, not because more than one kind of defect is contributing to (3.39), but generally from the influence of impurities (such as Y^{3+} in CaF_2) in controlling the defect populations, as described in § 3.2. It is evident that if one knows the charge-compensating defect which is introduced (e.g. that YF_3 gives F^- interstitials) then analysis of the ionic conductivity of doped crystals as a function of concentration and temperature by using (3.39) and equations such as (3.12) for the defect concentrations will give defect concentrations and mobilities and the corresponding activation energies.

Complications come from interactions among the defects as well as from the pairing of foreign ions with charge-compensating defects, so that this analysis, to be reliable, must have accurate input data as well as requiring considerable care in the numerical analysis. Examples showing the breadth of variation obtainable in inferred parameters even in well studied alkali halides are provided by Jacobs and Pantelis (1971).

This and other work shows that to obtain results accurate to a few tenths of 1 eV it is necessary to take careful account of these interaction effects. In particular, it is generally insufficient to divide up the ln σT vs. T curves into linear portions by eye, although of course such a procedure is often used to give a rough analysis. Independent data on diffusion coefficients is also of great value in making a confident analysis. Unfortunately among the fluorite compounds only a few analyses can be regarded as fully satisfactory. Further difficulties at the present time come from uncertainties in the nature of the doped crystal. We have already referred to the tendency of the fluorites to oxidise in moist air and water vapour. The effects of this on the ionic conductivity were first reported by Ure (1957) and it is clear that conductivity cells should be designed so that the crystals are kept under an inert atmosphere. Quite apart from this, there are some intrinsic complexities. Thus, for example, the solution of NaF in CaF_2 is generally assumed to give substitutional Na^+ ions and F^- vacancies, but it is also possible that it could give interstitial Na^+ ions and F^- interstitials as suggested on empirical grounds by Barsis and Taylor (1968*b*) for BaF_2 doped with NaF. Of course, these are only two extreme possibilities; and both substitutional and interstitial Na^+ ions could be present simultaneously. In fact as we shall see later (§ 3.5) such a partition seems quite possible on the basis of the defect-energy calculations which have been made by Franklin (1967*b*).

For these various reasons, defect characteristics inferred from present ionic conductivity measurements are still somewhat uncertain. Tables (3.2), (3.5), and (3.6) collect together some Frenkel-defect formation energies and defect activation energies inferred from ionic conductivity and other transport measurements. The values included are a selection from published results, being those which we judge to be the more reliable. In general the confidence one can have in them is less than one now has for some alkali and silver halides. However, there is reasonable agreement between the different results, although in some cases, the spread is obviously unacceptably wide. Those for the F^- interstitial in CaF_2 range from just over 0·5 eV (the nmr value) to 1·6 eV (ionic conductivity measurements of Ure, 1957), corresponding to a range of $\sim 10^4$ to 10^{-13} jumps/sec at room temperature! In most cases the anion interstitial has been inferred to be less mobile than the corresponding vacancy, just the opposite situation to that found in most solids.

TABLE 3.5

Experimentally determined values of the activation energy (u) of migration of the anion interstitials in some fluorite compounds.

Substance	u_{mi}(eV)	Reference
CaF_2	0·53–1·64	a, b, c, d
SrF_2	0·71–1·0	d, e, f
BaF_2	0·62–0·79	g, h
CdF_2	~1	i
UO_2	1.3	j

[a] Ure (1957); [b] Lysiak and Mahendroo (1966); [c] Twidell (1970); [d] Bollman *et al.* (1970); [e] Knowles and Mahendroo (1970); [f] Barsis and Taylor (1966); [g] Miller and Mahendroo (1968); [h] Barsis and Taylor (1968*b*); [i] Tan and Kramp (1970) but see Kessler and Caffyn (1972); [j] Belle (1969).

In addition to energies of Frenkel defect formation, these analyses also give the corresponding entropies but as these are of lesser theoretical interest at the present time, we have not tabulated them. For CaF_2 and SrF_2 the entropies of formation are about 10k. The entropy of activation of the F^- vacancy is about 1–2k while that of the interstitial is about 10k. Further details can be found in the references cited. The free volumes of Frenkel defect formation have been obtained by Lallemand (1971) from measurements of the influence of pressure on ionic conductivity; for CaF_2 and SrF_2 he obtained a value v_F of about

TABLE 3.6

Experimentally determined values of the activation energy (u) of migration of anion vacancies in some fluorite compounds.

Substance	u_{mv}(eV)	Reference
CaF_2	0·52–0·87	a, b, c
SrF_2	0·94–1·0	c, d
$SrCl_2$	0·34–0·46	d, e
BaF_2	0·56	f
CdF_2	0·39–0·51	g, h, i
CeO_2	<0·8	j, k

[a] Ure (1957); [b] Keig and Coble (1968); [c] Bollman *et al.* (1970); [d] Barsis and Taylor (1966); [e] Hood and Morrison (1967); [f] Barsis and Taylor (1968*a*); [g] Kessler and Caffyn (1972); [h] Tan and Kramp (1970); [i] Süptitz *et al.* (1972); [j] Lay and Whitmore (1971); [k] Tuller (1974).

25 per cent of the molecular volume, a low value consistent with Frenkel defects.

Ionic conductivity measurements can also yield energies of association between aliovalent ions and their charge-compensating defects, since the formation of such electrically neutral pairs removes carriers according to a definite dependence on temperature and impurity concentration (eqn 3.27). Both the data and the analysis have to be accurate if the inferred association energies are to be believed, however. A good example is provided by the work of Barsis and Taylor (1968*a*,*b*) on BaF_2. They obtained 0·56 eV as the association energy of O_s^{2-}–F_v^- pairs and 0·44 eV as that of Gd^{3+}–F_i^- pairs. Such values imply, for example, that at a total impurity fraction of 0·1 mole per cent, the equilibrium proportion which are paired at temperatures of 250 °C is ~90 per cent. As we have already seen the defect mobilities are such that equilibrium in this reaction is very rapidly obtained at these temperatures. The equilibrium fraction at lower temperatures rapidly approach 100 per cent. A collection of published association energies for CaF_2 and SrF_2 is contained in the paper by Bollman *et al.* (1970); some of these are clearly unreliable as these authors point out. Their own work yields 0·88 eV for O^{2-}–F_v^- pairs and 0·60 eV for Y^{3+}–F_i^- pairs in CaF_2; the last value is consistent with some other published values for RE^{3+}–F_i^- pairs. Both values, as one would expect, are somewhat larger than in BaF_2, giving higher degrees of association for the same concentrations and temperatures. It is much harder to accept the value of only 0·16 eV which Bollman *et al.* obtain for Y^{3+}–F_i^- pairs in SrF_2.

3.3.3. *Diffusion*

Mostly the values given in Tables 3.2, 3.5, and 3.6 have been obtained from ionic conductivity measurements, but some diffusion measurements have also been made. These are more limited than in other systems since the available isotopes of F and O are less convenient. Thus F^{18} has a half life of only 1·7 h while O^{18} is stable so that mass spectrometric determinations rather than counting methods must be used. Cation diffusion measurements can be made by conventional radiotracer methods. They give information about the minority defects, i.e. the cation vacancies. Where the relative rates of diffusion of anions and cations have been measured, the ratios D_{anion}/D_{cation} are found to be very large, thus confirming that the defects are predominantly in the anion sub-lattice.†

† See for example Belle (1969) and Matzke (1969) for UO_2, Hood and Morrison (1967) for $SrCl_2$ and Ure (1957) for CaF_2.

The basic expression for the diffusion coefficient of a tracer measured under conditions of negligibly small chemical concentration gradient is

$$D_{\mathrm{T}} = \tfrac{1}{6} \sum_{i} \Gamma_{i,\mathrm{T}} s_{i,\mathrm{T}}^{2} f_{i,\mathrm{T}}, \tag{3.41}$$

i.e. a sum of terms over all the defects i contributing to the motion of the tracer T, e.g. over anion interstitials and anion vacancies, in the case of anion self-diffusion. $\Gamma_{i,\mathrm{T}}$ is the frequency of *tracer* displacements each of magnitude $s_{i,\mathrm{T}}$ as caused by defect, i. A broad similarity between (3.41) for D_{T} and (3.40) for σ will be evident. The detailed relation, however, depends upon the net charge carried by the defects and details of the mode of defect movement. Firstly, there is the relation of the defect displacement distance s_i to that of the tracer $s_{i,\mathrm{T}}$. For vacancies they are evidently equal, but for an interstitialcy mechanism they are different (Fig. 3.4 thus gives $s_i = 2s_{i,\mathrm{T}}/\sqrt{3}$). Secondly, there is the relation of Γ_i and $\Gamma_{i,\mathrm{T}}$. In general for self-diffusion of anions and cations and for the diffusion of substitutional ions (e.g. Y^{3+} in CaF_2), $\Gamma_{i,\mathrm{T}}$ is the product of the probability that the tracer ion T has a defect in a neighbouring position times the frequency of tracer displacements caused by jumps of such neighbouring defects. For example, the vacancy contribution to the diffusion of fluorine tracers F* in CaF_2 is governed by the fraction of F* which have a vacancy in a neighbouring position and by the frequency of vacancy jumps which displace the tracer ($\frac{1}{6}$ of Γ_{v} for the vacancy); and similarly for interstitials.

Although as we have already indicated there are no significant (dynamical) correlations between the jumps of the defect itself, motion of a *tracer* T *via the agency of a defect* (vacancy or interstitialcy) does necessarily involve statistical correlations between successive steps of the tracer. This is illustrated qualitatively in Fig. 3.7. The consequence of this is that the usual simple random-walk argument (which led to eqn (3.35) for example) must be modified to take account of this correlation between the direction of successive moves of a particular ion. General reviews of this aspect of diffusion have been given recently by Manning (1968) and LeClaire (1970); here it will be sufficient to note that the factors $f_{i,\mathrm{T}}$ ($\leqslant 1$) express this feature of defect-induced diffusion. These factors are functions of the relative probabilities of the defect, i, making jumps in the different directions defined by the presence of the tracer, T. For self-diffusion via vacancies in the fluorite lattice (or indeed any cubic lattice) these relative probabilities are all equal to the reciprocal of the coordination number in the appropriate sub-lattice (i.e. to $\frac{1}{6}$ for anions, $\frac{1}{12}$ for cations) so that f is a pure number.

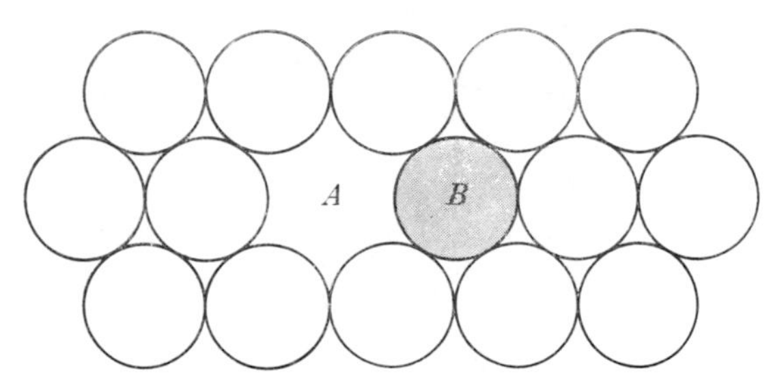

FIG. 3.7. A schematic diagram showing the configuration of a lattice after the shaded atom (impurity or isotope of the host) has just moved from A to B by jumping into a vacancy formerly at B. Site A is now vacant. As a consequence the next jump of the *shaded atom* is more likely to be back to A than to the the other neighbouring sites. With a vacancy mechanism there is thus a statistical correlation between the directions of two successive jumps of the shaded or labelled atoms. A similar conclusion follows with an interstitialcy mechanism (Fig. 3.4, p. 124).

Values of f for this example and for interstitials are given in Table 3.7. Substitutional foreign ions, in general, modify the jump frequencies of defects in their vicinities and in these cases f is a function of the ratios of the several frequencies and thus becomes a function of temperature (and pressure). For cations diffusing via vacancies the same models can be used as have been developed for the corresponding process in the NaCl structure since in both cases the motion is occurring on a f.c.c. sub-lattice. For detailed expressions for impurity correlation factors, both cation and anion, we again refer the reader to the reviews by Manning (1968) and LeClaire (1970).

TABLE 3.7

Correlation factors $f_{i,\mathrm{T}}$ for self-diffusion in the fluorite lattice.

Defect, i	Correlation factor $f_{i,\mathrm{T}}$
Anion vacancy	0·653
Cation vacancy	0·781
Anion interstitial	0·986
Cation interstitial	
(a) collinear	0·800
(b) perpendicular	1

The mechanism for interstitial anion motion is assumed to be as shown in Fig. 3.4. The mechanism of cation interstitial motion is also assumed to be of this same indirect or interstitialcy type, but there are two possibilities corresponding to the two displacements being (a) collinear, (b) perpendicular (after Compaan and Haven, 1956, 1957).

Measurements of self-diffusion by tracer methods have been made in the alkaline earth fluorides, CdF_2, $SrCl_2$, and UO_2, and the results are incorporated into Tables 3.5 and 3.6. In general, the results are in qualitative agreement with the models we have described. In particular, the anion self-diffusion coefficient is increased when the concentration of anion interstitials is increased by doping or non-stoichiometry (Matzke, 1970, Belle, 1969). Furthermore the ionic conductivity and the self-diffusion coefficients are in approximately the correct relation to one another (Hood and Morrison, 1967; Matzke, 1970). The precise ratio $D_T/(Ne^2\sigma/kT)$ as obtained for $SrCl_2$ at high temperatures (600–700 °C) is $\sim$0·6 which is consistent with diffusion predominantly via anion vacancies (0·56), although interstitialcy diffusion would give a value not greatly different (0·74). At lower temperatures, the ratio increased to between 2 and 6, depending on the specimen. These values suggest the presence of another mechanism contributing to the diffusion, but not to the conductivity; electrically neutral O_s^{2-}–Cl_v^- pairs would be a possibility.

Cation self-diffusion has been measured in CaF_2 (Matzke and Lindner 1964), SrF_2 (Baker and Taylor, 1969) and UO_2 (see, e.g. the review by Belle, 1969). The absolute magnitudes of the diffusion coefficients are much less than those of the anions and the activation energies are high ($\sim$4 eV), corresponding to the small number of intrinsic cation vacancies (eqns (3.12) and (3.15)). In the case of UO_2 the effect of doping with foreign ions and the effect of variable stoichiometry are qualitatively in agreement with the expectations of §§ 3.2.1 and 3.2.2. (Belle, 1969.) However, Reimann and Lundy (1969) from studies on single crystals of UO_2 disputed that true bulk diffusion was measured in these other experiments, which were carried out on polycrystals. Certainly it is under conditions of very low bulk diffusion that one expects grain-boundary and dislocation-enhanced diffusion to be most visible.

The above discussion all refers to diffusion under conditions of negligibly small chemical concentration gradient. This condition is easy enough to satisfy in self-diffusion experiments but with aliovalent impurity ions which necessarily change the defect population, the requirements are particularly stringent since the condition is really one of negligibly small gradient in defect concentration and mobility. When the condition is not satisfied, as it may well not be for aliovalent ions, then the diffusion process is more complex; for example, electrical diffusion potentials arise and D is no longer independent of concentration. The analysis of the experiment is then also more complex.

Such situations have been successfully studied in the alkali halides,† but there are few studies on fluorite structures of the same degree of sophistication. An exception is provided by the work of Lay (1970) on the 'chemical' diffusion of oxygen in UO_{2+x}, i.e. diffusion under a gradient of composition. This agreed with the self-diffusion coefficient of oxygen when analysed using the defect models so far described.

Lastly in this section, we summarize briefly the results obtained on rare-gas diffusion. The rare gases Ar, Kr and Xe are produced in CaF_2, SrF_2, and BaF_2 respectively by irradiation with fast neutrons as a consequence of the (n, α) reactions with the alkaline earth nuclei. Kr and Xe are produced in UO_2 by the fission of U^{235} nuclei by slow neutrons. The diffusion of these gases out of the crystals has been studied notably by Lagerwall (1962; CaF_2), Schmeling (1967; BaF_2), Felix and Lagerwall (1971; alkaline earth fluorides) and Miekeley and Felix (1972; UO_2).‡ Although some features of these results (e.g. the influence of non-stiochiometry on Xe diffusion in UO_2) are not yet understood, the broad features are consistent with the expectations of the model of interstitial gas diffusion. In this model we recognize the mobility of gas atoms when they are in interstitial positions, but assume negligible mobility otherwise, e.g. when trapped into vacancies and vacancy aggregates. Their macroscopic diffusion coefficient, D_G, is then given by their diffusion coefficient as an interstitial $D_{G,i}$

$$D_{G,i} = \tfrac{1}{6}\Gamma_{G,i}s_{G,i}^2,$$

multiplied by the fraction of time they are free as interstitials, i.e. $[G_i]/[G]$ as given by (3.28). In CaF_2 the observed activation energy for Ar diffusion is approximately 3 eV at high temperatures ($\gtrsim 900$ °C), but 6 eV below this; this suggests that the activation energy for migration of interstitial Ar atoms may be ~ 3 eV and that the energy of trapping into the radiation-induced vacancy defects is also about 3 eV. Lattice calculations bear out this interpretation (Norgett, 1971*a*,*b*), as we shall see in § 3.5. However, it may also be possible to account for the high-temperature diffusion in terms of gas atoms in pairs of F^- vacancies.

3.3.4. *Nuclear magnetic relaxation*

Valuable information about the diffusive movements of the atoms and ions in solids (and liquids) comes also from the motional narrowing of

† See especially the work of Fredericks and co-workers; for example Mannion *et al.* (1968), Allen *et al.* (1967) for NaCl and Keneshea and Fredericks (1963) for KCl. For theory see Howard and Lidiard (1964) or Adda and Philibert (1966).

‡ Earlier references for this material will be found in this paper.

nuclear magnetic resonance lines and measurements of nuclear magnetic relaxation rates. The important quantities are the spin-lattice relaxation time T_1, the spin-spin relaxation time T_2 and the spin-lattice relaxation time measured in the presence of an rf field (spin-lattice relaxation time in a rotating frame, $T_{1\rho}$). There are several mechanisms of relaxation and we refer the reader to the book by Abragam (1961) for a general review of these and the conditions under which they are separately important. Here we are interested in relaxation resulting from ionic jumps from one site to another via the change in nuclear interaction which these displacements cause. This mode of relaxation usually dominates the nuclear magnetic resonance line width when the mean time between jumps $\tau \ll \omega_{\mathrm{i}}^{-1}$ where $\hbar\omega_{\mathrm{i}}$ measures the strength of the coupling between different nuclei (dipolar or quadrupolar interactions). This condition specifies the motionally-narrowed region of nuclear magnetic resonance. Measurements of T_1 and T_2 in this region can be particularly successful for the study of diffusive motions when the main or only interactions are nuclear dipole-dipole couplings, although in view of their sensitivity to paramagnetic impurities (especially T_1, Abragam, 1961) it should be noted that this success depends upon having specimens of high purity. The theory of T_1 and T_2 in the motionally-narrowed region has been developed by Eisenstadt and Redfield (1963) following earlier work by Torrey (1953, 1954) and shows that the inference of jump frequencies is clear and direct. For example, if one measures both T_1 and T_2 for the anion nuclei and the cation nuclei, one has four pieces of information but only two unknowns, the mean anion and cation jump frequencies, w_{a} and w_{c}, respectively given by

$$\left.\begin{aligned} N_{\mathrm{a}}w_{\mathrm{a}} &= \sum_i \Gamma_{i,\mathrm{a}}n_i \\ N_{\mathrm{c}}w_{\mathrm{c}} &= \sum_i \Gamma_{i,\mathrm{c}}n_i \end{aligned}\right\} \tag{3.42}$$

where $N_{\mathrm{a}} = 2N_{\mathrm{c}}$ is the total number of anions per unit volume and $\Gamma_{i,\mathrm{a}}$ and $\Gamma_{i,\mathrm{c}}$ are the frequencies of anion and cation moves caused by defect i. A good check upon the consistency of the theory and the experiment is thus allowed. Good illustrations of this are provided by Eisenstadt's own work on LiF (Eisenstadt, 1963) and by the recent very complete study of NaI by Hoodless, Strange, and Wylde (1971) which also included $T_{1\rho}$. When quadrupolar couplings between nuclei are also important (in general when $I > \frac{1}{2}$) then the analysis is more complicated. In particular, quadrupolar interactions can lead to additional faster relaxations visible at the higher temperatures so that the free decay of a (transverse) spin polarization is no longer a simple exponential or first-order decay as it would otherwise be in the motionally-narrowed region. Even so it is still possible to obtain reliable information about ion migration as shown by Eisenstadt's work on NaCl (Eisenstadt, 1964). Lastly it should be noted that measurements of $T_{1\rho}$, allow one to study much slower diffusion rates than can be followed by conventional means (see Ailion and Slichter, 1965). The theory of Eisenstadt and Redfield (1963) has been extended to $T_{1\rho}$ by Look and Lowe (1966) and by Hoodless *et al.* (1971). These last authors showed how the jump of Na^+ ions in NaI doped with CdI_2 could be accurately obtained even in the low-temperature region where the CdI_2 was precipitating from solution. The work, like that of Eisenstadt on LiF, is particularly valuable in demonstrating that close agreement can be obtained between ionic jump frequencies inferred from nmr and those obtained from ionic conductivity and transport measurements. This point is very relevant to the alkaline earth fluorides since as we have already noted, conventional tracer diffusion measurements are not convenient. The full potential of the nmr method has not yet been used, although some studies of the alkaline earth fluorides have been made by Mahendroo and co-workers.

CaF_2 is a convenient material for studying these processes since Ca is more than 99 per cent Ca^{40} ($I = 0$) while F is entirely F^{19} which has $I = \frac{1}{2}$ (so that complicating quadrupolar interactions are also absent). Lysiak and Mahendroo (1966) measured T_1 in both pure CaF_2 and CaF_2 doped with Sm. The high-temperature region showed both the usual extrinsic and intrinsic regions of diffusion and the activation energies obtained were roughly the same as those found in ionic conductivity measurements. However, even though the Sm-doped crystals possessed only a single region (extrinsic region) this had a lower activation energy (0·5 eV) than the ionic conductivity analyses had assigned to free F^- interstitial migration, although it is not very different from the values for F^- vacancy migration and for the jumps of interstitial F^- bound to trivalent cations (section 3.4). Perhaps these experiments emphasize that while measurements by this technique can be particularly useful when tracer methods are difficult, their real value depends on a careful characterization of the system by other techniques as well. In this case more needs to be known about the state of the Sm ions in the conditions of the experiment (e.g. the proportions of Sm^{2+} and Sm^{3+} as functions of temperature, association between Sm^{3+} and interstitial F^-, incorporation of oxygen from the air, etc.). Similar experiments were carried out by Miller and Mahendroo (1968) on BaF_2 (both pure and doped with Eu^{3+}) and by Knowles and Mahendroo (1970) on SrF_2 (pure and doped with Gd^{3+}). As with CaF_2, the activation energies obtained for fluorine diffusion in the different regions are less than the corresponding values for the ionic conductivity; the lower values in Tables 3.5 and 3.6 are from these sources. This divergence may again reflect the difficulty of accurately specifying the defect state of these materials (defect association, presence of oxygen, etc.) and the present rather unsophisticated analysis of the transport and nmr data. For example, when defect association is significant then the ionic conductivity, tracer diffusion coefficients, and nuclear relaxation times all provide different information (eqns 3.40–3.42); the ionic conductivity measures the (drift) motion only of defects having a net charge while the nuclear relaxation times include contributions from all moving defects of the appropriate chemical species. To achieve the necessary range and precision it is important also to extend the existing nuclear relaxation measurements to include T_2 and $T_{1\rho}$ as well as T_1.

3.3.5. *Spectroscopic measurements of defect reactions*

The formation of associates of impurity ions with charge compensating or other defects will, in general, change the optical and paramagnetic resonance spectra of these ions. Thus under suitable conditions, such measurements can be used to follow the kinetics of association reactions as well as the equilibrium number of pairs. A successful example from the fluorites is provided by work by Twidell (1970) on the formation of Er^{3+}–F_i^- pairs in CaF_2. He found that this followed a first-order reaction rate equation, as might be expected, with an activation energy of 1·0 eV, presumably that of the F^- interstitial. The full potential of these spectroscopic techniques for following the kinetics and thermodynamics of such reactions has not yet been applied, however.

3.4. Dielectric and mechanical relaxation processes

Defect pairs, such as those formed by association of F^- interstitials with trivalent rare-earth ions in CaF_2, will not contribute to the (d.c.) ionic conductivity, σ, since they have no net electric charge. However, they do have an electric dipole moment and their reorientation as a result of thermally activated jumps can therefore be biased by an applied electric field. Thus in thermodynamic equilibrium there will be slightly more dipoles aligned parallel to the field direction than antiparallel to it (assuming for simplicity that the field is along one of the axes defined by the crystallographically equivalent orientations of the defect, e.g. ⟨100⟩). This contributes an additional defect term P_d, to the electric polarization ($P_d \sim \mu^2 E/kT$). However, if the electric field E is suddenly changed, P_d only approaches the new equilibrium value as $\sim\exp(-t/\tau)$, where τ^{-1} is some, often simple, combination of the frequencies of those jumps which reorient the complex defect. This lag of the polarization behind the field is, of course, a consequence of the randomness of these thermally activated jumps.

These complex defects are elastic as well as electric dipoles and can therefore also couple to elastic stress fields. Again, following a sudden change of stress, the defect contribution to the strain will relax exponentially towards its equilibrium value, though in general with a different time constant, τ.

In general, when several crystallographically inequivalent configurations (e.g. nearest and next-nearest-neighbour pairs) of a defect are possible and when as a consequence several distinct defect jump frequencies must be considered, there will be a number of distinct relaxation modes and a number of different relaxation times, τ. These relaxation modes can be analysed by their symmetry, just as vibration modes in molecules or point defects can be. Only modes of T_{1u} symmetry couple to electric fields and only those of A_{1g}, T_{2g}, or E_g symmetry couple to uniaxial elastic stresses (and then for E_g and T_{2g} only for stresses in specified directions). Thus a complete set of electric and elastic relaxation measurements can often go a long way towards identifying the symmetry of a complex defect and can provide detailed information about the various jump frequencies. Additionally, since dipole moments can be estimated reasonably accurately, the electric experiments also give guidance on absolute defect concentrations while, of course, for changes in relative concentrations following some defect reaction, e.g. precipitation, both electric and elastic methods can be valuable.

The experimental objective then is the determination of the strengths and relaxation times of these various relaxation modes. The most commonly applied techniques are oscillatory, i.e. a periodic stress or an alternating field is applied and the in-phase and out-of-phase components of the response are measured.† Each active relaxation mode

† The electrical measurements are made by a.c. bridge methods while the anelastic measurements will generally be made by setting the specimen into a resonance vibration and measuring the damping either from the frequency width of the resonance or from free decay.

contributes a Debye-like term to this response and thus the variation of response with frequency, ω, allows the determination of the relaxation times, τ. Such experiments can be conveniently carried out over a range of temperature and activation energies for the component jumps thereby inferred.

Measurements of electrical relaxation in ionic crystals have also been made by two other techniques; (i) measurements of the fast polarization following the sudden application of a steady electric field and (ii) ionic thermocurrent measurements (release of a frozen-in polarization on warming, analogous to thermoluminescence). In the first of these the polarization is analysed into a sum of exponential rises on applying the field and decays on removing it; these give the relaxation times directly. The method has been applied successfully to impurity-vacancy pairs in alkali halides (Dreyfus, 1961) but not yet to defects in fluorite crystals. In the ionic thermocurrent method (ITC), the defect polarization at a higher temperature is first frozen in by cooling the crystal in the electric field to sufficiently low temperatures that defect jumps cannot occur. The field is then removed and the crystal slowly warmed; at temperatures where defect mobility returns, a current pulse will be seen as the frozen-in polarization 'melts' away. The activation energies are directly related to the temperatures of these current pulses. This method was first described by Bucci *et al.* (1966) and has been applied successfully to trivalent ion–F_i^- pairs in the alkaline earth fluorides.

In both electric and elastic relaxation experiments, care must be taken to separate the response coming from the defects one wants to study from other sources of relaxation. Thus contributions to mechanical relaxation may arise from dislocations while large contributions to dielectric loss come from conduction processes. Firstly, the conductivity $\sigma(\omega)$ gives a contribution to $\tan\delta$ of $4\pi\sigma(\omega)/\epsilon\omega$ but this is generally fairly easily subtracted off since its frequency dependence is explicitly known and very different from the Debye-type loss (see Fig. 3.8). More troublesome are contributions from interfacial polarization at electrodes which can give rise to large contributions having the Debye frequency dependence (see, e.g. Daniel (1967) for a general discussion). These effects can arise, for example, from small air-gaps between the crystal and the electrodes (Miliotis and Yoon, 1969) or, in CaF_2 especially, from highly conducting surface layers resulting from oxidation (Franklin *et al.*, 1967). Combined studies of both dielectric and mechanical relaxation can thus be particularly valuable. In addition, spin-resonance methods can be used directly to measure certain relaxation

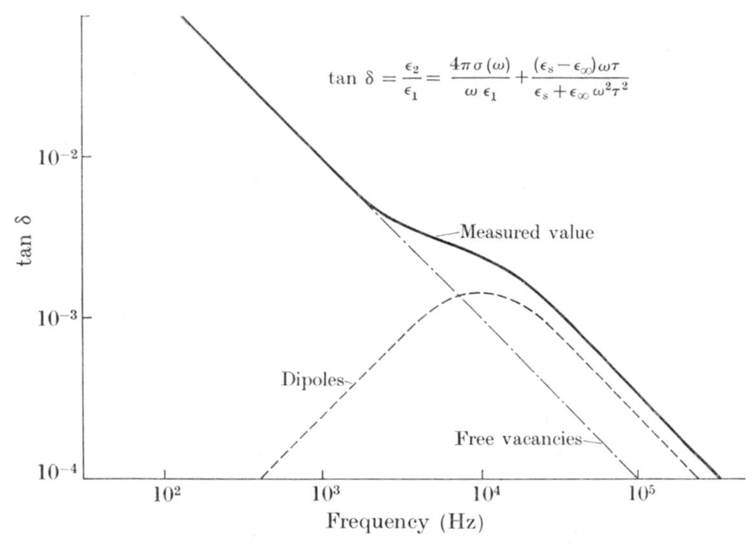

FIG. 3.8. A schematic log-log plot of tan $\delta \equiv \epsilon_2/\epsilon_1$ vs. frequency ω for a doped crystal containing dipolar impurity-vacancy (or interstitial) pairs showing how the observed value is a superposition of (i) a conductivity loss term, due to free vacancies and (ii) a dipolar Debye term coming from impurity-vacancy pairs. The scales of this diagram are for illustration only. The frequency at which the dipolar loss is a maximum is $\sim\tau^{-1}$ and thus depends exponentially upon temperature. The absolute magnitude of tan δ depends on defect concentration. Whether the inflection is visible in any actual case depends on the relative population of free- to bound-vacancies and other factors. Further, as explained in the text, impurity-vacancy and impurity–interstitial pairs may possess more than one mode of relaxation in which case there will be more than one Debye-like term in tan δ. Such terms may also arise in other ways e.g. from electrode polarization.

rates when the spectra of different configurations of the same complex can be separately measured. An elegant application of this is the work of Symmons (1970) on NaCl:Mn^{2+}, who followed the redistribution of charge-compensating vacancies between nearest and next-nearest neighbour positions to the Mn^{2+} ions following a sudden temperature change. (This is the symmetric A_{1g} relaxation mode which, although it couples to hydrostatic stresses, has not yet been seen in anelastic relaxation, presumably because the coupling is rather weak.) No equivalent studies in fluorite compounds have yet been reported, although epr has been used to monitor the concentration of defects of specified symmetry (see Chapter 6). Lastly we would draw attention to the value of measurements of lifetime broadening of epr lines as a means for the direct determination of defect relaxation times. This technique first demonstrated by Watkins (1959) on Mn^{2+}-vacancy pairs in alkali halides has also been used by Franklin and Marzullo (1970) to study Gd^{3+}-F_i^- pairs in CaF_2.

3.4.1. *Theory*

In this section we shall summarize the important part of the theory describing these relaxation effects.† The starting point of the theory is the phenomenological equations which replace the equilibrium constitutive equations between electric displacement and field and between strain and stress. For simplicity, we write these in scalar form, adequate for cubic materials when uniaxial stress is applied along a symmetry axis (⟨100⟩, ⟨110⟩, or ⟨111⟩) and sufficient for dielectric properties generally in this class of crystals. We shall use the notation of dielectric theory although with relatively minor changes the equations also apply to anelastic relaxation. Thus we denote the field (or stress) by E and the displacement (or strain) by D.‡

If we write a time-varying field $E(t)$ as an integral over its Fourier components

$$E(t) = \int_{-\infty}^{\infty} E(\omega)\mathrm{e}^{-\mathrm{i}\omega t}\,\mathrm{d}\omega, \tag{3.43}$$

then the corresponding displacement $D(t)$

$$D(t) = \int_{-\infty}^{\infty} \epsilon(\omega)E(\omega)\mathrm{e}^{-\mathrm{i}\omega t}\,\mathrm{d}\omega, \tag{3.44}$$

where $\epsilon(\omega)$ is the complex dielectric constant

$$\epsilon(\omega) = \epsilon_1(\omega)+\mathrm{i}\epsilon_2(\omega). \tag{3.45}$$

If the only contributions to $\epsilon(\omega)$ come from the perfect lattice and from relaxing defects, e.g. dipolar defects, then we may take $\epsilon(\omega)$ to be given by the Debye equation

$$\epsilon(\omega) = \epsilon_{\infty}+\sum_j \frac{\delta\epsilon_j}{1-\mathrm{i}\omega\tau_j} \tag{3.46}$$

where the summation is over all relaxing modes, j. In particular (3.46) assumes negligible d.c. conductivity due to unpaired defects; if this is not correct, an additional term

$$\frac{4\pi\mathrm{i}\sigma(\omega)}{\omega} \tag{3.47}$$

† For general background to relaxation phenomena in dielectrics see, e.g. Fröhlich (1958). For more detailed accounts of the theoretical methods used for point defects see, e.g. the reviews by Nowick and Heller (1963, 1965) and by Nowick (1967, 1972).

‡ It may be noted that Nowick (1972) pursues the analogy in terms of electric polarization P in place of D which leads to equations for the susceptibility in place of our equations for the dielectric constant. Since experimental results are generally represented in terms of the dielectric constant, we prefer the more usual formulation in terms of D (Fröhlich, 1958).

must be added to $\epsilon(\omega)$. In the anelastic relaxation experiments, creep processes (which are the analogue of d.c. conductivity) will be completely negligible and eqn (3.46) as an equation for the elastic compliance $s(\omega)$ is completely adequate. As far as defect-induced relaxation processes are concerned, ϵ_∞ is the limiting high-frequency dielectric constant, although in reality it will be dispersive if we go to optical and infrared frequencies. It is not difficult to show from (3.43) and (3.45) that the sudden application of a constant field E_0 gives an initial displacement $D_0 = \epsilon_\infty E_0$ which relaxes to the equilibrium displacement D_s according to

$$D(t) = D_0 + \sum_j E_0\, \delta\epsilon_i (1 - e^{-t/\tau_j}); \tag{3.48}$$

whence

$$D_s = \epsilon_s E_0 \tag{3.49}$$

with the equilibrium, or static, dielectric constant given by

$$\epsilon_s = \epsilon_\infty + \sum_j \delta\epsilon_j. \tag{3.50}$$

Likewise, if an alternating field of a single frequency ω is applied, the displacement

$$D(t) = \epsilon(\omega) E_0 e^{-i\omega t} = (\epsilon_1 + i\epsilon_2) E_0 e^{-i\omega t}, \tag{3.51}$$

lags in phase behind the applied field by an angle δ, the loss angle, such that

$$\tan\delta = \frac{\epsilon_2}{\epsilon_1} = \left\{\sum_j \left(\frac{\delta\epsilon_j \omega\tau_j}{1+\omega^2\tau_j^2}\right)\right\} \Big/ \left\{\epsilon_\infty + \sum_j \left(\frac{\delta\epsilon_j}{1+\omega^2\tau_j^2}\right)\right\}. \tag{3.52}$$

As is well known, the rate of absorption of energy from the electric field is proportional to $\tan\delta$. In the presence of a background conductivity $\sigma(\omega)$ due to unassociated mobile defects, it will be necessary to add $4\pi\sigma/\omega$ to the numerator ϵ_2 in (3.52). However, owing to the unimportance of creep in the usual anelastic relaxation experiments, (3.52) as it stands gives the loss angle in anelastic relaxation when the ϵ's are replaced by the corresponding compliances.

When only a single mode is significant, then

$$\tan\delta = \frac{(\epsilon_s - \epsilon_\infty)\omega\tau}{\epsilon_s + \epsilon_\infty \omega^2\tau^2}. \tag{3.53}$$

Furthermore, under this same condition it is not difficult to show that a plot of ϵ_2 vs. ϵ_1 is a semicircle whose centre lies on the axis of ϵ_1 (known as a Cole–Cole plot). In the presence of more than one relaxation time, the centre of such a plot will generally lie below the axis of ϵ_1 (Daniel, 1967).

For defect relaxations of the kind we are concerned with here $\epsilon_s - \epsilon_\infty \ll \epsilon_s$ and thus $\tan\delta \ll 1$; for a typical dielectric relaxation $\tan\delta$ may be $\sim 10^{-3}$ for a defect fraction of 10^{-4}. The magnitude of the maximum in $\tan\delta$, i.e. $\epsilon_s - \epsilon_\infty$ can be calculated from equilibrium statistical mechanics when the defect structure and its coupling to the external field is given. The relaxation times and thus the frequency ω_{max} at which this maximum occurs can, of course, only be obtained from an analysis of the defect jump rates; conversely, measurements of ω_{max} yield direct information about defect jump rates.

We turn now to the evaluation of the relaxation times τ_j in terms of the various defect jump frequencies. Let the different configurations of a defect be denoted by $u = 1, \ldots, n$. Then the rate equations describing the reorientation of these defects are

$$\frac{dn_u}{dt} = \sum_{v \neq u} n_v w_{vu} - \sum_{v \neq u} n_u w_{uv} \tag{3.54}$$

where n_v is the concentration of defects in configuration of type v and w_{vu} is the rate of jumps from configurations of type v to those of type u. These jump frequencies depend on the applied field. However, as we are only interested in linear responses to the field, we may expand the n_u and w_{uv} to first order in the applied field, by calling upon the rate process expression (3.30). Then

$$n_u = n_u(0) + \delta n_u, \tag{3.55}$$

$$w_{uv} = w_{uv}^{(0)} \left(1 - \frac{\delta E_{uv}^{s} - \delta E_u}{kT}\right) \tag{3.56}$$

in which δE_{uv}^{s} is the change in potential energy of the system at the saddle point on the path from u to v while δE_u is the energy change in the u configuration resulting from the application of the field. Expansion of (3.54) to first order in small quantities with use of the principle of detailed balance, i.e. of

$$n_u^{(0)} w_{uv}^{(0)} = n_v^{(0)} w_{vu}^{(0)} \tag{3.57}$$

then gives

$$\frac{d}{dt}(\delta n_u) = \left(\sum_{v \neq u} \delta n_v w_{vu}^{(0)} - \sum_{v \neq u} \delta n_u w_{uv}^{(0)}\right) + \sum_{v \neq u} n_v^{(0)} w_{vu}^{(0)} \frac{(\delta E_v - \delta E_u)}{kT}. \tag{3.58}$$

It should be noted that the second term on the right-hand side, the driving term, only depends on the energy differences between the various configurations. This is a consequence of the use of (3.56) together with the principle of detailed balance. It has the consequence that these relaxation phenomena are independent of any details of the path taken by a defect in moving from one site to another. The linearity of this equation (3.58) ensures that it is consistent with the phenomenological equations (3.43) to (3.52) above. By the properties of inhomogeneous linear differential equations the relaxation times τ_j can be found from the homogeneous parts of (3.58) above, i.e. from this equation in the absence of the driving terms in δE_u. These relaxation times are, in fact, easily seen to be the solutions of the determinantal equation

$$\det \|w_{vu}^{(0)} + \tau^{-1}\,\delta_{vu}\| = 0, \tag{3.59}$$

in which it is understood that

$$w_{uu}^{(0)} = -\sum_{v \neq u} w_{uv}^{(0)}. \tag{3.60}$$

TABLE 3.8

Possible relaxation normal modes and their symmetry type for defects of specified symmetry in cubic crystals. Also given are the symmetry coordinates of the electric field and stress components to which these modes may couple. (The usual subscript notation of Voigt is used for the stress components.)

Mode, Symmetry-type and degeneracy	Possible for defects of symmetry: Tetragonal	⟨110⟩ Orthorhombic	Trigonal	Symmetry coordinates: Electric field	Stress, σ
A_{1g} $(\times 1)$	Yes	Yes	Yes	None	$\sigma_1+\sigma_2+\sigma_3$
E_g $(\times 2)$	Yes	Yes	No	None	$(2\sigma_1-\sigma_2-\sigma_3,\ \sigma_2-\sigma_3)$
T_{2g} $(\times 3)$	No	Yes	Yes	None	$(\sigma_4,\ \sigma_5,\ \sigma_6)$
T_{1u} $(\times 3)$	Yes	Yes	Yes	$(E_x,\ E_y,\ E_z)$	None
A_{2u} $(\times 1)$	No	No	Yes	None	None
T_{2u} $(\times 3)$	No	Yes	No	None	None

In any particular system, it is especially convenient to make use of the symmetry of the defect in solving these equations and in specifying which relaxation modes can be stimulated by a given field. Certain general results can be stated for the three simplest defects, i.e. those of trigonal, tetragonal or orthorhombic symmetry. These results are gathered together in Table 3.8. From this we see, for example, that a tetragonal defect, i.e. one with a ⟨100⟩ axis, may have a triply-degenerate T_{1u} relaxation mode which will couple to an electric field, a doubly-degenerate E_g mode which will couple to uniaxial elastic stresses applied in either ⟨100⟩ or ⟨110⟩ directions (but not ⟨111⟩) and a non-degenerate A_{1g} mode which will couple to any hydrostatic stress.

As will be apparent already from preceding sections, two defects are of particular concern in fluorite crystals; (1) an anion interstitial associated with a substitutional foreign cation, e.g. RE^{3+} in CaF_2 and (2) an anion vacancy associated with a substitutional foreign cation, e.g. Na^+ in CaF_2. Since the origin of the association energy in these examples is the electrostatic attraction of defects bearing opposite

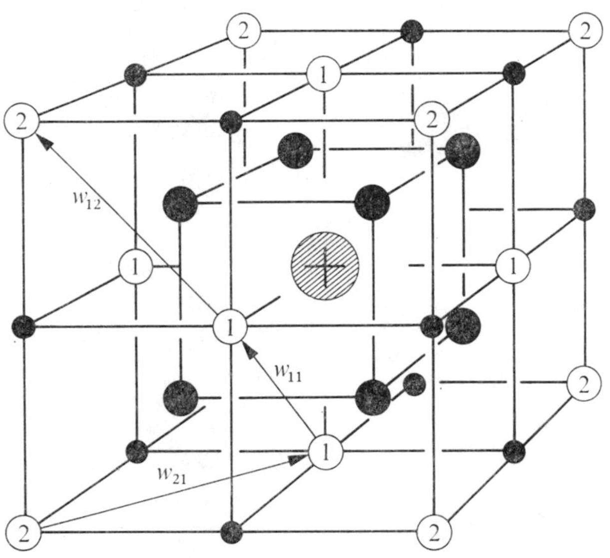

Fig. 3.9. A section of the fluorite lattice the centre of which is occupied by a trivalent cation shown shaded. The large filled circles denote the nearest anion neighbours of the impurity while the small filled circles are the cations. The open circles are interstitial sites (first and second neighbours to the impurity) which may be occupied by an anion interstitial. Jump frequencies w_{11}, w_{12}, and w_{21} are for the (net) displacements shown; actually the jumps are almost certainly interstitialcy jumps. (For clarity some of the further sites are not shown.)

effective charges, it is clear that the interaction energy may be appreciable not only at the nearest-neighbour separation, but at next-nearest-neighbour and more distant separations as well. We shall here give the results for these two defects assuming that only nearest and next-nearest configurations are significant.

(1) Interstitial anion-foreign cation pair. We denote the (six equivalent) nearest-neighbour positions of the anion by a subscript 1 and the (eight equivalent) next-nearest-neighbour positions by a subscript 2 (Fig. 3.9). Then w_{ij} represents the jump frequency from any particular i position to any particular j position. The relaxation times of the

modes are then (Franklin, Shorb and Watchman, 1964):

$$\begin{aligned}
A_{1g}: \tau^{-1} &= 4w_{12}+3w_{21} \\
A_{2u}: \tau^{-1} &= 3w_{21} \\
E_{g}: \tau^{-1} &= 4w_{12}+6w_{11} \quad \text{(2-fold degenerate)} \\
T_{1u}: \tau_{\pm}^{-1} &= 2w_{11}+2w_{12}+\tfrac{3}{2}w_{21}\pm \\
&\quad \pm[(2w_{11}+2w_{12}-\tfrac{3}{2}w_{21})+4w_{12}w_{21}]^{\frac{1}{2}} \\
&\quad \text{(each 3-fold degenerate)} \\
T_{2g}: \tau^{-1} &= 3w_{21}
\end{aligned} \tag{3.61}$$

The T_{1u} modes couple to electric fields while the A_{1g}, E_g, and T_{2g} modes couple to elastic stress fields as given in Table 3.8. Although no explicit calculations of the activation energies for these jumps have been made, it would seem reasonable to assume $w_{21} \gg w_{11} \gg w_{12}$, in view of the Coulomb attraction between the defects. In this case we see that the relaxation times of one of the two T_{1u} modes and of the E_g modes are largely determined by w_{11} while those of the other T_{1u} mode and of the T_{2g} mode are largely determined by w_{21}. Such simplifications obviously assist the analysis of measured relaxation times into component jump frequencies.

It will be noticed that we have taken $w_{22} = 0$ since we would not normally assume direct jumps to occur from one second neighbour position to another. (Figs. 3.3 and 3.4.) However, it is possible that this could happen by a kind of double interstitialcy mechanism; thus in Fig. 3.4 if the displaced normal ion, instead of moving on to an interstitial site as shown, moved downwards and so displaced the next normal ion on to an interstitial site, this would accomplish just the kind of displacement needed to give $w_{22} \neq 0$. In this case the expressions (3.61) for the inverse relaxation times of the A_{1g}, A_{2u}, T_{1u}, and T_{2g} modes will be altered by the addition of terms in w_{22}.

(2) Anion vacancy-foreign cation pair. In this case there are eight equivalent nearest-neighbour positions (giving a defect of trigonal symmetry) but twenty-four equivalent next-nearest positions and if both sets are included, the results are rather complicated owing to the low symmetry of the complex in its next-nearest-neighbour configuration (when it has a ⟨311⟩ axis). When only nearest-neighbour configurations are significant the defect is trigonal and gives rise to a (threefold-degenerate) electrically active T_{1u} mode, a (threefold-degenerate) elastically active T_{2g} mode (responsive to ⟨111⟩ and ⟨110⟩ uniaxial

stress but not to $\langle 100 \rangle$) and a non-degenerate A_{2u} mode which does not couple to uniform electric or elastic fields. In principle a trigonal defect also has a symmetric A_{1g} mode active under any hydrostatic stress, but such a mode can only arise if next-nearest or other more distant separations of the defect pair are possible; the relaxation time then depends on the frequencies of jumps to and from these more distant sites. For only nearest-neighbour jumps we have,

$$\begin{aligned} A_{2u}: \tau^{-1} &= 6w_{11} \\ T_{1u}: \tau^{-1} &= 2w_{11} \quad \text{(dielectric)} \\ T_{2g}: \tau^{-1} &= 4w_{11} \quad \text{(anelastic } \sigma\langle 111 \rangle\text{).} \end{aligned} \tag{3.62}$$

Expressions such as these enable the frequencies of the dielectric and anelastic loss maxima to be related to detailed models of the movements of the defect complexes. The absolute magnitudes of the loss factors, i.e. the $\delta\epsilon_j$, are relatively easily found from statistical thermodynamics when, as in the example of impurity-vacancy pairs above, we are concerned with only one active mode for then the quantity $\epsilon_s - \epsilon_\infty$ gives the desired magnitude. A simple calculation then shows that the relaxation of the dielectric constant $\delta\epsilon$ for cubic crystals is

$$\delta\epsilon = (\epsilon_s - \epsilon_\infty) = \frac{4\pi n_p \mu^2}{3v_m kT}, \tag{3.63}$$

where n_p is the concentration of complexes, μ their dipole moment, and v_m the molecular volume. Eqn (3.63) leaves out any consideration of internal field effects. While one would have to include a Lorentz internal field factor $\{(\epsilon_s + 2)/3\}^2$ for *point* dipoles, such a factor is not appropriate for extended dipoles such as defect pairs. Indeed there is no such general expression applicable to all defects, but detailed examination of specific examples by Ninomiya (1959) and by Boswarva and Franklin (1965) allows one to conclude that the correction to (3.63) is small. This is also borne out by detailed comparisons of relaxation strength with independent assessments of defect concentrations (see, e.g. Burton and Dryden, 1970).

The calculation of the corresponding expression for the relaxation of an elastic compliance $s_s - s_\infty$ is less familiar, but has been reviewed in detail by Nowick and Heller (1963). It is broadly analogous to that for dielectric relaxation except that in place of the dipole-moment vector $\boldsymbol{\mu}$ one has an elastic-defect tensor G which specifies the elastic strain generated in the lattice by the defect. Since any strain tensor can be diagonalized by a suitable orthogonal transformation, it follows that G has at most three principal values G_1, G_2, and G_3. For tetragonal or trigonal defects the two components perpendicular to the symmetry axis are equal and there are only two independent coefficients G_1 and G_2. For these two cases one has

$$\delta s = s_s - s_\infty = \beta \frac{v_m n_p (G_1 - G_2)^2}{kT}, \tag{3.64}$$

where $\beta = \frac{2}{3}$ for tetragonal defects and $\frac{4}{9}$ for trigonal defects. (For other more complicated cases see the reviews by Nowick and Heller 1963, 1965.) The calculation of G_1 and G_2 for specified defects can be made in principle by methods such as those described by Tewary (1969) and Hardy (1968), although it should be noted that their evaluation for ionic crystals has encountered some difficulties (see Faux and Lidiard, 1971 for their discussion of the elastic strain, i.e. G, due to vacancies). In practice the estimation of electric dipole moments is simpler and the approximation of taking μ as the product of effective charge times separation appears to be quite good.

When more than one mode contributes to either the dielectric or the anelastic relaxation, then one must partition the $\epsilon_s - \epsilon_\infty$ and $s_s - s_\infty$ given by (3.63) and (3.64) according to the relative strengths of these modes as obtained from the kinetic analysis. Details of such evaluations have been given for several defects in alkali halides; but as previously noted, some of these (e.g. Franklin *et al.* (1964)) are directly applicable to defect pairs in fluorite lattices. These analyses show that, for physically reasonable ranges of the relative jump frequencies, the mode of longest τ is generally dominant over others of the same type. In practice this makes it difficult, though not impossible, to obtain information about all modes. The accuracy with which dielectric loss measurements can be separated into a sum of Debye terms depends very much on the accuracy with which the background conductivity loss, $4\pi\sigma/\epsilon\omega$, can be separated off. In anelastic loss experiments, the background is not as large, but its frequency variation is less well known and therefore it is more difficult to subtract accurately. The ionic thermocurrent technique of dielectric relaxation is thus a particularly valuable addition to the experimental methods available, since it has no conductivity background. In this technique one measures the depolarizing current $i(T)$ released by slowly warming up a polarized crystal from a low temperature at which all τ_j are very long. Each relaxation mode gives a peak on a plot of $i(T)$ versus temperature.

Three features of the peak are immediately useful. Firstly at the beginning of a peak, $i(T)$ rises as

$$i(T) = \text{const. } \tau^{-1}. \tag{3.65}$$

Since one jump frequency w_{ij} will often dominate in the expression for τ we can write τ as

$$\tau = \tau_0 \exp(+E_\text{m}/kT) \tag{3.66}$$

and the expression for the initial value of $i(T)$ becomes

$$i(T) = \text{const. } \tau_0^{-1} \exp(-E_\text{m}/kT). \tag{3.67}$$

The activation energy E_m is then readily obtained. Secondly, if the heating rate $\mathrm{d}T/\mathrm{d}t \equiv b$ is constant, then the temperature of the maximum in $i(T)$ is

$$T_\text{m} = (bE_\text{m}\tau(T_\text{m})/k)^{\frac{1}{2}}, \tag{3.68}$$

from which the pre-exponential constant τ_0 is determined. Lastly, the area under the peak gives the contribution of that mode to the relaxation,

$$\delta\epsilon = \frac{1}{Eb}\int_\text{peak} i(T)\,\mathrm{d}T. \tag{3.69}$$

The technique is capable of some refinements; for example Müller and Teltow (1972) have shown the advantages of heating the specimen so that T^{-1} (rather than T) increases linearly with time. The method has proved a sensitive and valuable tool in practice.

3.4.2. *Experiment*

Experimental studies of defect relaxation processes in fluorite crystals have been made by applying the above methods to the alkaline earth fluorides (mainly CaF_2) and to the oxides CeO_2 and ThO_2. These have mainly aimed to obtain information about impurity ions paired with anion interstitials or anion vacancies. This work has once again shown the value of obtaining results by a combination of all the above techniques as well as by epr.

A number of relaxation experiments on CaF_2 doped with various trivalent ions (Y^{3+}, rare-earth ions) have been carried out, largely but not exclusively by ITC techniques. Although some of these results can be very satisfactorily interpreted in terms of the impurity–anion interstitial pair model described by eqn (3.61) (see Franklin and Crissman 1971) more characteristic activation energies are observed than are required in this model. Assignments of observed activation energies to particular defect jumps therefore remain rather uncertain. Nevertheless a variety of results obtained by ITC (Royce and Mascarenhas, 1970; Stott and Crawford 1971*a*,*b*; Kunze and Müller 1972; Wagner and Mascarenhas 1972; Stiefbold and Huggins 1972; Kitts and Crawford 1973; Kitts 1973), by dielectric relaxation and epr line broadening (Franklin and Marzullo 1970) and by anelastic relaxation (Franklin and Crissman 1971) point to the activation energy for the nearest neighbour interstitial jumps w_{11} as being close to 0·4 eV for many impurity ions. ITC measurements have also shown the existence of another lower activation energy ~0·15 eV (Stott and Crawford 1971*a*,*b*) which Franklin and Crissman (1971) show to be consistent with their anelastic relaxation data if we assign it to w_{21}. Both values, 0·15 and 0·4 eV, are, of course, considerably lower than that for free interstitial motion (Table 3.5). It may however be premature to assign the two ITC peaks to different relaxation modes of the same centre, since it is apparently possible to vary their relative strengths in a way inconsistent with the expected thermal equilibrium distribution of interstitial F^- ions between nearest and next-nearest positions (Stott and Crawford 1971*a*,*b*).

Recent ITC measurements on all three alkaline earth fluorides doped with Gd^{3+} (Kitts 1973) also disclose a further peak whose size increases from CaF_2 to BaF_2 and for which the activation energy varies from ~0·7 eV (CaF_2) to ~0·6 eV (BaF_2). On the grounds that epr shows the Gd^{3+}–F_i^- pairs to be predominantly next-nearest neighbours (trigonal spectra) in BaF_2 it is suggested that this activation energy corresponds to direct w_{22} jumps, which as we have already indicated would require a kind of double interstitialcy mechanism. We should then expect to see such an activation energy also characterising an elastically active T_{2g} mode, but anelastic relaxation experiments on BaF_2 have not yet been done.

In view of the importance of the hydrolysis of CaF_2 it is necessary to confirm that the above results do relate to the RE^{3+}–F_i^- centres and not to the Tr_1 and Tr_2 centres formed when doped CaF_2 is hydrolysed. This confirmation is provided by the recent work of Kitts and Crawford (1973) and Kitts (1973) who studied the changes in the ITC properties of CaF_2:Gd^{3+} before and after treatment known to produce Tr_1 and Tr_2 centres. The Tr_2-centre (Gd_s^{3+}–O_s^{2-} pair) is not expected to contribute to these relaxation processes since it cannot reorient thermally without the intervention of another defect but a new and distinct ITC peak seen in the oxygen-contaminated crystals is almost certainly due to the

Tr_1-centres. The activation energy of 0·5 eV is very close to that for the reorientation of Na_s^+–F_v^- pairs in CaF_2, possibly as one would expect from the structure of the Tr_1-centre (§ 6.5.9).

Lastly it is necessary to refer to processes having larger activation energies, ~1 eV, observed in older experiments by Southgate (1966) and by Chen and McDonough (1969). The dielectric losses seen by Chen and McDonough are too large to be due to defect pairs and may be due to surface and electrode polarization (see e.g. Franklin *et al.* 1967). The anelastic relaxations seen by Southgate (1966), however were assumed to be due to Y_s^{3+}–F_i^- pairs. Considered on their own, Southgate's results are consistent with that assignment, but this interpretation is not consistent with more recent work referred to above. In conclusion it seems clear that further work on these systems, preferably using selective techniques such as anelastic relaxation, is necessary before final assignments of all the known activation energies can be made.

Other relaxation experiments have been made on crystals doped with cations of a lower valency. The first of these were by Wachtman (1963) on ThO_2 doped with Ca^{2+} ions, who studied both dielectric and anelastic relaxations. His results were consistent with the model of Ca_s^{2+}–O_v^{2-} pairs; in particular the ratio of relaxation times τ_{diel}/τ_{anel} was close to 2 as required by (3.62). Similar results have also been found recently for Ca-doped CeO_2 (Lay and Whitmore, 1972). However, these two experiments were necessarily carried out on polycrystalline specimens so that the selection rule for anelastic relaxation due to trigonal defects (absence of a $\sigma\langle 100\rangle$ relaxation) could not be verified. Johnson, Tolar, Miller and Cutler (1969) were, however, able to confirm the absence of a $\sigma\langle 100\rangle$ relaxation in their studies of CaF_2 doped with Na^+ ions. Their work yielded a clear verification of the Na_s^+–F_v^- defect model and in doing so, showed that charge compensation of substitutional Na^+ cations occurs by anion vacancies and not by Na^+ interstitials (§ 3.2.2), since these would lead to tetragonal defects for which eqn (3.61) would have applied rather than (3.62). The activation energies for anion vacancy jumps around a cation impurity given by these measurements are not very different from those for free vacancy movements.

3.5. Theory of defect parameters (mainly energies)

Although the preceding sections have used only rather general statistical defect models and relied upon experiment to give energies and other characteristic defect parameters, it will be clear already from some examples that a mechanical theory of these parameters would be a valuable addition. Such a theory is useful because it allows us to narrow down the range of possible models to be considered, but for this it must be quantitatively accurate, since it is numerical values which are sought rather than relations. These calculations have developed in ways which originate from the work of Mott and Littleton (1938) and Mott and Gurney (1948). Some substantial steps towards improving the reliability of these calculations have been taken in the last few years and we shall therefore review this work fairly broadly. In § 3.5.1, we consider generally the methods of calculation, i.e. the physical models and the mathematical and computational methods employed. In § 3.5.2 we review the results so far obtained by these methods and models, relating them as far as possible to the experimental results of previous sections.

3.5.1. *Methods of calculation*

There are two aspects to be considered, firstly the physical models to be used and secondly the actual mathematical and computational means by which the desired quantities are evaluated.

Physical models. The obvious starting point for the calculation of defect properties in solids in or near thermodynamic equilibrium is provided by the Born–Oppenheimer and quasi-harmonic approximations (Chapter 2). These provide the only practical general scheme of calculation at the present time. We therefore begin with the familiar expression for the Helmholtz free energy of a crystal, namely,

$$F(T, V) = \Phi(V)+kT\sum_j \ln\left(2\sinh\frac{\hbar\omega_j}{2kT}\right). \tag{3.70}$$

Here the potential energy term $\Phi(V)$ and the lattice frequencies $\omega_j(V)$ are functions of the crystal volume, V. The relation of T and V to external pressure, P, is given by combining (3.70) with the thermodynamic equation

$$P = -\left(\frac{\partial F}{\partial V}\right)_T. \tag{3.71}$$

At high temperatures where $\hbar\omega_j \ll 2kT$ this becomes

$$P = -\left(\frac{\mathrm{d}\Phi}{\mathrm{d}V}\right)-\frac{kT}{V}\sum_j \frac{\mathrm{d}(\ln\,\omega_j)}{\mathrm{d}(\ln\,V)} \tag{3.72}$$

or, in the Grüneisen approximation in which all the derivatives

$\mathrm{d}(\ln \omega_j)/\mathrm{d}(\ln V)$ are equal,

$$\frac{\mathrm{d}(\ln \omega_j)}{\mathrm{d}(\ln V)} \equiv -\gamma, \tag{3.73}$$

$$P = -\left(\frac{\mathrm{d}\Phi}{\mathrm{d}V}\right) + \frac{3NkT\gamma}{V}, \tag{3.74}$$

where N is the number of atoms (ions) in the solid.

These equations can be used to describe both perfect solids and solids containing defects. By comparison, thermodynamic quantities of defect formation, migration etc. can therefore be derived. Thus at high temperatures, where $\hbar\omega_j \ll 2kT$, eqn (3.70) allows us to derive an expression for the internal energy change between two configurations as simply the difference in potential energies,

$$\Delta U = U_2 - U_1 = \Phi_2 - \Phi_1, \tag{3.75}$$

while the entropy change is

$$\Delta S = S_2 - S_1 = -k \sum \ln(\omega_j^{(2)}/\omega_j^{(1)}). \tag{3.76}$$

These equations are the basis of almost all detailed defect calculations which have been made to date. At a given pressure and temperature, the volume of the crystal will also change as the defect state of the crystal changes; such volume changes determine the effects of pressure upon defect concentration, mobility, etc. Expressions can be obtained from (3.74). Owing to the importance of defect energies in determining the temperature dependence of defect processes, the simple form (3.75) is especially convenient, since it requires the evaluation of only the potential energy of the static solid in its two defect configurations.

Before these energies or other quantities can be calculated for any given defect, it is obviously necessary to determine a suitable potential function which allows us to represent the potential energy as a function of atomic positions $\mathbf{r}_1, \mathbf{r}_2, \ldots, \mathbf{r}_N$. The general approach is to construct a semi-empirical model using as much macroscopic data on the solid as possible. For the substances of interest here one can obtain useful insights on the basis of the classic ionic model (Chapter 1), i.e. we consider the solid as a collection of ions interacting through their electrostatic charges (Coulomb interactions) and by the overlap of their electron clouds at close separations. These overlap or exchange repulsions are generally represented by the Born–Mayer form, i.e. by $b \exp(-r/\rho)$, where r is the separation of a pair of ions. Explicit quantum-mechanical and semi-classical calculations give general support to such a form. A recent example of such a priori calculations is the F^-–F^- interaction shown in Fig. 3.10. The ions are also assumed to be polarizable electrically and deformable as a result of overlap; this is often represented by using a *shell model* in which the ion is regarded as made up of a core (of charge X) surrounded by a shell (of charge Y) which may be displaced harmonically relative to the core

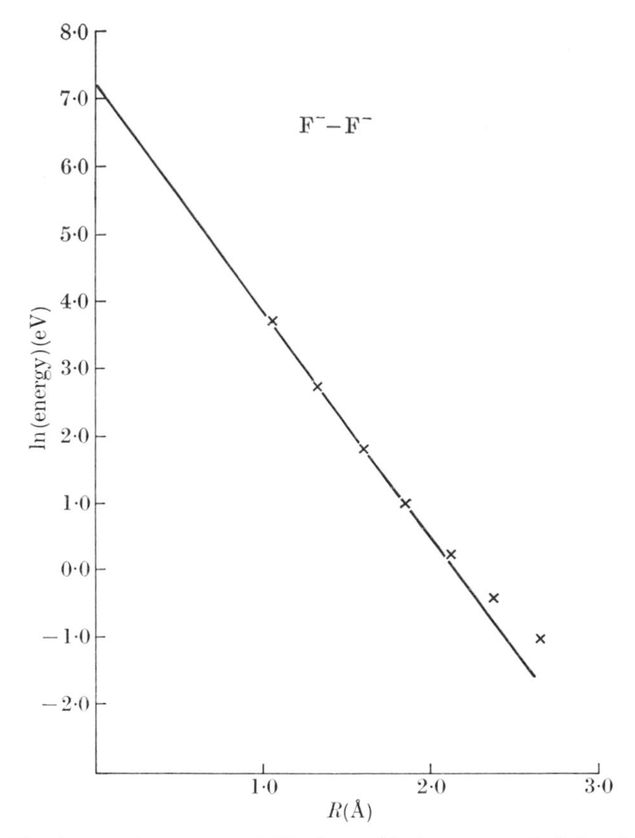

Fig. 3.10. The interaction energy (eV) of two F^- ions, corrected for the monopole-monopole electrostatic terms, as a function of internuclear separation (in Å). The crosses × represent values computed by means of the IBMOL V Hartree-Fock program. The energy is here plotted as its natural logarithm and the conformity of the points (×) to a straight line confirms the Born–Mayer form for the overlap potential (after Catlow and Hayns, 1972).

(spring constant k). Such shell models are extensively used to describe lattice dynamical behaviour (Chapter 2). In defect calculations where one needs to go beyond the harmonic approximation for those inter-ionic interactions in the strongly distorted region around the defect, we assume that the overlap repulsions act between the shells and have the exponential Born–Mayer dependence upon their separation. These models provide satisfactory descriptions of many physical properties, but obviously contain parameters which must be determined empirically. There are many different ways of weighting the empirical data but for defect calculations, particularly those relating to defects carrying a net electric charge, it is important to give good descriptions of the cohesive energy and of the dielectric constants and optical vibration frequencies such as the reststrahl and Raman frequencies (see Chapter 2). It has long been recognized (Jost, 1933) that the electrical polarization energy is a very important term in the energies of charged defects. For such defects the accuracy of the calculation is strongly dependent on, among other things, the accuracy with which ϵ_0 is described. In general, it is also important to describe the elastic properties well. Since the models we have described are all two-body central-force models we may not expect necessarily to describe all three elastic constants correctly; thus in the alkali halides three-body forces are necessary to account for the observed departures from the Cauchy condition $c_{12} = c_{44}$. In the fluorite structure, for which this condition is not a requirement, it does nevertheless appear that, for the compounds of principal interest, the differences $c_{12}-c_{44}$ are well described by the shell models which have been developed. Indeed their extension via the Born–Mayer overlap repulsion allows the description also of the third-order elastic constants (Catlow and Norgett, 1973) (see § 2.4.4).

We have been able here to describe the basic physical model only rather briefly and for a fuller description we refer the reader to the previous two chapters and to earlier reviews (Barr and Lidiard, 1970; Hardy and Flocken, 1970; Lidiard and Norgett, 1972). However, it should be noted that shell models, though undoubtedly the preferred empirical models, have been employed in defect calculations only relatively recently. A considerable amount of earlier work was carried out on so-called polarizable-point-ion models. These allow the ions to be polarizable electrically but not mechanically as a result of ionic overlap; nor do they allow the Born–Mayer interactions between the ions to depend on their polarization. Such models suffer from several intrinsic defects, of which the most dramatic is the possibility of polarization catastrophes, i.e. of negative divergencies in the energy associated with the dipole-dipole interactions. As Tosi and Doyama (1967) and Faux (1971) have shown, these will necessarily be found in any search for a lowest energy configuration if two ions i and j approach closer than a critical separation,

$$r_{\text{crit}} = (4\alpha_i\alpha_j)^{\frac{1}{6}} \tag{3.77}$$

where α_i, α_j are the electronic polarizability of ions i and j. This is not to say that calculations based on polarizable-point-ion models are altogether without value. The relaxations of a lattice around a defect are generally started from the perfect lattice positions and in many of these calculations stable configurations for which all r_{ij} are greater than the critical separation are found, so that useful predictions can then still be obtained with polarizable-point-ion models.

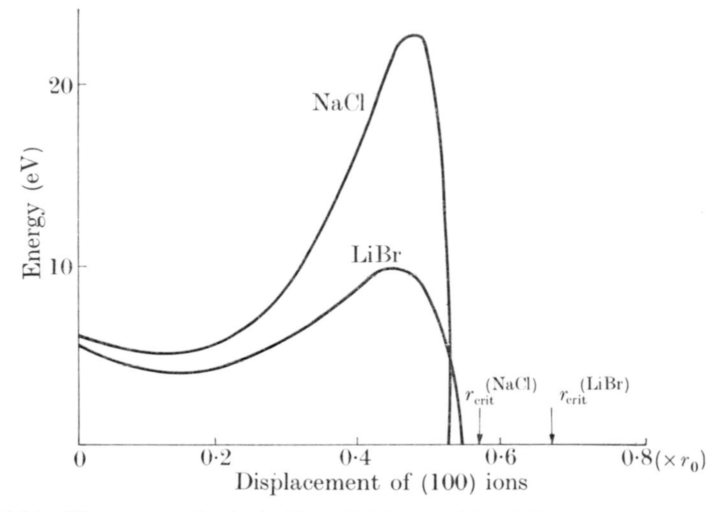

FIG. 3.11. The energy of polarizable-point-ion models of NaCl and LiBr crystals containing a cation vacancy shown as a function of the outward radial displacement of the nearest neighbours of the vacancy (measured in units of the anion-cation nearest-neighbour distance). The remainder of the crystal is allowed to relax appropriately. These curves show the onset of the polarization catastrophe at r_{crit} as well as the minima usually calculated. (After Faux 1971).

Of course these configurations are metastable with respect to the singularities but these are unphysical and may be assumed not to affect the solution greatly providing the matastability barrier is large; see Fig. 3.11 for an example from the alkali halides. These instabilities may be avoided by making the (realistic) assumption that the polarizability falls as the ions overlap (Quigley and Das, 1967) but this variation of the model has not been applied by others so far. Even if instabilities are avoided in this way or are not otherwise evident the polarizable point-ion model will often over-estimate the static dielectric constant and when it does so the model will yield too large an energy of relaxation around charged defects. This consequence has been clearly demonstrated in the alkali halides (Faux and Lidiard, 1971) but the same effect is apparent in some of the calculations for the alkaline earth fluorides (see the discussion in § 3.5.2). In general it is preferable to use shell models.

Mathematical methods. Having decided upon a suitable physical model, it now remains to evaluate the relaxation and polarization of the lattice in response to the forces which the defect exerts upon it, i.e. to find the configuration of minimum potential energy, Φ. For defects carrying a net charge the polarization of the lattice falls off only slowly as r^{-2} and important contributions to the energy of the equilibrium configuration come from regions distant from the defect. There can be

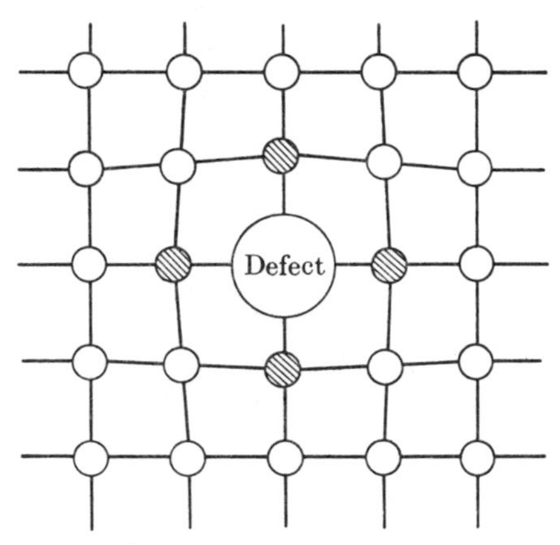

FIG. 3.12. Schematic diagram showing the division of the lattice into two regions (I and II) for the purposes of calculating lattice distortion, energy of relaxation, etc. In this figure region I is shown as the defect and its immediate neighbours while region II is the rest of the lattice. In any given calculation the choice of region I is guided by the method used but should be large enough that the results are insensitive to the exact boundary between I and II.

no hope, therefore, of relaxing just a few ions in the immediate vicinity of the defect and assuming that the rest are undisplaced and unpolarized. Even for defects without a net charge (e.g. neutral defects or pairs of oppositely charged defects) it may be necessary to relax many shells.

The general strategy of these calculations is thus to describe the relaxation of the more distant ions in either a harmonic or a continuum approximation and to minimize the energy function explicitly for only a finite number of ions in an inner region surrounding the defect ('Region I') where displacements and forces are large and anharmonic (Fig. 3.12). At present, for computers having large stores, reasonable upper limits to the size of this inner region are several hundred or a thousand ions, but good results can often be obtained with far fewer than this so that these calculations are not excluded when only smaller computers are available.

We express this strategy formally by writing the total potential energy of the defect solid as

$$\Phi = \Phi_{\mathrm{I}}(\mathbf{x})+\Phi_{\mathrm{I,II}}(\mathbf{x}, \boldsymbol{\xi})+\Phi_{\mathrm{II}}(\boldsymbol{\xi}), \tag{3.78}$$

where we use the vectors $\mathbf{x}$ and $\boldsymbol{\xi}$ to denote all the displacements (of both cores and shells or equivalently displacements and moments) in regions I and II respectively. We can re-arrange terms in $\boldsymbol{\xi}$ between $\Phi_{\mathrm{I,II}}$ and Φ_{II} so that the harmonic approximation for Φ_{II} takes the purely quadratic form

$$\Phi_{\mathrm{II}}(\boldsymbol{\xi}) = \tfrac{1}{2}\boldsymbol{\xi}^{\mathrm{T}}\mathbf{A}\boldsymbol{\xi}, \tag{3.79}$$

corresponding to the energy of a distorted region II filled with a perfect undistorted and unpolarized inner region. The term $\Phi_{\mathrm{I,II}}(\mathbf{x}, \boldsymbol{\xi})$ then represents the change in the interaction between regions I and II resulting from the presence of the defect and the distortions $\mathbf{x}$ in the inner region. Whenever we need to speak precisely of the interaction between the defect and the lattice we shall have in mind a decomposition such as this. The task now is to minimize the energy function Φ with respect to the $\mathbf{x}$ and the $\boldsymbol{\xi}$. Evidently minimization with respect to the $\mathbf{x}$ variables,

$$\frac{\partial\Phi}{\partial\mathbf{x}} = 0, \tag{3.80}$$

can only be carried out numerically since Φ_{I} and $\Phi_{\mathrm{I,II}}$ are complicated, anharmonic functions of $\mathbf{x}$. Methods are considered later. The object of separating off the harmonic term Φ_{II} however is to allow analytic methods to be used to obtain the solutions $\boldsymbol{\xi}$ which minimize Φ. If we now minimize Φ with respect to the displacements $\boldsymbol{\xi}$ we obtain as the equilibrium condition for region II

$$\mathbf{F} \equiv -\frac{\partial\Phi_{\mathrm{I,II}}}{\partial\boldsymbol{\xi}}(\mathbf{x}, \boldsymbol{\xi}) = \mathbf{A}\boldsymbol{\xi}, \tag{3.81}$$

which can be solved formally as long as it is adequate to expand the left-hand side only as far as the linear terms in $\boldsymbol{\xi}$; thus

$$-\frac{\partial\Phi_{\mathrm{I,II}}}{\partial\boldsymbol{\xi}} \equiv \mathbf{F}^{(0)}(\mathbf{x})+\mathbf{F}^{(1)}(\mathbf{x})\boldsymbol{\xi} \tag{3.82}$$

i.e.

$$(\mathbf{A}-\mathbf{F}^{(1)}(\mathbf{x}))\boldsymbol{\xi} = \mathbf{F}^{(0)}(\mathbf{x}) \tag{3.83}$$

or

$$\boldsymbol{\xi} = \mathbf{G}\mathbf{F}^{(0)}(\mathbf{x}), \tag{3.84}$$

where

$$\mathbf{G} = (\mathbf{A} - \mathbf{F}^{(1)}(\mathrm{x}))^{-1} \tag{3.85}$$

is the perturbed static Green function.

At this point it should be said that the methods used in practice divide into two kinds, (i) those which aim to solve these equations for region II explicitly (Kanzaki, 1957, Tewary, 1969) and (ii) those which use only the continuum limit (Mott and Littleton, 1938). Obviously with those methods based on the full harmonic approximation we can expect to take the boundary between regions I and II much closer to the defect. Here therefore the major effort goes into finding the solution in region II, generally leaving only a relatively minor problem in finding the solution of (3.80) for the inner region I. On the other hand with the second class of methods, the description of region II is relatively straightforward, but since region I is necessarily larger more effort must then go into the solution of (3.80). The emphasis is thus on quite different mathematical and numerical techniques. So far one or two general program packages, designed to allow the ready evaluation of the energies of a wide range of defects, have been developed on the basis of the continuum approach. In the next few pages we therefore discuss some of the details of these methods. We do so because these topics have as yet been rather little reviewed. The reader who is not immediately interested in the methods of these calculations may omit the following pages and turn next to the discussion of results of particular calculations (§ 3.5.2), referring back only as necessary.

Methods of explicit solution of equations (3.81)–(3.85) have been given by Kanzaki (1957), Hardy (1960, 1962) and by Tewary (1969). In Kanzaki's method, eqn (3.83) is Fourier-transformed over the crystal lattice (using periodic boundary conditions). The resulting block-diagonalization of $\mathbf{A}$ then renders the calculation of the transformed displacements

$$\mathbf{Q}(\mathbf{q}) = \sum_{l} \boldsymbol{\xi}_l \exp(-i\mathbf{q}\cdot\mathbf{R}_l) \tag{3.86}$$

relatively easy provided that the force $\mathbf{F}$ is of short range, i.e. has non-zero components only for ions l near to the defect. Furthermore the nature of the solutions obtained for particular defects shows that the harmonic description (3.79) should in many cases be valid very close to the defect itself, so that region I can be taken to contain very few atoms or ions. In practice, calculations by this method have taken region I to be simply the defect itself, thus eliminating the $\mathbf{x}$-variables from the problem—although this appears to go beyond the validity of (3.79) to some extent (Tewary, 1969). Despite these advantages, the method also has its drawbacks. Firstly, for charged defects the evaluation of the Fourier transform of the force $\mathbf{F}$ is not simple owing to the long range of the Coulomb term although it is practicable by an extension of the Ewald technique of lattice summation (Karo and Hardy, 1971). Secondly, in all cases, the inversion of the calculated $\mathbf{Q}(\mathbf{q})$ to get the actual displacements $\boldsymbol{\xi}$ involves numerical summations over the Brillouin zone. The method has not so far been applied very widely, although it has been used to evaluate the energies of formation of Frenkel and Schottky defects in alkali halide lattices (Karo and Hardy, 1971 and Schultze and Hardy, 1972*a*,*b*). No applications to fluorite lattices have yet been reported. The method would not appear to be well suited to the development of general program 'packages'. It is however obviously convenient for those properties such as X-ray and neutron scattering intensities which require the transformed displacements $\mathbf{Q}(\mathbf{q})$ directly (see e.g. Kanzaki, 1957 and MacDonald, 1972 for examples).

The alternative approach due to Tewary does not make a Fourier transformation but calculates the static-lattice Green function directly, using symmetry analysis to reduce the size of the matrices. However, as presently developed, it only appears to be practicable for localized forces and defects of reasonably high symmetry. It has been applied to vacancies and interstitials (including dumb-bell configurations) in cubic lattices but does not yet appear to be convenient for charged defects in ionic lattices. It is undoubtedly capable of further development and might form the basis of a convenient general program.

In summary then, those methods which aim to treat Region II in a full harmonic approximation, while elegant, are not, as presently developed, very convenient for defects of low symmetry and, due to the long range of Coulomb fields, are complicated for defects having a net charge. Most calculations have therefore been carried out by using a continuum, or $\mathbf{q} \to 0$, description of Region II while making Region I as large as possible so as to reduce to a minimum the error involved in using this limiting description of the outer region. This is the approach taken in all the calculations for fluorite lattices which have been published so far. In essence, it is that suggested originally by Mott and Littleton (1938).

In materials such as cubic crystals having no non-zero piezoelectric or electrostrictive constants one can separate the limiting $\mathbf{q} \to 0$ distortion due to a defect into elastic and electric terms,

$$\boldsymbol{\xi} = \boldsymbol{\xi}_{\text{elas}} + \boldsymbol{\xi}_{\text{elec}}. \tag{3.87}$$

The forms of ξ_{elas} and ξ_{elec} can be obtained by formal analysis based, for example, on the Kanzaki formulation (see, e.g. Hardy and Lidiard, 1967) but the results can also be understood from simple physical arguments, which we now present. We deal first with ξ_{elec} since this is often the most important term.

The basic physical idea is that a defect of net charge q causes a macroscopic polarization $\mathbf{P}$ which at distance $\mathbf{r}$ from the defect is

$$\mathbf{P} = \frac{q}{4\pi}\left(1 - \frac{1}{\epsilon_0}\right)\frac{\mathbf{r}}{r^3}, \tag{3.88}$$

and that this can be partitioned among the ions to give the actual moments and displacements. Thus in the polarizable-point-ion model of a fluorite lattice we should obtain electronic moments

$$\boldsymbol{\mu}_{c,a} = \frac{1}{4\pi}\left(1 - \frac{1}{\epsilon_0}\right) v_m \frac{\alpha_{c,a}}{(\alpha_c + 2\alpha_a + 4\alpha)} \frac{q\mathbf{r}}{r^3}, \tag{3.89}$$

where α_c and α_a are the electronic polarizabilities of the cations and anions respectively while α is the anion displacement polarizability and 2α is the cation displacement polarizability.† A simple calculation of the balance of forces in a polarized lattice allows one to relate α to the non-Coulombic interactions. As long as these do not extend beyond next-nearest neighbours we obtain

$$\alpha = Z^2e^2/R, \tag{3.90}$$

where

$$R = \frac{8}{3}\left(\phi''(r_0) + \frac{2}{r_0}\phi'(r_0)\right), \tag{3.91}$$

in which the anions are assumed to carry a charge $-Ze$ and the cations $2Ze$. ϕ is the non-Coulombic potential between nearest neighbours; the separations of anions (next-nearest neighbours) and of cations are not altered in a uniform polarization and the corresponding interactions do not therefore enter into (3.91). In the polarizable-point-ion model the displacement moments at the anion and cation sites are thus given by an expression like (3.89) except that α and 2α, respectively, appear in the numerator in place of the electronic polarizabilities. However, as we have previously noted, this polarizable-point-ion model generally

† That the cation displacement polarizability is twice that of the anions can be understood very simply. Firstly the cation charge is twice that of the anions, from which one expects a factor 2^2 since α is proportional to the square of the ionic charge; but secondly since each cation is surrounded by 8 anions whereas each anion is surrounded by only 4 cations the restoring force (via ϕ) on the cations is twice that on the anions. The net result is that the cation displacement polarizability is twice that of the anions.

overestimates the static dielectric constant ϵ_0 when α is evaluated from (3.90) and (3.91) and ϕ is obtained from a Born–Mayer potential fitted to the compressibility.

A similar calculation considering the balance of forces in a shell model allows one to derive the analogous equations for this case. The molecular polarizability, i.e. the total polarizability per unit cell is

$$\alpha_{\mathrm{m}} = e^2 \left(\frac{4Z^2}{R}+\frac{X_+^2}{k_+}+\frac{2X_-^2}{k_-}\right), \tag{3.92}$$

X_+e and X_-e being the charge on the cores of cations and anions respectively. The effective electronic polarizabilities are

$$\alpha_{\mathrm{c,a}} = -e^2X_\pm Y_\pm/k_\pm, \tag{3.93}$$

where the shell charges $eY_\pm$ are expected generally to be negative although the 'best fit' does not invariably lead to negative values (cf. Table 2.5). These expressions are effective or crystal-ion polarizabilities and not the free-ion polarizabilities, which are $e^2Y_\pm^2/k_\pm$ and thus always positive; the difference arises because the polarization of an ion in the crystal changes its interaction with its neighbours, because this interaction acts through the shells. The effective displacement polarizability α in the shell model is thus

$$\alpha = \frac{e^2Z}{2}\left(\frac{2Z}{R}+\frac{X_+}{k_+}-\frac{X_-}{k_-}\right). \tag{3.94}$$

As before, the displacement polarizability of the anions is α, that of the cations 2α. Eqns (3.92)–(3.94) allow us to partition the total polarization moment per cell into its various components. The actual displacements of the ions are, of course, the corresponding displacement moments divided by the total ionic charge, i.e. by $2Ze$ for cations, by $-Ze$ for anions.

These electrical polarization terms provide a most important description of the outer Region II around charged defects or combinations of charged defects. Their importance comes, as we have already noted, from the long range of electrical forces. However, the description is not yet complete since the defects are also sources of elastic distortion. The full description of these elastic displacements caused by the defect is more complicated, firstly because the elastic strength of the defect must be calculated a priori, even for a simple defect such as a vacancy, whereas the electric strength is just its net charge; secondly because whereas cubic solids are electrically isotropic their elastic anisotropy is often considerable. Of the substances of concern here, only BaF_2 is at all closely isotropic elastically.† In general the elastic properties of a point defect in an elastic continuum are specified by a tensor G_{ij} which gives the components F_i of the body force exerted by the defect on the continuum as

$$F_i = -\sum_j G_{ij} \frac{\partial}{\partial x_j} \delta(r), \tag{3.95}$$

where $\delta(r)$ is the Dirac δ-function. The displacements in a cubic elastic solid due to such a defect are in principle obtained by solving the equations of elasticity, which in tensor notation are

$$\sum_{j,k,l} c_{ijkl} \frac{\partial^2 \xi_k}{\partial x_l \partial x_j} = -F_i = \sum_j G_{ij} \frac{\partial}{\partial x_j} \delta(r) \tag{3.96}$$

by (3.95). For a cubic solid, of course, one has only three independent elastic constants $c_{11} \equiv c_{xxxx}$ etc., $c_{12} = c_{xxyy}$ etc., and $c_{44} = c_{xyxy}$ etc. It is then straightforward to solve (3.96) for the Fourier transform of the displacement, i.e. for $\mathbf{Q}(\mathbf{q})$:

$$\boldsymbol{\xi}_{\text{elas}} = \frac{1}{N} \sum_q \mathbf{Q}(\mathbf{q}) \exp(i\mathbf{q} \cdot \mathbf{r}) \tag{3.97}$$

where N is the number of lattice cells in the volume V (i.e. the periodicity volume). However, in general $\boldsymbol{\xi}$ can only be determined explicitly numerically. Tabulated Green functions for this problem have been given by Lie and Koehler (1968) and by Tewary and Bullough (1971). In the special case of elastic isotropy, however, one does have a simple solution for vacancies or interstitials in sites of high symmetry for which $G_{ij} = G\delta_{ij}$; namely

$$\boldsymbol{\xi}_{\text{elas}} = \frac{G}{4\pi c_{11}} \frac{\mathbf{r}}{r^3}. \tag{3.98}$$

To obtain the complete description of region II, solutions such as (3.98) or the equivalent for the anisotropic case should be added to the polarization displacements already described.

It remains to specify G_{ij} in terms of the interionic forces. By considering the equation of equilibrium of a solid containing the defect and selecting the limiting solution by the technique of Fourier-transforming this equation and taking the limit $\mathbf{q} \to 0$ one may show that

$$G_{ij} = \sum_l F_i(\mathbf{R}_l + \boldsymbol{\xi}_l) R_{lj} \tag{3.99}$$

† Perfect elastic isotropy requires $2c_{44} = c_{11} - c_{12}$, cf. Table 2.1. (p. 47)

where $F_i(\mathbf{R}_l+\boldsymbol{\xi}_l)$ is the ith component of the force exerted by the defect (region) on the ion l at its displaced position $\mathbf{R}_l+\boldsymbol{\xi}_l$ (see e.g. Tewary, 1969). The summation is over all ions l of the perfect lattice. For (3.99) to apply, it is only necessary that $\Phi_{II}(\boldsymbol{\xi})$ is harmonic and it may thus often be a good enough approximation to take the defect coordinates $\mathbf{x}$ as simply those for the defect, e.g. interstitial or vacancy itself; it is *not* necessary to take $\mathbf{x}$ as the whole Mott-Littleton region I. When this approximation is adequate then (3.99) simplifies considerably. Further simplifications follow for a vacancy resulting from the condition of stability of the lattice (3.74). These are described and applied in several papers; for monatomic lattices see e.g. Kanzaki (1957) and Hardy (1968), and for alkali halide lattices see Faux and Lidiard (1971). The extension of these formulae to fluorite structures is straightforward but we shall not enter into these details here, since no detailed applications to this class have yet been made.

In principle, therefore, we now have a complete description of the displacements and electronic moments in region II as the sum of (i) Mott-Littleton terms induced by the net charge(s) on the defect and (ii) elastic terms determined by the G_{ij} tensor. It will be seen from (3.99) that G_{ij} actually depends on the (total) displacements $\boldsymbol{\xi}_l$ and thus in a detailed calculation it will have to be determined self-consistently. In practice this is not a severe complication since it is generally sufficient to use (3.99) expanded to first order in $\boldsymbol{\xi}_l$ so self-consistency is a linear condition. For charged defects it seems to make little difference whether $\boldsymbol{\xi}_l$ or only $\boldsymbol{\xi}_{\mathrm{elec},l}$ is used in (3.99). As far as the energies of defects in the alkali halides are concerned, it probably makes little difference whether $\boldsymbol{\xi}_{\mathrm{elas}}$ is included or omitted altogether as long as region I is large enough. However, in materials containing divalent ions or ions with even higher charges this may not be true since the forces which are being balanced are so much larger. The point requires further investigation.

Numerical methods. The above approach gives the equilibrium displacements in Region II. The equilibrium displacements in Region I can only be found by minimizing Φ with respect to $\mathbf{x}$ numerically. In this field of minimization, there is a choice of several well-tried methods and for an up-to-date introduction to the extensive literature on this topic we refer the reader to a recent review by Fletcher (1970). These methods divide into three main groups (i) those which evaluate only the function itself (search procedures), (ii) those which also evaluate the first derivatives of the function (e.g. steepest descents), and (iii) those which evaluate second derivatives as well (variable metric or modified Newton–Raphson methods). We shall comment briefly on the use of these methods in defect calculations.

In the simplest search procedure, one seeks the minimum of the energy function by a cycle of linear searches in the variables of the problem, $x_1,\ldots,x_n$. This is equivalent to the successive relaxation of equivalent shells of atoms or ions as applied to point defect studies by a number of authors. However the convergence of this method is known to be bad, although computation times can be reduced ($\sim\frac{1}{2}n$ times) by not evaluating the total energy function at each step but only that part involving the shell of ions being relaxed. Various improvements of this direct search procedure have therefore been developed, of which the most widely used is probably that of Powell (1964). The general idea here is that of a linear transformation to new variables at the end of each cycle, which, by incorporating the results of that cycle, defines new, 'less strongly coupled' variables. This 'conjugate direction' method retains the advantage of not requiring derivatives, but converges satisfactorily. Even so, the required number of evaluations of the energy increases as the square of the number of variables, n, and for long-range forces a reasonable practical limit may be reached with $\sim$10–20 variables. These search procedures are convenient for small computers as they do not require large storage.

Methods which utilize derivatives of the energy function, i.e. the forces on the ions, in general converge faster than search methods, the number of iterations being reduced by a factor $\sim n$. The calculation of the function derivatives at any point specifies a local optimum direction of search. The choice of a suitable step length in this direction may be made according to several criteria. Thus a linear search may be made to minimize the energy and the various shells of ions may be relaxed independently or collectively; then the method is simply the application of the method of steepest descent. It is known, however, that the convergence of the steepest-descent method is poor and it is also well established in general that it is possible to gain substantially improved convergences within the limitations of this class of methods by using the conjugate-gradient method of Fletcher and Reeves (1964). This method has advantages in circumstances where a large store is not available. It has not yet been much applied in the theory of defects; but one example is provided by the study of dislocation core structure by Norgett, Perrin, and Savino (1972).

Lastly, we turn to methods derived from the Newton–Raphson method. These incorporate information on the first and second derivatives at each iteration and converge particularly rapidly. The basic equation relating the ith iteration to the $(i+1)$th is

$$\mathbf{x}^{(i+1)} = \mathbf{x}^{(i)} - \mathbf{H}^{(i)}\mathbf{g}^{(i)} \tag{3.100}$$

where $\mathbf{g}$ is the vector of derivatives of the energy and $\mathbf{H}$, the "Hessian" matrix, is the inverse of the matrix of second derivatives $\mathbf{W}$. A straightforward application of (3.100) is, however, ruled out since the calculation and inversion of the matrix of second derivatives at each iteration is far too time consuming. Fortunately this is not necessary. General methods designed to refine $\mathbf{H}$ during iteration from any well-conditioned approximation to the Hessian have been developed. These even avoid the initial calculation of the second derivatives (Fletcher and Powell, 1963). However, in a crystal lattice with pairwise interactions, the initial evaluation of $\mathbf{H}$ is not difficult, since each second derivative concerns only one pair of atoms or ions. A convenient adaptation of the method specially suited to defect calculations has been devised by Norgett and Fletcher (1970). It has been applied successfully to defects in fluorite structures but is capable of much wider application. The method has the important advantage that the number of iterations required to converge is only slowly dependent on the number of variables although, of course, since forces as well as energy must be calculated, each iteration takes somewhat longer (typically about two times) than in a search method which evaluates the energy alone. Comparison with search methods is thus very favourable (e.g. 10–20 times faster for 12 variables). The penalty is that considerable storage for the large Hessian matrix is required, so that the method is not suitable for small computers. Given that this storage requirement can be met these variable-metric methods are speedy and especially suitable for large numbers of variables and for models involving long-range forces, i.e. for ionic crystals.

Previously, with the slower search methods the large amount of computing necessary has meant that it has been possible to study only a limited range of simple symmetric defects and furthermore, that only a rather small region I could be used. This led to uncertainties about the accuracy of the Mott-Littleton approximation since this could not be tested by varying the size of Region I at all widely. Also it has been usual to regard each type of defect system as an isolated problem and to program it accordingly. This phase of theory is now ending. The application of these fast minimization methods means that it is now feasible to construct general programs that will calculate characteristic defect energies and configurations for a wide range of cases. The first general program based on these fast methods (although not the first general defect program) is one due to Norgett called HADES.† This is a comprehensive modular program for ionic crystals which will evaluate the relaxation about an arbitrary defect configuration made up of vacancies, interstitials and substitutional atoms or ions. The program allows a choice of polarizable-point-ion or shell models and is designed to require the specification of only essential crystal data (lattice and defect type, physical constants, size of region I etc.). This HADES program shows very clearly what is now possible in this field, namely the accurate calculation of the energies and configurations of defects in specified models; any uncertainties about the validity of such calculated results then repose clearly in the physical model employed.

3.5.2. *Particular calculations on point defects in fluorite lattices*

To date there have been relatively few calculations of the energies or other properties of point defects in fluorite structures. The most extensive work has been by Franklin (1967*a*,*b*, 1968), Norgett (1971*a*,*b*) Catlow (1973*a*,*b* and with Norgett, 1973) and Keeton and Wilson (1973). Calculations by Chakravorty (1971) and by Tharmalingam (1971) should also be noted. We discuss these various calculations under headings: (i) basic intrinsic defects, (ii) activation energies for motion, (iii) foreign ions, (iv) rare-gas atoms and (v) complex defects.

Intrinsic defects. The fundamental question is, 'what are the intrinsic, thermally-produced point defects in these materials or, in other words, which defects among the several possibilities have the lowest (effective) formation energies?' The results of the relevant calculations are summarized in Table 3.9. We make the following observations.

(i) All the calculations agree that in the alkaline earth fluorides anion Frenkel defects are the dominant intrinsic defects, i.e.

$$\tfrac{1}{2}h_{\mathrm{Fa}} < \tfrac{1}{3}h_{\mathrm{S}} < \tfrac{1}{2}h_{\mathrm{Fc}}. \tag{3.101}$$

The reasons for this appear to be (a) that the relatively large space at the centre of the empty F^- cubes allows the interstitial F^- ion to have a net negative energy there due to the polarization of the lattice which it

† Harwell Defect Evaluation System. This is based on the Norgett and Fletcher (1970) variable-metric method.

causes, (b) while this is also true for cation interstitials, in that case the double charge increases the magnitude of both the cation vacancy formation energy (positive) and the cation interstitial formation energy (negative) and thus of the Frenkel formation energy (positive) relative to the corresponding quantities for the anions.

(ii) The calculations listed in Table 3.9 have, except for those of Catlow and Norgett (1973), all been made with polarizable-point-ion models. To the extent that these over-estimate the actual static dielectric constant ϵ_0 these values will be somewhat too low. With such

TABLE 3.9

Calculated (internal) energies of formation of Frenkel and Schottky defects.

Substance	Energies of formation of defects (eV)			References
	Anion Frenkel	Schottky	Cation Frenkel	
CaF_2	1·7–2·8	4·3–8·6	6·0–9·2	a, b, c, d, e, f
SrF_2	1·9–2·5	4·4–8·1	7·0–8·6	a, e, g
BaF_2	1·6–2·3	4·8–6·9	7·5–8·8	a, e, f, g
UO_2	4·9–5·5	—	—	c

It is not possible to discuss each of the calculations in detail, but we observe that the most sophisticated calculations both in terms of mathematical techniques (HADES) and of physical model (shell models) are those of Catlow and Norgett (1973). The ranges of values given, however, correspond to all 'reasonable' values given by other authors as well, i.e. we have only excluded values which are clearly inapplicable for reasons which are understood (see section 3.5.1). Generally the results of Catlow and Norgett lie at the upper end of the ranges given.

[a] Franklin (1967*a*); [b] Franklin (1968); [c] Tharmalingam (1971); [d] Norgett (1971*a*); [e] Catlow and Norgett (1973); [f] Keeton and Wilson (1973); [g] Norgett (1971*b*).

models, as we have previously noted, the error increases as the inner region I is enlarged since the correct empirical value of ϵ_0 is generally assumed in the Mott–Littleton expressions used to describe region II. Thus the lowest values are those of Norgett (1971*a*,*b*) and of Keeton and Wilson (1973) all of whom included many shells in region I while the values of Franklin (1967*a*, 1968), who used only 2 shells, are higher and in better agreement with the experimental values (Table 3.2). The most reliable calculations should be those of Catlow and Norgett (1973); these gave formation energies of anion Frenkel defects of 2·6–2·7 eV for CaF_2, 2·2–2·4 eV for SrF_2 and 1·6–1·9 eV for BaF_2, which compare well with the values in Table (3.2). The calculations used three different shell models which were evaluated by means of the HADES program.

(iii) The calculations assumed that the interstitial position of lowest energy was at the centre of the empty anion cube. In view of the results from the neutron diffraction studies (§ 3.1), and the fact that the Madelung electrical potential at the cube-centre position is actually negative (i.e. repulsive for a negative ion), it is important to test this assumption. Catlow made an extensive test on a CaF_2 model using the HADES program and found that the cube-centre position was definitely a minimum with respect to displacements in all three crystal directions $\langle 100 \rangle$, $\langle 110 \rangle$, $\langle 111 \rangle$. There is no tendency to form dumb-bell interstitials, as happens in some other systems. This conclusion is likely to follow for other substances as well although it does depend on the nature of the anion-anion interaction; thus in the oxides the O^{2-}–O^{2-} interaction appears to be such that the stable interstitial configuration is slightly off-site (Catlow, 1973*b*).

Activation energies. Published results are summarized in Table 3.10.

TABLE 3.10

Calculated activation energies for migration of free anion vacancies and interstitials.

Substance	Activation energy		Reference
	Anion vacancy	Anion interstitial	
CaF_2	0·15–0·64	0·69–1·56	a, b, c
SrF_2	0·30	0·69	b
BaF_2	0·37–0·4	0·68	b, c

The values for the interstitials are for migration via the interstitialcy mechanism; direct interstitial migration requires higher activation energies, more than twice as large.

[a] Chakravorty (1971); [b] Catlow and Norgett (1973); [c] Keeton and Wilson (1973).

The first calculations of F^- vacancy and interstitial activation energies in CaF_2 were by Chakravorty (1971). The potential models used were those of Franklin (1968: his Set 3) and of Axe (1965), both embodied in the polarizable-point-ion model. Although these calculations were only approximate mathematically, their interest lay in two qualitative features in the results; namely (i) that the F^- vacancy activation energy was less than that of the F^- interstitial and (ii) that the activation energy for the interstitialcy or indirect interstitial mechanism was considerably lower than that for direct $\langle 110 \rangle$ migration from one interstitial site to another. These two qualitative results are preserved in the more extensive shell-model calculations of Catlow and Norgett

(1973). This work, which included a very careful search for the saddle-point configuration, shows that the F^- interstitialcy activation energy is 0·7 eV in all three alkaline earth fluorides. This value appears quite insensitive to the model since it is also obtained when the potentials used by Chakravorty are evaluated with the HADES program and values very close to it (0·6–0·8 eV) are also obtained in the extensive polarizable-point-ion calculations of Keeton and Wilson (1973). By contrast direct ⟨110⟩ interstitial migration would require above 3 eV.

Rather surprisingly the vacancy migration energies given by these more extensive calculations (both shell models and polarizable-point models) are too low, ranging from 0·4 eV for BaF_2 down to only 0·2 eV for CaF_2. It has been confirmed by both Catlow and Norgett (1973) and Keeton and Wilson (1973) that the higher activation energy obtained by Chakravorty for vacancy motion is a consequence of his restricted relaxation.

Foreign ions. Since many studies of defects in ionic crystals depend upon the incorporation of foreign ions ('dopants') to change intrinsic defect concentrations in controlled ways, it is obvious that theory could be useful in predicting the nature of the charge compensation which occurs. For example, while there is considerable empirical evidence that doping CaF_2 with YF_3 and other trivalent rare-earth fluorides introduces F^- interstitials, it has generally been assumed that doping with NaF leads to F^- vacancies, even though the empirical evidence for this is far less definite. The only extensive examination of this question is that by Franklin (1967*b*) who calculated the energies of solution of YF_3 and NaF in CaF_2. These calculations bring one immediately up against uncertainties over the interactions between the foreign ions and those of the host lattice, where one has generally far less information. For example, in the absence of an analysis of YF_3 itself, there is no reliable way of estimating the important Y^{3+}–F^- interactions in CaF_2/YF_3 solid solutions. At present one is obliged to fall back on the concept of ionic radii. In fact, in this example these indicate that Y^{3+} is very like Ca^{2+} apart from its extra charge; Franklin (1967*b*) therefore assumed that the Born–Mayer overlap interactions between Y^{3+} and other ions were identical to those between Ca^{2+} and these ions. Similar ideas were used to guide the choice of the interactions of Na^+ ions with their neighbours, although for the Na^+–F^- interactions one has an analysis of NaF itself. The calculations of the energies of solid solution of NaF and YF_3 were carried out by the same method as Franklin used for the intrinsic defects; namely region I included the first two shells of neighbours around each defect, while region II was described by a Mott–Littleton approximation for the electrical polarization and displacements supplemented by an isotropic elastic displacement field (eqn 3.98), the strength of which was fixed by requiring continuity with the second-neighbour displacements. However, as we have previously noted for the intrinsic defects, the inclusion of this elastic term may not make much difference to the calculated energies. All the calculations were carried out with polarizable-point-ion

TABLE 3.11

Calculated (internal) energies (eV) *of solution of* NaF *and* YF_3 *in* CaF_2

NaF		
	Na^+ substitutional, Na^+ interstitial	1·3
	Na^+ substitutional, F^- vacancy	1·5
	Na^+ substitutional, Ca^{2+} interstitial	2·7
	Na^+ interstitial, F^- interstitial	3·7
YF_3		
	Y^{3+} substitutional, F^- interstitial	~few × 0.1

For both it is assumed that the solution is in equilibrium with pure solute in its own intrinsic form. For NaF different modes of solution are included (see text). (After Franklin, 1967*b*.)

models but were repeated for the different choices of Born–Mayer potentials also used in Franklin's other calculations. Possibly the best results are those in Table 3.11 which gives the calculated energies of solid solution in CaF_2 of YF_3 and NaF relative to their normal solid phases.† It will be seen from these results that the energy of solution of YF_3 is expected to be very small, in qualitative agreement with the observed high solubilities found experimentally.

It will also be seen that the lowest energy of solution of NaF is obtained if we assume that this occurs with substitution of Na^+ ions on cation sites accompanied by either interstitial Na^+ ions or F^- vacancies; the predicted difference in energy between these two possibilities is too small to be relied upon. It would appear possible from the closeness of the results for these two cases that the preferred mode of solution might change in going from one system to another, e.g. from CaF_2/NaF

† The possibility of the formation of intermediate solid phases should be kept in mind when considering the precipitation of solutes at low temperatures. Thus in alkali halides doped with divalent cations intermediate, but metastable, Suzuki phases, e.g. $NaCl{\cdot}6CdCl_2$, may precipitate first.

to BaF_2/KF, or might be influenced by allowance for pairing between complementary defects. However, it should be recalled that the energies appearing in Table 3.11 are those appearing in the reaction constant for the *solubility product* for the defects in question considered in thermodynamic equilibrium with the solute phase, i.e. with precipitates. This solubility product should be considered as being satisfied simultaneously with those governing Frenkel and Schottky disorder and any other important defect reactions. As a result there are really only two independent reactions (not four), which we can take as (i) the transfer of a molecule of NaF from the NaF phase and its incorporation in the CaF_2 phase in any *one* of the four possible ways listed in Table 3.11, and (ii) the reaction for the partition of the Na^+ ions between substitutional and interstitial sites, i.e.

$$Na_i^+ + Ca_v^{2+} \rightleftharpoons Na_s^+, \tag{3.102}$$

a sort of impurity-Frenkel reaction. This latter reaction must be considered also when no NaF precipitate phase is present. The equilibrium equation corresponding to (3.102) is

$$\frac{[Na_i^+][Ca_v^{2+}]}{[Na_s^+]} = K_N \equiv K_{NO} \exp(-h_N/kT), \tag{3.103}$$

while the values in Table 3.11 give $h_N = 4{\cdot}7$ eV. Although this energy is large, it does not necessarily follow that the fraction of Na^+ interstitials is negligible, since $[Ca_v^{2+}]$ may be very small. This is in fact true at low concentrations of Na^+ ions when these are almost all incorporated substitutionally and charge compensation is effected by formation of F^- vacancies. Setting $[F_v^-] \simeq [Na_s^+]$ (negligible intrinsic disorder) and using the equation (3.15) of Schottky disorder, we obtain from (3.103)

$$[Na_i^+] = \frac{K_N}{K_S}[Na_s^+]^3 \tag{3.104}$$

and, by Franklin's results,

$$\frac{K_N}{K_S} \sim \exp\left(+\frac{1{\cdot}1\, eV}{kT}\right), \tag{3.105}$$

apart from entropy factors. At 1100 K, K_N/K_S is thus $\sim 10^5$ and therefore for Na^+ fractions of only 10^{-3} $[Na_i^+]$ is already 10 per cent of the total Na^+ fraction. Beyond this, say, when the Na^+ fraction is 10^{-2}, we expect the Na^+ interstitials to be dominant. Thus even when no precipitate phase is present, the interpretation of conductivity and ionic transport measurements on systems such as CaF_2/NaF may be more complex than is often supposed. These complications do not arise in this way in other systems. Thus in the alkali halides the energies of interstitial ions are probably sufficiently high that such complications due to partition of doping ions between substitutional and interstitial sites can be safely ignored. However, in the silver halides AgBr and AgCl which show predominantly cation Frenkel disorder, it is found that Cu^+ ions enter interstitial sites more readily than Ag^+ ions do, resulting in the addition of Ag^+ vacancies (Süptitz and Teltow, 1967).

We have gone into this detail in order to bring out the value of the calculations on foreign ions in suggesting the models and conditions likely to be important in analysing experimental data. This value lies in showing these likely possibilities rather than in the exact reproduction of energies already inferred experimentally.

Rare-gas atoms. While the equilibrium solubilities of the rare gases in solids generally are very low, and the fluorite compounds are no exception, these gases are nevertheless produced in significant quantities under fast neutron irradiation and they exert important effects on the subsequent behaviour of the irradiated solid, particularly at high temperatures. In the alkaline earth fluorides the most important reactions are the (n, α) reactions with the alkaline earths. Norgett (1971*a*,*b*) and later Keeton and Wilson (1973) calculated the activation energies of rare-gas interstitials as well as the energies of trapping into anion and cation vacancies (cf. §§ 3.2.4 and 3.3.3 describing the diffusion of these gases).

As with Franklin's calculations of solid solubilities, a central question is the magnitude of the Born–Mayer overlap interactions between the gas atoms and the ions of the solvent matrix. Norgett estimated this in two ways; (i) by replacing the ions by their isoelectronic rare-gas equivalents and appealing to the Thomas–Fermi–Dirac calculations of Abrahamson (1964, 1969) on the interaction potentials of rare gases, (ii) by regarding the gas atom as an ion of zero charge and extrapolating from the empirical Born–Mayer potentials obtained for the alkali halides by Fumi and Tosi (1964) by using the geometric-mean rule. Keeton and Wilson (1973) calculated their gas-ion potentials by a semi-classical method due to Wedepohl (1967). These alternative potentials were embodied in polarizable-point-ion models, but subsequent repetition by Norgett of his calculations with shell models indicates relatively small change in the predicted characteristics. Norgett evaluated the various energies by an early version of the HADES Program; for

TABLE 3.12

Calculated activation energies for free migration of interstitial rare-gas atoms through fluorite lattices.

System	Absolute energy (eV)	Activation energy (eV)	Reference
CaF_2:He	0·5	1·3	a
CaF_2:Ne	1·6	1·5	a
CaF_2:Ar	3·0–3·6	1·4–2·5	a, b
SrF_2:Kr	3·8–4·0	2·2–2·4	c
BaF_2:Xe	3·6–4·6	2·6–3·0	c

Absolute energies of the interstitial atom in the lattice are also given (relative to 'a state of rest at infinity'). The values quoted correspond to different choices for the interaction between a gas atom and ions of the solids.
[a] Keeton and Wilson (1973); [b] Norgett (1971*a*); [c] Norgett (1971*b*).

those defects bearing a net charge, e.g. a gas atom in a vacancy, region II was described in the Mott-Littleton approximation but with neglect of elastic terms. This means that for the neutral defects, e.g. a rare-gas interstitial, region II was taken to be rigid and unpolarized. Keeton and Wilson (1973) made the same assumption for *all* their defects, charged and uncharged, but then attempted to extrapolate to an infinite region I by assuming a variation linear in R_{I}^{-1} for charged defects and R_{I}^{-2} for neutral defects. The first of these choices follows from the expression for the polarization energy of a dielectric continuum and the results obtained this way for the charged defects agree well with those obtained by others using the Mott-Littleton method. On the other hand the elastic energy stored in an isotropic elastic continuum subject to the distortion (3.98) falls off as R_{I}^{-3} so that the values given by Keeton and Wilson (1973) for neutral defects may be too low. The results are given in Tables 3.12 and 3.13.

TABLE 3.13

Effective activation energies, Q, for diffusion of rare-gas atoms through pure fluorite lattices containing thermal equilibrium concentrations of Frenkel and Schottky defects according to Norgett (1971*a*,*b*).

System	Effective activation energy, Q (eV)
CaF_2:Ar	2·5–2·5
SrF_2:Kr	3·0–3·1
BaF_2:Xe	3·3–3·7

These activation energies Q allow for the trapping and detrapping of gas atoms in vacancies. Trapping into cation vacancies is important even though they are minority defects. The models and methods used are as in Tables 3.12 and 3.4.

Several features should be noted. Firstly, the absolute energies of the interstitial gas-atoms are high (~3–4 eV) and their equilibrium thermodynamic solubilities are thus extremely low. This is still true when one allows also for their trapping into vacancies; the trapping energies are only 1–2 eV while one must add the effective formation energies of the vacancies, so that the net energies of solution remain much the same as for the free interstitial atoms. Secondly, it should be noted that the calculated activation energies for free interstitial migration are high, like the experimental results, and not small (~few tenths of an eV) as they are in the alkali halides. Thirdly, when one takes account of the trapping of gas atoms into thermally produced (intrinsic) anion and cation vacancies, one obtains the effective activation energies, Q, for gas diffusion given in Table 3.13. These are in reasonably good agreement with experiment (Felix and Lagerwall, 1971 and Schmeling, 1967). As these results are undoubtedly sensitive to the gas-ion potential this is very encouraging. These calculations thus not only give a good description of an interesting class of impurities and thereby confirm the interpretations of the experiments, but they also explain the contrasting behaviour of the alkali halide systems (cf. Norgett and Lidiard, 1968).

Complex defects. In view of the high concentrations of defects which can be introduced into fluorite lattices there is no doubt about the importance of defect interactions and the likelihood of defect clustering. These interactions can now be studied quantitatively by means of programs such as HADES. Catlow (1973*a*) has studied models of the alkaline earth fluorides containing Y^{3+} and other trivalent cations using this program. Firstly he has verified the $\langle 100 \rangle$ tetragonal symmetry of the Y^{3+}—F_i^- nearest-neighbour pairs as required by many experimental results (§ 3.4.2 and chapter 6). Of much greater interest are his results on the interactions of two such pairs. If these are brought together in a (100)-plane as shown in Fig. 3.13 and the lattice is constrained to retain this symmetry on relaxation (ions A in particular being allowed to relax only normal to the (100) plane) then the energy of interaction is relatively small, being practically zero in $CaF_2{:}Y^{3+}$ and about $\frac{1}{2}$ eV in $SrF_2{:}Y^{3+}$ and $BaF_2{:}Y^{3+}$. However if this symmetry constraint is relaxed a further considerable energy lowering follows; the two interstitials pull together along the $\langle 110 \rangle$ direction shown (primarily as a result of their attraction to the trivalent cations) while the normal anions marked (A) above and beneath the (100)-plane are pushed away in directions close to $\langle 111 \rangle$ towards the vacant cube-centre positions lying above and below the trivalent ions. The other anions

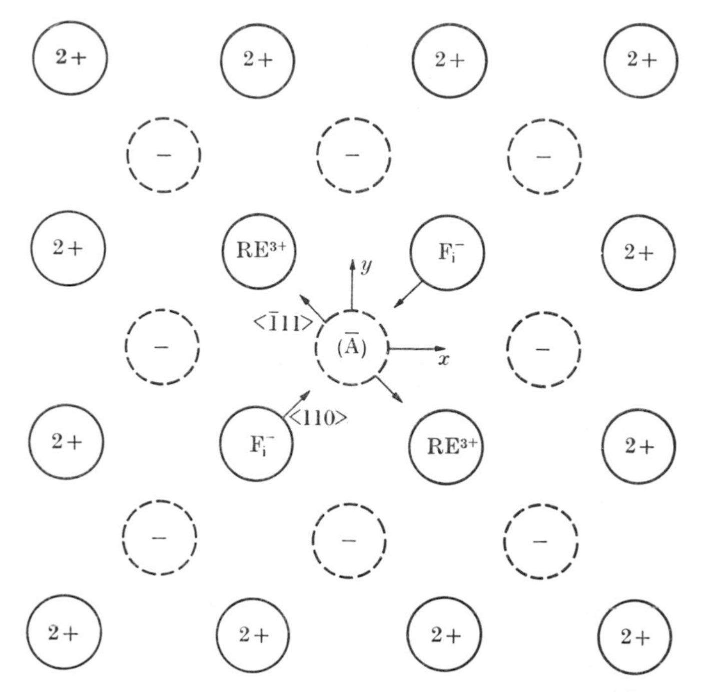

FIG. 3.13. A (100) section of a fluorite lattice showing the principal relaxations arising when two RE^{3+}–F_i^- pairs are brought together in the configuration shown. Ions initially lying in a (100) cation plane are shown as full circles while the anions on neighbouring planes are shown as broken circles. The origin of the coordinates given in Table (3.14) lies midway between the two RE^{3+} sites with the z-axis rising out of the figure. The two anions A above and beneath the cation plane relax towards the centres of the empty anion cubes above and below the two RE^{3+} ions; there are thus two equivalent configurations for the relaxed cluster.

are displaced to a very much smaller extent. The calculated displacements of these anions from normal sites are given in Table 3.14; it will be seen that they are remarkably close to those inferred from the neutron diffraction experiments of Cheetham *et al.* (1971). Possibly they would appear even closer if the neutron diffraction analysis were refined by allowing for the smaller displacements of the other anions. The binding energy of two pairs into this cluster was calculated to be 1·2 eV in CaF_2:Y^{3+} and BaF_2:Y^{3+} and 1·6 eV in SrF_2:Y^{3+}. Their concentration will thus be appreciable even in quite lightly doped crystals at relatively high temperatures. There can thus be little doubt that these calculations provide the explanation for the structures inferred from the neutron diffraction experiments on CaF_2:YF_3 and on UO_{2+x} (§ 3.1) and in so doing show that they are fully consistent with the highly-symmetric interstitial configurations found in dilute systems

TABLE 3.14

Relaxed positions of the ions of the defect cluster formed by bringing together two Y^{3+}–F_i^- nearest-neighbour pairs in CaF_2 as shown in Fig. 3.14.

Ion	Position	
	Calculated	Experimental
⟨110⟩ Interstitials	(0, 0·32, 0·32) and (0, −0·32, −0·32)	(0, 0·26, 0·26) and (0, −0·26, −0·26)
⟨111⟩ Interstitials	(−0·26, 0·26, 0·75) and (0·26, −0·26, −0·75)	(−0·34, 0·34, 0·84) and (0·34, −0·34, −0·84)
Y^{3+} Ions	(−0·47, 0·47, 0·05) and (0·47, −0·47, −0·05)	—

The other anions neighbouring the Y^{3+} ions are displaced to a much smaller extent than the anions marked A in this figure ($\sim$5% of the normal F^-—F^- separation). In the experimental determination of ion positions these and the displacements of the Y^{3+} are neglected (Cheetham *et al.*, 1971). The coordinates are given in units of the normal F^-—F^- nearest-neighbour distance. (After Catlow, 1973a.)

by spectroscopic, resonance and relaxation methods. They provide one of the clearest illustrations of the value of these defect calculations.

Acknowledgements. I would like to acknowledge the benefit obtained from discussions and correspondence on the content of this chapter with Drs. C. R. A. Catlow, A. D. Franklin, M. J. Norgett, and J. H. Strange and Professor A. Kessler.

Bibliography

For comprehensive books on point defects and atomic and ionic transport processes we refer the reader to

(i) Y. Adda and J. Philibert, *La Diffusion dans les solides* (Presses Universitaires de France, Paris 1966).

(ii) C. P. Flynn, *Point defects and diffusion* (Clarendon Press, Oxford, 1972).

(iii) L. Slifkin and J. W. Crawford, (eds.) *Point defects in solids*, Vol. 1 (Plenum Press, New York 1972).

For a detailed and general treatment of the kinetic theory of diffusion in crystals see,

(iv) J. R. Manning, *Diffusion kinetics for atoms in crystals* (Van Nostrand, Princeton 1968).

References

ABRAGAM, A. (1961). *The principles of nuclear magnetism*, Clarendon Press, Oxford, p. 291.

ABRAHAMSON, A. A. (1964). *Phys. Rev.* **133**, A990.

—— (1969). *Ibid* **178**, 76.

ADDA, Y. and PHILIBERT, J. (1966). *La diffusion dans les solides*, Presses Universitaires de France, Paris.

AILION, D. C. and SLICHTER, C. P. (1965). *Phys. Rev.* **137,** A235.
ALLEN, C. A., IRELAND, D. T., and FREDERICKS, W. J. (1967). *J. chem. Phys.* **47,** 3068.
ALLNATT, A. R. (1965). *Ibid* **43,** 1855.
—— (1967). *Adv. Chem. Phys.* **11,** 1.
—— and LOFTUS, E. (1973). *J. chem. Phys.* **59,** 2541 and 2550.
ANDERSON J. S. (1946). *Proc. R. Soc.* **A185,** 69.
—— (1970). *The Chemistry of extended defects in non-metallic oxides* (Eds L. Eyring and M. O'Keefe), North-Holland, Amsterdam, p. 1.
ARONSON, S., RULLI, J. E., and SCHANER, B. E. (1961). *J. chem. Phys.* **35,** 1382.
ASHBEE, K. H. G. and FRANK, F. C. (1970). *Phil. Mag.* **21,** 211.
AXE, J. D. (1965). *Phys. Rev.* **139,** A1215.
BAKER, J. M., DAVIES, E. R., and HURRELL, J. P. (1968). *Proc. R. Soc.* **A308,** 403.
BAKER, M. and TAYLOR, A. (1969). *J. Phys. Chem. Solids* **30,** 1003.
BARR, L. W. and LIDIARD, A. B. (1970). *Physical chemistry—an advanced treatise*, Vol. X, p. 151. Academic Press, New York.
BARSIS, E. and TAYLOR, A. (1966). *J. chem. Phys.* **45,** 1154.
—— —— (1968*a*). *Ibid* **48,** 4357.
—— —— (1968*b*). *Ibid* **48,** 4362.
BELLE, J. (1969). *J. nucl. Mat.* **30,** 3.
BOLLMAN, W., GÖRLICH, P., HAUK, W., and MOTHES, H. (1970). *Phys. Stat. Sol.* **2,** 157.
BONTINCK, W. (1958). *Physica* **24,** 650.
BOSWARVA, I. M. and FRANKLIN, A. D. (1965). *Phil. Mag.* **11,** 335.
BROWN, L. M. (1961). *Phys. Stat. Sol.* **1,** 585.
BUCCI, C., FIESCHI, R., and GUIDI, G. (1966). *Phys. Rev.* **148,** 816.
BURTON, C. H. and DRYDEN, J. S. (1970). *J. Phys. C.* **3,** 523.
CATALANO, E., BEDFORD, R. G., SILVEIRA, V. G. and WICKMAN, H. H. (1969). *J. Phys. Chem. Solids* **30,** 1613.
CATLOW, C. R. A. (1973*a*). *J. Phys. C,* **6,** L64.
——(1973*b*). D. Phil thesis University of Oxford.
—— and HAYNS, M. R. (1972). *J. Phys. C,* **5,** L237.
—— and NORGETT, M. J. (1973). *J. Phys. C.* **6,** 1325.
CHAKRAVORTY, D. (1971). *J. Phys. Chem. Solids* **32,** 1091.
CHEETHAM, A. K., FENDER, B. E. F., and COOPER, M. J. (1971). *J. Phys. C.* **4,** 3107.
CHEN, J. H. and MCDONOUGH, M. S. (1969). *Phys. Rev.* **185,** 453.
COMPAAN, K. and HAVEN, Y. (1956). *Trans. Faraday Soc.* **52,** 786.
—— —— (1957). *Ibid* **54,** 1498.
CORISH, J. and JACOBS, P. W. M. (1973). *J. Phys. C,* **6,** 57.
CROATTO, U. and BRUNO, M. (1946). *Gazz. Chim. Ital.* **76,** 246.
—— and MAYER, A. (1943). *Ibid* **73,** 199.
DANIEL, V. (1967). *Dielectric relaxation.* Academic Press, New York.
DE CONINCK, R. and DEVREESE, J. (1969). *Phys. Stat. Sol.* **32,** 823.
DE VRIES, K. J. (1965). Ionic conductivity and non-stoichiometry in $PbCl_2$. Thesis, Utrecht.
DREYFUS, R. W. (1961). *Phys. Rev.* **121,** 1675.
DWORKIN, A. S. and BREDIG, M. A. (1968). *J. phys. Chem.* **72,** 1277.
EISENSTADT, M. (1963). *Phys. Rev.* **132,** 630.
—— (1964). *Ibid* **133,** A 191.
—— and REDFIELD, A. G. (1963). *Ibid* **132,** 635.

ESHELBY, J. D., NEWEY, C. W. A., PRATT, P. L., and LIDIARD, A. B. (1958). *Phil. Mag.* **3,** 75.
ETSELL, T. H. and FLENGAS, S. N. (1970) *Chem. Rev.* **70,** 339.
EVANS, A. G. and PRATT, P. L. (1969). *Ibid* **20,** 1213.
FAUX, I. D. (1971). *J. Phys. C,* **4,** L211.
—— and LIDIARD, A. B. (1971). *Z. Naturforsch.* **26a,** 62.
FELIX, F. and LAGERWALL, S. Y. T. (1971). *Phys. Stat. Sol.* (a) **4,** 73.
FLETCHER, R. (1972). *Comp. Phys. Commun.* **3,** 159.
—— and POWELL, M. J. D. (1963). *Comput. J.* **6,** 163.
——and REEVES. C. M. (1964). *Ibid* **7,** 149.
FLYNN, C. P. (1972). *Point defects and diffusion* (Oxford University Press).
—— and STONEHAM, A. M. (1970). *Phys. Rev.* **B1,** 3966.
FONG, F. K. (1966). *Prog. Solid State Chem.* **3,** 135.
FRANKLIN, A. D. (1967*a*). *Proc. Brit. Ceram. Soc.* No. 9, 15.
—— (1967*b*). *J. Amer. Ceram. Soc.* **50,** 648.
—— (1968). *J. Phys. Chem. Solids* **29,** 823.
—— (1972). In *Point defects in solids,* Vol. 1, p. 1. (Eds. J. H. Crawford and L. M. Slifkin). Plenum Press, New York.
—— and CRISSMAN, J. (1971). *J. Phys. C,* **4,** L239.
—— and MARZULLO, S. (1970*a*). *Ibid* **3,** L171.
—— —— (1970*b*). *Proc. Brit. Ceram. Soc.* No. 19, 135.
——, ——, and WATCHMAN, J. B. (1967). *National Bureau of Standards J. of Res.* **71A,** 355.
——, SHORB, A. and WATCHMAN, J. B. (1964). *Ibid* **68A,** 425.
FRENKEL, J. (1946). *Kinetic theory of liquids.* Clarendon Press, Oxford.
FRÖHLICH, H. (1958). *Theory of dielectrics,* 2nd edn. Clarendon Press, Oxford.
FUMI, F. G. and TOSI, M. P. (1964). *J. Phys. Chem. Solids* **25,** 31 and 45.
HARDY, J. R. (1960). *Ibid* **15,** 39.
—— (1962). *Ibid* **23,** 113.
—— (1968). *Ibid* **29,** 2009.
—— and FLOCKEN, J. W. (1970). *C R C Critical Reviews in Solid State Sciences* **1,** 605.
—— and LIDIARD, A. B. (1967). *Phil. Mag.* **15,** 825.
HOOD, G. M. and MORRISON, J. A. (1967). *J. appl. Phys.* **38,** 4796.
HOODLESS, I. M., STRANGE, J. H., and WYLDE, L. E. (1971). *J. Phys. C,* **4,** 2742.
HOWARD, R. E. and LIDIARD, A. B. (1964). *Rep. Prog. Phys.* **27,** 161.
IPPOLITOV, E. G., GARASHINA, L. S., and MAKLASKLOV, A. G. (1967). *Inorg. Mater.* **3,** 59. (Original Russian version *Izv. Akad Nauk SSSR, Neorganicheskie Materialy* **3,** 73).
JACOBS, P. W. M. and PANTELIS, P. (1971). *Phys. Rev.* **B4,** 3757.
JOHNSON, H. B., TOLAR, N. J., MILLER, G. R., and CUTLER, I. B. (1969). *J. Phys. Chem. Solids* **30,** 31.
JOST, W. (1933). *J. chem. Phys.* **1,** 466.
KANZAKI, H. (1957). *J. Phys. Chem. Solids* **2,** 24 and 37.
KARO, A. M. and HARDY, J. R. (1971). *Phys. Rev.* **B3,** 3418.
KEETON, S. C. and WILSON, W. D. (1973) *Ibid* **B7,** 834.
KEIG, G. A. and COBLE, R. L. (1968). *J. appl. Phys.* **39,** 6090.
KENESHEA, F. J. and FREDERICKS, W. J. (1963). *J. chem. Phys.* **38,** 1952.
KESSLER, A. and CAFFYN, J. E. (1972). *J. Phys. C,* **5,** 1134.
KINGSLEY, J. D. and PRENER. J. S. (1962). *Phys. Rev. Lett.* **8,** 315.
KITTEL, C. (1971) *Introduction to solid state physics,* 4th edn. (Wiley, New York) especially Chapter 19.
KITTS, E. L. (1973) Ph.D. thesis, University of North Carolina.
—— and CRAWFORD, J. H. (1973), *Phys. Rev. Lett.* **30,** 443.

KLIEWER, K. L. (1965). *Phys. Rev.* **140, A** 1251.
—— (1966). *J. Phys. Chem. Solids* **27,** 705 and 719.
—— and KOEHLER, J. S. (1965). *Phys. Rev.* **140, A** 1226.
KNOWLES, J. T. and MAHENDROO, P. P. (1970). *Phys. Lett.* **31A,** 385.
KOEHLER, J. S., LANGRETH, D. and VON TURKEVICH, B. (1962). *Phys. Rev.* **128,** 573.
KRÖGER, F. A. (1964). *Chemistry of imperfect crystals.* North-Holland, Amsterdam.
——(1966). *Z. phys. Chem.* **49,** S178.
KUNZE, I. and MÜLLER, P. (1972) *Phys. Stat. Sol.* (a) **13,** 197.
KVIST, A. (1972). *Physics of electrolytes* (ed. J. Hladik), Vol. 1, Academic Press, London and New York, p. 319.
LAGERWALL, T. (1962). *Nukleonik* **4,** 158.
LALLEMAND, P. (1971). Thesis, University of Paris.
LAY, K. W. (1970). *J. Amer. Ceram. Soc.* **53,** 369.
—— and WHITMORE, D. H. (1971). *Phys. Stat. Sol.* (b) **43,** 175.
LECLAIRE, A. D. (1970). *Physical chemistry—an advanced treatise,* Vol X, Academic Press, New York, p. 261.
LEHOVEC, K. (1953). *J. chem. Phys.* **21,** 1123.
LIDIARD, A. B. (1954). *Phys. Rev.* **94,** 29.
—— (1957). *Handbuch der Physik,* Vol. XX, p. 246. Springer-Verlag, Berlin.
—— (1971). *Theory of imperfect crystalline solids,* p. 339. International Atomic Energy Agency, Vienna.
—— and NORGETT, M. J. (1972). *Computational solid state physics* (eds. F. Herman, N. W. Dalton and T. R. Koehler), Plenum Press, New York, p. 385.
LIE, K. H. C. and KOEHLER, J. S. (1968). *Adv. Phys.* **17,** 421.
LOOK, D. C. and LOWE, I. J. (1966). *J. chem. Phys.* **44,** 2995.
LYSIAK, R. J. and MAHENDROO, P. P. (1966). *Ibid* **44,** 4025.
MACDONALD, R. A. (1972). *J. Phys.* **F2,** 209.
MANNING, J. R. (1968). *Diffusion kinetics for atoms in crystals.* Van Nostrand, Princeton.
MANNION, W. A., ALLEN, C. A. and FREDERICKS, W. J. (1968). *J. chem. Phys.* **48,** 1537.
MATZKE, H. (1969). *J. nucl. Mat.* **30,** 26.
—— (1970). *J. mat. Sci.* **5,** 831.
—— and LINDNER, R. (1964). *Z. Naturforsch.* **19a,** 1178.
MIEKELEY, W. and FELIX, F. W. (1972). *J. nucl. Mat.* **42,** 297.
MILIOTIS, D. and YOON, D. N. (1969). *J. Phys. Chem. Solids* **30,** 1241.
MILLER, J. R. and MAHENDROO, P. P. (1968). *Phys. Rev.* **174,** 369.
MOTT, N. F. and GURNEY, R. W. (1948). *Electronic processes in ionic crystals,* 2nd edn. Clarendon Press, Oxford.
—— and LITTLETON, M. J. (1938). *Trans. Faraday Soc.* **34,** 485.
MÜLLER, P. and TELTOW, J. (1972) *Phys. Stat. Sol.* (a) **12,** 471.
NABARRO, F. R. N. (1967). *Theory of crystal dislocations.* Clarendon Press, Oxford.
NINOMIYA, T. (1959). *J. Phys. Soc. Japan* **14,** 30.
NORGETT, M. J. (1971*a*). *J. Phys. C,* **4,** 298.
——— (1971*b*). *Ibid* **4,** 1284.
—— and FLETCHER, R. (1970). *Ibid* **3,** L190.
—— and LIDIARD, A. B. (1968). *Phil. Mag.* **18,** 1193.
——, PERRIN, R. C. and SAVINO, E. J. (1972). *J. Phys. F,* **2,** L73.
NOWICK, A. S. (1967). *Adv. Phys.* **16,** 1.
—— (1972). In *Point defects in solids* (eds. J. H. Crawford and L. M. Slifkin). Vol. 1, p. 151, Plenum Press, New York.
—— and HELLER, W. R. (1963). *Adv. Phys.* **12,** 251.
—— —— (1965). *Ibid* **14,** 101.

POWELL, M. J. D. (1964). *Comput. J.* **7,** 155.
QUIGLEY, R. J. and DAS, T. P. (1967). *Phys. Rev.* **164,** 1185.
RANON, U. and LOW, W. (1963) *Phys. Rev.* **132,** 1609.
REDDY, T. Rs., DAVIES, E. R., BAKER, J. M., CHAMBERS, D. M., NEWMAN, R. C., and ÖZBAY, B. (1971) *Phys. Lett.* **36A,** 231.
REIMANN, D. K. and LUNDY, T. S. (1969). *J. Amer. Ceram. Soc.* **52,** 511.
ROYCE, B. S. H. and MASCARENHAS, S. (1970). *Phys. Rev. Lett.* **24,** 98.
SCHMELING, P. (1967). *Phys. Stat. Sol.* **20,** 127.
SASHITAL S. R. and VEDAM, K. (1972). *J. appl Phys.*, **43,** 4396.
SCHULTZE, P. D. and HARDY, J. R. (1972*a*). *Phys. Rev.* **B5,** 3270.
—— —— (1972*b*). *Ibid* **B6,** 1580.
SHORT, J. and ROY, R. (1963). *J. Phys. Chem.* **67,** 1860.
SIERRO, J. (1961). *J. chem. Phys.* **34,** 2183.
SIMKOVICH, G. (1963). *J. Phys. Chem. Solids* **24,** 213.
SLIFKIN, L. M., MCGOWAN, W., FUKAI, A., and KIM, J. S. (1967). *Photographic Science and Engineering* **11,** 79.
SOUTHGATE, P. D. (1966). *J. Phys. Chem. Solids* **27,** 1623.
STEELE, B. C. H. (1972) in *Solid State Chemistry*. Vol. 10, p. 117. Butterworths, London.
STIEFBOLD, D. R. and HUGGINS, R. A. (1972) *J. Solid State Chem.* **5,** 15.
STOCKBARGER, D. C. (1949). *J. opt. Soc. Amer.* **39,** 731.
STONEHAM, A. M. and DURHAM, P., 1973, *J. Phys. Chem. Solids* **34,** 2127.
STOTT, J. P. and CRAWFORD, J. H. (1971*a*). *Phys. Rev. Lett.* **26,** 384.
—— —— (1971*b*). *Phys. Rev.* **B4,** 668.
SÜPTITZ, P., BRINK, E., and BECKER, D. (1972). *Phys. Stat. Sol.* (b) **54,** 713.
—— and TELTOW, J. (1967). *Ibid* **23,** 9.
SYMMONS, H. (1970). *J. Phys. C*, **3,** 1846.
TAN, Y. T. and KRAMP, D. (1970). *J. chem. Phys.* **53,** 3691.
TELTOW, J. (1949). *Ann. Phys. Lpz*, **5,** 71.
TEWARY, V. K. (1973). *AERE Report* T.P. 388. *Adv. Phys.* **22,** 757.
—— and BULLOUGH, R. (1971). *J. Phys. F*, **1,** 554.
THARMALINGAM, K. (1971). *Phil. Mag.* **23,** 181.
THORN, R. J. and WINSLOW, G. H. (1966). *J. chem. Phys.* **44,** 2632.
TIEN, T. Y. and SUBBARAO, E. C. (1963). *Ibid* **39,** 1041.
TORREY, H. (1953). *Phys. Rev.* **92,** 962.
—— (1954). *Ibid* **96,** 690.
TOSI, M. P. and DOYAMA, M. (1967) *Ibid* **160,** 716.
TUBANDT, C. (1932). *Handbuch der Experimental physik* **12** (1), 383.
TULLER, H. L. (1974) Ph.D. thesis, Columbia University (New York).
TWIDELL, J. W. (1970). *J. Phys. Chem. Solids* **31,** 299.
URE, R. W. (1957). *J. chem. Phys.* **26,** 1363.
VERWEY, J. F. and SCHOONMAN, J. (1967). *Physica* **35,** 386.
VINEYARD, G. (1957). *J. Phys. Chem. Solids* **3,** 121.
WACHTMAN, J. B. (1963). *Phys. Rev.* **131,** 517.
WAGNER, J. and MASCARENHAS, S. (1972) *Phys. Rev.* **B6,** 4867.
WATKINS, G. D. (1959). *Phys. Rev.* **113,** 79 and 91.
WEDEPOHL, P. T. (1967). *Proc. Phys. Soc.* **92,** 79.
WILLIS, B. T. M. (1963). *Proc. R. Soc.* **A274,** 122 and 134.
—— (1964). *Proc. Brit. Ceram. Soc.* **1,** 9.
—— (1965). *Acta Cryst.* **18,** 75.

4
COLOUR CENTRES

4.1. Introduction

MANY alkali halides provide an attractive medium for colour-centre research because of the ease with which they can be grown in reasonably pure single-crystal form and because of the ease with which they may be coloured by ionizing radiations. In the past most colour-centre research has been carried out in these materials (for reviews see Schulman and Compton, 1962; Compton and Rabin, 1964; Fowler, 1968; Crawford and Slifkin, 1972). Colour-centre work in other materials such as alkaline earth oxides (Henderson and Wertz, 1968; Hughes and Henderson, 1972) and alkaline earth fluorides† has expanded in recent years because of the increasing availability of good single crystals. This work has, of course, benefited considerably from concepts developed in alkali halide research. Some of the earliest investigations of colour centres in alkaline earth fluorides were carried out in Przibram's laboratory on naturally occurring crystals (Przibram, 1956). Przibram and his collaborators made a survey of both the natural and radiation-induced colours and fluorescences of fluorite minerals from various sources. The first detailed correlation between colour-centre bands in alkaline earth fluoride and alkali halide crystals was made by Mollwo (1934) who investigated the optical properties of additively-coloured natural CaF_2 crystals.‡ Mollwo found that additive coloration of CaF_2 in calcium vapour followed by rapid quenching produced two main absorption bands which he called the α and β bands (Fig. 4.6). [This notation is potentially confusing because the bands labelled α and β in alkali halides do not have a similar origin (see p. 104 of Schulman and Compton). We shall, however, retain Mollwo's notation here because of its general usage in the past.]

Colour-centre research in alkaline earth fluorides remained rather quiescent after Mollwo's (1934) pioneer work but was revived by the availability of furnace-grown single crystals. The earliest description of optical bands produced by X-irradiation of synthetic single crystals

† In this monograph alkaline earth fluorides are not taken to include BeF_2 and MgF_2. These materials do not have the fluorite structure.

‡ For a general discussion of additive coloration see e.g. Markham (1966, p. 58). Experimental techniques suitable for alkaline earth fluorides have been described by van Doorn (1961–62) and Phillips and Duncan (1971).

of CaF_2 was given by Smakula (1950) and further reports quickly followed (Barile, 1952; Smakula, 1953). Smakula (1950) found a four-band spectrum in his irradiated crystals unrelated to the α and β bands of Mollwo (1934) and this complicated the problems of interpretation. Lüty (1953) subsequently showed that the four-band spectrum could be produced by additive coloration at relatively low temperatures (*c.* 550 °C). However, if the crystals were quenched after additive coloration at higher temperatures (*c.* 700 °C) Mollwo's α and β bands were predominant. Lüty (1953) concluded that Smakula's four-band spectrum was associated with unidentified impurities always present in CaF_2. Subsequent investigations showed that the four-band spectrum was associated with the presence of yttrium impurity in the crystals. Further discussion of this spectrum will be given in Chapter 7.

It was quickly noticed that the colourability of undoped alkaline earth fluorides exposed to ionizing radiations at room temperature was very small by comparison with most alkali halides and that the purer the crystal the smaller was the colourability. Effects of heat and chemical treatment on the X-ray colourability were investigated by Schulman, Ginther, and Kirk (1952) and by Adler and Kveta (1957) who found that such treatments could give rise to marked increases in colourability. Bontinck (1958a) found that neutron bombardment of CaF_2 produced a spectrum different from Smakula's (1950) four-band spectrum and he observed an absorption band at 520 nm which he identified with Mollwo's (1934) β band. Bontinck's investigations led him to suggest that Mollwo's β band was due to an electron trapped in an anion vacancy, i.e. an F centre, and that the α band was due to an F centre near an interstitial fluorine ion. These suggestions did not coincide with Mollwo's (1934) view that the α band was due to an unperturbed F centre. This confused situation was briefly reviewed by Przibram (1959).

Further optical measurements of irradiation effects were reported by Scouler and Smakula (1960) who investigated effects of 2·5 meV electrons on CaF_2 doped with yttrium or sodium and by Messner and Smakula (1960) who investigated effects of 2·5 meV electrons and 150 kV X-rays on undoped CaF_2, SrF_2, and BaF_2. Descriptions of the optical properties of coloured alkaline earth fluoride crystals were given by Karras (1961*a*, *b*) and the general situation at this time was reviewed by Görlich, Karras and Lehmann (1961*a*, *b*). It was clear from this latter review that, although much experimental observation was available, there was no detailed understanding of the structure of colour centres in alkaline earth fluorides. Further descriptions of the optical properties of coloured crystals followed from the Jena school (Görlich and Karras, 1962; Görlich, Karras and Lehmann, 1963; Görlich, Karras and Koch, 1965).

Detailed understanding of the structure of colour centres in alkaline earth fluorides began with the application of epr techniques (Hayes and Twidell (1962); Arends (1964)) and has been greatly extended in recent years by the use of endor methods and high-resolution optical techniques, by the investigation of the effects of uniaxial stress, electric fields and magnetic fields on optical spectra and by theoretical studies. We now know that the α band in additively-coloured CaF_2 is a complex superposition of bands arising from F centres, M centres,† and higher F-aggregate centres, and that the shape of the band depends on details of coloration, and crystal purity. The β band is predominantly an absorption band of M centres. Impurities can give rise to absorption in both the α and β band regions in additively-coloured and X-irradiated crystals. The properties of additively-coloured SrF_2 and BaF_2 are somewhat similar to those of CaF_2. Close study of the optical properties of additively-coloured CaF_2 and SrF_2 has revealed sharp zero-phonon lines at low temperatures. Study of these lines has led to identification of R centres (composed of three F centres).

The self-trapped hole (V_K centre) has been shown to exist in crystals X-irradiated at low temperatures and its structure has been established. The self-trapped exciton has been identified. The structure and some properties of the H centre (a hole trapped at an interstitial anion site) have also been determined.

Although colour-centre work in crystals with the fluorite structure has been largely confined to alkaline earth fluorides some investigations have also been reported for $SrCl_2$, ThO_2, CeO_2, and PbF_2.

The determination of the structure of intrinsic colour centres in alkaline earth fluorides has led to investigations of the events that occur in the radiolysis of these crystals. These and other recent developments will be reviewed in this chapter.

4.2. The F centre

4.2.1. *Magnetic resonance investigations*

The epr spectrum of F centres was identified in CaF_2 and BaF_2 by Arends (1964) who carried out his investigations on additively-coloured

† An M centre is composed of a pair of F centres and is sometimes described as an F_2 centre. This latter notation is not altogether suitable for fluoride crystals and it will not be used here.

crystals. The epr spectrum of F centres in additively-coloured SrF_2 was subsequently described by den Hartog and Arends (1967*a*). A description of the endor spectrum of the F centres in CaF_2 was given by Hayes and Stott (1967) and in SrF_2 and BaF_2 by Stoneham, Hayes, Smith, and Stott (1968). As a result of the magnetic resonance investigations the model for the F centre shown in Fig. 4.1 was established. Approximate values for the peak position of the F band in CaF_2, SrF_2, and BaF_2 were obtained by Arends (1964) and den Hartog and Arends (1967*a*) by investigating effects of optical bleaching on the intensity

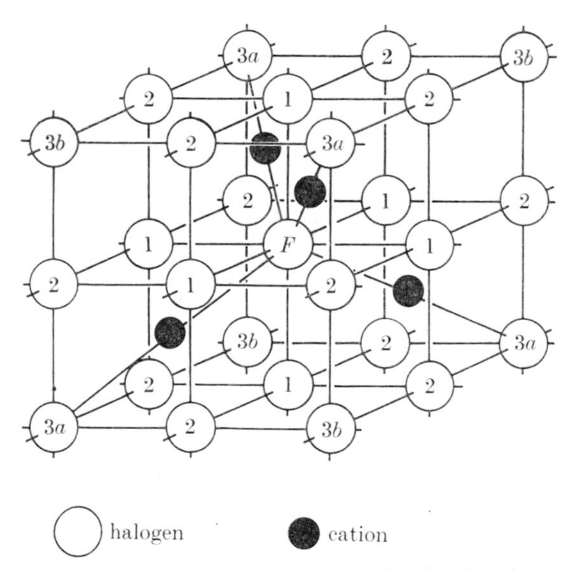

FIG. 4.1. Schematic representation of the fluorite lattice showing the first three fluorine shells and the first cation shell (represented by filled circles) centred on an F centre. The fluorines in the first shell are labelled 1, in the second shell 2 and in the third shell 3*a* and 3*b* (after Hayes, 1970).

of F centre epr spectra. More precise values for the peak position of the F band and values for the spin-orbit coupling constant in the excited ^2P state of the F band were obtained, using magneto-optical methods, by Cavenett, Hayes, Hunter, and Stoneham (1969). The Stark effect of the F centre in SrF_2 was investigated by Bartram, Harmer, and Hayes (1971). A detailed assessment of the theory of the F centre in alkaline earth fluorides will be given in § 4.2.3.

The F centre in alkaline earth fluorides is a single electron trapped in a fluorine vacancy. The nearest neighbours are four cations at the corners of a regular tetrahedron (Fig. 4.1) and the point symmetry of the F center is T_d. This is in contrast with the O_h point symmetry

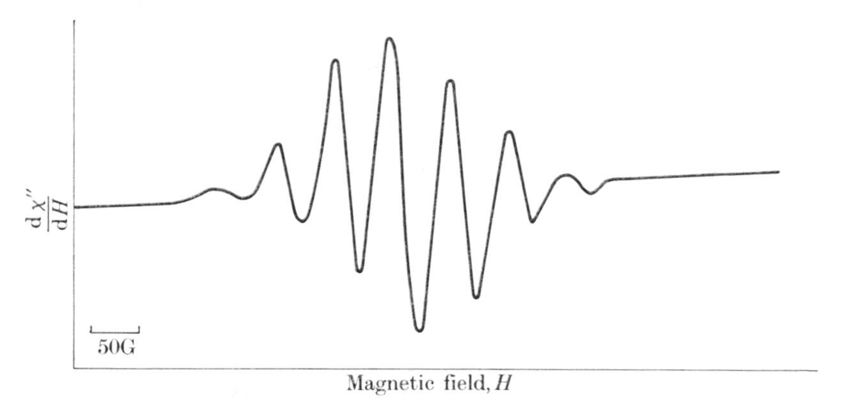

FIG. 4.2. First derivative of the epr absorption line of F centres in CaF_2 at 300 K. The external magnetic field is aligned along [110] (after Arends, 1964).

of the F centre in alkali halides. The second-nearest neighbours are six fluorines at the corners of a regular octahedron and in CaF_2 and SrF_2 they give rise to an isotropic seven-line hyperfine structure in the epr spectrum (Fig. 4.2).

In CaF_2 the concentration of positive ions with nuclear moments is too small (^{43}Ca with $I = \frac{7}{2}$ is 0·13% abundant) to give measurable hyperfine structure. In the case of SrF_2 the hyperfine interaction with ^{87}Sr ($I = \frac{9}{2}$, 7% abundant) is not observable indicating that the extent of the hyperfine structure is within the envelope of fluorine hyperfine structure. The Zeeman and fluorine interaction terms for the F centre in alkaline earth fluorides are contained in the spin-Hamiltonian

$$\mathscr{H} = g\beta\mathbf{H}\cdot\mathbf{S}+\sum_l(\mathbf{S}\cdot\mathbf{A}^l\cdot\mathbf{I}^l-g_F\beta_n\mathbf{H}\cdot\mathbf{I}^l), \tag{4.1}$$

where $S = \frac{1}{2}$, $I^l = \frac{1}{2}$ is the spin of the lth fluorine nucleus and g_F is the fluorine nuclear g-factor. The g-values for the F centre in the various crystals are given in Table 4.1. Values for the components of the

TABLE 4.1
F centre g-values

CaF_2	SrF_2	BaF_2
1·998[a] ±0·001	1·989[c]	1·953[a] ±0·001
1·9978[b] ±0·0005	1·986[d] ±0·001	1·9542[d] ±0·0005

[a] Arends (1964). [b] Hayes and Stott, (1967) [c] den Hartog and Arends (1967*a*) [d] Stoneham *et al.* (1968).

fluorine hyperfine tensor **A**, determined by endor methods, are given in Table 4.2. The hyperfine tensor for shell 1 fluorines has orthorhombic symmetry because of the T_d point symmetry of the F centre, and has a cube edge and two face diagonals as principal axes. In Table 4.2, A_3 for shell 1 fluorines is the principal value of A referred to the F centre-fluorine bond axis and A_1 and A_2 refer to face diagonals perpendicular to the bond axis. The hyperfine tensor for second-shell fluorines also has orthorhombic symmetry. The component A_3 for

TABLE 4.2

Fluorine hyperfine parameters (MHz) *for the F centre in* CaF_2, SrF_2, *and* BaF_2 (*Stoneham et al.*, 1968)

Shell	h.f. Parameters	CaF_2	SrF_2	BaF_2
1	A_1	160·82±0·05	106·73±0·08	46·73±0·01
	A_2	165·22±0·05	109·37±0·08	46·90±0·01
	A_3	221·6±0·1	148·70±0·11	67·87±0·01
2	A_1	9·17±0·01	8·38±0·02	4·15±0·01
	A_2	8·91±0·01	7·90±0·02	3·94±0·01
	A_3	14·40±0·01	12·74±0·02	7·12±0·01
3*a*	$A_1 = A_2$	18·33±0·01	25·71±0·04	37·19±0·01
	A_3	27·33±0·01	37·49±0·04	53·58±0·01
3*b*	$A_1 = A_2$	2·82±0·01	2·75±0·02	1·68±0·01
	A_3	5·52±0·01	4·96±0·02	3·32±0·01
4	$A_1 = A_2$	2·53±0·01	1·41±0·02	0·51±0·01
	A_3	4·72±0·01	3·01±0·02	1·56±0·01

these fluorines refers to a principal axis which is displaced in a {110} plane from the line joining the fluorine to the F centre. This displacement is presumably in the direction of the nearest cation in this plane (see Fig. 4.1) and is found to be 3° in CaF_2, 7·7° in SrF_2 and 13° in BaF_2 (Hayes and Stott; Stoneham *et al.*, 1968). The hyperfine tensor for shell 3 fluorines has trigonal symmetry but there are two different types of third-shell fluorines labelled 3*a* and 3*b* (Fig. 4.1 and Table 4.2) characterized by the presence and absence respectively of a cation between the F centre and the fluorine. In the case of fourth-shell fluorines departures from axial symmetry are too small to be measurable.

Inspection of Table 4.2 shows that departures from axial symmetry of the fluorine hyperfine tensor for shell 1 and shell 2 fluorines are small. For comparison with theory we shall, as an adequate approximation, ignore departures from axial symmetry and write the fluorine hyperfine interaction for all shells as the sum of an isotropic (contact) term and an anisotropic term:

$$A = A_s + A_p(3\cos^2\theta - 1). \tag{4.2}$$

Values of A_s and A_p for the four nearest fluorine shells are given in Table 4.3 for CaF_2, SrF_2, and BaF_2.

The appearance of the epr spectrum of the *F* centre in BaF_2 is quite different from that in CaF_2 and SrF_2 (Arends 1964; Stoneham *et al.*, 1968). There are two Ba nuclei, ^{135}Ba (6·59% abundant) and ^{137}Ba (11·32% abundant) each with $I = \frac{3}{2}$, and hyperfine structure due to these nuclei is observable in the spectrum. The strong central structure in the spectrum does not have the simple isotropic hyperfine structure from six equivalent fluorines found with CaF_2 and SrF_2. This difference in the case of BaF_2 is due primarily to the fact that the hyperfine interaction with $3a$ fluorines is comparable with that of first-shell fluorines (Table 4.2). The hyperfine interaction with $3a$ fluorines has the unusual feature that it increases from CaF_2 to SrF_2 to BaF_2 whereas all other interactions decrease. The interaction of the F centre electron in BaF_2 with the nearest-neighbour Ba nuclei was also investigated by endor methods and the quadrupole parameter Q' was determined (Stoneham *et al.*, 1968).

The measured endor parameters provide a sensitive check of theoretical wavefunctions for the *F* centre ground state. The *F* centre in the fluorite structure is interesting because of its tetrahedral, rather than cubic, symmetry. The anisotropy of the ground-state wavefunction associated with the T_d symmetry is particularly evident in the difference between the hyperfine parameters of $3a$ and $3b$ fluorines

Although the various theories of the *F* centre will be reviewed in § 4.2.3 we shall make some preliminary comments here. There are three main classes of theory: continuum models based on effective-mass theory, molecular-orbital models, and pseudo-potential approaches. The continuum approaches have several weaknesses and will not be used here. In particular, they have built-in spherical symmetry, and so are unsuitable for calculating the anisotropic terms in the wavefunction. The difficulty with molecular-orbital approaches (e.g. den Hartog, 1970) is that the coefficients which are most important in the hyperfine structure carry rather little weight in the energy, so that a formidable calculation is needed to get useful results. The pseudo-potential methods prove to be most convenient; they use a variational method with trial wavefunctions centred on the vacancy.

The pseudo-potential methods predict a pseudo-wavefunction $|\phi\rangle$. The spin-resonance parameters must be calculated using the true wavefunction $|\tilde{\phi}\rangle$, related to $|\phi\rangle$ by orthogonalization to the core orbitals $|c\rangle$ of the lattice ions:

$$|\tilde{\phi}\rangle = N[|\phi\rangle - \sum_c |c\rangle\langle c|\phi\rangle]. \tag{4.3}$$

It is generally acceptable to put N equal to 1 in (4.3). The errors are usually only a few per cent, and do not affect relative magnitudes. The sum over core orbitals can be written in terms of projection operators $P_{Ic} \equiv |Ic\rangle\langle Ic|$, where I labels the ion and c the orbital on the ion. The orthogonalization is simplified by three assumptions in all the calculations described here. Firstly, the core orbitals are assumed to be free-ion Hartree–Fock orbitals. Secondly, the overlaps between core orbitals on different sites are nearly always assumed to be zero. Thirdly, the pseudo-wavefunction $|\phi\rangle$ is usually assumed to vary slowly over the core orbitals to simplify the overlap integrals $\langle c|\phi\rangle$ and similar matrix elements.

The true core-wavefunctions in the crystal will differ from free atomic functions for two reasons. Firstly, the true core-orbitals should be orthogonalized to each other. The effects will be small if the overlaps are sufficiently small. Secondly, the ion experiences a crystal field due to the other ions. There is some evidence that this perturbation does not change the wavefunctions greatly (Watson, 1958). The X-ray diffraction data analysed by Maslen (1967) verify that free-ion Hartree-Fock orbitals are a good approximation for CaF_2 (see § 1.3.1).

The approximation that the core orbitals do not overlap is probably adequate for the alkaline earth fluorides. This is not true of the alkali halides (Wood, 1970), where omission of the core overlaps leads to a function $|\tilde{\phi}\rangle$ which falls off too rapidly with distance. If the overlaps are small we may rewrite the expression for $|\phi\rangle$ using a modified projection operator:

$$\sum_{I,c} \rightarrow \sum_{I,c} P_{I,c} - \sum_{I,c} \sum_{J \neq I,d} P_{I,c} P_{J,d},$$

where $P_{Ic}P_{Jd} \equiv |Ic\rangle\langle Ic|Jd\rangle\langle Jd|$ involves the overlap of core orbitals on different sites.

The theoretical pseudo-wavefunctions are of the form

$$|\phi\rangle = \lambda\,|s\rangle + \mu|f\rangle, \tag{4.4}$$

where $|s\rangle$ is spherically symmetric and $|f\rangle$ transforms like (xyz/r^3). We now give the results for various detailed models, but defer discussion of the validity of the models until later (see § 4.2.3).

First we consider the predictions when $|\phi\rangle$ is assumed to be spherically symmetric. Stoneham *et al.* (1968) calculated isotropic and anisotropic fluorine hyperfine constants for *F* centres in CaF_2, SrF_2, and BaF_2 using the point-ion wavefunctions of Bennett and Lidiard (1965). The results are given in Table 4.3. The general features are predicted

TABLE 4.3

Observed and calculated hyperfine constants (MHz) *for the F centre in* CaF_2, SrF_2, *and* BaF_2

		A_s			A_p	
Crystal	Shell	Experiment[a]	Theory[a]	Theory[b]	Experiment[a]	Theory[a]
CaF_2	1	182·51	245·5	279·16	24·55	26·7
	2	10·85	16·4	38·75	1·81	2·58
	3a	21·33	2·33	43·92	3·00	0·87
	3b	3·72	2·33	1·10	0·90	0·87
	4	3·26	0·47	1·99	0·32	0·49
SrF_2	^{87}Sr	$\leqslant 100$	71·3	—	—	—
	1	121·6	195	225·17	13·55	20·4
	2	9·7	11·8	30·72	1·53	1·96
	3a	29·6	1·55	36·60	3.93	0·70
	3b	3·49	1·55	1·20	0·74	0·70
	4	1·94	0·3	1·52	0·53	0·4
BaF_2	^{135}Ba	258·9	363·9	—	15·5	—
	^{137}Ba	289·9	388·0	—	17·5	—
	1	53·8	149·0	174·70	7·0	15·5
	2	5·1	7·7	23·09	1·1	1·45
	3a	42·6	0·89	28·52	5·4	0·55
	3b	2·2	0·89	1·20	0·5	0·55
	4	0·86	0·15	1·09	0·35	0·33

[a] Stoneham *et al.* (1968). [b] Bartram *et al.* (1971).
The fluorine shells are numbered 1 to 4. In the case of SrF_2 and BaF_2 the hyperfine constants for first shell-cations are given. All theoretical results use $N^2 = 1$.

rather well, apart from the differences between the 3*a* and 3*b* fluorines. Further, when the overlaps of the core orbitals are included, two other effects are predicted. One is a part of the difference in hyperfine constants for the 3*a* and 3*b* shells. The other is the anisotropy of the hyperfine constants for the nearest-neighbour fluorine ions when the magnetic field is rotated in the $\{100\}$ plane. The anisotropy is dominated by terms arising from the overlap of the F^- ion and the two cations which are nearest neighbours both to it and to the *F* centre. The observed anisotropy can be expressed as $2{\cdot}83 \times 10^{-2}$ (a.u.)$^{-3}$. The predicted value from the spherical wavefunction, with allowance for core overlaps, is $0{\cdot}5\times10^{-2}$ (a.u.)$^{-3}$; this, however, is very approximate as it is the difference between two large terms.

Clearly, the $|f\rangle$ term in (4.4) should not be neglected and two calculations have included it. The earlier, by Stoneham *et al.* (1968), used a combination of perturbation theory and a wavefunction $|f\rangle$ from effective-mass theory. This was not a variational wavefunction; the spherically-symmetric part used was just the Bennett-Lidiard result. The predictions gave general order-of-magnitude agreement for the differences between the $3a$ and $3b$ fluorine hyperfine constants in CaF_2. Table 4.4 compares the results with experiment, and gives the relative

TABLE 4.4

Third-shell hyperfine constants (MHz) *for* CaF_2

	$(A_s^{3a}-A_s^{3b})/2$	$(A_p^{3a}-A_p^{3b})/2$
Experiment	8·79	1·05
Theory		
f admixture.	14·43 (21·41*)	−0·113
Core orthogonalization.	1·35	0·625
Total	15·78	0·512

Theory is from Stoneham *et al.* (1968) apart from one value (*) taken from Bartram *et al.* (1971).

contributions of $|f\rangle$ admixture and core-orbital orthogonalization to these terms. The weakness of this treatment of the anisotropy of the wavefunction is particularly apparent from the values of $(A_p^{3a}-A_p^{3b})$, where the contribution has the wrong sign because the $|f\rangle$ wavefunction at the $3a$ and $3b$ fluorines is still increasing, rather than decreasing, with distance. It would appear that a variational calculation is needed which varies both the $|s\rangle$ and $|f\rangle$ terms, and which includes their admixture. This has been done by Bartram, Harmer, and Hayes (1971) whose results for A_s in CaF_2, SrF_2, and BaF_2 are also given in Table 4.3. Bartram *et al.* concentrate on the $|f\rangle$ admixture part of the anisotropy, and do not include the terms from core-orbital orthogonalization. The general agreement for A_s is very good. This agreement would probably also hold for A_p which was not calculated; since the $|f\rangle$ term is much more compact it should give a contribution of the correct sign to $(A_p^{3a}-A_p^{3b})$.

In summary, the agreement between experiment and theory is good; it is better than for F centres in the alkali halides. It is especially satisfying that the anisotropy in the wavefunction is predicted with reasonable accuracy, since these terms are a very sensitive check of the theory.

The g-factors have also been discussed theoretically. Thus, den Hartog and Arends (1967*a*) have shown that the change Δg from the free electron g-value is roughly linear in the ratio of cation spin-orbit splitting to F-band energy, and they argue that this indicates roughly the same anisotropic admixture in the CaF_2, SrF_2, and BaF_2 centres. Quantitative estimates for CaF_2 (den Hartog and Arends, 1967*b*) using the Bennett–Lidiard wavefunctions give Δg contributions of $-0{\cdot}0030$ from the nearest Ca^{2+} ions and $-0{\cdot}0046$ from the nearest F^- ions. The total $-0{\cdot}0076$, is in acceptable agreement with the experimental value of $-0{\cdot}004$.

4.2.2. *Magneto-optical and Stark effects*

Approximate estimates of the peak position of the F band were made in CaF_2 and BaF_2 (Arends, 1964) and in SrF_2 (den Hartog and Arends, 1967*a*) by investigating effects of optical bleaching on the intensity of epr spectra in additively-coloured crystals. More precise values were obtained by Cavenett *et al.* using magneto-optical methods. The F band, which arises from a ${}^2S(\Gamma_1)$–${}^2P(\Gamma_5)$ type transition is broad (Fig. 4.3) due to strong coupling of the F centre electron to lattice vibrations; this coupling partly quenches the spin-orbit interaction in the ${}^2P$ state. It was shown by Henry, Schnatterly, and Slichter (1965) that magneto-optical effects in such strongly coupled systems may be conveniently investigated by determining the changes in moments of the band shape induced by the applied magnetic field. They showed that the change in zeroth moment of an F band induced by an external field H for right (+) and left (−) circularly polarized light is

$$\langle \Delta \mathrm{A} \rangle_{\pm} = 0 \tag{4.5}$$

and the change in the first moment is

$$\langle \Delta E \rangle_{\pm} = \pm \langle x| \, L_z \, |y\rangle(\beta H + \lambda \langle S_z \rangle) \tag{4.6}$$

where $\langle x| \, L_z \, |y\rangle$ is the matrix element of orbital angular momentum in the ${}^2P$ state of the F centre, λ is the spin-orbit coupling constant in the ${}^2P$ state and $\langle S_z \rangle$ is the thermal-average value of the spin in the ${}^2S$ ground state. Eqn (4.5) means that the area under the F band envelope does not change; it does not apply if H admixes other electronic states into the F band states. The change in the first moment, given by eqn (4.6) is the change in the centre of gravity of the F band induced by H.

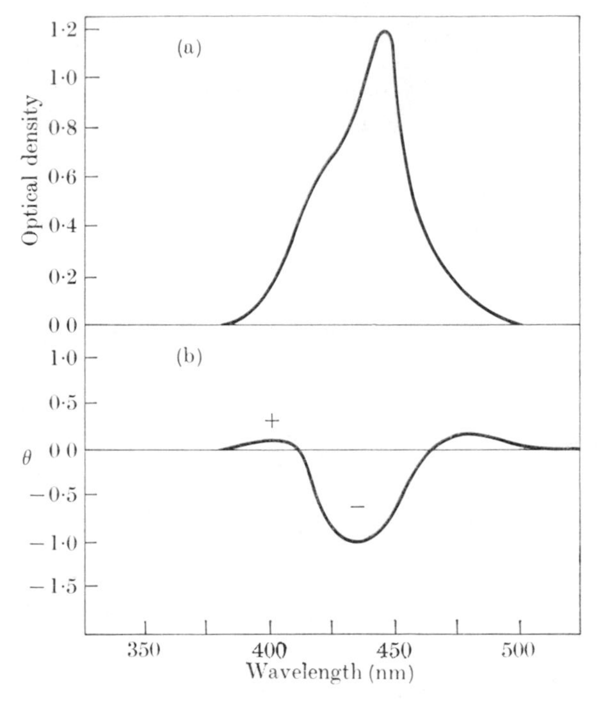

FIG. 4.3. (a) Optical density at 4 K of additively-coloured SrF_2. (b) Faraday rotation in degrees, at 4 K and 7·0 T, associated with the optical density in (a) (after Hayes, 1969).

The first term on the right hand side of eqn (4.6) is temperature independent and, because it is small, it was not determined with accuracy for alkaline earth fluorides. The second term in eqn (4.6) is temperature dependent and gives rise to large magneto-optical effects at low temperatures when λ is large. A measurement of the optical density of the F band and the associated Faraday rotation (or magnetic circular dichroism) pattern gives $\langle \Delta E_{\pm} \rangle$ and hence $\langle x| \, L_z \, |y\rangle\lambda = \lambda'$. The measured value of λ' provides a check on wavefunctions for the ^2P state of the F centre (see § 4.2.3). Measurements of changes in higher moments induced by H give information about electron-lattice interaction for the F centre (Henry *et al.*) but are difficult to determine with precision.

We have already mentioned that the α band in additively-coloured alkaline earth fluorides arises from a superposition of bands due to F centres, M centres, higher F-aggregate centres and impurities. It is not possible to produce F centres in additively-coloured crystals without some of the other centres present and hence it is not possible to determine with accuracy the contribution of F centres to the optical density in the α band. It is therefore not possible to determine λ' with precision in additively-coloured crystals. However, it is possible to obtain an accurate estimate of the peak position of the F band from the minimum of the Faraday rotation pattern (Fig. 4.3) and the F band peak positions determined in this way for additively-coloured CaF_2, SrF_2, and BaF_2 are given in Table 4.5.

F centres are produced by X-irradiation of undoped alkaline earth fluoride crystals at 77 K or lower temperatures (§ 4.9) and the F band is not superimposed in crystals treated in this way by F-aggregate bands. However, detailed investigation of the radiation-produced F band in SrF_2 (see § 4.9.1) suggests the presence of a fluorine interstitial

TABLE 4.5

Values of peak position and λ' for the F band at 4 K in CaF_2, SrF_2, and BaF_2 (after Cavenett et al., 1969)

Crystal	Peak Position nm	eV	λ' (cm^{-1})
CaF_2	376	3·30	~ -25
SrF_2	435	2·85	-66
BaF_2	611	2·03	-90

near the F centre. The presence of a fluorine interstitial will not appreciably affect the value of the spin-orbit coupling parameters λ' in the ^2P state of the F centre. Values of λ' obtained by Cavenett *et al.* from radiation-produced F bands in CaF_2, SrF_2, and BaF_2 are given in Table 4.5.

We have already seen the pronounced effects of the anisotropy of the potential at the F centre site on the endor measurements on the $^2S(\Gamma_1)$ ground state. This anisotropy has consequences for the excited $^2P(\Gamma_5)$ state also. The magnitude of the spin-orbit energy in this state does not provide a sensitive test of anisotropy. The Stark effect is more profoundly affected since the lack of inversion symmetry allows a linear Stark splitting whose magnitude is a measure of the admixture of p and d orbitals by the anisotropic potential. Again, because of the strong phonon coupling it is convenient to investigate the effect of the Stark field using the method of moments. The matrix of the electric field perturbation may be written in the form (Bartram *et al.*, 1971).

$$KeF\begin{pmatrix} 0 & \cos\theta & \sin\theta\sin\phi \\ \cos\theta & 0 & \sin\theta\cos\phi \\ \sin\theta\sin\phi & \sin\theta\cos\phi & 0 \end{pmatrix},$$

where θ and ϕ are polar angles of the applied field intensity $\mathbf{F}$, e is the electronic charge, and K may be calculated from the Γ_5 wave functions (see § 4.2.3). Assuming that $\mathbf{F}$ is parallel to $\mathbf{z}$ the linear Stark effect gives

$$\begin{aligned} \langle \Delta A \rangle &= 0, \\ \langle \Delta E_n \rangle &= 0, \\ \langle \Delta E_z^2 \rangle &= 0, \\ \langle \Delta E_x^2 \rangle &= \langle \Delta E_y^2 \rangle = (KeF)^2, \end{aligned} \tag{4.7}$$

where $\langle \Delta A \rangle$, $\langle \Delta E \rangle$, and $\langle \Delta E^2 \rangle$ are changes induced by $\mathbf{F}$ in the zeroth, first and second moments of the F band and $\eta = x, y, z$ defines the polarization of the absorbed light. Using an a.c. Stark spectrometer it was shown by Bartram *et al.* (1971) that, for the F band in SrF_2,

$$(KeF/hc) \leqslant (0{\cdot}5 \pm 0{\cdot}2)F \text{ cm}^{-1} \tag{4.8}$$

where F, in units of 10 kV/cm, is the local electric field. A Lorentz local-field correction was used. The value of K will be increased by about 80% if this correction is omitted (see e.g. Smith and Dexter, 1972). A calculation of the magnitude of the linear Stark effect in SrF_2 will be described in the following section.

The quadratic Stark effect was also experimentally investigated for the F band in SrF_2 and found to be less than the limiting sensitivity of experiment (Bartram *et al.*, 1971). This result was used to put a bound on the dipole matrix element between the $2s$ and $2p$-like excited states of the F centre ($\langle 2s|\, z\, |2p\rangle \leqslant 0{\cdot}4 \times 10^{-8}$ cm) if the two states are exactly degenerate. Alternatively, using a theoretical estimate of $1{\cdot}5 \times 10^{-8}$ cm for the matrix element it follows that the $2s$ and $2p$-like states must be separated by more than 0·7 eV.

4.2.3. *Theoretical investigations*

Theories of the F centre in the fluorite structure have concentrated on two types of method. On the one hand there are pseudo-potential methods. These include as a special case the point-ion model (Gourary and Adrian, 1960) in which the defect electron moves in a potential obtained by replacing all ions by appropriate classical point charges. (Point-ion calculations for many alkaline earth fluorides have been given by Bennett and Lidiard (1965)). On the other hand there are continuum methods based on effective-mass theory with some allowance for lattice polarization. The continuum treatments can only be considered semiquantitative. They use effective-mass theory well outside its regime of validity, they assume an isotropic lattice for a problem in which asymmetric distortions and asymmetric crystal fields are important, and they tend to use the wrong sort of macroscopic parameters. For example, the F centre electron primarily overlaps the cations, whereas the dielectric constant weights the anions more highly. We therefore concentrate on the pseudo-potential calculations and use the continuum work for different information. A list of calculations discussed here is given in Table 4.6.

TABLE 4.6

Theoretical models for the F centre in the fluorite structure

I	Pseudo-potential calculations with no anisotropy terms
	(a) Point-Ion, Variational (Bennett and Lidiard, 1965)
	(b) Point-Ion, Integration of Schrödinger equation (Bennett, 1969).
	(c) Point Ion plus Ion-Size terms, Variational (Bartram *et al.*, 1968).
II	Calculations with Anisotropic Admixture
	(a) Perturbation admixture (Stoneham *et al.*, 1968; Cavenett *et al.*, 1969).
	(b) Point-Ion, Variational (Bartram *et al.*, 1971)
III	Calculations based on continuum models
	(a) Simpson-Pekar model, using approach of Smith (1957) as corrected by Markham (1966) (Stoneham, unpublished, 1967)
	(b) Self-consistent Polaron model (Bennett, 1971*a*)

The defect-electron wavefunction must be orthogonal to the other occupied orbitals of the crystal. On the one hand the variational trial function can be explicitly orthogonalized to the core orbitals. Alternatively, the orthogonality constraint can be replaced by an extra term in the Hamiltonian, so that one may obtain the energy levels using an unconstrained trial function. This second approach is the basis of the pseudo-potential method. The Hamiltonian contains the usual crystal potential V and a term representing the orthogonality constraint, V_c. The sum of the two gives the pseudo-potential. The eigenfunction obtained is the pseudo-wavefunction $|\phi\rangle$, which must be orthogonalized to the occupied orbitals (see eqn (4.3)) to give the true wavefunction

TABLE 4.7

Trial wavefunctions for the F centre in the fluorite structure

The Γ_1 ground state $|g\rangle$ is a combination of functions transforming as 1 (s-like) and xyz/r^3 (f-like), whereas the Γ_4 excited state $|e\rangle$ combines functions transforming as z/r (p-like) and xy/r^2 (d-like):

$$|g\rangle = s\,|s\rangle + c|f\rangle$$
$$|e\rangle = \bar{s}\,|p\rangle + \bar{c}|d\rangle$$

where all functions are normalized. The radial dependence of the components is either of the form given in equation (4.10) or of the form

$$\begin{array}{ll} s & (1+\alpha r)\exp(-\alpha r) \\ p & r\exp(-\beta r) \\ d & r^2\exp(-\gamma r) \\ f & r^3\exp(-\delta r) \end{array}$$

where r is in units of the nearest-neighbour distance.

TABLE 4.8

Energy levels and transition energies for the F centre given by the models of Table 4.6

	Expt.	I(a)	I(b)	I(c)	II(b)	III(a)	III(b)	
CaF_2								
Ground state				−7·66	−9·41	−6·64		
Excited state				−4·08	−5·90	−2·79		
Transition energy	3·30	3·19	3·5	3·59	3·51	3·85	3·26	2.80*
			3.07					0·735*
SrF_2								
Ground state				−7·99	−9·11	−6·38		
Excited state				−5·08	−5·85	−2·82		
Transition energy	2·85	2·91		2·91	3·26	3·56	2·45	
BaF_2								
Ground state				−8·60	−8·77	−6·09		
Excited state				−6·52	−5·76	−2·84		
Transition energy	2·03	2·61		2·08	3·01	3·25	1·74	

All values are in eV. Energy zeros are the vacuum level apart from III, where energies are measured from the bottom of the conduction band. Where two figures are quoted, the upper refers to absorption and the lower to emission, and undistorted lattice values are not available. Values marked * used numerical integration of the Schrödinger equation.

$|\tilde{\phi}\rangle$ needed in predicting observables. There are two particular advantages of the pseudo-potential method. The first is that $|\phi\rangle$ is usually a much smoother function than $|\tilde{\phi}\rangle$, so that it is easier to guess a good trial function. The second advantage is that there is an important cancellation between V and V_c. At large distances from a lattice ion the F centre electron experiences the appropriate point-ion potential. As the F centre electron moves closer to the nucleus the ion's electrons screen less and less of the nuclear charge and, as a consequence, a strong attractive term appears in V. This correction to the point-ion potential is largely cancelled by V_c since the orthogonality constraint is repulsive in just the region of the core orbitals. To a good approximation one can write

$$V+V_c = V_{\mathrm{PI}}+V_{\mathrm{IS}}, \tag{4.9}$$

where V_{PI} is the point-ion potential, and V_{IS} encompasses fairly small ion-size corrections.

The trial wavefunctions used in the variational calculations fall into two classes. In the first the F centre ground state is spherically symmetric and the excited state is a simple p-function. In the second the admixture of more anisotropic components by the crystal field is included. It is important to note that in the first case, where simple s- and p-functions occur, the expectation value of the potential energy only involves the spherically symmetric part of the potential energy. If $V_{\mathrm{av}}(r)$ is the average over angles of a potential $V(\mathbf{r})$ with at least tetrahedral symmetry, then the expectation values of V_{av} and V are identical within s or within p functions. We shall not discuss work which ignores this result and makes other approximations instead (Feltham and Andrews (1965); Fleming (1966)).

We now outline the different calculations, starting with those which ignore the anisotropic admixtures; wavefunctions and energies are summarized in Tables 4.7 to 4.10. Bennett and Lidiard (1965) used the point-ion model for nine crystals with the fluorite structure and found that the best variational trial functions were of the form

$$\begin{aligned} \Psi_{1s}(\mathbf{r}) &\sim \sin(\xi r/a)/(r/a) & (r<a) \\ &\sim e^{\eta} \sin \xi . \exp(-\eta r/a)/(r/a) & (r>a); \\ \eta &= -\xi \cot \xi \end{aligned} \tag{4.10a}$$

for the ground state, and

$$\begin{aligned} \Psi_{2p}(\mathbf{r}) &\sim j_1(\xi r/a) e^{-\eta} \cos\theta & (r<a) \\ &\sim j_1(\xi)(r/a) e^{-\eta r/a} \cos\theta & (r>a); \\ \eta &= 3-\xi^2/(1-\xi \cot \xi) \end{aligned} \tag{4.10b}$$

TABLE 4.9

Wavefunction parameters† for the ground and excited states of the F band

Model	I(a)	I(c)	II(b)	II(a)	III(a)
CaF_2 ground state	$\xi = 2{\cdot}31$ $\eta = 2{\cdot}11$ $(\alpha = 2{\cdot}81)$	$\xi = 2{\cdot}24$ $\eta = 1{\cdot}76$	$\alpha = 2{\cdot}631$ $\delta = 4{\cdot}192$ $c = -0{\cdot}486$	$\delta = 1{\cdot}12$ $c = -0{\cdot}15$	$\alpha = 2{\cdot}847$
excited state	$\xi = 3{\cdot}16$ $\eta = 3{\cdot}06$ $(\beta = 2{\cdot}21)$	$\xi = 3{\cdot}07$ $\eta = 2{\cdot}78$	$\beta = 2{\cdot}43$ $\gamma = 3{\cdot}23$ $\bar{c} = -0{\cdot}613$	$\gamma = 1{\cdot}49$ $\bar{c} \leqslant 0{\cdot}5$	$\beta = 2{\cdot}369$
SrF_2 ground state	$\xi = 2{\cdot}33$ $\eta = 2{\cdot}21$ $(\alpha = 2{\cdot}86)$	$\xi = 2{\cdot}22$ $\eta = 1{\cdot}67$	$\alpha = 2{\cdot}649$ $\delta = 4{\cdot}210$ $c = -0{\cdot}506$	$\delta = 1{\cdot}18$	$\alpha = 2{\cdot}904$
excited state	$\xi = 3{\cdot}20$ $\eta = 3{\cdot}19$ $(\beta = 2{\cdot}28)$	$\xi = 3{\cdot}06$ $\eta = 2{\cdot}76$	$\beta = 2{\cdot}507$ $\gamma = 3{\cdot}264$ $\bar{c} = -0{\cdot}625$	$\gamma = 1{\cdot}58$	$\beta = 2{\cdot}442$
BaF_2 ground state	$\xi = 2{\cdot}35$ $\eta = 2{\cdot}32$ $(\alpha = 2{\cdot}94)$	$\xi = 2{\cdot}17$ $\eta = 1{\cdot}44$	$\alpha = 2{\cdot}687$ $\delta = 4{\cdot}259$ $c = -0{\cdot}526$	$\delta = 1{\cdot}27$	$\alpha = 2{\cdot}982$
excited state	$\xi = 3{\cdot}24$ $\eta = 3{\cdot}33$ $(\beta = 2{\cdot}36)$	$\xi = 3{\cdot}06$ $\eta = 2{\cdot}77$	$\beta = 2{\cdot}535$ $\gamma = 3{\cdot}296$ $\bar{c} = -0{\cdot}636$	$\gamma = 1{\cdot}69$	$\beta = 2{\cdot}536$

† The wavefunction parameters ξ and η are defined in equation (4.10). Other coefficients c, $\bar{c}$, α, β, γ, and δ are defined in Table 4.7. The various models used are described in Table 4.6.

TABLE 4.10

Contributions to the energy of the ground $|g\rangle$, and excited, $|e\rangle$, states of the F band (Bartram et al. 1968). Results are in Ry units (1 Ry = 13·6 ev).

	CaF_2		SrF_2		BaF_2	
Contribution	$\|g\rangle$	$\|e\rangle$	$\|g\rangle$	$\|e\rangle$	$\|g\rangle$	$\|e\rangle$
Point ion	−0·535	−0·301	−0·511	−0·299	−0·479	−0·294
Ion size	−0·018	+0·027	−0·057	−0·049	−0·115	−0·149
Ionic polarization	−0·009	−0·040	−0·011	−0·044	−0·017	−0·050
Lattice polarization	−0·002	+0·014	−0·007	+0·020	−0·020	+0·015

for the excited state. Subsequently Bennett (1968) integrated the radial Schrödinger equation numerically and verified that these trial functions were quite accurate. Bartram, Stoneham, and Gash (1968) included the leading terms beyond the point-ion model; these ion-size corrections give a substantial improvement for both alkali halides and alkaline earth fluorides. The improvement is especially marked for BaF_2, where the large Ba^{2+} ion has a large ion-size effect.

Two calculations include anisotropic components. In one (Stoneham *et al.*, 1968; Cavenett *et al.*, 1969) perturbation theory was used to admix anisotropic terms into the Bennett-Lidiard wavefunctions. This approach did not estimate effects on transition energies but attempted to analyse the anisotropic contributions to spin-resonance and magneto-optic properties. The second calculation (Bartram *et al.*, 1971) used the point-ion model, and involved a proper variational calculation of the anisotropic terms. The parameters of the $|s\rangle$ and $|p\rangle$ terms and their $|f\rangle$ and $|d\rangle$ admixtures were varied, whilst including the correct estimated admixture. Both calculations agree that the anisotropic $|f\rangle$ and $|d\rangle$ terms fall off very rapidly with distance.

For completeness, results of two continuum calculations are also discussed (Table 4.6). One (Stoneham, 1967) is based on the well known methods of Simpson (1949) and Pekar (Pekar, 1951; Pekar and Deigen, 1948). The specific form adopted is described at length by Smith (1957), whose expressions for the energy function have been corrected by Markham (1966). The second calculation (Bennett, 1968) makes use of Haken's (1963) treatment of the lattice and electronic polarization. Both methods suffer from the weaknesses of continuum models described earlier. Bennett's (1968) model suffers also from the weaknesses of the Haken method, which proves a bad approximation even in the cases in which it should be at its best (Stoneham, 1972). Bennett's (1968) model also involves an effective mass which must be guessed, since no measurements are available. It is interesting to note that numerical integration of the Schrödinger equation gives very different results to the variational method in Bennett's (1968) continuum model, contrary to the results of the point-ion model.

We now compare the theories with each other and with experiment. Experiment shows clearly that the transition energies, ΔE, do not obey the empirical Mollwo–Ivey law (Ivey, 1947). This law, which works rather well for the alkali halides, relates ΔE to the lattice parameter a:

$$\Delta E \sim E_0 a^{-n}, \tag{4.11}$$

where E_0 and n are constants. This result is only expected to be valid for a potential which scales with the lattice parameter, i.e. $V \equiv V(r/a)$ (Hayes and Stoneham, 1969), and large deviations from it suggest that ion-size corrections are important. The deviations here are largest for BaF_2 for which only two models predict reasonable transition energies. These are the pseudo-potential treatment of Bartram *et al.* (1968), and Bennett's (1968) continuum model. Since the continuum model works less well for SrF_2, and has the other difficulties mentioned earlier, it is probable that its success in the case of BaF_2 is fortuitous.

The various contributions to the energy of the ground and excited state of the F band found by Bartram *et al.* (1968) are shown in Table 4.10. The point-ion contributions dominate but there are significant

TABLE 4.11

Outward displacements of the nearest neighbours of the F centre as a percentage of the nearest-neighbour distance

Model as in Table 4.6	CaF_2	SrF_2	BaF_2
I(b)	4·2	4·1	4·1
I(c)	1·5	2·5	5·5
III(b)	0·5	−0·4	−1·4

ion-size terms. By contrast, the polarization contributions are not very large. All theories which include lattice distortion agree that it is small, and that the nearest neighbours move by just a few per cent of the lattice spacing (Table 4.11). The theories differ in detail, however. This is not surprising in view of the rather complete cancellation between the outward forces on the ions arising because a compact anion has been replaced by a diffuse defect electron, and the inward forces arising from the elimination of the short-range interaction between the anion removed and its neighbours.

The $|s\rangle$ and $|p\rangle$ radial wavefunctions prove rather insensitive to the model, even to the effect of the ion-size corrections. The inclusion of anisotropic $|f\rangle$ and $|d\rangle$ terms in the wavefunction does not affect the other terms much, although the transition energies do not agree so well with experiment. There are three checks of these wavefunctions: the spin-resonance parameters, considered earlier, the Stark effect and the spin-orbit splitting of the excited state.

Both the linear Stark effect and the spin-orbit splitting are determined by the unrelaxed excited state of the F centre. The linear Stark effect is determined by the admixture of d-like orbitals into the p-like excited state. The constant K used to describe the Stark effect in § 4.2.1 is related to the parameters $\bar{s}$, $\bar{c}$, β, and γ of Table 4.7 by

$$K/a = -2\bar{c}\bar{s} \,.\, 192\sqrt{\tfrac{2}{3}}\beta^{\frac{5}{2}}\gamma^{\frac{7}{2}}(\beta+\gamma)^{-7}, \tag{4.12}$$

where a is the distance of the nearest cation from the F centre. Comparison of K_{theory} (eqn. 4.12) with experiment requires a local-field correction (§ 4.2.1): the predicted second-moment change induced by the applied electric field (4.7) involves the local field, F_{LOC}, whereas one measures the applied field F. Thus the measured splitting parameter K should be compared with values of $(K_{\text{theory}} \,.\, F_{\text{LOC}}/F)$. In units of Bohr radii the

TABLE 4.12

Contributions (cm^{-1}) *to the spin-orbit coupling constant of the F centre (after Cavenett et al. 1969)*

Contribution	CaF_2	SrF_2	BaF_2
Nearest-neighbour shell			
ion-ion term	−11·9	−61·2	−159.6
	(−33·3)	(−168)	(−501)
ion-vacancy term	−0·1	+0·8	−0·2
Vacancy-centred term	+2·7	+3·3	+3·8
Next nearest-neighbour shell			
ion-ion term	−9·8	−13·3	−13·5
ion-vacancy term	+0·7	+0·7	+0·7
Total (p-like function)	−18·4	−69·7	−168·8
Experiment (see Table 4.5)	~ −25	−66	−90

The results are for pure p-like functions, apart from the bracketed values which are for pure d-like functions.

coefficients K_{theory} of Bartram *et al.* (1971) for CaF_2, SrF_2, and BaF are 2·03, 2·12, and 2·36. Using the Lorentz local-field correction to give an approximate value of F_{LOC}, the predicted linear Stark effect in SrF_2 is consistent with the experimental limit (4.8).

The spin–orbit splitting of the excited $2p(^2p)$-like state of the F centre in some fluorites has been calculated by Cavenett *et al.* (1968) using the methods of Smith (1965). The theory correctly predicts that the six-fold degenerate level splits so that the quartet state lies below the doublet state, and gives reasonable values of the splitting. Experiment and theory are compared in Table 4.12. The predictions use the point-ion wavefunctions of Bennett and Lidiard (1965), although very similar results can be obtained using wavefunctions with ion-size corrections. It is essential to orthogonalize the point-ion wavefunction to the core orbitals. This leads to three types of contribution to the spin-orbit splitting: those involving just the vacancy-centred function, those linear in overlap with the ion cores (ion-vacancy terms), and those quadratic in the overlaps (ion–ion terms). The largest contributions are always from the ion–ion terms. The nearest-neighbour cations dominate in all except CaF_2; here the nearest-neighbour anions, which are further from the vacancy than the cations, contribute significantly. The results are not too sensitive to admixtures of d-like orbitals into the p-like trial functions; this can be verified from the results for a purely d-like wavefunction also given in Table 4.12.

4.3. M and M^+ centres

4.3.1. *General optical properties of M centres*

F centres in additively-coloured alkaline earth fluoride crystals readily aggregate forming more complex centres. The simplest aggregate, the M centre, is composed of two F centres. In a series of investigations by Beaumont and Hayes (1969), Hayes, Lambourn, and Smith

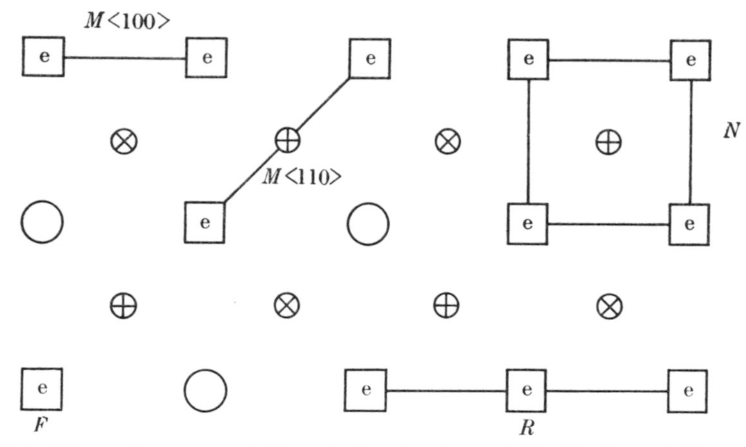

FIG. 4.4. Schematic representation of the structure of F, M, R, and N centres in fluorites. ⊗ and ⊕ are cations above and below the plane of the paper. ○ are anions.

(1970) and Beaumont, Harmer, and Hayes (1972*a*) it was shown that two types of M centre can exist in CaF_2 and SrF_2 (Fig. 4.4). One type is composed of two nearest F centres on a crystal cube edge and it will be described in future as the $M\langle 100\rangle$ centre. The other type is composed of two nearest F centres on a face diagonal and it will be described as the $M\langle 110\rangle$ centre. Investigations of M centres have also been carried out in BaF_2 (Beaumont and Hayes, 1969) but the results are much less detailed than in CaF_2 and SrF_2.

The β band, which occurs in CaF_2 at 521 nm and in SrF_2 at 595 nm, is due predominantly to absorption of M centres; absorption associated with impurities also contributes (Hayes *et al.*, 1970). Under high resolution partly resolved vibronic structure is observed on the low-energy side of the band. This structure is associated with a zero-phonon line in CaF_2 at 539·6 nm and in SrF_2 at 605·3 nm (Fig. 4.5); these lines have widths of 0·7 and 6·6 cm^{-1} respectively at 20 K. Both $M\langle 100\rangle$ and $M\langle 110\rangle$ centres contribute to absorption in the β band and this band will sometimes be referred to as the M band. The zero-phonon lines are transitions of the $M\langle 110\rangle$ centres. The integrated zero-phonon line intensity bears a constant ratio to the integrated intensity of the M band for a wide range of coloration densities and this ratio is about 20 times greater for SrF_2 than for CaF_2. Both the M band and the zero-phonon line are destroyed by F band light above 200 K. Bleaching with M band light at room temperature reduces the intensity of the zero-phonon line and the M band in constant proportion and it is not possible to change this proportion by using narrow-band excitation in different parts of the M band.

Excitation in either the F or M band regions gives a broad fluorescence band peaking in CaF_2 at 585 nm and in SrF_2 at 644 nm with weak vibronic structure on the high energy side (Fig. 4.5). This vibronic structure is associated with the $M\langle 110\rangle$ zero-phonon line which also occurs in fluorescence. The broad fluorescence band is again a superposition of bands of $M\langle 100\rangle$ and $M\langle 110\rangle$ centres and will in future be referred to as the M emission band. From the strength of the vibronic absorption in Fig. 4.5 it appears that the absorption of $M\langle 100\rangle$ centres predominates in CaF_2 whereas the absorption of $M\langle 110\rangle$ centres predominates in SrF_2.

4.3.2. *Polarization effects in absorption and fluorescence of M centres*

The optical absorption of cubic crystals containing anisotropic defect centres is isotropic when the defect centres are in equilibrium with the lattice. It is possible, however, to induce optical anisotropy in such crystals by bleaching with polarized light (Feofilov, 1961; Feofilov and Kaplyanskii, 1962; Compton and Rabin, 1964). This anisotropy arises from different populations, induced by the polarized bleaching light, of the various allowed orientations of the anisotropic defects. The different populations are produced either by preferential

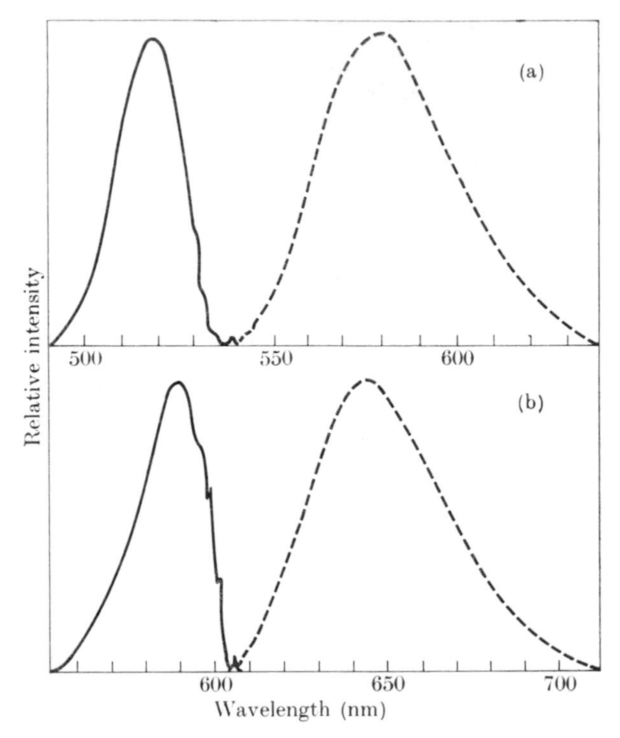

FIG. 4.5. M band absorption (full line) and emission (broken line) at 20 K for (a) CaF_2 and (b) SrF_2. The emission is induced by excitation in the F-band region (after Beaumont *et al.*, 1972*a*).

destruction of the centres in a particular orientation by the light or by redistribution of the centres among the various orientations after preferential excitation of the centres in some orientations by the light. The latter mechanism requires reorientation of the centres in optically excited states. The production of preferential alignment may require thermal energy. It is possible, for example, to preferentially align M centres in CaF_2 at 77 K but not at 20 K.

Let us assume, for simplicity, that the anisotropic centres have axial symmetry and that their optical absorption bands are due to electric-dipole transitions of the π-type (allowed for the electric vector parallel to the defect axis) or of the σ type (allowed for the electric vector perpendicular to the defect axis). If we limit our considerations to centres aligned along the $\langle 100 \rangle$, $\langle 110 \rangle$, and $\langle 111 \rangle$ directions it is possible, by carrying out only two polarized bleaching experiments, to distinguish between the various directions:

(i) If excitations with linearly polarized light with the electric vector $\boldsymbol{\epsilon}$ in the orientation $\boldsymbol{\epsilon} \parallel [0\bar{1}1]$ causes anisotropy when the absorption spectrum is subsequently measured with polarized light with $\boldsymbol{\epsilon} \parallel [0\bar{1}1]$ and with $\boldsymbol{\epsilon} \parallel [011]$, the centres must be aligned along $\langle 110 \rangle$ or $\langle 111 \rangle$; the cube edge orientation is excluded since the electric vector of the exciting light makes equal angles with the [001] and [010] orientations.

(ii) If excitation with polarized light with $\boldsymbol{\epsilon} \parallel [001]$ causes anisotropy when the absorption spectrum is subsequently measured with light with $\boldsymbol{\epsilon} \parallel [010]$ and with $\boldsymbol{\epsilon} \parallel [001]$ the centres must be oriented along $\langle 110 \rangle$ or $\langle 100 \rangle$; they cannot be oriented along $\langle 111 \rangle$ since [001] polarized light makes equal angles with all $\langle 111 \rangle$ directions.

It therefore follows that if anisotropy is induced with light with $\boldsymbol{\epsilon} \parallel [001]$ but not with $\boldsymbol{\epsilon} \parallel [0\bar{1}1]$ the centres must be aligned along $\langle 100 \rangle$. This result is unambiguous even when we allow orientations different from the three considered above. If anisotropy is induced by light with $\boldsymbol{\epsilon} \parallel [001]$ and with $\boldsymbol{\epsilon} \parallel [0\bar{1}1]$ the centres are oriented along $\langle 110 \rangle$. This result is not unambiguous, however, if we allow orientations different from the three considered here. It is not possible, for example, to distinguish between $\langle 110 \rangle$ and $\langle 121 \rangle$ orientations using polarized bleaching techniques (Compton and Rabin, 1964). If anisotropy is induced by light with $\boldsymbol{\epsilon} \parallel [0\bar{1}1]$ but not by light with $\boldsymbol{\epsilon} \parallel [001]$ the centres are aligned along $\langle 111 \rangle$ and this is, again, a unique determination.

Fig. 4.6 shows the effect of bleaching additively-coloured CaF_2 in the α band at 77 K with polarized light with $\boldsymbol{\epsilon} \parallel [001]$. Smakula's four-band spectrum (see Fig. 7.1) contributes noticeably to the absorption in Fig. 4.6 and, in addition, another unidentified impurity gives rise to a sharp peak at 552 nm and a band at 384 nm. It is apparent that bleaching in the α band region makes the crystal optically dichroic in both the α and β band regions. Very slight dichroism is produced in the α band but none in the β band by bleaching with light with $\boldsymbol{\epsilon} \parallel [0\bar{1}1]$. These results indicate that the centres which give rise to the β band also absorb in the α band region and that the electric dipoles responsible for the β band are oriented along cube edges. Excitation in the β band is very inefficient in producing dichroism. Since the dichroism in the α band region is opposite in sign from that in the β band region (Fig. 4.6) it follows that the dipole responsible for the dichroic absorption in the α band is perpendicular to that giving rise to the dichroism in the β band. Similar results have been obtained for SrF_2 by Beaumont and Hayes (1969) and for M centres in alkali halides by Okamoto (1961).

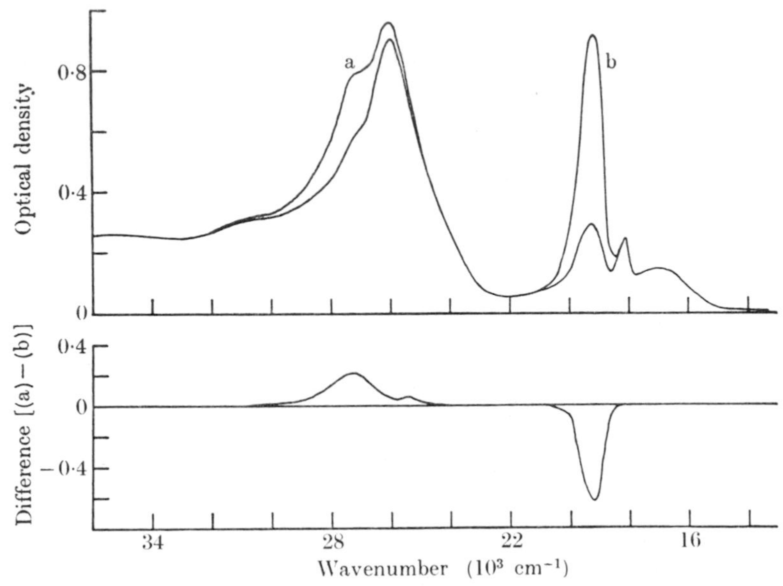

FIG. 4.6. Optical absorption at 77 K of additively-coloured CaF_2 after bleaching with linearly polarized light at 77 K with $\epsilon \parallel [001]$. Curve (a) was measured after the bleach with light with $\epsilon \parallel [100]$ and curve (b) with $\epsilon \parallel [001]$. The induced dichroism [(a)–(b)] is also indicated (after Beaumont and Hayes, 1969). The main absorption bands at about 26×10^3 and $19{\cdot}2 \times 10^3$ cm^{-1} are generally referred to as the α and β bands.

Since the experiments described above can only give orientations of dipoles relative to crystal axes and not to axes of the defect centre they cannot distinguish between π and σ transitions. This distinction may be made with other techniques (§§ 4.3.2 and 4.3.4). It will be shown later that the dipole moment of the $M\langle 110\rangle$ centre contributing to the β band is parallel to a cube edge of the crystal and perpendicular to the molecular axis of the centre. It will also appear that the dipole moment of the $M\langle 100\rangle$ centre contributing to the β band is parallel to the molecular axis and hence to a crystal cube edge. This similarity of behaviour makes the determination of the separate optical properties of the two centres very difficult.

The two main peaks in the dichroism of M centres in CaF_2 and SrF_2 are given in Table 4.13; the higher energy peak is sometimes called the M_F band because it occurs in the F band region.

Pumping in the F band region of CaF_2 with linearly polarized light at 77 K also induces dichroism in the 539·6 nm zero-phonon line and its magnitude is similar to that in the M band. This indicates that about the same fraction of $M\langle 100\rangle$ and $M\langle 110\rangle$ centres is preferentially aligned by the bleaching light. Pumping at 20 K in the F band region does not

TABLE 4.13

Main peaks (nm) *in the dichroism of M centres in* CaF_2 *and* SrF_2 (*after Beaumont and Hayes*, 1965)

	M_F band	M band
CaF_2	366	521
SrF_2	427	595

produce preferential alignment of either the $M\langle 100\rangle$ or $M\langle 110\rangle$ centres indicating that the alignment is thermally activated. The activation energy for thermal reorientation of M centres in the ground electronic state in CaF_2 has been measured by Beaumont and Hayes (1969) by measuring the decay with increasing temperature of optically induced dichroism in the β band. The behaviour could be described by a single activation energy $\Delta E = 0{\cdot}98 \pm 0{\cdot}01$ eV and this is not surprising in view of the preponderance of $M\langle 100\rangle$ centres. A value of $\Delta E = 1{\cdot}0$ eV has been found for M centres in KCl (Ishii, Tomiki, and Ueta, 1958, 1959).

Beaumont *et al.* (1972*a*) investigated in some detail the dichroism induced in the M band and the 605·3 nm line in SrF_2 by bleaching with linearly polarized light with $\boldsymbol{\epsilon} \parallel [001]$ in the F band region at 77 K (Fig. 4.7). The dichroism in the M band is greater than in the zero-phonon line and the band peaks at higher energy for $\boldsymbol{\epsilon} \parallel [001]$ than for $\boldsymbol{\epsilon} \parallel [100]$. These results are consistent with the presence of superimposed bands of $M\langle 100\rangle$ and $M\langle 110\rangle$ centres in the M band region, the absorption of the former being at somewhat higher energy and more dichroic. Beaumont *et al.* (1972*a*) estimated that the peak of the $M\langle 100\rangle$ band occurred at 575 nm and the $M\langle 110\rangle$ band at 590 nm with the latter about four times more intense than the former. Using this result it is possible to estimate the Huang–Rhys factor S for the $M\langle 110\rangle$ centre from the relationship

$$S = \ln\{\mathrm{OD_{band}}/\mathrm{OD_{zp}}\}, \tag{4.13}$$

where $\mathrm{OD_{band}}$ is the integrated optical density of the $M\langle 110\rangle$ band and $\mathrm{OD_{zp}}$ is the integrated optical density of the zero-phonon line. The large estimated value of $S \simeq 7{\cdot}5$ is consistent with the compact nature of the $M\langle 110\rangle$ centre (M centres in alkali halides do not have zero-phonon lines of measurable intensity). The fact that $M\langle 100\rangle$ centres do not exhibit a zero-phonon line suggests an even larger S-value and this is consistent with the even more compact nature of these centres.

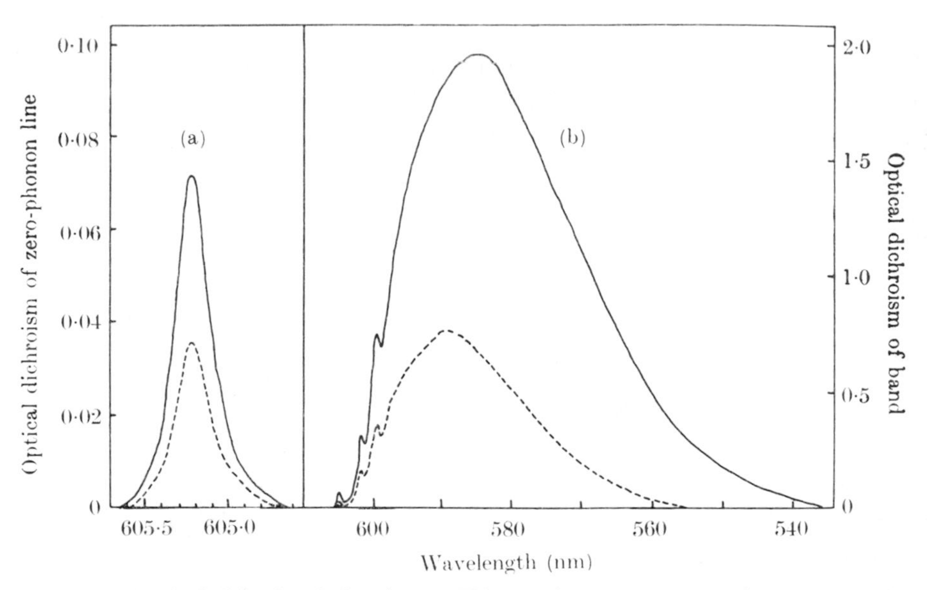

FIG. 4.7. Optical dichroism induced at 77 K in (a) the 605·3 nm zero-phonon line and (b) the M absorption band in SrF_2 by pumping with polarized light with $\epsilon \parallel [001]$ in the F-band region. Absorption was measured at 20 K with $\epsilon \parallel [001]$ (broken curves) and $\epsilon \parallel [100]$ (solid curves). Note different scales for (a) and (b) (after Beaumont *et al.*, 1972*a*).

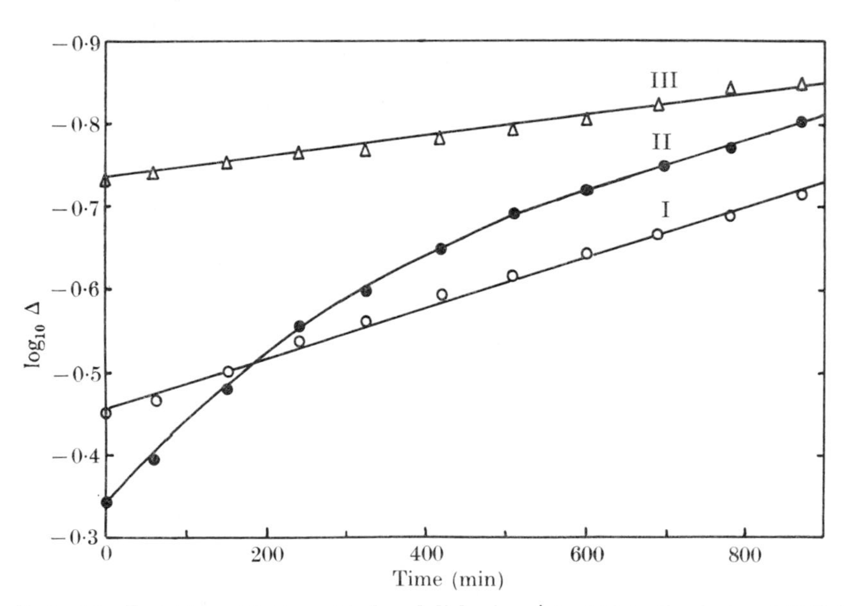

FIG. 4.8. Variation with time of induced dichroism Δ in SrF_2 in the dark at 293 K for I 605·3 nm line of $M\langle 110\rangle$ centre, II M band, III 769·2 nm zero-phonon line of R centre (after Beaumont *et al.*, 1972*a*).

The thermally induced decay in darkness of the dichroism $\triangle$ in the M band and the 605·3 nm line in SrF_2 at 293 K is shown in Fig. 4.8: $\triangle$ is the difference between the full and dashed curves in Fig. 4.7. Trace I shows that the dichroism of the zero-phonon line decays exponentially with a lifetime of 1400 min. Initially the decay of dichroism in the M band is faster than in the zero-phonon line (trace II) and has a lifetime of 430 min. However, after 600 min. the decay rate in the band is constant and is the same as in the zero-phonon line. These results indicate that the $M\langle 100\rangle$ centres reorient about three times faster than the $M\langle 110\rangle$ centres in SrF_2 at 293 K. The reorientation may lead to interconversion of $M\langle 110\rangle$ and $M\langle 100\rangle$ centres and hence help to establish thermodynamic equilibrium between the centres. The fact that the intensity ratio of the zero-phonon line and the M band is always the same after additive coloration suggests that such an equilibrium exists (trace III in Fig. 4.8 refers to R centres and is discussed in § 4.4).

The emission band of the M centres in CaF_2 is yellow in colour (Fig. 4.5). The fluorescence investigated by Feofilov (1953) in additively-coloured CaF_2 was red in colour and its origin has not been established. Beaumont *et al.* (1972*a*) found that the regions of excitation for the emission bands in CaF_2 and SrF_2 occurred in the same regions as the dichroic peaks (Fig. 4.6) and had the same polarization properties. The dipole moment of the emission band is parallel to the polarization direction for excitation in the β band region and perpendicular to the polarization direction for excitation in the α band region. It seems that after excitation in the α band region the M centres decay non-radiatively to the excited states of the β band and then decay radiatively to the ground states with the Stokes shift shown in Fig. 4.5. The excitation spectrum for the $M\langle 110\rangle$ zero-phonon line was also investigated in CaF_2 by Beaumont *et al.* (1972*a*) and found to be similar to that of the broad M emission band.

Both the $M\langle 100\rangle$ and $M\langle 110\rangle$ centres contribute to the dichroic peaks and the excitation bands in the α and β band regions. It is not possible to determine the separate contributions of the two types of centre to these optical spectra but it would appear that the $M\langle 100\rangle$ contribution dominates in CaF_2.

Beaumont and Hayes (1969) found that the lifetime of the M emission band in CaF_2 is $3{\cdot}6\times10^{-8}$ s for both α and β band excitation and is independent of temperature in the range 77–300 K. This short lifetime is characteristic of strong absorption lines and relatively small Stokes shifts (2100 cm^{-1} in this case) and is similar to the value of 6×10^{-8} s found for the M centre in KCl (Swank and Brown, 1963).

4.3.3. *Effects of uniaxial stress and electric fields*

Because of the narrowness of the $M\langle 110\rangle$ zero-phonon lines the effects of uniaxial stress are readily resolved and fit closely the patterns

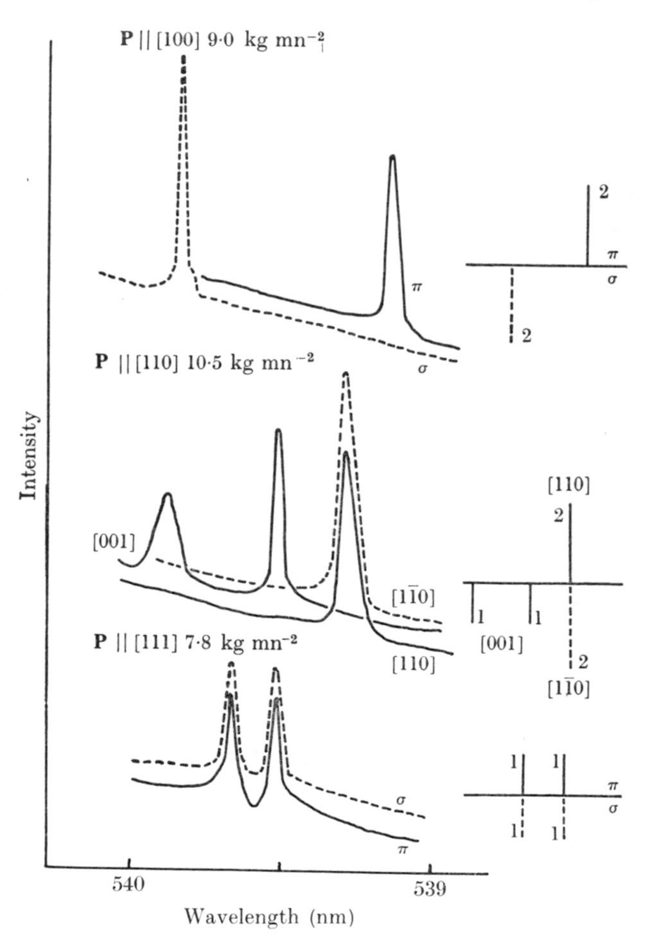

FIG. 4.9. Measured and predicted stress patterns for the 539·6 nm zero-phonon line of the $M\langle 110\rangle$ centre in CaF_2. π and σ indicate absorptions measured with $\boldsymbol{\epsilon} \parallel \mathbf{P}$ and $\boldsymbol{\epsilon} \perp \mathbf{P}$. For $\mathbf{P} \parallel [110]$ the orientation of $\boldsymbol{\epsilon}$ is indicated by a lattice direction (after Beaumont *et al.*, 1972*a*).

expected for centres with orthorhombic I symmetry (Fig. 4.9) (Kaplyanskii (1964) discusses these patterns). The zero-phonon line transitions occur between orbital singlet states and the stress raises orientational degeneracy of the centres. The stress Hamiltonian is

$$\mathscr{H} = A_1\sigma_{zz}+A_2(\sigma_{xx}+\sigma_{yy})+2A_3\sigma_{xy}, \tag{4.14}$$

where the σ's are the components of the piezospectroscopic tensor and the A-values are determined from experiment. The stress results show that the dipole moment of the transition giving rise to the zero-phonon line is oriented along $\langle 001\rangle$ and is perpendicular to the molecular axis of the $M\langle 110\rangle$ centre; the other two principal axes of the dipole moment are in the perpendicular plane and are along $\langle 110\rangle$ and $\langle 1\bar{1}0\rangle$. The measured parameters of eqn (4.14) at 20 K are given in Table 4.14 for

TABLE 4.14
Parameters (cm^{-1} (kg/mm^2)$^{-1}$) *of the piezo-spectroscopic tensor for the* $M\langle 110\rangle$ *centre zero-phonon line* (*Beaumont et al.*, 1972c)

	A_1	A_2	A_3
CaF_2	2·36±0·12	−0·72±0·04	0·70±0·04
SrF_2	3·22±0·16	−1·13±0·06	0·84±0·05

the 539·6 nm line in CaF_2 and the 605·3 nm line in SrF_2.

Both the 539·6 and 605·3 nm zero-phonon lines show large linear Stark effects due to raising of orientational degeneracy of the $M\langle 110\rangle$ centres by the applied electric field. The point groups in the fluorite structure with orthorhombic I symmetry are D_{2h} (the point group of the $M\langle 100\rangle$ centre), D_2 and C_{2v}. A linear Stark effect cannot be associated with D_{2h} because it has inversion symmetry. Even though D_2 lacks inversion symmetry a linear Stark effect cannot occur with this group because the Stark operator does not have the required transformation properties. It therefore follows from the results of the combined stress and Stark effects that the point symmetry of the $M\langle 110\rangle$ centre is C_{2v}. In this symmetry the C_2 axis may be parallel to $\langle 100\rangle$ or $\langle 110\rangle$ and there are similar possibilities for the orientation of the optical dipole moment **d** of the zero-phonon line transition. Analysis of the Stark patterns enables a unique choice to be made between these various combinations of possibilities.

The electric field Hamiltonian is

$$\mathscr{H} = \mathbf{D}\cdot\mathbf{F} \tag{4.15}$$

where **D** is the effective static dipole moment of the transition and **F** is the applied electric field. Comparison of observed and predicted Stark patterns (Fig. 4.10) shows that both the C_2 axis and **d** are aligned along ⟨100⟩. Measurements with **F** ∥ [100] gave, for those centres with the C_2 axis along [100], in units of cm^{-1} $(V/cm)^{-1}$

$$D_{100} = 6 \cdot 0 \pm 0 \cdot 5 \times 10^{-5}$$

for the 539·6 nm line in CaF_2 and

$$D_{100} = 5 \cdot 5 \pm 0 \cdot 5 \times 10^{-5}$$

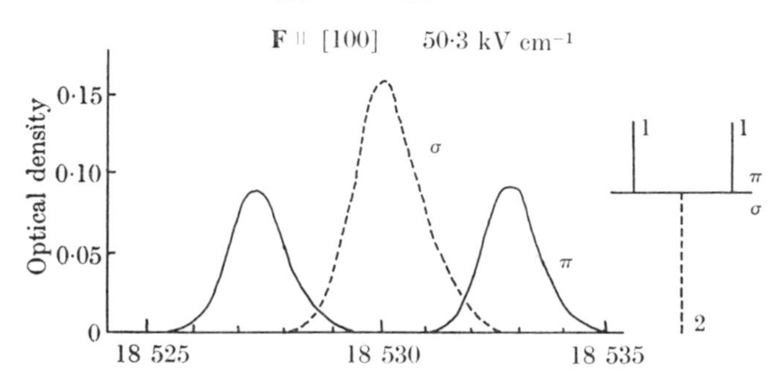

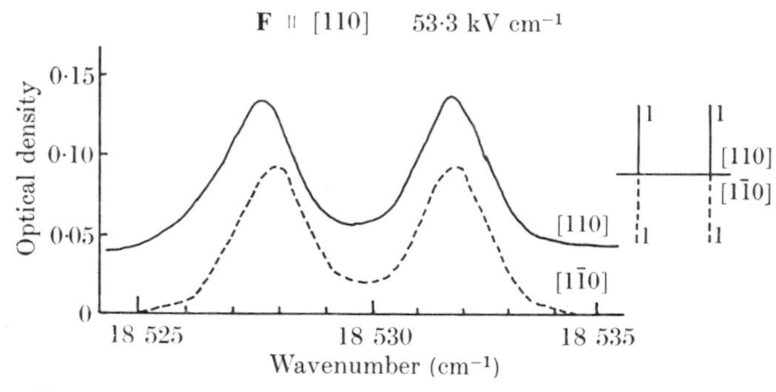

FIG. 4.10. Measured and predicted Stark patterns for the 539·6 nm line in CaF_2. π and σ indicate absorption measured with $\boldsymbol{\epsilon} \parallel \mathbf{F}$ and $\boldsymbol{\epsilon} \perp \mathbf{F}$. For **F** ∥ [110] the orientation of $\boldsymbol{\epsilon}$ is indicated by a lattice direction (after Beaumont *et al.*, 1972*a*).

for the 605·3 nm line in SrF_2. The Stark measurements also show (Beaumont *et al.*, 1972*a*) that the vibronic structure in the M absorption band is associated with the M⟨110⟩ zero-phonon line.

4.3.4. *Metastable states and paramagnetic resonance*

The main features of the optical behaviour of M⟨100⟩ centres may be explained using a simple model based on the electronic structure of the hydrogen molecule. The ground electronic state of such a system is $(1\sigma^2)^1\Sigma_g$ and the first excited state is $(1\sigma^2)^3\Sigma_u$. Calculations of the singlet-triplet splitting by Stoneham (unpublished, 1972), following

Berezin (1972), estimate that the triplet lies about 0·5 eV above the ground state for CaF_2, SrF_2, and BaF_2, Extending the H_2 model, allowed electric-dipole transitions are expected from the $^1\Sigma_g$ ground state to $(1\sigma\, 2p\sigma)^1\Sigma_u$ and $(1\sigma\, 2p\pi)^1\Pi_u$ excited states. It will be shown later that the absorption of the $M\langle 100\rangle$ centres in the α band region is σ-polarized and using the H_2 model it can be assigned to transitions of the $^1\Sigma_g \rightarrow {}^1\Pi_u$ type. The absorption of $M\langle 100\rangle$ centres in the β band region is π-polarized and may be assigned to $^1\Sigma_g \rightarrow {}^1\Sigma_u$ type transitions. It should be emphasized that the point symmetry of the $M\langle 100\rangle$ centre (D_{2h}) is lower than that of the H_2 molecule ($D_{\infty h}$) and that the degeneracy of a π state will be raised in the defect centre.

The electronic structure of the $M\langle 110\rangle$ centres cannot be simply related to that of the H_2 molecule. It appears that the cation nearest the paired F centres is an intrinsic part of the centre and that its effect is strong enough to bring one of the $^1\Pi_u$ levels of the H_2 model to a sufficiently low energy to give a σ polarized transition in the β band region. The near coincidence of absorption of $M\langle 100\rangle$ and $M\langle 110\rangle$ centres in the β band region appears to be fortuitous.

Effects arising from populating excited states of M centres in CaF_2 by optical pumping methods have been investigated by optical and epr methods (Hayes *et al.*, 1970). The optical pumping reduced the absorption of M centres in the α and β band regions due largely to transfer of population from the ground state of the $M\langle 100\rangle$ centres to the metastable $^3\Sigma_u$ type state. From the recovery rate of M band intensity it was concluded that the metastable state had a lifetime of $1{\cdot}7 \pm 0{\cdot}15$ s. The temperature dependence of this lifetime was measured in the region 4 to 226 K and was found to fit the expression

$$\frac{1}{\tau} = \frac{1}{\tau_0} + \nu \exp\left(-\frac{\Delta E}{kT}\right) \tag{4.16}$$

with

$$\tau_0 = 2{\cdot}6 \text{ s}, \quad \nu = 0{\cdot}52 \text{ s}^{-1} \quad \text{and} \quad \Delta E = 0{\cdot}006 \text{ eV}.$$

The corresponding values for the M centre in KCl are (McCall and Grossweiner, 1967)

$$\tau_0 = 69 \text{ s}, \quad \nu = 3 \text{ s}^{-1} \quad \text{and} \quad \Delta E = 0{\cdot}064 \text{ eV}.$$

Using filtered light it appeared that population of the $^3\Sigma_u$ type state could be produced by excitation in either the α or β band regions with comparable efficiency.

When additively-coloured crystals are optically pumped at 77 K in the resonant cavity of an epr spectrometer a metastable light-induced epr spectrum with orthorhombic symmetry is observed (Fig. 4.11) which fits the spin Hamiltonian

$$\mathscr{H} = \beta\mathbf{H}.\mathbf{g}.\mathbf{S}+D\{S_z^2-\tfrac{1}{3}S(S+1)\}+E(S_x^2-S_y^2), \tag{4.17}$$

with

$$S = 1, \quad g_x = g_y = g_z = 1{\cdot}999\pm0{\cdot}002,$$

$$D/g\beta = 615\pm5\ \text{G}, \qquad E/g\beta = 70\pm5\ \text{G}.$$

This spectrum is due to the ${}^3\Sigma_u$ state of the $M\langle100\rangle$ centre. The zero-field splitting D comes primarily from the magnetic dipole-dipole interaction between the two electrons. In the point-dipole limit the

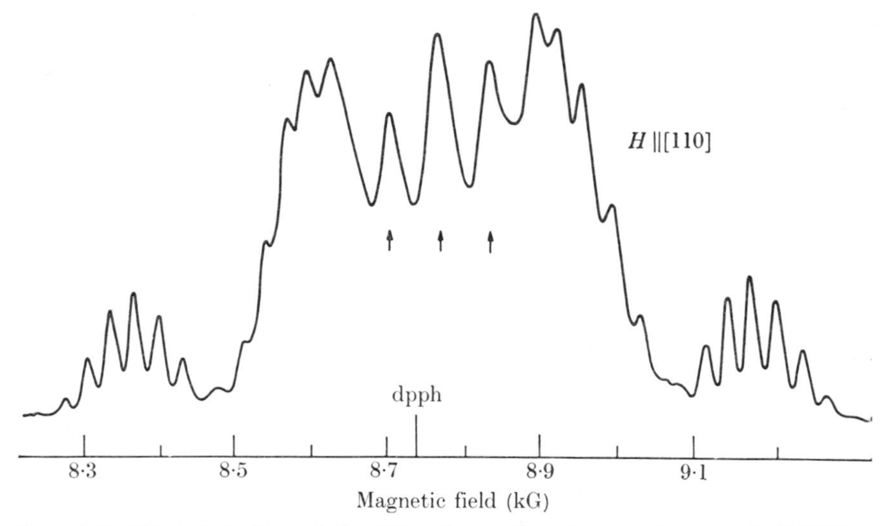

FIG. 4.11. First derivative of the K-band epr dispersion spectrum, with $H \parallel [110]$, of the triplet state of the $M\langle100\rangle$ centre in CaF_2 at 77 K. The arrows indicate some F-centre lines (after Hayes *et al.*, 1970).

observed value corresponds to a dipole separation of about 76·5% of the separation of the lattice sites. If 1s-like hydrogenic wavefunctions are used instead of point dipoles most of this discrepancy is removed.

The z direction in the spin-Hamiltonian (4.17) coincides with a cube edge and the x and y directions with face diagonals. The arrangement of neighbouring calcium ions (Fig. 4.4) results, for each cube edge direction, in two distinguishable $M\langle100\rangle$ centres with x axes (and y axes) rotated 90° relative to each other and there are therefore six distinguishable $M\langle100\rangle$ centres in the magnetic unit cell. The $M\langle100\rangle$ centre has 10 nearest-neighbour fluorines which are not all equivalent. The line groupings centred on 8·36 and 9·17 kG in Fig. 4.11 are due to resolved fluorine hyperfine structure. The fluorine hyperfine structure for centres aligned along [001] is best resolved with $H \parallel [110]$ since 8 of the nearest fluorines are then magnetically equivalent. Because of the complexity of the epr spectrum, arising primarily from the anisotropy of the fine structure, it is not possible to carry out a detailed analysis of the fluorine hyperfine structure; endor measurements would be required.

Neglecting the inequivalence of the fluorines the isotropic component of the fluorine hyperfine structure may be approximately described by the expression $\frac{1}{2}AS\sum I$, where $S = 1$, $I = \frac{1}{2}$ is the spin of a fluorine nucleus and the summation covers the nearest fluorines. Comparison of this expression with experiment gives $A \sim 174$ MHz which is, not surprisingly, comparable with the measured isotropic hyperfine interaction of 182·51 MHz for the nearest-neighbour fluorine hyperfine structure of the F centre in CaF_2 (Table 4.3).

The lifetime of the epr spectrum at 77 K was found to be $1{\cdot}6 \pm 0{\cdot}2$ s in agreement with the value of 1·7 s quoted earlier for the recovery of the M band intensity at 77 K; the agreement supports the view that this recovery is due primarily to depopulation of the $^3\Sigma_u$ type state of the $M\langle 100\rangle$ centre.

Wavelengths in the region 300–500 nm are effective in producing the $^3\Sigma_u$ epr spectrum and effects arising from polarization of the exciting light are readily observable. When the d.c. magnetic field is aligned along [100] it is parallel to the molecular axis of those $M\langle 100\rangle$ centres aligned along [100] and perpendicular to the axis of the centres aligned along [010] and [001]. It is possible to distinguish between these two groups of $M\langle 100\rangle$ centres because of the anisotropy of the fine structure of the $^3\Sigma_u$ epr spectrum. If the crystal is illuminated in the α band region at 77 K with polarized light with $\boldsymbol{\epsilon} \parallel [100]$ only the epr spectrum of the centres aligned along [010] and [001] is found; the epr spectrum of the centres aligned along [100] is missing. It follows, if one assumes that population of the $^3\Sigma_u$ state is due to excitations of $M\langle 100\rangle$ centres in the α band region, that the transitions in this region are σ-polarized and hence that the transitions of the $M\langle 100\rangle$ centres in the β band region are π-polarized. Pumping in the α band region will also excite F centres but no detectable effects were found arising from transfer of energy from F centres to $M\langle 100\rangle$ centres. This result is in contrast with the situation in KCl where transfer of energy from F centres to M centres is effective in populating the metastable triplet state (Schneider, 1966; McCall, and Grossweiner, 1967; Haarer and Pick, 1967).

The lifetime for recovery of the intensity of the 539·6 nm zero-phonon line of $M\langle 110\rangle$ centres after removal of the optical pump is

~0·4 s (Beaumont *et al.*, 1972*a*). However, no metastable epr spectrum associated with $M\langle 110\rangle$ centres has been found in CaF_2 and this may be due in part to the relatively low concentration of these centres.

In SrF_2 the M band amplitude recovers with a lifetime of $0{\cdot}12 \pm 0{\cdot}06$ s at 20 K on removal of the pumping light and the amplitude of the 605·3 nm zero-phonon line recovers with a lifetime of ~0·4 s (Beaumont *et al.*, 1972*a*). The preponderance of $M\langle 110\rangle$ centres in SrF_2 suggests that these lifetimes should be similar and the apparent discrepancy appears to be due partly to an increase in width of the zero-phonon arising from heating effects of the lamp. This gives a reduction of amplitude which recovers with a thermal time constant.

No metastable epr spectrum of $M\langle 100\rangle$ centres has been found in SrF_2. However, a metastable epr spectrum of $M\langle 110\rangle$ centres has recently been observed in SrF_2 which is not inconsistent with the model of Fig. 4.4 for these centres (Call, Hayes and Smith, 1974).

In concluding this section it should be noted that in additively-coloured CaF_2 crystals the number of F centres and M centres are comparable even after rapid quenching (den Hartog and Flim, 1972). For this reason it was suggested by these authors that the M centres may be formed preferentially at impurities. However, identical M centres are produced in irradiated crystals (Hayes and Lambourn, 1973) in a manner which does not require assumptions about impurities (see § 4.9.2). It has also been pointed out earlier in this section that the measured D-value of the triplet state of $M\langle 100\rangle$ centres in CaF_2 is consistent with the model of an unperturbed M centre. The mechanisms of additive coloration are not understood in any detail and there seems no strong reason to suggest at the present time that the M centres largely responsible for the intensity of the β band in CaF_2 are other than normal M centres. However, the situation in SrF_2 is less clearcut (Call, Hayes, and Smith, 1974).

4.3.5. *Optical properties of M^+ centres and reorientation of M centres*

It was demonstrated by Collins (1973*a*) that M^+ centres can be produced in additively-coloured CaF_2 by X-irradiating the crystals at 77 K or by optically bleaching in the α band at 77 K. An absorption band of these centres occurs on the low energy side of the M band and Collins (1973*a*) gives the peak position as 545 nm. The M^+ centres may also be made as an intermediate step in the conversion of F to M centres after heavy irradiation at 77 K (Hayes and Lambourn, 1973b; see also § 4.9.2) and in this situation the M^+ band may be measured with greater precision. Hayes and Lambourn (1973) give the peak position of the M^+ band at 77 K as 552 nm. The polarization properties of the band are consistent with a transition of the type $^2\Sigma$–$^2\Sigma$ expected on the basis of a H_2^+ model. Collins (1973*a*) also finds weak absorption bands of the M^+ centre in CaF_2 at 360 and 300 nm and a broad emission band of the type $^2\Sigma$–$^2\Sigma$ at 645 nm. The position of the emission band is determined by the Stokes shift associated with the 552 nm absorption band. It seems likely that the M^+ centres in CaF_2 are of the type $M^+\langle 100\rangle$.

Hayes and Lambourn (1973b) find that M^+ centres in irradiated SrF_2 have a broad absorption band peaking at 658 nm at 20 K; a zero-phonon line occurs at 699·4 nm which may be associated with this band. An investigation of effects of uniaxial stress and electric fields on this line would be helpful in establishing whether the M^+ centres in SrF_2 are predominantly of the type $M^+\langle 110\rangle$ (see § 4.3.3).

Reorientation of optically excited M centres in alkali halides is a two-photon process involving firstly formation of M^+ centres by light absorbed in the M_F band (Schneider, 1970; Collins and Schneider, 1972). The M^+ centres then absorb light in the same spectral range and reorient during de-excitation. The reoriented M^+ centres subsequently trap electrons giving rise to reoriented M centres. Collins (1973*b*) has shown that a similar mechanism operates in CaF_2. This mechanism requires electron traps in the crystal to form a population of M^+ centres during optical excitation and in the alkali halides F centres act as traps, forming F^- centres (an anion vacancy containing two electrons). The F^- centre has not been observed in alkaline earth fluorides. However, it has been discussed theoretically by Bennett (1971*b*) who uses a point-ion model and finds a photoionization energy of about 1·3 eV for CaF_2, SrF_2, and BaF_2.

Collins (1973*b*) finds that the introduction of electron-trapping impurities into CaF_2 increases the reorientation efficiency of M centres. He also finds that an M^+ centre excited in the α band region has a probability close to $\frac{2}{3}$ of reorienting in the process of de-excitation. Assuming complete randomization in the process of de-excitation this is the maximum possible value for $\langle 100\rangle$ oriented centres.

4.4. R and R^- centres

4.4.1. *Optical properties of R centres*

Some of the larger F-aggregate centres in alkaline earth fluorides show narrow zero-phonon lines at low temperatures. In particular, a zero-phonon line occurs at 677·4 nm in additively coloured CaF_2 whose width at half height is 0·7 cm^{-1} at 20 K. A corresponding line occurs in SrF_2 at 769·2 nm whose width at half height is 3·9 cm^{-1} at 20 K. These zero-phonon lines are associated with broad asymmetric

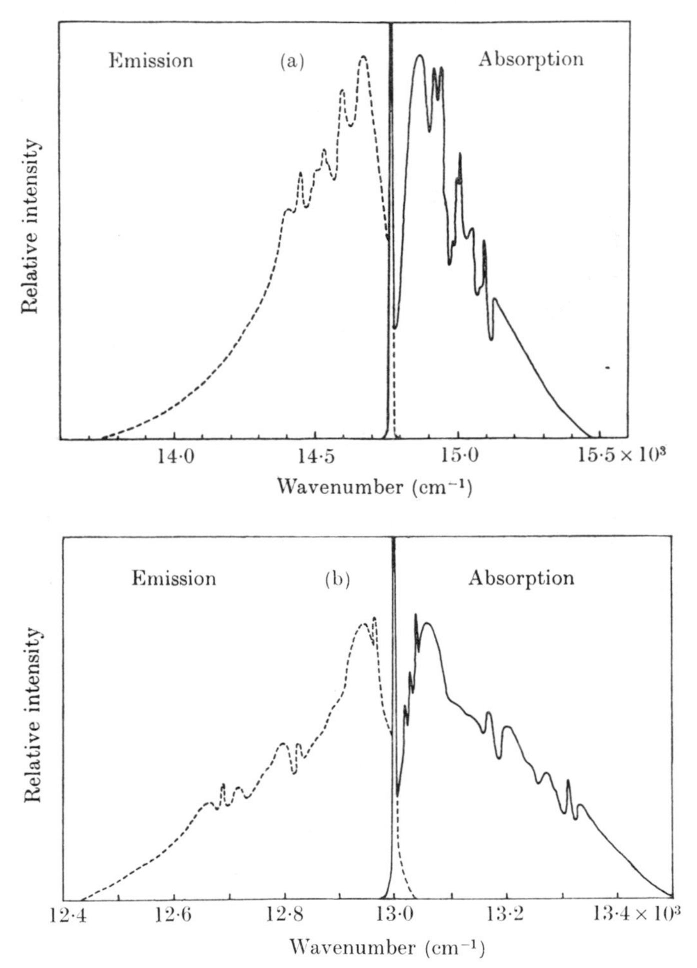

FIG. 4.12. Absorption and emission of the R centre at 20 K in (a) CaF_2 and (b) SrF_2 (after Beaumont *et al.*, 1972*b*).

absorption bands at higher energies which show partly resolved phonon structure at low temperatures (Fig. 4.12). Optical excitation of the crystals at 20 K at energies greater than that of the zero-phonon line gives rise to a broad fluorescence band (Fig. 4.12). An investigation of these optical spectra by Beaumont, Harmer, and Hayes (1972*b*,*c*) led them to suggest that linear R centres, composed of three nearest-neighbour F centres aligned along $\langle 100 \rangle$ (Fig. 4.4) were responsible for the spectra.

The ground electronic configuration of the R centre is analogous to the ground configuration $1s1s1s$ ($^2\Sigma_g^+$, $^2\Sigma_u^+$, $^4\Sigma_u^+$) of the linear H_3 molecule discussed by Hirschfelder, Eyring and Rosen (1936). The linear F_3 model has D_{2d} point symmetry and in this symmetry the electronic states of H_3 (which has point symmetry D_{2h}) are classified as $^2\Gamma_1$, $^2\Gamma_4$, and $^4\Gamma_4$. Calculations by Hirschfelder *et al.* (1936) for H_3 give the $^2\Gamma_4$ state lowest and the $^2\Gamma_1$ state next highest. All the measured properties of the R centre zero-phonon line are consistent with a transition of the type $^2\Gamma_4$–$^2\Gamma_1$.

Beaumont *et al.* (1972*b*) showed that if a freshly prepared crystal of additively-coloured CaF_2 is bleached at room temperature with light in the M band region (~520 nm; see § 4.3) the intensity of the 677·4 nm line of the R centre may be increased by a factor of about 5, to a maximum value. Subsequent bleaching in the broad R band (at ~665 nm; see Fig. 4.12) at room temperature produces a slow decrease in the intensity of the R band and a small increase in the intensity of the M band. Similar changes occur in additively-coloured SrF_2 subjected to the same bleaching operation.

Beaumont *et al.* (1972*b*) measured the excitation spectrum for the fluorescence of the R centre using linearly polarized pumping light. Measurements were made at 20 K to prevent centres realining under the influence of the polarized light (see the next paragraph). The use of polarized pumping light enables one to distinguish between π- and σ-polarized bands in the excitation spectrum (Fig. 4.13). In crystals containing M centres excitation in the M band gives a strong R fluorescence due to energy transfer from M to R centres. However, M centres may be preferentially destroyed by optical bleaching in the M band at room temperature and Fig. 4.13 was obtained after optical destruction of M centres. The R fluorescence may also be generated by pumping in the broad R band (Fig. 4.12), which absorbs π-polarized light, but this part of the excitation spectrum is not included in Fig. 4.13.

Crystals of CaF_2 and SrF_2 containing R centres may be made optically dichroic in the region of the R absorption band by pumping crystals in the F band region at 77 K with linearly polarized light. Light of wavelength longer than the F band does not produce the dichroism. Because of overlap of π- and σ-polarized bands of the R centre in the F

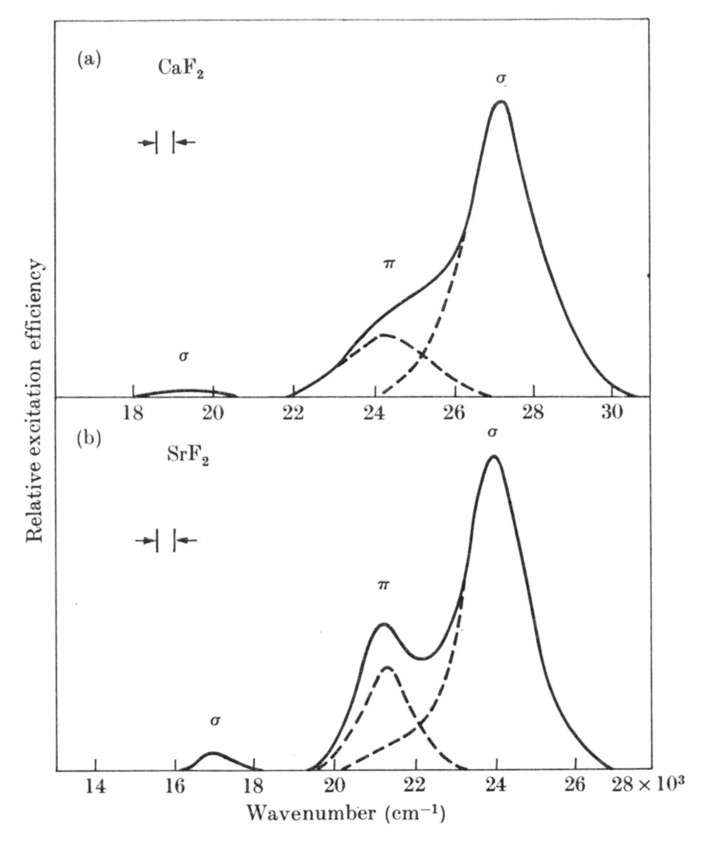

FIG. 4.13. Excitation spectrum for the *R* centre zero-phonon line fluorescence in (a) CaF_2 and (b) SrF_2 showing polarization of the bands (after Beaumont *et al.*, 1972*b*).

band region (Fig. 4.13) the sign of the dichroism depends on the particular filter combination used for optical pumping. The dichroism arises largely from preferential removal of *R* centres from a specific cube edge orientation by the exciting light and subsequent realignment in either of the other two possible orientations. In this process the total number of *R* centres is conserved. No realignment of *R* centres can be produced by optical excitation at 20 K indicating that some thermal energy is required for the reorientation. Dichroism induced at 77 K is destroyed in the dark by thermally induced disorientation above 270 K. It is also possible to induce dichroism in the *R* band by preferential destruction of *R* centres in the temperature range 150–270 K by bleaching in the *F* band region with linearly polarized light with $\epsilon \parallel \langle 100 \rangle$.

The Huang-Rhys factor S for the R band was found by Beaumont *et al.* (1972*b*) to be

$$S = 3{\cdot}1 \pm 0{\cdot}5 \quad \text{for CaF}_2,$$

$$S = 2{\cdot}6 \pm 0{\cdot}5 \quad \text{for SrF}_2.$$

These values of S are comparable with those obtained for R centres in alkali halides (Fitchen, 1968).

The phonon structure in the R band in CaF_2 is compared in Fig. 4.14

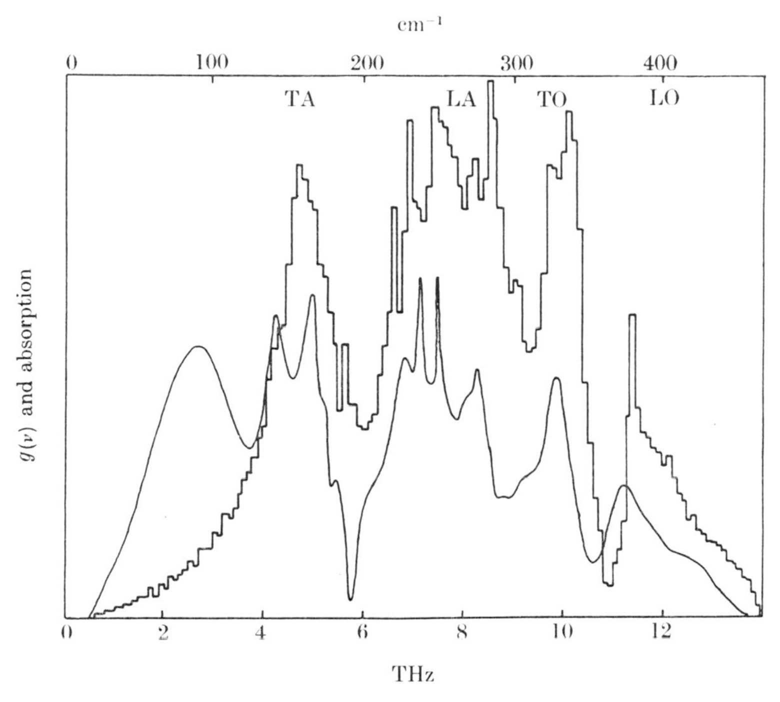

FIG. 4.14. Comparison of the phonon structure in the R centre absorption band in CaF_2 at 1·5 K with a single-phonon density of states $g(\nu)$ (histogram) given by Elcombe and Pryor (1970) (after Beaumont *et al.*, 1972*b*).

with a calculated single-phonon density of states obtained from a shell model fit to neutron diffraction measurements (Elcombe and Pryor, 1970; see § 2.3). It is apparent that, in general, there is good agreement between the positions of the phonon peaks in the R band and the peaks in the density of states. The selection rules for vibronic transitions of a defect at a lattice point in the fluorite structure have been considered by Loudon (1964) (see also Hughes, 1967). For a $\Gamma_4 \rightarrow \Gamma_1$ transition coupling only to phonons of Γ_1 symmetry is allowed for transitions originating on $n = 0$ vibrational levels. This situation is not very restrictive for vibronic transitions and similarity to the full single-phonon density of states is not surprising.

In both absorption and fluorescence a peak occurs in the broad band separated by 85 cm^{-1} from the zero-phonon line in CaF_2 and 60 m^{-1} in SrF_2 (Fig. 4.12). There is no corresponding peak in the single-phonon density of states; the peak in the density of states for TA phonons occurs at 161 cm^{-1} in CaF_2 (Fig. 4.14) and at $\sim$120 cm^{-1} in SrF_2 (see § 2.3). It seems that the low-energy peaks may be due to resonance modes associated with the defect. Such modes occur commonly in the optical bands of *F*-aggregate centres in the alkali halides (Pierce, 1964).

In both CaF_2 and SrF_2 some relatively weak satellite lines occur near the *R* zero-phonon line due to *R* centres perturbed by other defects; these are readily visible in CaF_2 (Fig. 4.12). The ratio of the intensities

TABLE 4.15

Parameters (cm^{-1}(kg/mm^2)$^{-1}$) *of the piezospectroscopic tensor for the zero-phonon line of R* (*Beaumont et al.*, 1972*c*) *and* R^- (*Beaumont et al.*, 1972*d*) *centres*

	R centre		R^-	
	A_1	A_2	A_1	A_2
CaF_2	$1{\cdot}61\pm0{\cdot}08$	$-0{\cdot}48\pm0{\cdot}02$	$0{\cdot}97\pm0{\cdot}05$	$-0{\cdot}27\pm0{\cdot}01$
SrF_2	$1{\cdot}56\pm0{\cdot}08$	$-0{\cdot}53\pm0{\cdot}03$	$1{\cdot}50\pm0{\cdot}07$	$-0{\cdot}82\pm0{\cdot}04$

of these lines to the main zero-phonon line is highly sample dependent and can be varied by optical or thermal treatment (Beaumont *et al.*, 1972*b*).

Because of the narrowness of the zero-phonon lines of the *R* centre the line splittings produced by application of uniaxial stress are readily resolved. The *R* centre zero-phonon line has been assigned to a transition of the type $^2\Gamma_4 \rightarrow {}^2\Gamma_1$ (Beaumont *et al.*, 1972*c*) and no orbital degeneracy is involved. In this case application of uniaxial stress raises orientational degeneracy only and for a defect with a fourfold (z) axis the components of the piezospectroscopic tensor referred to the principal axes of the centre are

$$A_{zz} = A_1, \qquad A_{xx} = A_{yy} = A_2, \qquad (4.18)$$
$$A_{xy} = A_{yz} = A_{zx} = 0.$$

so that only two independent parameters are required to describe the stress-induced shifts. The values of A_1 and A_2 for the *R* centre zero-phonon line in CaF_2 and SrF_2 are given in Table 4.15. The uniaxial stress measurements show that the zero-phonon lines are due to transitions between orbital singlet states in centres with tetragonal symmetry and that the electric dipole moment of the transition is parallel to the tetragonal axis.

The Stark effect of the R centre zero-phonon line has also been investigated by Beaumont *et al.*, 1972*c*. They found, using a d.c. electric field, that the zero-phonon line of the R centre in CaF_2 and SrF_2 shows a quadratic red shift but no measurable linear effect. The linear R model lacks inversion symmetry. However, a linear Stark effect cannot occur for a $\Gamma_4 \rightarrow \Gamma_1$ transition in D_{2d} symmetry because, in this symmetry, the electric dipole operator does not have the required symmetry properties (Kaplyanskii and Medvedev, 1967).

The ground state of the R centre is expected to be of the type $^2\Gamma_4$ and the centre should be paramagnetic in its ground state with a g-value close to 2. The g-value of the F centre is also close to 2 (Table 4.1) and its epr spectrum should overlie that of the R centre. It has not yet been found possible to prepare crystals of CaF_2 or SrF_2 containing R centres in concentration comparable with that of F centres and no ground state epr of R centres has been identified.

In addition to spin doublets the R centre also gives rise to a set of quartet spin states. Optical excitation from the ground spin-doublet may result in the transfer of some energy to the quartet system. Relaxation from the quartet system to the ground spin-doublet is expected to be relatively slow because of the change in spin multiplicity (see § 4.3 for a discussion of such effects for M centres). Beaumont *et al.* (1972*b*) found that the zero-phonon line of the R centre in SrF_2 is reduced in intensity by 15 per cent on exposure to an unfiltered SP 500 W mercury lamp at 20 K and that the intensity recovers with a lifetime of about 0·65 s when the pumping light is removed. The corresponding effect in CaF_2 was too small to be observed. It appears that about 15 per cent of the R centres in SrF_2 may be maintained in quartet states. These states are paramagnetic but the possibility of observing epr presents sensitivity problems.

4.4.2. *Optical properties of R^- centres*

The spectrum of zero-phonon lines in additively-coloured CaF_2 and SrF_2 is enriched by optical pumping in the F band region at low temperatures (Harmer, 1971). In particular a zero-phonon line appears in CaF_2 at 732·7 nm whose width at half height is 0·6 cm^{-1} at 20 K; a corresponding line occurs in SrF_2 at 835·0 nm whose width at half height is 5·5 cm^{-1}. These lines were assigned by Beaumont *et al.* (1972*d*) to the linear R^- centre which has the same point symmetry (D_{2d}) as the linear R centre. The lines are readily produced at temperatures down to 4 K and since ionic motion is unlikely at these temperatures it seems that an ionized form of an existing centre is involved.

The electronic structure of the R^- centre is analogous to that of the linear H_3^- molecule. The electronic structure of this molecule has been investigated by Macias (1968) who showed that the lowest electronic state is $^1\Sigma_g^+$. In D_{2d} symmetry this is classified as $^1\Gamma_1$ and the zero-phonon line of the R^- centre is probably a transition of the type $^1\Gamma_1 \rightarrow {}^1\Gamma_4$. Such a transition is allowed for π-polarized light and is consistent with results of investigations by Beaumont *et al.* (1972*d*).

The R^- zero-phonon line cannot be observed in crystals which do not contain R centres in measurable concentration. There is a linear relationship between the maximum obtainable intensity of the R^- zero-phonon line and the R centre concentration. It appears that R centres trap electrons ionized into the conduction band by light in the F band region, forming R^- centres. Application of an electric field of about 50 kV/cm during optical pumping at 20 K increases the obtainable concentration of R^- centres by a factor of about 5. This arises from increased ionization of optically excited states in the F band region by the electric field. However, the maximum concentration of R^- centres obtainable in this way is too small to allow observation of any vibronic structure which may be associated with the zero-phonon line.

On warming a crystal containing R^- centres it is found that the centres disappear over a period of about 5 minutes at 285 K in CaF_2 and at 145 K in SrF_2. The R^- centres are also destroyed by optical excitation at the energy of the zero-phonon line. In CaF_2 the R^- centres are destroyed in a few seconds at 150 K and in about 30 minutes at 20 K by light from a 100 W quartz-iodine lamp passed through a monochromator with a band width of 0·8 nm. Light of wavelength longer than the zero-phonon line does not destroy the centres. The crystals may be made optically dichroic in the R^- zero-phonon line by using linearly polarized bleaching light at 20 K. Bleaching with light with $\boldsymbol{\epsilon} \parallel [100]$ at the energy of the zero-phonon line preferentially destroys the R^- centres aligned along [100] and hence gives rise to dichroism.

As in the case of the R centre the R^- zero-phonon line shows a well-resolved uniaxial stress pattern characteristic of transitions between orbital singlet states in a centre with tetragonal symmetry (Beaumont *et al.*, 1972*d*). These measurements show that the dipole moment of the transition is of the π type. The piezospectroscopic tensor has the same form as that for the R centre and the tensor parameters measured at 20 K for CaF_2 and SrF_2 are given in Table 4.15.

The R^- zero-phonon line in CaF_2 shows a small quadratic Stark shift but no linear Stark effect could be detected. Since R and R^- have the same point symmetry it would seem that the absence of an observable linear Stark effect for both systems may be due to symmetry restrictions rather than accidental effects and this is consistent with the model proposed for the centres.

4.5. Other centres, including colloids, in additively-coloured crystals

4.5.1. *Centres displaying zero-phonon lines*

Additively-coloured CaF_2 and SrF_2 crystals show many zero-phonon lines in addition to those described in §§ 4.3 and 4.4. Most of these lines appear to be associated with impurities so far unidentified. An extensive account of the lines has been given by Beaumont, Harmer, Hayes, and Spray (1972) (see also Tanton, Stettler, Shatas, Williams, and Mukerji, 1968 and Runciman, Stager, and Crozier, 1963). Of the unidentified lines the 598·8 nm line in CaF_2 and the 679·9 and 689·0 nm lines in SrF_2 appear to be the most interesting. These lines have associated vibronic structure on the high energy side related to the single-phonon density of states of the host crystal. The Huang-Rhys factor S is found to have the relatively low value of $\sim$0·7 for the 598·8 nm line and $\sim$1·1 for both the 679·9 and 689·0 nm lines. Uniaxial stress measurements show that the lines arise from allowed electric dipole transitions between orbital singlet states in centres with tetragonal symmetry. It was tentatively suggested by Beaumont, Harmer, Hayes, and Spray that a complex of four F centres (N centre; Fig. 4.4) is a possible origin of the 598·8 nm line in CaF_2 and the 679·9 nm line in SrF_2 and that the 689·0 nm line in SrF_2 could arise from this complex with an impurity in the nearest anion site. However, these authors emphasized the difficulty of establishing the structure of intrinsic F aggregate centres in CaF_2 involving four or more F centres, primarily because of the low concentrations obtainable.

4.5.2. *Colloids*

Slow cooling of alkaline earth fluorides after heavy additive coloration produces large metallic precipitates (up to 10 μm in size) in the crystals.

The structure of these precipitates was investigated by den Hartog (1969) and den Hartog, Tinbergen, and Perdok (1970). Small metallic precipitates (of the order of 10 nm in size), known as colloids, may be conveniently produced by rapidly quenching additively-coloured crystals and subsequently heating them to temperatures between 200 and 350 C. In a CaF_2 crystal treated in this way den Hartog (1969) found a colloid absorption band peaking at 550 nm; the half-width was about 0·4 eV and did not vary appreciably between 300 and 77 K. A corresponding band occurred in SrF_2 at 620 nm whose half-width was about 0·6 eV. Earlier investigations of colloids in CaF_2 were made by Mollwo (1934) and by Mayerl (1951). The decoration of dislocations by precipitated metal in additively coloured CaF_2 was investigated by Bontinck and Dekeyser (1956) and by Bontinck and Amelinckx (1957). A similar effect in CaF_2 crystals irradiated with 1·5 MeV protons was studied by Kubo (1966).

The epr of conduction electrons in the small colloids was investigated by den Hartog (1969) who found a line in CaF_2 with $g = 1{\cdot}999$. The width of this resonance varied from sample to sample over the range 0·07 to 0·14 G but the g-value was sample independent. In SrF_2 the g-value was 2·002 and the sample dependence of the line width was 0·4 to 3·0 G. It appears that relaxation of the conduction electrons, and hence the resonance line width, depends to some extent on particle size. The integrated intensity of the resonance line in SrF_2 was found to be independent of temperature. In BaF_2 the g-value of the colloid resonance was 2·003 and the linewidth was about 1 G. No theoretical investigation of the colloid g-values has yet been made.

4.6. Trapped holes and trapped excitons

4.6.1. *The self-trapped hole (V_K centre) and the self-trapped exciton*

In some ionic crystals, holes may be stabilized at particular lattice sites due to strong electron-lattice interaction. These holes are described as self-trapped when no other lattice defects are involved. The self-trapped hole is paramagnetic and its epr spectrum in CaF_2 was described by Hayes and Twidell (1962). The corresponding centre in the alkali halides (Castner and Känzig, 1957) is generally referred to as the V_K centre or $\{X_2^-\}$ centre where X is a halogen. We shall use the former notation here. The centres may be produced by X-irradiating undoped CaF_2 crystals at 77 K. However, Hayes, and Twidell (1962) found that if the crystals are doped with the heavier trivalent rare-earth ions such as thulium, the initial rate of production of the centres could be increased by more than an order of magnitude. This increase is due to trapping of electrons by trivalent rare-earth ions forming divalent rare-earth ions with a consequent reduction in electron-hole recombination. A proportion of the divalent rare-earth ions formed in this process survive to room temperature and above (see § 4.6.3, Chapter 5, and § 5.7.2 for further discussion.)

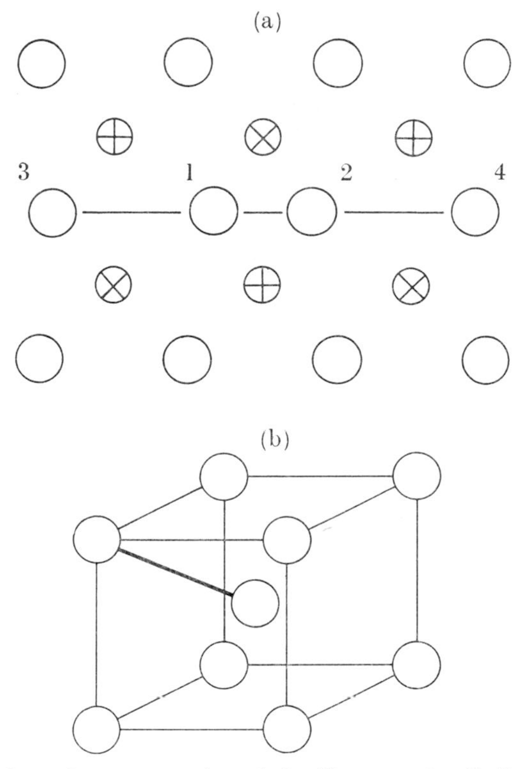

FIG. 4.15. (a) Schematic representation of the V_K centre in alkaline earth fluorides. ⊗ and ⊕ are cations above and below the plane of the paper. ○ are anions. The hole in the V_K centre is primarily located in fluorines 1 and 2 and spreads slightly to fluorines 3 and 4. (b) Schematic representation of the *H* centre in alkaline earth fluorides. The hole in the *H* centre is located primarily on an interstitial fluorine and a lattice fluorine.

Analysis of the epr spectrum showed that the V_K centre in alkaline earth fluorides was a molecular-type centre with its principal axis aligned along $\langle 100 \rangle$ (Fig. 4.15(a)). The point symmetry, D_{2h}, is the same as that of the V_K centre in the alkali halides where the alignment is, however, along $\langle 110 \rangle$. The epr of the V_K centre in SrF_2 and BaF_2 was investigated by Tzalmona and Pershan (1969) and by Beaumont,

Hayes, Kirk, and Summers (1970) and in BaF_2 by Kazumata (1969). Departures from axial symmetry are too small to be readily detectable in the epr spectrum and an adequate description of the spectrum is given by the spin-Hamiltonian

$$\mathscr{H} = g_{\parallel}\beta H_z S_z + g_{\perp}\beta(H_x S_x + H_y S_y) + A I_z S_z + B(I_x S_x + I_y S_y) + \\ + A' I'_z S_z + B'(I'_x S_x + I'_y S_y), \quad (4.19)$$

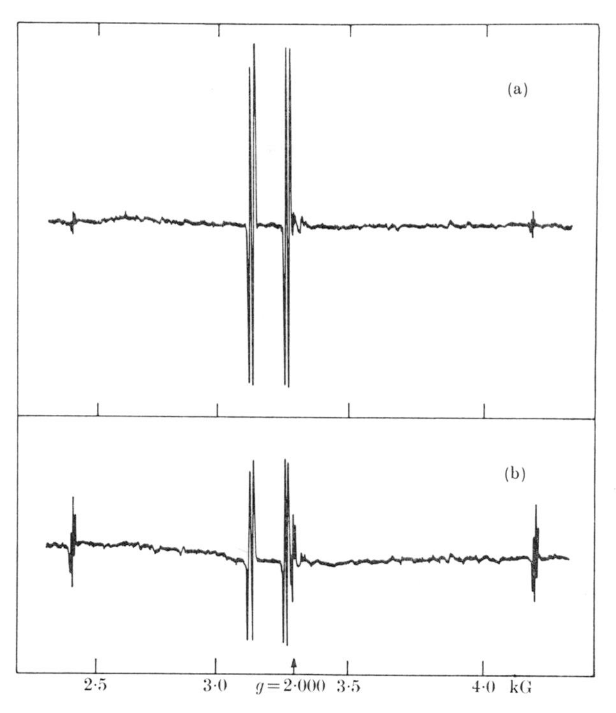

FIG. 4.16. First derivative of the X-band epr spectrum of the V_K centre in SrF_2: Tm at 77 K after bleaching at 77 K into the ultraviolet band of the centre with linearly polarized light with $\boldsymbol{\epsilon} \parallel [001]$. (a) $H \parallel [001]$, (b) $H \parallel [010]$ (after Beaumont *et al.*, 1970).

with $S = \frac{1}{2}$, $I = I_1 + I_2$, and $I' = I_3 + I_4$ where $I_1 = I_2 = \frac{1}{2}$ are the spins of fluorine nuclei 1 and 2 and $I_3 = I_4 = \frac{1}{2}$ are the spins of fluorine nuclei 3 and 4 (Fig. 4.15(a)). The hyperfine interaction with fluorines 1 and 2 is large indicating that the hole is localized almost entirely in the vicinity of fluorines 1 and 2 (the notation $[X_2]$ is used for this reason). The hyperfine interaction with fluorines 3 and 4 is small and is resolved in CaF_2 and SrF_2 (Fig. 4.16) but not in the larger BaF_2 lattice. In addition to the parameters of eqn (4.19) the epr linewidths are also given in Table 4.16.

TABLE 4.16
Spin-Hamiltonian parameters of self-trapped holes (V_K centres) in CaF_2, SrF_2, BaF_2, *and* $SrCl_2$

		$g_{\parallel}$	$g_{\perp}$	A (G)	B (G)	A'(G)	B'(G)	Line width (G)
CaF_2	a	2·001 ±0·001	2·020 ±0·001	899·9 ±1·0	48·5 ±2·0	15·1 ±0·3	⩽4	4·1
	b	2·002 ±0·001	2·019 ±0·001	898 ±1·0	43 ±10	—	—	6·0
SrF_2	c	2·002 ±0·002	2·020 ±0·002	899 ±2	47 ±3	6·3 ±0·2	—	—
	d	2·002 ±0·002	2·022 ±0·002	899·0 ±1·0	44·0 ±1·0	6·2 ±0·3	⩽3	3·5
BaF_2	c	2·000 ±0·002	2·020 ±0·002	899 ±2	46 ±2	—	—	—
	d	2·004 ±0·002	2·024 ±0·002	897·0 ±2·0	41·5 ±1·0	⩽2	⩽2	5·6
$SrCl_2$ ^{35}Cl	e	2·0020 ±0·0004	2·0362 ±0·0015	98·6 ±0·25	10·3 ±1·5	5·0 ±0·7	< 2·5	
$SrCl_2$ ^{37}Cl	e	2·0020 ±0·0004	2·0362 ±0·0015	82·1 ±0·5	—	4·1 ±1·0	<2·5	

[a] Hayes and Twidell (1969). [b] Kazumata (1969). [c] Tzalmona and Pershan (1962). [d] Beaumont, Hayes, Kirk and Summers (1970). [e] Bill, Suter, and Lacroix (1966).

An investigation of the interaction of the unpaired electron with more distant nuclei has been given for CaF_2 and BaF_2 by Marzke and Mieher (1969) using endor methods; the reader is referred to the original paper for details.

Alkaline earth fluorides containing V_K centres give rise to a strong optical absorption band in the ultraviolet. Estimates of the peak position of this band in CaF_2 were made by Merz and Persham (1967) and by Arkangel'skaya and Alekseeva (1966). An unambiguous identification of the optical absorption of the V_K centres in CaF_2, SrF_2, and BaF_2 in the ultraviolet and near infrared was made by Beaumont, Hayes, Summers, and Twidell (1968), Beaumont, Hayes, Kirk, and Summers (1970), Hayes (1969), and Hayes (1970) by correlating effects of polarized bleaching light on the optical and epr spectra. Neglecting departure from axial symmetry in the V_K centre the optical absorption spectrum can be described in terms of the energy levels of the free molecule ion F_2^- (Beaumont *et al.*, 1968, 1970). The ground state of the system is

$$2p\sigma_g^2\, 2p\pi_u^4\, 2p\pi_g^4\, 2p\sigma_u\, (^2\Sigma)_u,$$

and excited states in order of increasing energy are

$$2p\sigma_g^2\, 2p\pi_u^4\, 2p\pi_g^3\, 2p\sigma_u^2\, (^2\Pi_g),$$

$$2p\sigma_g^2\, 2p\pi_u^3\, 2p\pi_g^4\, 2p\sigma_u^2\, (^2\Pi_u),$$

$$2p\sigma_g\, 2p\pi_u^4\, 2p\pi_g^4\, 2p\sigma_u^2\, (^2\Sigma_g).$$

The transition $^2\Sigma_u \rightarrow {}^2\Sigma_g$ is allowed for electric dipole radiation with the electric vector $\boldsymbol{\epsilon}$ parallel to the molecular axis (π-type) and gives rise to the intense ultraviolet band. A weakly allowed transition occurs to the state $^2\Pi_{g\frac{1}{2}}$ for light with $\boldsymbol{\epsilon}$ perpendicular to the molecular axis (σ-type) and gives rise to a band in the near infrared (Table 4.17).

TABLE 4.17

Peak position (nm) *at* 77 K *of optical absorption bands of the self-trapped hole* (V_K *centre*)

	$^2\Sigma_u \rightarrow {}^2\Sigma_g$	$^2\Sigma_u \rightarrow {}^2\Pi_{\frac{1}{2}g}$
CaF_2[a]	320	~750
SrF_2[a]	326	~750
BaF_2[a]	336	~750
$SrCl_2$[b]	397	~800

[a] Beaumont *et al.* (1970).
[b] Hayes, Lambourn, Rangarajan, and Ritchie (1973).

Although the excited state $^2\Pi_u$ is not observed in absorption it does affect the ground state g-value (for a good description of the electronic structure of the V_K centre see Slichter, 1963).

Because of the anisotropy in the optical absorption of the V_K centre it is possible to produce preferential alignment of the centres using the methods already described for F-aggregate centres (§ 4.3). The self-trapped hole in alkaline earth fluorides is immobile at 77 K. Fig. 4.17(a) shows effects of bleaching into the ultraviolet band of the V_K centre in SrF_2:Tm at 77 K with linearly polarized light with $\boldsymbol{\epsilon} \parallel \{001\}$. In the process of de-excitation, realignment of the centre takes place with the result that the {001} orientation is depopulated and the population of the {010} and {100} orientations increases. The population changes are detected by measuring the optical absorption with polarized light

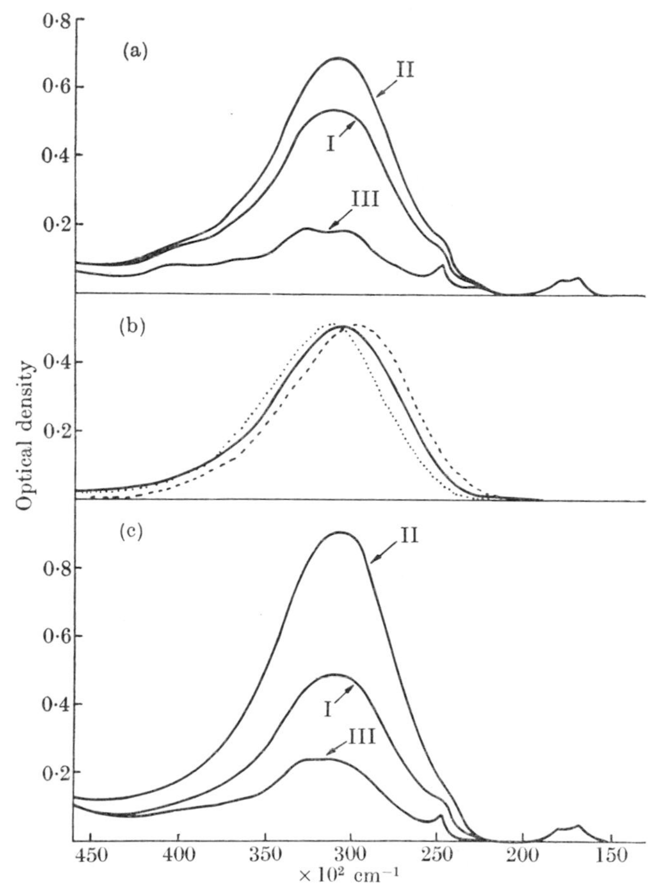

FIG. 4.17. (a) I, Optical absorption at 77 K of a SrF_2: Tm crystal after X-irradiation at 77 K. II. Optical absorption with polarized light with $\epsilon \parallel [010]$ by the crystal described in I above after bleaching in the 365 nm region with polarized light with $\epsilon \parallel [001]$. III Optical absorption of light with $\varepsilon \parallel [001]$ of the crystal described in II above. (b) Full line is the difference between II and III in (a) showing anisotropic absorption of the V_K centre. Dotted and dashed lines show the corresponding anisotropic absorption for CaF_2 and BaF_2. (c) As in (a) except that the bleaching light was in the region 750 nm and had $\epsilon \parallel [010]$. The sharp structure in (a) and (c) is due to Tm^{2+} ions (after Beaumont *et al.*, 1970).

with $\epsilon \parallel \{001\}$ and $\{010\}$ after the polarized bleach. A little consideration shows that for a π-polarized transition, the difference in areas I–III and II–I in Fig. 4.17(a) should be in the ratio 2 : 1 if no V_K centres are lost in the bleaching process. The observed ratio is 2·6 : 1 and this discrepancy is due primarily to destruction of about 20% of the V_K centres, presumably as a consequence of hole motion through the valence band. It should be emphasized that these investigations of effects of polarized bleaching light on the optical absorption spectrum of the V_K centre show only that the dipole responsible for the ultraviolet band is oriented parallel to a $\langle 100 \rangle$ crystal direction (see § 4.3).

The orientation of the dipole relative to the V_K molecular axis can be determined by investigating the effect on the epr spectrum of irradiating into the ultraviolet band with linearly polarized light. Fig. 4.16 shows the changes induced in the epr spectrum of the V_K centre in SrF_2:Tm by bleaching a crystal in the ultraviolet band at 77 K in a microwave resonance cavity with polarized light with $\boldsymbol{\epsilon} \parallel \{001\}$. It is apparent from intensity changes in the epr spectrum when the external magnetic field is rotated from $\{001\}$ to $\{010\}$ that the number of centres aligned along $\{010\}$ is about six times greater than the number aligned along $\{001\}$. This optically-induced magnetic anisotropy in the cubic crystal shows conclusively that the ultraviolet band is an excitation of the V_K centre and that the band is excited by light polarized parallel to the principal axis of the centre (π type). This experiment shows that the ultraviolet band has the polarization properties expected from a $^2\Sigma_u \rightarrow {}^2\Sigma_g$ transition.

The difference in area II–III in Fig. 4.17(a) gives the anisotropic absorption of the V_K centre in the ultraviolet band and this is shown in Fig. 4.17(b). This difference gives an accurate band shape for the $^2\Sigma_u \rightarrow {}^2\Sigma_g$ transition without interference from isotropic centres such as Tm^{2+} (Tm^{2+} absorbs to the long wavelength side of 400 nm (Fig. 4.17(a); see also §5.2.2)). The band shapes for the ultraviolet band in CaF_2 and BaF_2, obtained in the same way are also shown in Fig. 4.17(b). The peak positions of the bands are given in Table 4.17. The bands are very broad (about 8000 cm^{-1} at half height).

The $^2\Sigma_u \rightarrow {}^2\Pi_{g\frac{1}{2}}$ transition is at least 200 times weaker than the $^2\Sigma_u \rightarrow {}^2\Sigma_g$ transition and has not been detected in absorption. The peak position of the band has been detected, however, and the band has been shown to be of the σ-type by bleaching the crystals at 77 K with polarized infrared light with $\boldsymbol{\epsilon} \parallel [010]$ and by detecting preferential alignment of the V_K centres using the anisotropic absorption in the ultraviolet band (Fig. 4.17(c)). The peak position is determined by measuring the magnitude of the induced alignment as a function of infrared wavelength and is given in Table 4.17 for CaF_2, SrF_2, and BaF_2. It is readily shown, assuming excitation in a $^2\Sigma_u \rightarrow {}^2\Pi_{g\frac{1}{2}}$ transition, that the difference in areas II–I and I–III in Fig. 4.17(c) should be in the ratio 2:1. The observed ratio of 1·9:1 is in good agreement with expectation. This suggests that the $^2\Pi_{g\frac{1}{2}}$ state may be in the band gap of the crystal, leading to realignment with little migration and hence to smaller V_K losses.

These investigations, using polarized bleaching light, are similar to earlier investigations of V_K centres in alkali halides (Delbecq, Smaller, and Yuster, 1958; Delbecq, Hayes, and Yuster, 1961).

The identification of the self-trapped hole in alkaline earth fluorides gave rise subsequently (Hayes, Kirk, and Summers, 1969; Beaumont *et al.*, 1970) to the identification of the self-trapped exciton; this may be regarded as an electron trapped in the positive potential of a V_K centre (i.e. $e^- + V_K$). It was found that illumination of CaF_2:Tm at wavelengths longer than 540 nm, after X-irradiation at 77 K, gives rise to luminescence. Spectral decomposition of this luminescence showed a broad asymmetric band peaking at 280 nm. This band is due to recombination luminescence of the V_K centre; it results from capture

TABLE 4.18

Peak position (nm) *at* 77 K *of the emission band of the self-trapped exciton*

	X-ray fluorescence	Recombination luminescence photoluminescence	Recombination luminescence thermoluminescence
CaF_2	279	280	277
SrF_2	300	298	300
BaF_2	310	310	312

Errors in all peak positions ±4 nm
All measurements by Beaumont *et al.* (1970).

by these centres of electrons liberated by infrared light (a discussion of the photoconductivity of CaF_2 containing Tm^{2+} ions has been given by Kiss and Staebler, 1965); the luminescing centre is, in fact, the self-trapped exciton. It is readily demonstrated that the 280 nm band is associated with the V_K centres by optically aligning the centres before illuminating the crystals with infrared light and by investigating the polarization of the luminescence. The 280 nm band was also observed in the thermoluminescence of irradiated CaF_2:Tm. Similar results were obtained by Hayes *et al.* (1969) and by Beaumont *et al.* (1970) for SrF_2:Tm and BaF_2:Tm (Table 4.18). It was found by Beaumont *et al.* (1970) using flash-tube excitation, that the lifetime of the 280 nm fluorescence in CaF_2:Tm was shorter than 40 ns.

Extending the simple molecular-orbital description of the V_K centre the unstable ground state of the self-trapped exciton may be written in the form

$$2p\sigma_g^2\, 2p\pi_u^4\, 2p\pi_g^4\, 2p\sigma_u^2\, (^1\Sigma_g).$$

The next highest configuration is $2p\sigma_g^2\, 2p\pi_u^4\, 2p\pi_g^4\, 2p\sigma_u\, 3s\sigma_g\ (^1\Sigma_u, {}^3\Sigma_u)$. The recombination luminescence of the self-trapped hole in alkaline earth fluorides has been assigned to the allowed electric dipole transition

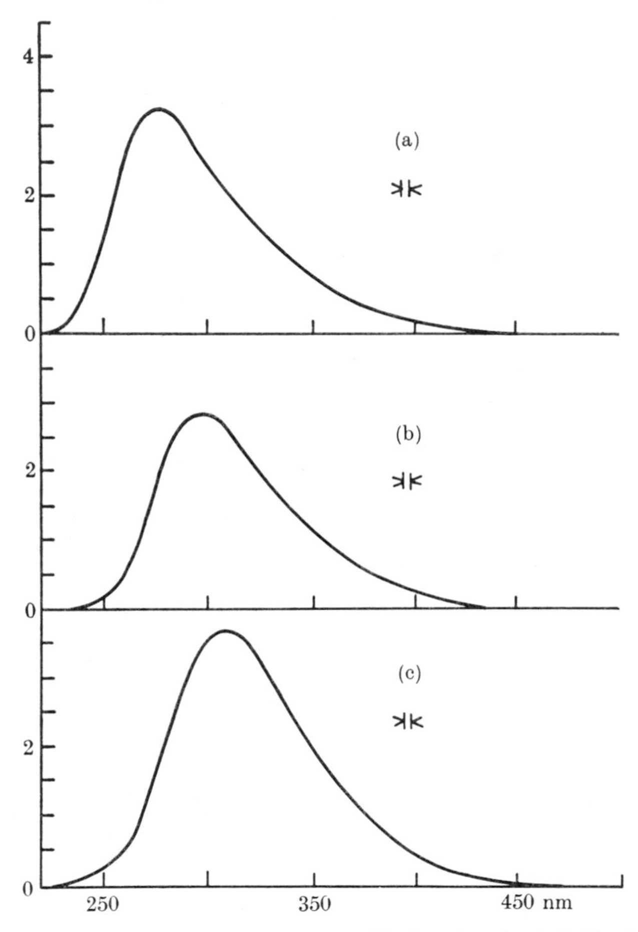

FIG. 4.18. X-ray fluorescence spectrum at 77 K of undoped (a) CaF_2, (b) SrF_2, and (c) BaF_2 (after Beaumont *et al.*, 1970).

$^1\Sigma_u \rightarrow {}^1\Sigma_g$; the polarization and lifetime measurements are consistent with this assignment.

Emission of the self-trapped exciton has also been detected (Beaumont *et al.*, 1970) through spectral decomposition of the X-ray fluoressence of undoped CaF_2, SrF_2, and BaF_2 (Fig. 4.18 and Table 4.18)†. X-ray fluorescence bands are not expected to be an exact replica of the

†It seems that centres other than the self-trapped exciton also contrbute to the intensity of the bands of Fig. 4.18 (P. S. Call. Private communication).

recombination luminescence bands because, in part, of the presence of the ultraviolet absorption band of V_K centres in crystals giving rise to recombination luminescence. The peak of this band occurs at 320 nm in CaF_2 and it will tend to reduce intensity on the low-energy side of the recombination luminescence band and hence the apparent width (see Table 4.18). The effect is more pronounced in photoluminescence than in thermoluminescence because the destruction of V_K centres is less rapid in the former case. Absorption by aligned V_K centres will also effect the polarization of the recombination luminescence to some extent.

The properties of the self-trapped exciton in alkaline earth fluorides parallel closely those of the self-trapped exciton in alkali halides (see e.g. Kabler and Patterson, 1967 and Murray and Keller, 1967).

4.6.2. *The H centre*

In alkali halides the *H* centre (a hole trapped at an interstitial anion site) is a prominent primary product of X-irradiation at helium temperatures and as such is important in consideration of radiation damage mechanisms (see § 4.9.1). Paramagnetic resonance measurements by Känzig and Woodruff (1958) showed that the *H* centre hole in some alkali halides is located equally on the interstitial anion and a nearest substitutional anion giving, in effect, a molecule ion of the type $[X_2^-]$ on a single anion site with its molecular axis along $\langle 110 \rangle$. The point symmetry (D_{2h}) of the *H* centres described by Känzig and Woodruff is the same as that of the self-trapped hole and the epr spectra of the two types of centre are somewhat similar. In KCl the *H* centre decays at about 42 K (Hayes and Hodby, 1966) and motion of the V_K centre sets in at about 173 K (Delbecq, Hayes, and Yuster, 1961).

H centres are not produced in observable concentration in undoped alkaline earth fluorides by irradiating for many hours with 50 kV X-rays at temperatures down to 4 K. They are readily formed, however, in crystals doped with the heavier trivalent rare-earth ions (Re^{3+}). In suitably grown crystals these ions introduce charge-compensating fluorine interstitials (F_i^-) and a fraction of these are not closely associated with the rare-earth ions. When such crystals are X-irradiated at 77 K large concentrations of V_K centres are produced (see § 4.6.1) and a relatively small concentration of *H* centres. However, when the crystals are heated to induce migration of V_K centres (see § 4.6.3) a fraction of the moving holes are retrapped at F_i^- sites producing additional *H* centres. The hole is located on the interstitial fluorine and a nearest substitutional fluorine giving a $\langle 111 \rangle$ oriented molecular ion (Fig.

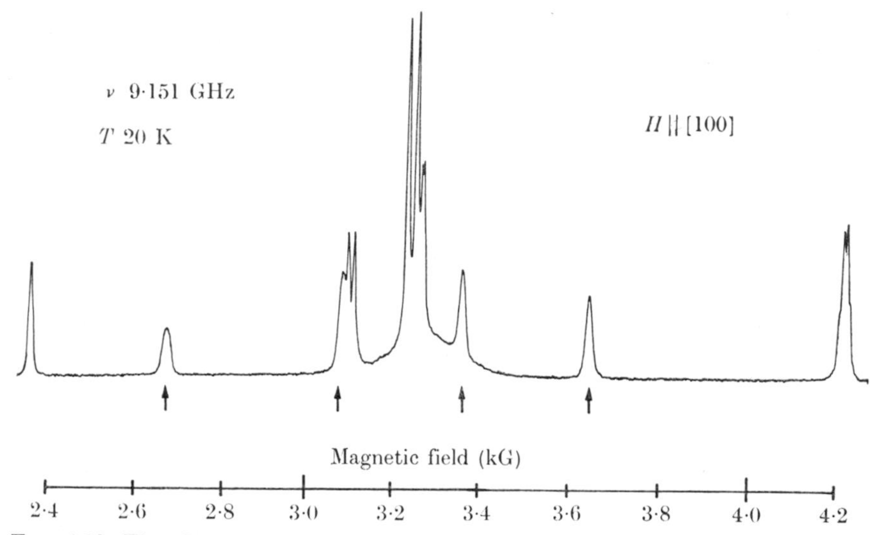

FIG. 4.19. First derivative of the x-band epr dispersion spectrum of SrF_2 at 20 K after electron irradiation at 20 K, warming to 130 K for 5 minutes and recooling to 20 K. The lines marked with an arrow are due to *H*-centres. The remaining lines are due to V_{KA} centres (after Hayes, Lambourn, and Stott, 1974).

4.15(b). In contrast with alkali halides the *H* centre in alkaline earth fluorides is more stable than the V_K centre and in CaF_2 decays at about 170 K.

H centres produced by X-irradiation of rare-earth doped crystals have been identified in CaF_2 by Hall, Leggeat, and Twidell (1969) and in CaF_2, SrF_2, and BaF_2 by Beaumont, Hayes, Kirk, and Summers (1970) (the latter authors used the notation V_H rather than *H*; see Fig. 4.20 below). In rare-earth doped CaF_2 a number of closely similar epr spectra of *H*-type centres are found due presumably to varying degrees of association of the fluorine interstitial with RE^{3+} ions or other impurities. However, thermal annealing experiments show that one of these epr spectra is more stable than all others and is present in all samples (Hall *et al.*, 1969) and this we assume to be the normal *H* centre. These effects are less noticeable in rare-earth doped SrF_2 and BaF_2.

The *H* centre spectrum may be fitted to the spin Hamiltonian

$$H = g_{\parallel}\beta H_z S_z + g_{\perp}\beta(H_x S_x + H_y S_y) + A_1 I_z^1 S_z + B_1(I_x^1 S_x + I_y^1 S_y) + A_2 I_z^2 S_z + B_2(I_x^2 S_x + I_y^2 S_y), \quad (4.20)$$

where $S = \frac{1}{2}$ and A_1 and $I^1 = \frac{1}{2}$ refer to the interstitial fluorine and A_2 and $I^2 = \frac{1}{2}$ to the substitutional fluorine involved in the centre. The parameters of eqn (4.20) are given in Table 4.19. In contrast with the

TABLE 4.19

Spin-Hamiltonian parameters of H centres in CaF_2, SrF_2, *and* BaF_2

	$g_{\parallel}$	$g_{\perp}$	A_1 (G)	B_1 (G)	A_2 (G)	B_2 (G)
CaF_2	[a] 2·0026 ±0·0005	2·039 ±0·001	1076·9 ±1	189 ±5	591·6 ±1	40 ±5
	[b] 2·0015 ±0·0005	2·038 ±0·001	1078·9 ±1·3	195 ±6.0	592·8 ±3·0	35 ±11
SrF_2	[b] 2·0044 ±0·0001	2·0353 ±0·0001	1070·8 ±3·0	147·9 ±5·0	606·9 ±3·0	41·8 ±7
BaF_2	[b] 2·0045 ±0·001	2·0306 ±0·001	1048·4 ±3.0	127·1 ±6.0	667·0 ±3.0	35·4 ±11

[a] Hall *et al.* (1969), X-irradiated rare-earth doped crystals.
[b] Hayes, Lambourn, and Stott (1974), electron-irradiated undoped crystal.

alkali halides the hyperfine interaction with the fluorine labelled 1 is almost twice as large as the hyperfine interaction with the fluorine labelled 2. The larger hyperfine interaction (A_1, B_1) is arbitrarily assigned here to the interstitial fluorine because it is expected, on electrostatic grounds, that the interstitial will provide the deeper hole trap. The large difference between A_1 and A_2 shows that the structure of the *H* centre in alkaline earth fluorides is closer to a purely interstitial neutral atom than the *H* centre in alkali halides.

H centres may be produced in undoped alkaline earth fluorides by heavy irradiation with 1 MeV electrons at 77 K (see also § 4.9). The parameters of the spin-Hamiltonian (eqn (4.20)) for the epr spectrum (Fig. 4.21) of *H* centres produced in this way are also given in Table 4.18. In some orientations of the external magnetic field the epr lines of *H* centres produced by electron irradiation show partly resolved structure suggesting perturbation by another defect produced by the radiation (Hayes, Lambourn, and Stott, 1974).

Beaumont *et al.* (1970) found evidence for the existence of a broad absorption band of the *H* centre peaking at about 308 nm in CaF_2 and SrF_2 and at about 330 nm in BaF_2. It was not found possible to produce preferential alignment of the *H*-centres by bleaching into this band with linearly polarized light at temperatures down to 4 K. This may be due to motion of the molecular bond about the interstitial site even at these low temperatures.

4.6.3. *Thermal annealing of V_K centres and formation of impurity-stabilized hole centres*

When alkaline earth fluorides are X-irradiated at 77 K the V_K centre is the predominant hole centre formed. When the irradiated crystals are warmed past the decay temperature of the V_K centre some

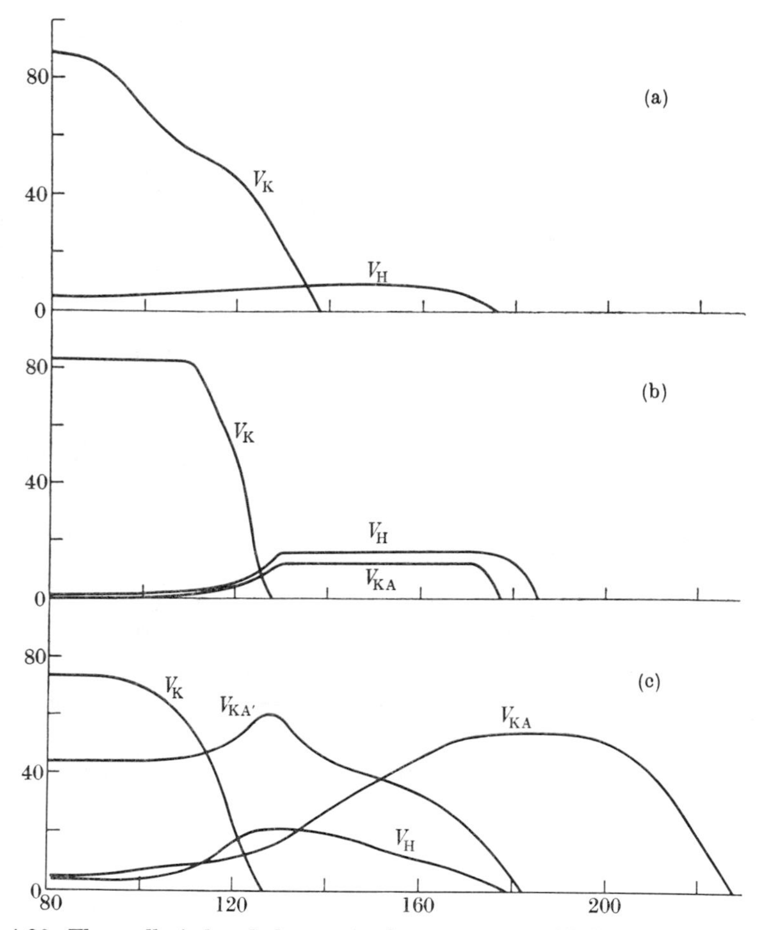

FIG. 4.20. Thermally induced changes in the epr spectra of hole centres in (a) CaF_2: Tm (b) SrF_2: Tm and (c) BaF_2 :Tm after X-irradiation at 77 K (after Beaumont *et al.*, 1970).

of the moving holes are retrapped at impurity sites which stabilize the holes to higher temperatures (Fig. 4.20). In this section we shall firstly give a brief description of the properties of impurity-stabilized hole centres which have been investigated by epr methods and we shall then describe in some detail the annealing behaviour of paramagnetic hole centres in thulium-doped crystals (the annealing behaviour of hole centres in undoped crystals is somewhat different). Two main categories of impurity-stabilized hole centres have been found.

I. Centres with ⟨100⟩ alignment which are similar in structure to the V_K centre. These may be classified into three sub-groups:

(i) Centres which have been observed in thulium-doped SrF_2 and BaF_2 by Tzalmona and Pershan (1969) and by Beaumont *et al.* (1970) and which have been described as V_{KA} centres by the latter authors. The thermal stability of these centres is shown in Fig. 4.20(b) and (c). The centres have a strong optical absorption band in the ultraviolet analogous to the V_K band (see § 4.6.1) and it is possible to optically align the V_{KA} centres by bleaching into this band at 77 K with linearly polarized light (Beaumont *et al.*, 1970). It is possible that the V_{KA} centre may be a V_K centre trapped at a cation impurity site; detailed confirmation of the structure is, however, lacking.

(ii) Centres which have been described as $V_{KA'}$ centres by Beaumont *et al.* (1970) and which have been observed only in thulium-doped BaF_2 (see Fig. 4.20(c)).

No suggestions have been made for the structure of this centre.

(iii) Centres stable at room temperature which have been observed in natural CaF_2 (Sierro, 1965) and in CaF_2 doped with NaF (Hall, Leggeat, and Twidell, 1969). It was suggested by Sierro that this centre was a V_K centre stabilized by a cation vacancy but later experiments by Hall *et al.* (1969) seem to be consistent with stabilization by a substitutional Na^+ ion.

II. A variety of hole centres with ⟨110⟩ alignment have been observed by epr methods in CaF_2 doped with yttrium or lutecium impurities by Hall *et al.* (1970). They are stable in the vicinity of room temperature. A model has been proposed for these centres by Hall *et al.* (1970) involving a hole trapped by a pair of nearest-neighbour charge-compensating interstitial fluorine ions associated with a pair of trivalent impurity ions. The formation in CaF_2 of impurity-interstitial aggregates has been discussed by Osiko (1965) (see also § 3.2.3).

It should be emphasized that the models proposed for impurity-stabilized hole centres in alkaline earth fluorides are tentative. It was shown by Bass and Mieher (1968), using endor methods, that the V_{KA} centre in NaF is a V_K centre trapped by a Li^+ impurity in a nearest-neighbour cation site. Experiments of this type have not been carried out in alkaline earth fluorides.

Paramagnetic resonance measurements show that when CaF_2:Tm is X-irradiated at 77 K an intense V_K spectrum, a weak H spectrum and a strong Tm^{2+} spectrum is obtained; the number of paramagnetic hole centres formed is approximately equal to the number of Tm^{2+} ions. An annealing curve obtained by monitoring the decay of the epr spectra of the hole centres is shown in Fig. 4.20(a). It is apparent that there are two kinks in the decay curve of the V_K centres, at 90 and 120 K. The decay at 90 K follows first-order kinetics approximately with an activation energy of 0·19 eV. The decay at 120 K follows second-order kinetics with an activation energy of 0·31 eV (Table 4.20). This result

TABLE 4.20

Activation energies (eV) *for motion of* V_K *centres*

System	Observed value[a]	Predicted values[b] 90°	180°	180°
		Model IIa	Model IIa	Model IIb
CaF_2 (90 K) (180° jump)	0·19	0·08	0·09	0·03
CaF_2 (120 K) (180° jump)	0·31	0·21	0·25	0·09
SrF_2 (110 K) (180° jump)	0·21	0·43	0·54	0·28
SrF_2 (120 K) (90° and 180° jumps)	0·30	0·50	0·61	0·32
BaF_2 (95 K) (90° and 180° jumps)	0·30	0·50	0·64	0·39
BaF_2 (115 K) (90° and 180° jumps)	0·38	0·65	0·84	0·51
$SrCl_2$ (105 K) (90° and 180° jumps)	0·25	—	—	—

[a] Values for fluorides are due to Beaumont *et al.* (1970) and for $SrCl_2$ are due to Hayes, Lambourn, Rangarajan, and Ritchie (1973).

[b] Norgett and Stoneham (1972*b*).

suggests that the V_K centres are formed preferentially near Tm^{2+} sites and that the first decay stage results in the capture of moving V_K centres by nearby Tm^{2+} ions. This stage is associated with relatively intense Tm^{3+} fluorescence arising from the capture of moving holes by Tm^{2+} ions in cubic sites (see § 7.2). However, only a small fraction of the Tm^{2+} and V_K centres is lost at this stage and an approximation to first-order kinetics is reasonable. It was also suggested by Schlesinger and Menon (1969) that holes are trapped near rare-earth ions in CaF_2

irradiated at low temperatures. The second stage of decay seems to correspond to a more general motion of V_K centres through the body of the lattice, giving rise primarily to paired (nonparamagnetic) hole centres and hence to second-order kinetics. A small fraction of the hole centres in the second stage is lost by combining with Tm^{2+} and a small fraction by conversion to H centres (Fig. 4.20(a)).

Since no paramagnetic hole centres were observed in crystals of CaF_2:Tm after decay of the H centres at about 175 K (Fig. 4.20(a)) and since about 90% of the Tm^{2+} ions survive to room temperature, it was concluded by Beaumont *et al.* (1970) that about 90% of the holes created at 77 K exist at room temperature in a non-paramagnetic form. Tm^{2+} ions are still observable up to 450 C indicating that non-paramagnetic hole centres are stable to this temperature. There has been some speculation about the structure of these high-temperature hole centres in CaF_2 (Merz and Pershan, 1967*a*; Kiss and Staebler, 1965; Arkangel'skaya, 1965) but the experimental information presently available is inadequate to provide reliable models.

The thermally induced motion of V_K centres was studied in greater detail by Beaumont *et al.* (1970) by firstly preferentially aligning the centres at 77 K using polarized bleaching light in the ultraviolet band with $\epsilon \parallel [001]$. It was found using epr methods that the centres preferentially aligned along [010] and [001] decayed at the same rate to complete destruction at about 138 K (see Fig. 4.20(a)) and it was concluded that only linear (180°) motion of V_K centres occurs in CaF_2 below the temperature of disappearance.

X-irradiation of SrF_2:Tm at 77 K produces an intense V_K spectrum and V_{KA} and H centres with an intensity about one-fiftieth of that of the V_K spectrum. The thermally-induced decay of the epr spectrum of these centres is shown in Fig. 4.20(b). It should be emphasized that curves of the type shown in Fig. 4.20 are fairly crude and the detailed appearance depends to some extent on heating rate. Nevertheless, there are discrepancies in temperature between Fig. 4.20(b) and the corresponding curve given by Tzalmona and Pershan which seem difficult to account for. It is apparent from Fig. 4.20(b) that there are two kinks in the decay curve of the V_K centres at 110 and 120 K. The decay stage at 110 K follows first-order kinetics fairly closely with an activation energy of 0·21 eV and the decay stage at 120 K follows second-order kinetics approximately with an activation energy of 0·30 eV (Table 4.20). This behaviour is similar to that found in CaF_2.

About 25% of the V_K centres in SrF_2 are converted on thermal annealing to V_{KA} and H centres (Fig. 4.20(b)). The initial concentration of V_K centres is approximately equal to that of Tm^{2+}. On being warmed to room temperature about 50% of the Tm^{2+} ions are converted to Tm^{3+} ions by mobile holes. No paramagnetic hole centres survive to room temperature so that it is apparent that about 50% of the holes created at 77 K exist at room temperature in a non-paramagnetic state.

A study of the effects of thermal annealing on the optically-induced preferential alignment of the V_K centres showed (Beaumont *et al.*, 1970) that only linear motion of the centres occurred in SrF_2:Tm at the 110 K decay stage. However, random (90°) jumps occurred at 120 K. The difference in activation energies for 180° and 90° jumps is about 0·1 eV.

The decay of optically-induced preferential alignment of V_{KA} centres was also investigated in SrF_2:Tm by Beaumont *et al.* (1970). It was found that these centres reorient without loss of intensity at 123 K with an activation energy of 0·32 eV. The V_{KA} centre decays at 184 K (Fig. 4.20(b)); this decay obeys second-order kinetics with an activation energy of 0·93 eV. The reorientation involves pivoting around the impurity which stabilizes the hole and comparison of the activation energies for reorientation and decay gives a defect-induced stabilization energy of 0·61 eV.

X-Irradiation of BaF_2:Tm at 77 K gave epr spectra of V_K, V_{KA}, $V_{KA'}$ and H centres with relative intensities shown in Fig. 4.20(c). Investigation of the epr spectra of optically aligned centres showed (Beaumont *et al.*, 1970) that random motion of the V_K centres occurred before a noticeable decay set in and that there is no measureable difference between the activation energies for 180° and 90° jumps. Tzalmona and Pershan suggest, however, that 180° motion of the V_K centre may be occurring in the early stages of decay in BaF_2:Tm. The $V_{KA'}$ centres do not occur in their BaF_2:Tm crystals.

The V_K centres in BaF_2:Tm were found to reorient with an activation energy of 0·30 eV at 95 K; decay at 115 K followed second-order kinetics with an activation energy of 0·38 eV (Table 4.21).

It was shown by Beaumont *et al.* (1970) that the V_{KA} centre in BaF_2 reorients without decay at 91 K with an activation energy of 0·24 eV and decays with second-order kinetics at 209 K with an activation energy of 0·85 eV. As in SrF_2 the reorientation is a pivotal motion about the associated impurity; the impurity stabilization energy of 0·61 eV is equal to that found in SrF_2. No detailed investigation of the decay of the $V_{KA'}$ centres has been reported.

The fraction of Tm^{2+} ions surviving on warming the irradiated crystals from 77 K to room temperature (90% in CaF_2, 50% in SrF_2, and 10% in BaF_2) suggests that interaction of holes with Tm^{2+} increases at the expense of pairing along the series and this may be due, in part, to smaller overlap of hole wavefunctions with increasing lattice size. It should be emphasized that details of decay and the type and concentration of impurity-stabilized hole centres formed depend to some extent on the origin and hence the impurity content of the crystals used.

4.6.4. *Theory of the V_K centre*

The polarization and distortion of the host lattice produced by the V_K centre make it very immobile, so that in many respects the self-trapped hole centre resembles a conventional lattice defect. The major approximation in the theory of this centre can be described as the 'molecule in a crystal' hypothesis. The spin-resonance data (§ 4.6.1) show that the hole is concentrated primarily on two of the fluorine ions. We assume that the electronic properties of the V_K centre are those of an F_2 molecular ion, and that the crystal environment merely changes the interatomic spacing R from the value for the free ion. The accuracy of this assumption can be tested in three ways. Firstly, Jette, Das, and Gilbert (1969) give expressions for the hyperfine constants and excited states of the F_2^- ion, based on the Hartree-Fock calculations of Gilbert and Wahl (1971). If the hypothesis is valid, the observed spin-resonance and optical data should define a unique separation; the separations obtained from the measurements are, in fact, quite close (Fig. 4.21). It is not certain whether the deviations which occur are due to inadequacy of the wavefunctions of the free molecular ion or to effects of the crystal other than changes of the equilibrium spacing. A second test of the hypothesis concerns the hole charge-density on ions neighbouring the F_2^- group. Hyperfine structure from the neighbouring axial ions has been detected in the epr spectrum (Table 4.16); it is small for CaF_2, less in SrF_2 and unresolved in BaF_2. Thus the hole is almost completely concentrated on the F_2^- group. The third test of the 'molecule in a crystal' hypothesis concerns the effect of crystal fields on the molecular ion. Unlike the free F_2^- ion, the V_K centre does not have axial symmetry so that there should be two distinct g-factors for magnetic fields normal to the axis, and transitions to π states may be split. However, these effects have not been seen; there are no appreciable deviations from axial symmetry in alkaline earth fluorides.

Calculations of the lattice distortion by the V_K centre have been made

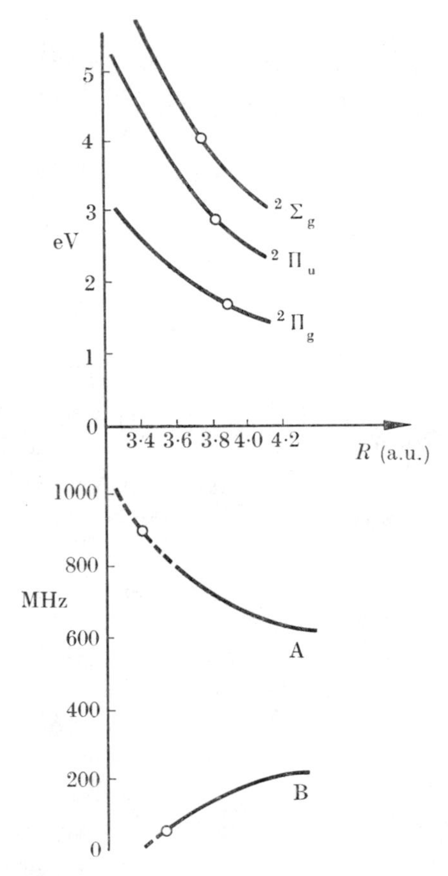

FIG. 4.21. Calculated energy levels and hyperfine constants of the F_2^- molecule ion (see § 4.6.1) plotted against internuclear distance R. Experimental points are for the V_K centre in CaF_2 (after Norgett and Stoneham, 1973*a*).

by Jette and Das (1969) for CaF_2 and by Norgett and Stoneham (1973*a*) for CaF_2, SrF_2, and BaF_2. We shall concentrate on the latter work, since it gives a fuller treatment, improving the interatomic potentials, the relaxation procedure, and the number of atoms allowed to move in calculating the local distortion. Both methods are similar in principle, and are in general agreement where comparison is possible.

The static equilibrium configuration of the V_K centre minimizes the sum of three energies: the interactions within the F_2^- molecular ion, the interaction of the F_2^- ion with the host, and the energy of the rest of the lattice. Two of these terms are easily treated. Thus the interactions within the F_2^- ion are taken from Hartree-Fock calculations for the free molecular ion by Gilbert and Wahl (1971). A shell model is used for the interactions between ions outside the V_K centre with parameters the same as in the perfect crystal (§ 2.3). However, the interactions between the F_2^- ion and the rest of the lattice present more problems. If the ion is regarded as $(F^{-\frac{1}{2}}\,F^{-\frac{1}{2}})$, then it is not clear what repulsive interactions and polarizabilities should be attributed to each $F^{-\frac{1}{2}}$ ion. Norgett and Stoneham (1972*a*) show that the results are not sensitive to the polarizability. The only property at all sensitive to the repulsive interaction is one of the motion energies, discussed later.

The lattice-relaxation calculations follow the usual prescription of treating a region close to the defect explicitly, and the remainder of the crystal in the Mott and Littleton (1938) approximation (see § 3.5). The innermost region was sufficiently large that the results did not vary significantly on changing its size. The most important parameters predicted are the separations of the two ions constituting the V_K centre. Experiment suggests separations of about 2·01 Å from optical data and about 1·85 Å from spin-resonance results (see Fig. 4.21). The predicted separations vary with model and host lattice over the range 1·84 Å to 1·96 Å. Thus the F_2^- molecule is only slightly distorted in its crystal from the free-molecule separation of 1·90 Å. In principle the endor data of Marzke and Mieher (1969) could yield displacements of other ions to compare with the predictions, but the necessary analysis has not been done.

The calculations also yield other results. Firstly, the relaxation energies verify that the hole should be self-trapped. The energy gain by lattice relaxation is much larger than that required to localize the hole on two anions. Secondly, one can estimate an effective mass and force constant for the relative motion of the two fluorine ions. From these one may estimate the linewidths of the optical transitions in excellent agreement with experiment. One may also verify that the characteristic frequency associated with this motion is too large to be associated with the temperature dependence of the hyperfine constants. Assmus and Dreybrodt (1969) suggest modes involving cation motion to account for this temperature dependence.

The self-trapped hole moves through the crystal lattice by a hopping motion which is qualitatively different from the propagative motion of electrons in metals. Its motion can be regarded as a succession of random independent jumps between equivalent sites in the lattice. This view of the V_K centre and its motion corresponds to the "small polaron" models of Yamashita and Kurosawa (1958) and of Holstein (1959). Appel (1968) has reviewed these approaches. Similar models have been proposed for a variety of hopping processes (e.g. Flynn, 1971). The strong coupling of the hole to the lattice phonons is an essential element in the theory.

The activation energy for the motion can be calculated in much the same way as the distortion near the defect. This has been done by Norgett and Stoneham (1973*b*). The lattice distortion in the initial site can be regarded as the response of a harmonic lattice with force-constant matrix $\mathbf{A}$ to forces $\mathbf{F}_1$, due to the presence of the hole. Then the lattice relaxation energy in the initial site is:

$$\Delta E_1 = -\tfrac{1}{2}\mathbf{F}_1 \cdot \mathbf{A}^{-1} \cdot \mathbf{F}_1. \tag{4.21}$$

If the forces appropriate to the final site are $\mathbf{F}_2$, then ΔE_1 and ΔE_2 are identical because the initial and final sites are distinct but equivalent. The activation energy in small polaron theory is a lattice activation energy

$$E_a = -\tfrac{1}{2}\bar{\mathbf{F}} \cdot \mathbf{A}^{-1} \cdot \bar{\mathbf{F}} - \Delta E_1, \tag{4.22}$$

where $\bar{\mathbf{F}} = \tfrac{1}{2}(\mathbf{F}_1 + \mathbf{F}_2)$, and can be calculated straightforwardly. Calculations of this sort have been done for the alkali halides (Song, 1970, 1971; Flynn, 1971). They give order-of-magnitude agreement with experiment, but all predict that E_a increases with the jump distance. This is not observed in the fluorite structure. There are, however, three major difficulties. Firstly, the forces $\mathbf{F}$ change significantly during the self-trapping lattice relaxation, and this non-linearity is important. Secondly, the lattice anharmonicity is important. Moreover, since the distortions very close to the V_K centre are large, it is essential to allow enough atoms to move in calculating the relaxation energies involved in E_a. Thirdly, the theory of the motion of small polarons predicts deviations from $\tau^{-1} \sim W \exp(-E_a/kT)$ below the Debye temperature. It is necessary to parameterize the deviations so that the apparent activation energy U and E_a can be compared. The largest errors in the theory at present arise from the difference between U and E_a at the temperatures of the experiments.

Results for activation energies U (Norgett and Stoneham, 1973*b*) are given in Table 4.20, where several models are shown indicating different choices of input parameter. These are effective activation energies, and so are temperature dependent.

Unfortunately, the 90° jump has a lower value of E_a than the 180° jump for all three substances and this is reflected in the U. This contradicts the experimental observation that 180° jumps are observed in CaF_2 at low temperatures, before 90° motion begins (§ 4.6.3). One possible explanation of this effect is the chemical binding of the V_K ions with the neighbouring axial fluorines. This would have the correct trend with crystal, since the binding should be largest for smaller lattice parameters. However, this explanation must be ruled out. The spin-resonance data suggest that the amplitude of the hole wave-function on these neighbours is so small that enhanced binding is probably negligible; this is confirmed by approximate molecular-orbital calculations.

An alternative possibility is that the bonding of the molecule contracts the charge clouds of the V_K ions more along the defect axis than perpendicular to it. This need not be a large effect in terms of a reduction in ionic radius to have a significant effect on the calculated difference in E_a for 90° and 180° jumps. In model IIb the axial interaction between the ions of the F_2^- molecule and the nearest lattice F^- ion is reduced arbitrarily by half. This corresponds to a reduction of only $\sim 10\%$ in the ionic radius. This anisotropy in the overlap interaction then gives the desired result that the 180° jumps have lower values of E_a. As expected, the effect is much greater for CaF_2 than for SrF_2 and BaF_2.

Agreement of predicted and observed activation energies is satisfactory. In fact, the agreement is better than the table suggests, since the discrepancies are greatly enhanced by the very non-linear form of the relation between E_a and U. This can be seen by using the relation in reverse to derive an 'experimental' E_a. The ratio of the experimental to the theoretical value for the 90° BaF_2 jumps proves to be $0{\cdot}92 \pm 0{\cdot}06$, the errors quoted being experimental. A very modest change in parameters can remove the inconsistency. One of the problems of the fluorite systems is that the experiments have been performed at temperatures where the relation of U and E is the weakest link in the calculation. It is for this reason that we do not quote values of the hopping integrals J; their variation with small changes in model is so large as to make them meaningless.

The trend from crystal to crystal appears to be correctly given. Thus the activation energies get higher along the sequence CaF_2, SrF_2, BaF_2, and the 90° jump becomes progressively more favoured in comparison with the 180° jump. Both these features show improvements over Flynn's attractive and very simple treatment, where a slight opposite trend in activation energies along the sequence is indicated.

4.7. Impurity anions

We have already emphasized that coloration effects in alkaline earth fluorides are strongly influenced by some impurities. Evidence for the existence of hydroxyl and carbonate ions in CaF_2 have been obtained by infrared methods (Bontinck, 1958b; Wickersheim and Hanking, 1959) but the effects of molecular impurities on the coloration properties of the crystals have not been explored. Some trivalent rare-earth ions strongly affect the colourability of crystals because of their electron-trapping properties and because of the charge-compensating defects introduced into the crystals by them; these effects are discussed elsewhere in this monograph (see § 4.6 and Chapter 7). In this section we shall concern ourselves with hydrogen and oxygen impurities. These impurities affect the behaviour of alkaline earth fluorides under ionizing radiations because they act as hole traps and under certain circumstances introduce anion vacancies into the crystals. Hydrogen-doped crystals colour quickly under X-irradiation at room temperature because of rapid generation of *F*-aggregate centres; an explanation of this effect is suggested in § 4.7.2.

4.7.1. *Paramagnetic resonance and ultraviolet absorption of hydrogen centres*

The investigations of hydrogen-doped crystals discussed here were carried out on materials prepared in the manner described by Hall and Schumacher (1962) (see § 2.5 for details). Using infrared methods it was shown by Elliott *et al.* (1965) that crystals treated in this way contain hydrogen dissolved as H^- ions in fluorine sites (H_s^- centres) and concentrations of H_s^- approaching $10^{20}\ cm^{-3}$ are readily achieved (§ 2.5).

The first detailed investigations of CaF_2:H crystals were carried out by Hall and Schumacher (1962) who showed, using epr methods, that X-irradiation at room temperature produced hydrogen atoms stable in interstitial sites (H_i^0 centres). It was subsequently shown by Bessent, Hayes, and Hodby (1965, 1967) and Bessent, Hayes, Hodby, and Smith (1969), again using epr methods, that X-irradiation at 77 K produced hydrogen atoms in fluorine sites (H_s^0 centres) and that the H_s^0 centres are converted to H_i^0 centres on warming to room temperature. The epr spectrum of both H_s^0 centres and H_i^0 centres (Fig. 4.22) may be described by the spin Hamiltonian

$$\mathscr{H} = g\beta\mathbf{H}\cdot\mathbf{S}+A\mathbf{I}\cdot\mathbf{S}-g_p\beta_n\mathbf{H}\cdot\mathbf{I}+\sum_n[\mathbf{S}\cdot\mathbf{A}_n^F\cdot\mathbf{I}_n^F-g_F\beta_n\mathbf{H}\cdot\mathbf{I}_n^F] \quad (4.23)$$

where $S = \frac{1}{2}$, $I = \frac{1}{2}$ is the spin of the proton ($I = 1$ for the deuteron), $I^{F}_{n} = \frac{1}{2}$ is the spin of the nth nearest-neighbour fluorine and g_n (or g_d) and g_F are the proton (or deuteron) and fluorine nuclear g-factors. The first three terms in eqn (4.23) give the energy of the hydrogen atom alone. The terms in the square bracket describe the hyperfine structure due to interaction with the six nearest-neighbour fluorines for H^0_s centres and the eight nearest-neighbour fluorines for H^0_i centres.

The hyperfine interaction with nearest-neighbour fluorines is well resolved and makes it easy to distinguish between H^0_s and H^9_i centres. In the former case a seven-line hyperfine structure characteristic of interaction with six equivalent fluorines is observed when the external magnetic field is aligned along (111) whereas, in the latter case, a nine-line hyperfine structure characteristic of interaction with eight equivalent fluorines is obtained when the external magnetic field is along $\langle 100 \rangle$ (Fig. 4.22).

As in the case of the F centre (§ 4.2) the point symmetry at the H^0_s site is T_d and we expect the hyperfine tensor for nearest-neighbour fluorines to have orthorhombic symmetry. However, departure from axial symmetry is too small to be observed in the epr spectrum and the hyperfine interaction with a single fluorine nucleus may be written

$$\mathbf{S} \cdot \mathbf{A} \cdot \mathbf{I}^{F} = A_{\|} I^{F}_{z} S_{z} + A_{\perp}(I^{F}_{x} S_{x} + I^{F}_{y} S_{y}) \tag{4.24}$$

for both H^0_s and H^0_i centres where the z direction coincides with the hydrogen-fluorine bond axis. The parameters $A_{\|}$ and $A_{\perp}$ may be decomposed:

$$\begin{aligned} A_{\|} &= A_s + 2A_p, & A_{\perp} &= A_s - A_p \\ &= A_s + 2(A_\sigma + A_d), & &= A_s - (A_\sigma + A_d), \end{aligned} \tag{4.25}$$

where A_s is the contact interaction through fluorine s orbitals, A_σ is the interaction through fluorine $p\sigma$ orbitals and A_d may be approximated by the classical dipole-dipole interaction $(g\beta\, g_F \beta_n)/R^3$ with R equal to the hydrogen-fluorine spacing. Equations (4.25) are a modified form of eqn (4.2). Values of the parameters of eqn (4.25) for H^0_s, D^0_s, H^0_i, and D^0_i centres in alkaline-earth fluorides are given in Table 4.21.

The hyperfine interactions of H^0_s and H^0_i centres in CaF_2 with second and further fluorine shells were investigated by endor methods by Hall and Schumacher (1962), by Bessent *et al.* (1967) and by Böttcher, Frieser, Milsch, Völkel, Wartewig, Welter, and Windsch (1970).

The effect of temperature on the magnitude of the proton and nearest-neighbour fluorine hyperfine interactions of H^0_i centres was investigated by Hall and Schumacher (1962) and by Welber (1964). A more detailed

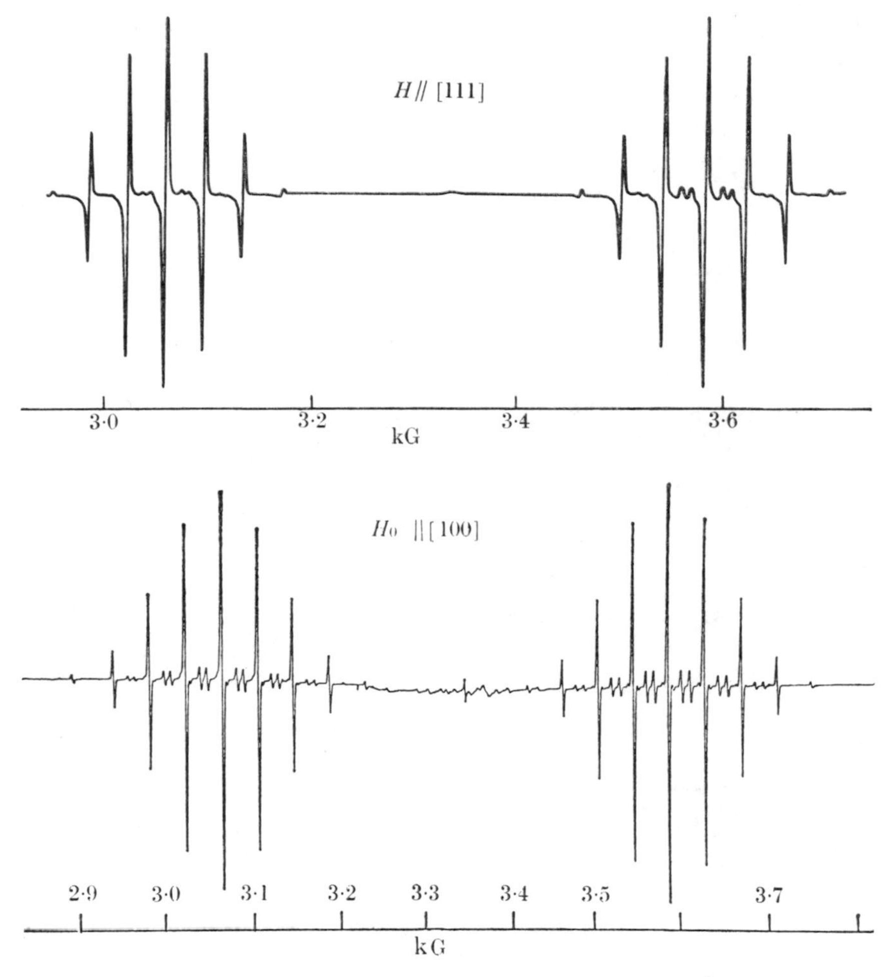

FIG. 4.22. First derivative of the epr absorption spectrum of (a) H_s^0 centres in CaF_2 (after Bessent *et al.*, 1967) and (b) H_i^0 centres in CaF_2 (after Hall and Schumacher, 1962).

investigation of these effects was carried out for both H_s^0 and H_i^0 centres by Hodby (1969) who found that, in general, with increasing temperature, the proton hyperfine splitting falls, the coupling with fluorine *p*-orbitals is unchanged and the interaction with fluorine *s*-orbitals increases. These effects were also investigated by Sligar and Blum (1971). The theory of the spin-resonance parameters of H_s^0 and H_i^0 has been discussed by several groups (Hall and Schumacher, 1962, Blum, 1967*a*, *b*, Bessent *et al.*, 1967 and Hodby, 1969). The analysis starts by considering a simple model in which the only effect of the

TABLE 4.21

Parameters of the epr *spectra of substitutional and interstitial hydrogen atoms in* CaF_2, SrF_2, *and* BaF_2

System	Temperature (K)	g	A (MHz)	$A_{\parallel}$ (MHz)	$A_{\perp}$ (MHz)	A_s (MHz)	A_{σ} (MHz)
$CaF_2:H_s^{0a}$	77	2·00235 ±0·00006	1439·44 ±0·3	168·11 ±0·5	52·76 ±0·3	91·21 ±0·4	34·77 ±0·3
$CaF_2:D_s^{0a}$	77	2·00249 ±0·0001	221·45 ±0·2	165·23 ±0·3	51·05 ±0·3	89·11 ±0·3	34·38 ±0·2
$CaF_2:H_i^{0b}$	77	2·00226	1464·19 ±0·2	173·99 ±0·3	69·09 ±0·3	104·06 ±0·3	29·30 ±0·2
$CaF_2:D_i^{0b}$	300	2·00242	224·92 ±0·2	171·94 ±0·3	67·69 ±0·3	102·44 ±0·3	29·08 ±0·2
$SrF_2:H_s^{0a}$	77	2·0028 ±0·001	1437·83 ±0·5	100·96 ±0·2	23·49 ±0·2	49·31 ±0·2	22·87 ±0·2
$SrF_2:D_s^{0a}$	77	2·00276 ±0·0001	221·29 ±0·2	98·86 ±0·3	21·76 ±0·3	47·46 ±0·3	22·74 ±0·2
$SrF_2:H_i^{0a}$	77	2·00286 ±0·0001	1444·16 ±0·2	127·14 ±0·3	42·57 ±0·3	70·76 ±0·3	23·65 ±0·2
$BaF_2:H_i^{0a,c}$	300	2·00232 ±0·00006	1424·5 ±0·10	83·44 ±0·2	20·35 ±0·2	41·38 ±0·2	17·16 ±0·2

g-value of free hydrogen atoms: 2·002256. A-value of free hydrogen atoms: 1420·40573 MHz.
[a] Bessent *et al.* (1969). [b] Hall and Schumacher (1962). [c] Welber (1964).

crystal on the hydrogen atom is to make its wavefunction Ψ_{H_0} orthogonal to the wavefunctions Ψ_c of the other crystal ions. Thus, for the free ion,

$$\Psi_{H_0} = \sqrt{(1/\pi a_0^3)}\exp(-r/a_0)$$

and, in the crystal,

$$\Psi_H = N[\Psi_{H_0} - \sum_c \Psi_c \langle \Psi_c \mid \Psi_{H_0} \rangle].$$

Other simplifications are possible; one might take $N = 1$ because the overlaps are small, or one might assume Ψ_H varies slowly over the cores of the neighbours. Hydrogen and fluorine hyperfine parameters attained to this order are compared with experiment in Table 4.22. The general orders of magnitude are satisfactory, but the agreement is not remarkable.

Further treatments go beyond this model in two respects. Firstly, covalent bonding of the hydrogen to its neighbours is included. This modifies the hyperfine parameters and g-factor. Secondly, the dynamics of the hydrogen are brought in to interpret the temperature dependence

TABLE 4.22

Comparison of calculated and observed epr *parameters* (MHz) *for hydrogen in* CaF_2, SrF_2, *and* BaF_2

		A^H		A_s^F		A_p^F	
		Theory	Expt.	Theory	Expt.	Theory	Expt.
H_s	CaF_2	1456[c]	1439	15·5[b] 31[c], 27[c*]	91		
	SrF_2	1443[c]	1438	7[b] 18[c]	49		
	BaF_2	1433[c]		9.4[c]			
H_i^0	CaF_2	1660[a] 1666[c] 1586[c*]	1464	129[a] 62[b] 129[c] 68[c*]	104	33[a]	34·93
	SrF_2	1403[c]	1444	31[b]	71		
	BaF_2	1411[c]	1425	18·5[b]	41		

[a] Hall and Schumacher (1962), using a free-hydrogen wavefunction.
[b] Bessent, Hayes, and Hodby (1967), using a free-hydrogen wavefunction and the assumption that ψ_{H_0} varies slowly over the cores.
[c] A. H. Harker (unpublished), using point-ion wavefunctions. Values marked * include the orthogonalization of the core orbitals to each other.
Experimental values are from Table 4.21. (A_p is determined from A_σ using eqn (4.25).)

of the hyperfine structure and isotope dependence of the central hyperfine constant.

Blum (1967*a*) discussed covalency for the H_i^0 centre. He assumed the covalency was weak, so one could concentrate on the hydrogen and one of its neighbouring fluorines, superposing the effects of the others at the end. Blum considers covalent admixtures of fluorine $2p$ orbitals. The hydrogen orbital Ψ_{H_0} and the fluorine $2p\sigma$ orbitals, which lie along the H-F axis, form bonding and antibonding orbitals:

$$\phi_a \simeq \Psi_{H_0} - \lambda\Psi_F,$$

$$\phi_b \simeq \Psi_F + (\lambda - S)\Psi_{H_0},$$

where S is the overlap $\langle\Psi_{H_0}|\Psi_F\rangle$ and λ is assumed to be spin-independent. The covalency causes some electron transfer from the $2p\pi$ orbitals to the antibonding orbitals, φ_a. This transfer requires admixture of states higher in energy by $(E_\pi - E_a)$. Blum (1967*a*) calculates the change in g-factor in terms of λ, S, and $\bar{E}$, where $\bar{E}$ is an effective promotion energy for states admixed by the spin-orbit coupling. The covalent contribution to the g-factor, Δg_{cov}, is conveniently expressed in terms of Δg, the deviation from the free-spin value with no covalency:

$$\Delta g_{\text{cov}} \simeq -\Delta g \frac{\lambda - S}{\lambda} \frac{\bar{E}}{E_{\pi} - E_{\text{a}}}. \tag{4.26}$$

The observed g-value has been used by Blum (1967*a*) to obtain an estimate of $E_{\pi} - E_{\text{a}}$ with values for overlaps given by Hall and Schumacher and with the following assumptions for other parameters:

$$\Delta g = 2.10^{-4}, \qquad \lambda - S = 0{\cdot}03, \qquad \bar{E} = 8 \text{ eV}.$$

He finds $(E_{\pi} - E_{\text{a}}) \simeq 1{\cdot}3$ eV. Thus the observed g-factor can be understood in terms of plausible covalency parameters. The covalency also affects the hyperfine constants, as shown in Table 4.23. Hodby has

TABLE 4.23

Calculated epr *parameters* (MHz) *for* $CaF_2 : H_i^0$, *including covalency* (*Blum*, 1967*a*)

	Calculated			
	No covalency No distortion	Covalency No distortion	Covalency 0·1 a.u. dilatation	Expt. from Table 4.21
A^{H}	1660	1566	1476	1464
A_s^F	129	129	103	104
A_p^{F}	33	44	40	35

argued that it is necessary to include admixture of fluorine $2s$ orbitals as well as fluorine $2p$ orbitals to fit the temperature dependence of the epr parameters at constant volume.

The temperature and pressure dependence of the epr parameters are discussed by Hall and Schumacher (1962), Blum (1967*b*), Hodby (1969), and Sligar and Blum (1971). The theory of these effects is primitive. In outline, the pressure dependence can be predicted in the way just described, with an assumption about λ as a function of lattice spacing. Blum (1967*a*) assumes λ is proportional to the overlap, S. He concludes that the local compressibility is reduced near the defect by about a factor of two. His results are shown in Table 4.23.

Hodby (1969) observes that, at constant volume, there is a decrease of A with temperature, instead of the increase expected on simple models. For this reason he includes $2s$ covalent admixtures. Sligar and Blum (1971) analyse these constant-volume data in terms of both orthogonalization to core orbitals and temperature-dependent charge transfer. They show that, as for the alkali halides (Spaeth, 1969), it is essential to include corrections for the hydrogen zero-point motion.

Harker (1972) has made point-ion calculations of the $^2S \rightarrow {}^2P$ transition of H_i^0 and H_s^0 centres in CaF_2, SrF_2, and BaF_2 and obtains transition energies of about 9 eV for H_i^0 and about 12 eV for H^0. Measurements of these transitions have not been reported.

The H_s^- centres give rise to a strong absorption band of the $^1S \rightarrow {}^1P$ type in the ultraviolet. The peak position of this band in CaF_2, SrF_2, and BaF_2 was measured by Beaumont, Gee and Hayes (1970) (Table 4.24). Singh, Galipeau, and Mitra (1970) have considered the $^1S \rightarrow {}^1P$ transition using a calculation similar to the point-ion theory of the F centre discussed in § 4.2.3. The main difference comes from the fact that the H_s^- ion has two electrons, and it is essential to include the correlation between them. This is done in two ways. First, the trial wavefunction allows for radial correlation:

$$\Psi_g \sim [\exp(-\alpha r_1)\exp(-\beta r_2)+\exp(-\alpha r_2)\exp(-\beta r_1)].$$

TABLE 4.24

Observed (Beaumont, Gee, and Hayes, 1970) *and calculated (Singh et al.,* 1970; *Bennett,* 1972 *(values due to this author in brackets)) optical properties of* H_s^- *in* CaF_2, SrF_2, *and* BaF_2

	Wavefunction parameters				Transition energy (eV)			Calculated oscillator strength	
	α	β	γ	δ	calc		obs		
CaF_2	0·997	0·516	1·017	0·474	(6·42)	6·82	7·65	(1·94)	1·80
SrF_2	1·003	0·495	1·015	0·461	(5·90)	6·37	7·04	(1·94)	1·80
BaF_2	1·012	0·473	1·012	0·448	(5·47)	5·89	6·00	(1·94)	1·80

For the free H^- ion this prescription gives about 60% of the deviation from the Hartree-Fock energy. Secondly, Singh *et al.* arbitrarily correct the ground-state energy by 0·0143 a.u., this being the difference between the exact ground-state energy of a free H^- ion and the value obtained from the prescription for Ψ_g above. The excited state, Ψ_x, is similar in form to Ψ_g:

$$\Psi_x \sim [\phi_s(r_1)\phi_p(r_2)+\phi_s(r_2)\phi_p(r_1)],$$

where ϕ_s and ϕ_p are hydrogenic 1s and 2p functions: $\phi_s \sim \exp(-\gamma r)$ and $\phi_p \sim r\exp(-\delta r)$. Results are shown in Table 4.24 for the energies, wavefunction parameters and oscillator strengths. The calculated energies are in good general agreement with the measurements.

Bennett (1972) has also used the point-ion model to treat H_s^- in CaF_2, SrF_2, BaF_2, and CdF_2. Instead of using a variational wavefunction with radial correlation Bennett integrated the one-electron Schrödinger equation numerically and used empirical exchange and correlation potentials. He also considered excited states in which one electron was excited to a $2s$ orbital. The calculated energy levels for CaF_2 are, in order of increasing energy, $^1S(1s^2)$, $^3P(1s, 2p)$, $^3S(1s, 2s)$, $^1P(1s, 2p)$, and $^1S(1s, 2s)$ with much the same order for the other crystals. Bennett's calculated results for the $^1S-^1P$ transition are given in Table 4.24. Optical transitions of H_s^- in alkaline earth fluorides have also been considered by Hagston (1970).

4.7.2. *Production and thermal annealing mechanisms of hydrogen centres*

The thermal annealing of hydrogen-doped crystals after X-irradiation at low temperatures was studied in detail by Bessent *et al.* (1969). High-temperature thermal annealing of H_i^0 centres in CaF_2 produced by room temperature X-irradiation has been investigated by Atwater (1964) and by Welber (1965). It was found by Bessent *et al.* (1969) that X-irradiation in CaF_2:H crystals at 120 K and lower temperatures led to formation of H_s^0 centres and perturbed F centres. The H_s^0 centres are converted to H_i^0 centres on warming past 135 K. On warming past 110 K the perturbed F centres change gradually to apparently normal F centres and perturbed M centres and this annealing is independent of the annealing of the H_s^0 centres. The reader is referred to Bessent *et al.* (1969) for a detailed description of the properties of the perturbed F and M centres. The H_i^0 centres, F centres and perturbed M centres decay gradually at room temperature but the decay time is sample dependent, varying from hours to weeks.

In one batch of CaF_2:H crystals investigated by Bessent *et al.* (1969) the H_i^0 centres were stable indefinitely at room temperature due presumably to the presence of impurities acting as deep electron traps. Correlated measurements were made in this material of the effects of high-temperature annealing on the epr spectrum of H_i^0 and the infrared vibrational spectrum of H_s^-. These measurements showed that up to 90% of the H_s^- centres may be destroyed by sufficiently long X-irradiation at any temperature between 20 and 300 K and that most of the H_s^- centres destroyed are converted to H^0 centres. On warming the irradiated crystals it was found that the H_s^- intensity recovered by roughly equal amounts in each of two stages centres at about 400 and 500 K. After completion of the second annealing stage the concentration of H_s^- centres was within 2% of its original value. The epr spectrum of the H_i^0 centres disappeared completely at the 400 K stage. The form in which the nonparamagnetic hydrogen, which reappears as H_s^- at 500 K, exists in the crystal has not been established but molecular hydrogen is a possibility. These investigations suggest that molecular hydrogen may be formed from association of H_i^0 centres at about 400 K and imply that H_i^0 centres are immobile below 400 K.

It was found by Bessent *et al.* (1969) that the production rate of H_i^0 centres and F centres in CaF_2:H above 135 K was about 100 times greater than the production rate of H_s^0 and perturbed F centres below 120 K. The production rate of H_s^0 centres in CaF_2:Er:H containing 0·5% molar concentration of Er^{3+} ions was about 100 times greater than in CaF_2:H crystals because of the electron-trapping properties of the Er^{3+} ion (see Twidell, 1970). It was also found that a preirradiation for 20 minutes of CaF_2:H at room temperature, forming H_i^0 centres,

TABLE 4.25

Activation energies (eV) *for the conversion of* H_s^0 *and* D_s^0 *to* H_i^0 *and* D_i^0 *in* CaF_2, SrF_2, *and* BaF_2 (*Bessent et al.*, 1969)

	CaF_2	SrF_2	BaF_2
Hydrogen	0·37	0·36	0·33
Deuterium	0·37	0·38	0·34

Error in all measurements ±0·02 eV.

increased the initial production rate of H_s^0 at 77 K by a factor of about 20 due to rapid conversion of H_i^0 to H_s^0 by the irradiation.

X-irradiation of SrF_2:H at 77 K produces H_s^0 centres and F centres in almost equal numbers. The H_s^0 centres convert to H_i^0 centres at 130 K and the F centres are stable to room temperature. The production rate of H_s^0 centres and F centres above 130 K is again about 100 times greater than the production rate of H_s^0 and F centres below 120 K. The behaviour of hydrogen in BaF_2:H is similar to that in SrF_2:H; the H_s^0 centres convert to H_i^0 centres at 125 K.

The conversion of H_s^0 to H_i^0 obeys first-order kinetics over a concentration range of 20:1 in CaF_2, SrF_2, and BaF_2 (Bessent *et al.*, 1969). The measured activation energies for the conversion are given in Table 4.25. It is apparent that the measured activation energies for hydrogen and deuterium are the same within experimental error and that the activation energies decrease only slightly from CaF_2 to SrF_2 to BaF_2. In all three materials there appears to be a 100% conversion of H_s^0 to H_i^0. Observation of the fluorine hyperfine structure in epr spectra during the conversion showed that the cube of eight fluorines surrounding H_i^0 was always intact.

The fact that H_i^0 in CaF_2 becomes mobile only at 400 K suggests that the conversion of H_s^0 to H_i^0 may involve motion of the hydrogen to only the nearest interstitial site. If this is the case it follows from the fact that the cube of fluorines surrounding H_i^0 is always intact that the conversion requires removal of the anion vacancy created by motion of the hydrogen. Endor measurements show (Bessent *et al.*, 1969) that there are no lattice defects within 3 fluorine-fluorine spacings of H_s^0 or of H_i^0. An activation energy of greater than 0·5 eV has been found for fluorine vacancy motion in alkaline earth fluorides (Table 3.6). The measured activation energies for conversion of H_s^0 to H_i^0 are smaller (*c.* 0·36 eV) than the measured activation energies for unassociated vacancies. This difference may not rule out the possibility of anion-vacancy motion from a hydrogen site; the observed first-order kinetics are consistent with this mechanism. It would seem that H_s^0 centres may oscillate between a substitutional and a nearest interstitial site but are trapped in interstitial sites only when the temperature of the lattice is high enough to cause anion vacancy removal from a hydrogen site.

The rapid conversion of H_i^0 to H_s^0 by irradiation at 77 K and the large increase in the production rate of H_s^0 when electron-trapping impurities are added to the crystal show that the generation of electron traps by the irradiation is the limiting factor in the production rate of H_s^0 centres in CaF_2 (see § 4.9). The rapid increase in the production rate of *F* and *F*-aggregate centres and H_i^0 centres above the conversion temperature of H_s^0 to H_i^0 is due to trapping of electrons in anion vacancies created by the conversion.

4.7.3. *Oxygen centres*

Investigations by Adler and Kveta (1957) and by Bontinck (1958*b*) show that CaF_2 reacts readily with water vapour at high temperatures and suggest that the resulting hydrolysis gives rise to a variety of defects which includes O^{2-} ions in fluorine sites and charge-compensating fluorine vacancies (see also Messier, 1968 and § 3.2.2). It has been suggested that absorption at about 200 nm in hydrolysed CaF_2 is associated with the presence of oxygen (Bruch, Görlich, Karras, and Lehmann, 1964; Feltham and Andrews, 1965; Bontinck, 1958). Charge compensation of Y^{3+} and trivalent rare-earth ions in CaF_2 by O^{2-} ions is fairly well established (see Chapter 6).

A first step in the understanding of the effects of ionizing radiation on CaF_2 containing oxygen was made by Bill and Lacroix (1966*a*) who investigated hydrolysed CaF_2 by epr methods. They found a spectrum after X-irradiation at room temperature which they assigned to O^- ions in fluorine sites. This is produced by capture of a hole by substitutional O^{2-} ions. The O^- spectrum showed axial symmetry and was described by the spin Hamiltonian

$$\mathscr{H} = g_{\parallel}\beta H_z S_z + g_{\perp}\beta(H_x S_x + H_y S_y) + {}^{\mathrm{F}}A_{\parallel} S_z I_z + \\ + {}^{\mathrm{F}}A_{\perp}(S_x I_x + S_y I_y) - g_{\mathrm{F}}\beta_n \mathbf{H}\cdot\mathbf{I}, \quad (4.27)$$

where $S = \frac{1}{2}$, $I = \frac{1}{2}$ is the spin of the fluorine nucleus and

$$g_{\parallel} = 2{\cdot}0016 \pm 0{\cdot}0008, \quad |{}^{\mathrm{F}}A_{\parallel}| = 53{\cdot}7 \pm 0{\cdot}4 \text{ G},$$

$$g_{\perp} = 2{\cdot}0458 \pm 0{\cdot}0008, \quad |{}^{\mathrm{F}}A_{\perp}| = 15{\cdot}0 \pm 0{\cdot}4 \text{ G}.$$

TABLE 4.26

Parameters of the epr *spectrum of* $^{17}O^-$ *in* CaF_2 *at* 4·5 K (*Bill*, 1969)

$g_{\parallel} = 1{\cdot}9955 \pm 0{\cdot}001$	$g_{\perp} = 2{\cdot}1047 \pm 0{\cdot}001$
${}^{\mathrm{F}}A_{\parallel} = 63{\cdot}6 \pm 0{\cdot}9$ G	${}^{\mathrm{F}}A_{\perp} = 15{\cdot}4 \pm 0{\cdot}9$ G
${}^{17}\mathrm{O}A_{\parallel} = 100 \pm 2$ G	${}^{17}\mathrm{O}A_{\perp} = 9 \pm 6$ G

Since the ground state of the free O^- ion is ^2P a Jahn-Teller distortion in the solid is not unexpected and the axial symmetry was assigned to this effect by Bill and Lacroix (1966*a*). However, since a hyperfine interaction was observed with only one fluorine on the *z*-axis it is possible that a nearest charge-compensating fluorine vacancy on the *z*-axis is partly responsible for lowering the symmetry.

An O^- centre was observed by Bill (1971) in CaF_2 which showed a genuine Jahn-Teller distortion. The crystals were hydrolysed in a manner described by Bill (1969) and subsequently X-rayed at room temperature. An axial epr spectrum of O^- was found with its principal (*z*) axis along $\langle 100 \rangle$ crystal directions. The spectrum showed a resolved fluorine hyperfine interaction with two equivalent fluorines on the *z*-axis. The presence of oxygen in the crystals was confirmed by doping the crystals with $^{17}O(I = \frac{5}{2})$. The measured parameters of the epr spectrum are given in Table 4.26. The distortion axis of a particular O^- oscillates between the three cube edge directions and a population difference between these directions may be induced by uniaxial stress at low temperatures in $\langle 100 \rangle$ or $\langle 110 \rangle$ directions. A similar (paraelastic) effect has been observed for O_2^- centres in alkali halides (Känzig, 1962).

Bill and Lacroix (1966*b*) found that X-irradiation at room temperature of hydrolysed yttrium-doped CaF_2 crystals gave an epr spectrum with orthorhombic symmetry which they suggested was due to a hole trapped at a complex involving oxygen and yttrium. Fedder (1970*a*, *b*) X-irradiated CaF_2 and SrF_2 after diffusing silver into the crystal at a high temperature and he found an epr spectrum which he assigned to Jahn-Teller distorted Ag^{2+} ions. Comparison with the work of Bill (1971) suggests that the spectrum observed by Fedder is due to O^- ions. It seems that Fedder's crystals were contaminated by oxygen during the high-temperature diffusion (Bill, 1972).

4.8. Colour centres in $SrCl_2$, ThO_2, CeO_2, and PbF_2

Colour-centre investigations in crystals with the fluorite structure have been largely confined to the alkaline earth fluorides. However, some investigations have been carried out in other materials and results are presented in this section.

4.8.1. *Strontium chloride*

Colour centres are not readily produced in $SrCl_2$ by X-irradiation at room temperature. They do, however, colour at low temperature. As in the case of CaF_2 and SrF_2 the predominant colour centres produced by X-irradiation between 4 and 77 K are V_K centres and perturbed F centres (see § 4.9.1). Hayes, Lambourn, Rangarajan, and Ritchie (1973) did not observe H centres in the crystals irradiated at temperatures down to 4 K. These authors found that the F band produced by irradiation at 77 K shows two partly-resolved peaks at 580 and 667 nm. Polarized bleaching experiments show that the splitting of the F band is due to a perturbation with symmetry close to trigonal; the higher energy component is a transition to an orbital doublet and the low energy component to an orbital singlet. The source of the perturbation has not been established but alkali impurities are a possibility (see Rzepka, Baltog, Lefrant, Yuste, and Taurel 1973).

Magneto-optical investigations give a value of λ' of about $-100\ cm^{-1}$ for the ^2P state of the F centre (Billardon, Duran, Lefrant, and Taurel 1973). This is numerically larger than the value of $\lambda' = -67\ cm^{-1}$ found for the ^2P state of the F centre in SrF_2 (Table 4.5) due presumably to a larger negative contribution to λ' from the first chlorine shell.

The epr spectrum of the crystals irradiated at 77 K shows a line with unresolved structure with a g-value of 1·992. This line is due to the perturbed F centres; it has a width of 40 G at half height.

The structure of the V_K centre in $SrCl_2$ is similar to that in CaF_2 (§ 4.6.1). The epr spectrum has been measured by Bill, Suter, and Lacroix (1966), Catton and Symons (1968) and Hayes, Lambourn, Rangarajan, and Ritchie (1973). Chlorine has two isotopes, ^{35}Cl and ^{37}Cl, each with $I = \frac{3}{2}$. The hyperfine interaction with these isotopes is resolved and the spin-Hamiltonian (4.19) may be used to describe the spectra (Table 4.16, p. 233). The positions of the $^2\Sigma_u \rightarrow {}^2\Sigma_g$ and $^2\Sigma_u \rightarrow {}^2\Pi_{g\frac{1}{2}}$ type transitions of the V_K centre were determined by Hayes, Lambourn, Rangarajan, and Ritchie (Table 4.17, p. 234) using techniques similar to those described in § 4.6.1.

Dichroism induced in the optical and epr spectra of the V_K centres by bleaching with linearly polarized light begins to disappear at about 105 K. Only 180° motion of the centres is observable at 103 K but 90° motion is pronounced at 107 K. Measurement of the time dependence of the decay of the V_K intensity at 103·1, 105·0, and 107·3 K gives a mean activation energy for the two types of motion of $0{\cdot}251 \pm 0{\cdot}013$ eV (Table 4.20, p. 244).

The motion of the V_K centres at 105 K does not result in the disappearance of paramagnetic hole centres. Some of the moving holes are trapped at a defect site giving a more stable trapped hole centre with an epr spectrum similar to that of the V_K centre. This centre was called a V_{KA} centre by Hayes, Lambourn, Rangarajan, and Ritchie (1973) because of behaviour analogous to that of the V_{KA} centre in SrF_2 (§ 4.6.3). The V_{KA} centre in $SrCl_2$ has an absorption band of the type $^2\Sigma \rightarrow {}^2\Sigma$ peaking at 379 nm and dichroism is readily induced in this band at 77 K by bleaching with polarized light. The dichroism disappears at 140 K although the V_{KA} intensity does not decay until about 290 K. As in the case of SrF_2 (§ 4.6.3) the axis of the V_{KA} centre tumbles about the stabilizing defect before motion through the lattice sets in.

Heating irradiated $SrCl_2$ beyond 77 K also changes the trapped electron centres. Up to about 110 K the decay of the F band matches that of the V_K centres. At temperatures higher than ~130 K the appearance of the spectra in the 500–800 nm region changes. Two bands characteristic of M-type centres grow at 525 and 750 nm (Matei, 1971), but the structure of these centres has not been explored in detail. They reach maximum intensity at ~180 K and decay at higher temperatures. This decay is accompanied by growth of a band peaking at 580 nm which is partly due to F-type centres. The 580 nm band reaches a maximum intensity at 250 K and decays with the V_{KA}

centres at $\sim$290 K. This complex annealing behaviour is unusual and may be in part a consequence of the impurity content of the crystals.

Effects of X-irradiation of hydrogen-doped $SrCl_2$ were examined by Baltog, Lefrant, Houlier, Yuste, Chapelle, and Taurel (1971). Later Lefrant, Jumeaux, and Taurel (1972) X-irradiated SrF_2 at 90 K after heating in hydrogen and found an epr absorption 40 G wide with $g = 1{\cdot}991$. On warming the crystals to 240 K the epr spectrum developed a resolved hyperfine structure with a line spacing of about 8 G and the optical absorption spectrum simplified to a single band at 580 nm. Lefrant *et al.* suggested that these changes arose from conversion of perturbed F centres to unperturbed F centres.

It was found by den Hartog, Mollema, and Schaafsma (1973) that crystals of $SrCl_2$ with a low impurity content contained F centres when rapidly quenched after additive coloration. The epr spectrum ($g = 1{\cdot}989$) had a resolved chlorine hfs with a line spacing of 7·5 G. These authors conclude that the peak of the unperturbed F band is at 630 nm, in disagreement with the results of Lefrant *et al.* It would appear that combined magneto-optical and endor measurements on F centres in $SrCl_2$ are desirable.

den Hartog *et al.* (1973) concluded that the presence of oxygen in $SrCl_2$ strongly enhances the coagulation of F centres during additive coloration.

4.8.2. *Thorium oxide and cerium oxide*

Kolopus, Finch, and Abraham (1970) found that crystals of ThO_2 grown from lead-based solvents showed several epr spectra after irradiation at room temperature with β or γ rays. They assigned one spectrum, which was isotropic, with

$$g = 1{\cdot}9666 \pm 0{\cdot}0006 \quad \text{and} \quad A = 13{,}400 \pm 50 \text{ G}(1{\cdot}230 \text{ cm}^{-1})$$

to Pb^{3+}; the ^{202}Pb isotope with $I = \frac{1}{2}$ is 22·6% abundant. They found a corresponding spectrum in CeO_2 grown from a PbF_2-based solvent with

$$g = 1{\cdot}9649 \pm 0{\cdot}008 \quad \text{and} \quad A = 13{,}130 \pm 50 \text{ G } (1{\cdot}20 \text{ cm}^{-1}).$$

In addition, in irradiated ThO_2, a second Pb^{3+} spectrum was found with trigonal symmetry with

$$g_{\parallel} = 1{\cdot}9704 \pm 0{\cdot}0007, \qquad g_{\perp} = 1{\cdot}9637 \pm 0{\cdot}0007,$$
$$A_{\parallel} = 1{\cdot}194 \pm 0{\cdot}010 \text{ cm}^{-1}, \qquad A_{\perp} = 1{\cdot}181 \pm 0{\cdot}010 \text{ cm}^{-1}.$$

This spectrum of Pb^{3+} showed a resolved hyperfine interaction with the nucleus of another ion with $I = \frac{1}{2}$. The spectrum was not found in crystals grown from a PbO-based solvent and Kolopus *et al.* concluded that the additional hyperfine structure was due to interaction with a fluorine nucleus in one of the eight nearest-neighbour oxygen sites. The super-hyperfine splitting could be fitted to an expression of the form $\Delta H = a+b(3\cos^2\theta-1)$ with $a = 14{\cdot}5$ G and $b = 3{\cdot}0$ G. Kopolus *et al.* suggest that the trigonal Pb^{3+} spectrum is the same as that found earlier by Neeley, Gruber, and Gray (1967) and assigned by the latter authors to the F centre.

An extensive study of photoelectronic processes in undoped and rare-earth doped ThO_2 was carried out by Rodine and Land (1971) using thermoluminescence techniques. Ultraviolet excitation was used to create free electrons and holes; these were subsequently trapped at lattice defects creating the thermoluminescent centres.

Wagner and Murphy (1972) measured epr of undoped CeO_2 crystals after irradiation at 77 K with light of energy greater than the band gap. They observed spectra of three different centres, labelled V_I, V_{II}, and V_{III}, with orthorhombic symmetry. The principal axes (x, y, z) of the g-tensor are parallel to [110], [001], and [110] respectively; the components of the g-tensor are:

	g_x	g_y	g_z
V_I	2·0311	2·0065	2·0330
V_{II}	2·0175	2·0054	2·0317
V_{III}	2·0268	2·0050	2·0415

The V_I and V_{II} centres do not show hyperfine structure but the V_{III} spectrum shows a resolved hyperfine structure with a nucleus with 100% $I = \frac{1}{2}$; the components are $A_x = 3{\cdot}1$, $A_y = 2{\cdot}3$, and $A_z = 2{\cdot}7$ in units of 10^{-4} cm^{-1}. All three centres bleach readily on exposure to room light. The V_I spectrum was observable at 4 K but disappeared above 15 K due to rapid spin-lattice relaxation. By contrast, the spectra of the V_{II} and V_{III} centres could only be observed above 10 K. The V_I, V_{II}, and V_{III} centres decay at 145, 110, and 117 K respectively. Wagner and Murphy suggest that the spectra may be due to O_2^{3-} holes associated with a stabilizing impurity ion. In the case of the V_{III} centre the Y^{3+} ion with $I = \frac{1}{2}$ is a possibility.

4.8.3. *Lead fluoride*

Rose and Schneider (1971) described an extremely complex epr spectrum produced by γ-irradiation at 77 K of β-PbF_2, the only lead halide with the fluorite structure. The spectrum, which occurs near $g = 2$ consists of a large number of highly-orientation-dependent narrow resonance lines due to interaction with lead and fluorine nuclei. Centres giving rise to the epr spectrum also give rise to an optical absorption band at 443 nm. The spectra decay on heating to 160 K. The complexity of the epr spectrum has prevented an unambiguous identification of the centres being made.

4.9. Radiation damage

4.9.1. *The primary defects*

It has been recognized for a considerable time that alkaline earth fluoride crystals with a low impurity content do not colour readily on

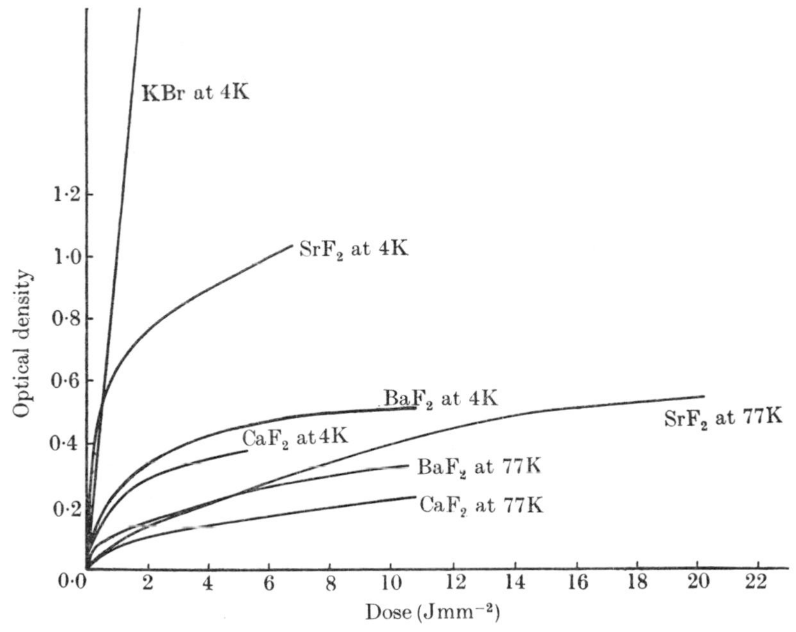

FIG. 4.23. Dependence of F centre optical density on irradiation dose for CaF_2, SrF_2, BaF_2 and KBr exposed to 70 KeV electrons at 4 and 77 K (after Call, Hayes, Stott, and Hughes, 1974).

exposure to X-rays at room temperature. However, much of the CaF_2 generally available contains appreciable concentrations of yttrium impurity and colour centres incorporating this impurity are readily formed at room temperature by X-rays (these centres are described in detail in Chapter 7). The situation at low temperatures is quite different. The crystals show a rapid first-stage coloration which saturates, followed by a slower second-stage growth (Fig. 4.23). Work carried out recently has been concerned to a considerable extent with the temperature dependence of coloration.

The observation of an F band in CaF_2 irradiated with 3 MeV electrons at helium temperatures was reported by Kamikawa, Kazumata, Kikuchi, and Ozawa (1966). Subsequently measurements on CaF_2 irradiated with 0·4 μs pulses of 15 MeV electrons at room temperature were described by Patterson and Fuller (1967) who found no evidence for a transient F band with a lifetime longer than 0·2 μs. This negative result suggested low efficiency for F centre production at room temperature rather than efficient production followed by rapid decay. The production of colour centres by neutron irradiation at low temperatures has been studied in CaF_2 by Kamikawa and Ozawa (1968) and in BaF_2 by Kazumata (1969).

It was shown by Mukerji, Tanton, and Williams (1967) that X-irradiation of CaF_2 at 77 or 4 K produced an F band which disappeared on warming the crystals to room temperature. Arends (1968) found that the F band produced by X-irradiation at 77 K was readily destroyed by optical bleaching at 77 K and could be quickly restored by subsequent X-irradiation at 77 K. This result indicated that anion vacancies required for F centre formation were generated by the original X-irradiation. Görlich, Karras, Symanowski, and Ullman (1968) found that the colourability of alkaline earth fluorides at low temperatures increases with decreasing temperature and at a fixed low temperature increases from CaF_2 to SrF_2. Görlich and Ullman (1972) have also investigated F centres produced by irradiation of mixed alkaline earth fluoride crystals.

More recently, an investigation of effects of temperature on X-ray induced coloration of SrF_2 has been carried out by Hayes and Lambourn (1973*a*). They found that the holes complementary to the F centres produced by irradiation at both 77 and 4 K are V_K centres. In alkali halides, by contrast, irradiation in the helium temperature range generally produces H centres rather than V_K centres (Känzig and Woodruff, 1958).

Fig. 4.24 shows the relative numbers of first-stage F centres produced in SrF_2 irradiated for a fixed time at various temperatures. The production rate falls by a large amount at 34 K over a range of a few deg ees. A further fall in production rate occurs at 67 K. Finally, when V_K motion sets in at $\sim$110 K the production rate falls to a very low value and does not recover at higher temperatures (Hayes and Lambourn (1973*a*) refer to the temperature ranges 4–34, 34–67, 67–110 K as ranges

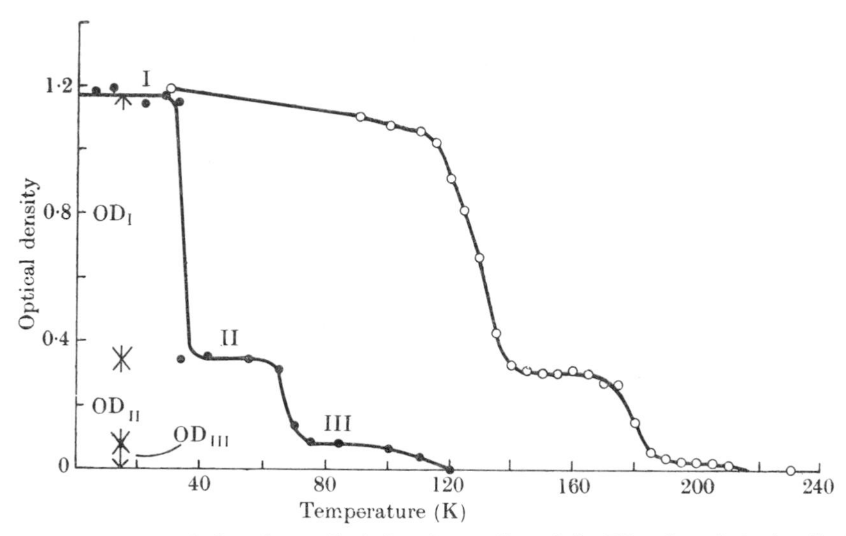

FIG. 4.24. Open circles: thermally-induced annealing of the F band made by irradiating SrF_2 at 30 K. Closed circles: peak optical density produced by X-irradiation for 60 minutes at different temperatures (after Hayes and Lambourn, 1973*a*).

I, II, and III). This result is again in contrast with the situation found in some alkali halides, e.g. KCl, where F centre production efficiency falls in the range 4–100 K, increases again in the range 100–200 K (Comins and Wedepohl, 1966) and falls at higher temperatures.

The F band produced by irradiation in range I shows partly resolved structure (Fig. 4.25*a*) and the shape of this band is the same for all temperatures of irradiation in this range. The F band produced by irradiation in range II shows less well resolved structure (Fig. 4.25*b*). The F band produced by irradiation in range III is about 5% narrower than the band produced by irradiation in range II; there is, however, no pronounced change in shape associated with the decay step at 67 K. It was suggested by Hayes and Lambourn (1973*a*) that the structure in the F band was due primarily to association with negatively charged fluorine interstitials (I centres). It is possible, however, that in the first stage impurities may also be directly involved (see below).

Excitation at 32 K with linearly polarized light in the low energy wing of the F band created by irradiation in range I produces a linear dichroism. This arises from preferential destruction of the perturbed F centres by excitation of electric dipoles close to $\langle 111 \rangle$ and is due to ionization of the F centres. The F band produced by irradiation in range II also shows dichroism characteristic of centres with trigonal symmetry.

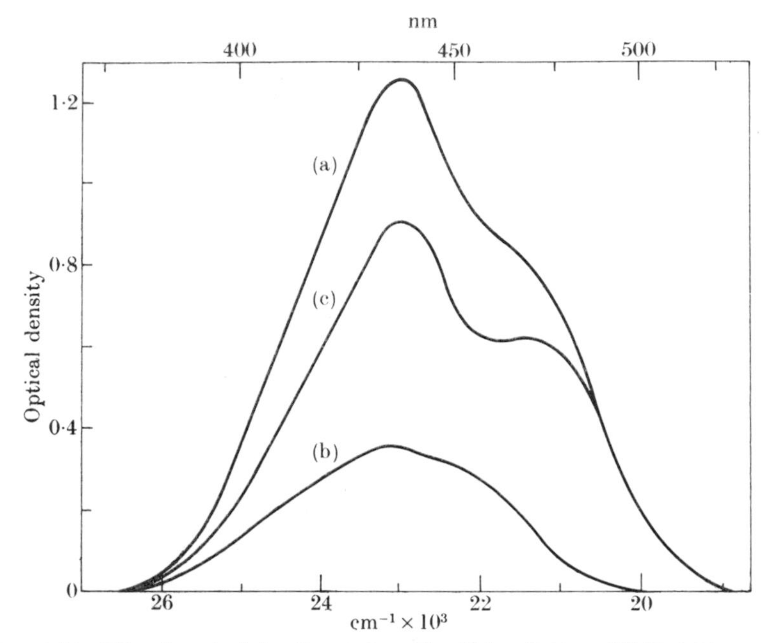

FIG. 4.25. F band optical density produced by X-irradiation of SrF_2 (a) in range I and (b) in range II for the same time (see Fig. 4.24). (c) is the difference between (a) and (b) (after Lambourn, 1972).

The intensity of the F band produced by irradiation in range I does not change appreciably with increasing temperature until the onset of V_K motion at $\sim$110 K (Fig. 4.24). Further heating to $\sim$175 K leads to liberation of holes from V_{KA} centres and further destruction of F centres (Fig. 4.24). The thermal-annealing curve for the F band produced by irradiation in range II is similar to that produced by irradiation in range I and again there is no change in shape of the band with increasing temperature.

If the F band in SrF_2 is destroyed by optical bleaching at the temperature of creation a short reirradiation at the same temperature reconstitutes the band. This holds for ranges I, II, and III and shows that the F–I pairs are converted to F⁺–I pairs by the bleaching and that the latter are stable. However, if the crystal is taken through the series of operations (1) X-irradiate in range I (2) optically bleach the F band (3) raise the temperature for a few minutes to range II (4) re-cool to range I and (5) re-irradiate for a minute, an F band occurs with the shape and intensity that would have been produced if the original irradiation had been carried out in range II. This result indicates that about 70% of the F band intensity produced by irradiation in range I

(OD_I in Fig. 4.23) is due to an *F* centre associated with a nearby fluorine interstitial and that when the *F* electron is removed and the crystal is heated to range II the interstitial rapidly recombines with the vacancy. A similar result is obtained if the series of operations described above is carried out for ranges II and III. It appears that the ranges I, II, and III are associated with F–I pairs of increasing separation and that the *F* band produced by irradiation in range I is a superposition of F bands associated with the three ranges with relative intensities OD_I, OD_{II}, and OD_{III} (Fig. 4.24). Removing the contributions OD_{II} and OD_{III} from the *F* band created in range I leaves the contribution OD_I (Fig. 4.25*c*) to the nearest F–I pair.

The epr spectra of *F* centres produced by irradiation in ranges I and II show an isotropic seven-line hfs characteristic of interaction with six fluorines. The *g*- and *A*-values of the epr spectra are given in Table 4.27

TABLE 4.27

epr *parameters of* F *centres in* X-*irradiated and additively-coloured* SrF_2

		g	A (MHz)
X-irradiated†	25 K	1·9860 ±0·0003	137±2
	41 K	1·9861 ±0·0003	120±1
Additively‡ Coloured		1·9860 ±0·0010	120±6

† Hayes and Lambourn (1973*a*).
‡ Stoneham *et al.* (1968) (see also den Hartog and Arends (1967)).

and the corresponding parameters for the unperturbed *F* centres found in additively coloured crystals are included for comparison. The *A*-value for the *F* centres produced in range I is noticeably larger than the unperturbed value and this larger hyperfine interaction with the nearest fluorines is confirmed by endor measurements (Hayes and Lambourn, 1973*a*). The epr spectrum produced by irradiation in range II is also present in crystals irradiated in range I; it makes the width of the epr lines produced by irradiation in range I somewhat larger than that found for irradiation in range II. The endor measurements show that the concentration of unperturbed *F* centres is very small in crystals irradiated in both ranges I and II.

The detailed temperature dependence of first-stage coloration has also been investigated for CaF_2 by Hayes and Lambourn (1973*b*) and for BaF_2 by Call, Hayes, Stott, and Hughes (1974) with results somewhat similar to those found in SrF_2. The latter authors also showed, using heavy irradiations with 70 keV electrons at low temperatures, the presence of a first and second stage in the coloration of CaF_2, SrF_2, and BaF_2 (Fig. 4.23). The range of 70 keV electrons in these materials is only about 25 μ and one can therefore irradiate to very high doses without the peak optical density of the F band becoming excessive. The F band in SrF_2 gradually changes with dose from the perturbed shape characteristic of the first stage (Figure 4.25) to a symmetrical shape, suggesting that the second-stage F centres are not perturbed.

Magnetic resonance measurements on irradiated alkaline earth fluorides at 20 K show an initial rapid production of V_K centres; this saturates and is associated with the first stage (Call, Hayes, Stott, and Hughes). A much slower production of H centres also occurs which continues without saturation up to high doses and is associated with the second stage. Thermal annealing of the V_K centres at temperatures ($\sim$130 K) at which H centres are still stable leads to an increase of H centre population. The original situation could be restored, however, by a brief reirradiation at 77 K showing that there is a radiation-induced equilibrium between V_K centres and H centres.

Recent work on irradiated $SrCl_2$ has shown coloration effects at low temperatures (§ 4.8.1) similar to those found in the first-stage coloration of SrF_2. However, these effects are not observed in zone-refined $SrCl_2$ (S. Lefrant and L. Taurel, private communication). In addition, no F centre production could be found in zone-refined $SrCl_2$ using total electron doses at 4 K adequate to produce a well-developed second stage in SrF_2 (Call, Hayes, Stott, and Hughes). These results suggest that the first-stage coloration in alkaline earth fluorides may be impurity controlled and it is possible that the perturbed first-stage F centres may be associated with an impurity ion as well as an interstitial. However, it seems that the second stage is characterized by production of isolated F centres and H centres (see also § 4.9.2) and is an intrinsic process.

4.9.2. *Formation of F-aggregate centres*

If a large concentration of F centres (peak optical density in the F band of about 3) is produced in CaF_2 or SrF_2 by irradiation with 1 MeV electrons at 77 K it is found that the F band can subsequently be reduced to an optical density of about 0·2 by optical bleaching. The residual F band is strongly resistant to further optical bleaching at 77 K and this stability suggested that F-aggregate centres could be formed in these crystals. (Hayes and Lambourn, 1973*b*.)

Curve I of Fig. 4.26 shows the optical absorption of SrF_2 after heavy electron irradiation at 77 K. The main absorptions are the F and V_K bands at 435 and 326 nm. There is an additional weak band of unknown origin at 625 nm which is not associated with F-aggregate formation. Curve II shows effects of warming the crystal to 150 K for 10 minutes

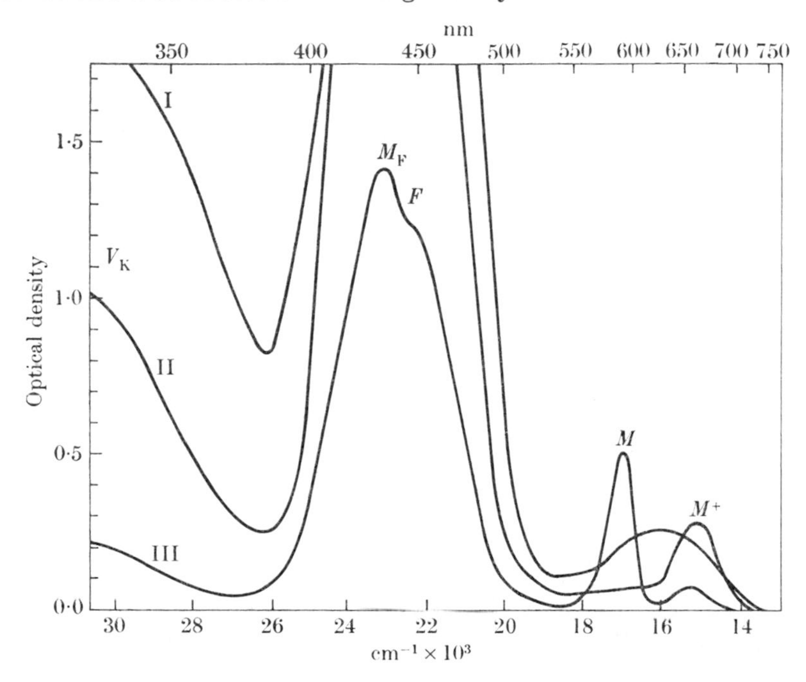

FIG. 4.26. I. Optical absorption at 77 K of SrF_2 after irradiation with 1 MeV electrons at 77 K. II, same as I except that the crystal was heated to 150 K and recooled to 77 K. III, same as II except that the crystal was optically bleached in the F-band region (after Hayes and Lambourn, 1973*b*).

and recooling to 77 K. The F and V_K bands have dropped in intensity, the 625 nm band has disappeared, and the M^+ band has appeared at 685 nm (see § 4.3.5). Somewhat similar behaviour has been observed in alkali halides (Seager, Welch, and Royce, 1972). Curve III shows effects of subsequently bleaching in the F band at 77 K. This produces a gradual reduction in intensity of the M^+ band and a growth of the M band at 595 nm and of the M_F band at 427 nm. High-resolution optical measurements show that R centres are also present weakly in the crystals; the highest optical density achieved at the peak of the zero-phonon line of the *R* centre (§ 4.4.1) is about 0·02. The behaviour of CaF_2 is very similar to that of SrF_2 (Hayes and Lambourn, 1973*b*).

The residual *F* band and also the *M* and M^+ bands produced in irradiated SrF_2 are stable to about 180 K. Above 180 K the intensity of the bands falls off gradually with increasing temperature and about 20% survives to room temperature.

The formation of M^+ centres by annealing irradiated CaF_2 and SrF_2 to 150 K appears to be due to migration of F^+ centres, converting *F* centres to M^+ centres. Because interaction between vacancies and interstitials can lead to mutual destruction it seems unlikely that the *F* centres converted in this way are associated with interstitials. This result supports the view that heavy irradiation produces isolated *F* centres in observable concentration (see § 4.9.1).

4.9.3. *Mechanisms of defect formation*

It seems that the radiolysis of some alkali halides can be reasonably well accounted for by non-radiative electron-hole recombination (Pooley, 1966*a*; Hersch, 1966; Keller and Patten, 1969). However, there is still some difficulty in satisfactorily accounting for the displacement of anions with the energy available from this mechanism (Smoluchowski, Lazareth, Hatcher, and Dienes, 1971). The primary damage products in alkali halides appear to be *F* centres and *H* centres (Ueta, Kondo, Hirai, and Yoshinari, 1969). The *F* centres and *H* centres are many lattice spacings apart and this may be due to a replacement sequence involving a halogen atom moving in a closely packed ⟨110⟩ direction (Pooley, 1966*b*; Torrens and Chadderton, 1967). The replacement sequences will be of varying length and it is possible that *F* centres and *H* centres separated by a few lattice spacings may transform into F^+ and *I* centres by a tunnelling process; the electric field of a nearby interstitial anion prevents vacancies trapping electrons and forming *F* centres (Lüty, 1961).

The first-stage coloration in alkaline earth fluorides may arise from stabilization of vacancy-interstitial pairs at impurity sites. This stage saturates in SrF_2 at a concentration of about $2\times10^{18} F$ centres cm^{-3}. The second stage has been extended without saturation to a similar concentration (Figure 4.23). It should be emphasized that many of the *H* centres produced in the second stage appear to be perturbed by other lattice defects (§ 4.6.2) and hence it seems that even production of second-stage *F* centres may be indirectly controlled by impurities.

The efficiency of second-stage coloration in alkaline earth fluorides is about two orders of magnitude less than the colourability of KBr in similar circumstances (see Figure 4.23). Call, Hayes, Stott, and Hughes point out that this is not inconsistent with conclusions based on consideration of ion sizes and efficiency of collision replacement sequences (see e.g. Townsend 1973). They also discuss the absence of second-stage coloration in zone-refined $SrCl_2$.

In alkali halides the (non-paramagnetic) hole centres created by irradiation at room temperature appear to be trapped at anion interstitial complexes (Sonder and Sibley 1972; Hobbs, Hughes, and Pooley 1973). Holes may be stabilized at room temperature and above in irradiated alkaline earth fluorides doped with rare-earth ions (§ 4.6.3), possibly by complexes of charge-compensating interstitial anions. About 20 per cent of second-stage F centres survive on warming alkaline earth fluoride crystals to room temperature (§ 4.9.2) suggesting that some of the radiation-produced interstitials aggregate or are trapped above their mobility temperature and that many of the traps or aggregates are stable at room temperature.

In conclusion, it now seems that there is a general understanding of the way in which presently available crystals of alkaline earth fluorides respond on exposure to ionising radiation at low temperatures. However, no direct observations have yet been made of F^+ and I centres in crystals irradiated at low temperatures and experimental investigations of the primary damage process of the type described by Ueta, Kondo, Hirai, and Yoshinari (1969) are required. Further progress will depend to a considerable extent on the availability of purer crystals but with the high-melting-point fluorides purification presents major problems.

References

Adler, H. and Kveta, I. (1957). *S.B. Österreich Akad. Wiss. Math.-Naturw. Klasse*. Abt. II **166**, 199.

Appel, J. (1968). *Solid State Physics* (ed. F. Seitz and D. Turnbull) Academic Press Inc., New York, **21**, 193.

Arends, J. (1964). *Phys. Stat. Sol.* **7**, 805.

—— (1968). *Sol. St. Comm.* **6**, 421.

Arkangel'skaya, V. A. (1965). *Opt. Spectrosc.* **18**, 46.

—— and Alekseeva, L. A. (1966). *Ibid*, **21**, 50.

Assmus, W. and Dreybrodt, W. (1969). *Phys. Stat. Sol.* **34**, 183.

Atwater, H. A. (1964). *J. chem. Phys*, **40**, 606.

Baltog, I., Lefrant, S., Houlier, B., Yuste, M., Chapelle, J. P., and Taurel, L. (1971). *Phys. Stat. Sol.* **48**, 345.

Barile, J. (1952). *J. chem. Phys.* **20**, 297.

Bartram, R. H., Harmer, A. L., and Hayes, W. (1971). *J. Phys. C., Sol. St. Phys.* **4,** 1665.
——, Stoneham, A. M., and Gash, P. (1968). *Phys. Rev.* **176,** 1014.
Bass, I. L. and Mieher, R. L. (1968). *Ibid* **175,** 421.
Beaumont, J. H., Gee, J. V., and Hayes, W. (1970). *J. Phys. C, Sol. St. Phys.* **3,** L152.
——, Harmer, A. L., and Hayes, W. (1972*a*). *J. Phys. C, Sol. St. Phys.* **5,** 1475.
——, ——, —— (1972*b*). *Ibid* **5,** 257.
——, ——, —— (1972*c*). *Ibid* **5,** 266.
——, ——, —— (1972*d*). *Ibid* **5,** 275.
——, ——, ——, and Spray, A. R. L. (1972). *Ibid* **5,** 1489.
—— and Hayes, W. (1969). *Proc. R. Soc.* **A309,** 41.
——, ——, Kirk, D. L., and Summers, G. P. (1970). *Ibid* **A315,** 69.
——, ——, Summers, G. P., and Twidell, J. W. (1968). *Solid State Commun.* **6,** 903.
Bennett, H. S. (1968). *Phys. Rev.* **169,** 729.
—— (1969). *Ibid* **184,** 919.
—— (1971*a*). *Ibid* **B3,** 2763.
—— (1971*b*). *Ibid* **B4,** 1327.
—— (1972). *Ibid* **B6,** 3936.
—— and Lidiard, A. B. (1965). *Phys. Lett.* **18,** 253.
Berezin, A. A. (1972). *Phys. Stat. Sol.* (*b*) **49,** 51.
Bessent, R. G., Hayes, W., and Hodby, J. W. (1967). *Proc. R. Soc.* **A297,** 376.
——, ——, ——, and Smith, P. H. S. (1969). *Ibid.* **A309,** 69.
Bill, H. (1969). *Helv. Phys. Acta* **92,** 771.
—— (1971). *Solid State Commun.* **9,** 477.
—— (1972). *Phys. Rev.* B, **6,** 4359.
—— and Lacroix, R. (1966*a*). *Phys. Lett.* **21,** 257.
—— —— (1966*b*). *Ibid* **22,** 250.
——, Suter, H., and Lacroix, R. (1966). *Ibid* **22,** 241.
Billardon, M., Duran, J., Lefrant, S., and Taurel, L. (1973) *Phys. Stat. Sol*(b). **58,** 673.
Blum, H. (1967*a*). *J. chem. Phys.* **46,** 650.
—— (1967*b*). *Phys. Rev.* **161,** 213.
Bontinck, W. (1958*a*). *Physica,* **24,** 639.
—— (1958*b*). *Ibid* **24,** 650.
—— and Dekeyser, W. (1956). *Ibid* **22,** 595.
—— and Amelinckx, S. (1957). *Phil. Mag.* **2,** 94.
Böttcher, R., Frieser, A., Milsch, B., Völkel, G., Wartewig, S., Welter, M., and Windsch, W. (1970). *Phys. Stat. Sol.* **38,** K 97.
Bruch, H., Görlich, P., Karras, H., and Lehmann, R. (1964). *Ibid* **4,** 685.
Call, P. J., Hayes, W., and Smith, P. H. S., (1974). *J. Phys. C, Sol. St. Phys.* **7,** 1214.
——, ——, Stott, J. P., and Hughes, A. E. (1974) *J. Phys. C., Sol. St. Phys.,* to be published.
Catton, R. C. and Symons, M. C. R. (1968). *J. chem. Soc.* **A,** 2155.
Castner, T. G. and Känzig, W. (1957). *J. Phys. Chem. Solids* **3,** 178.
Cavenett, B. C., Hayes, W., Hunter, I. C., and Stoneham, A. M. (1969). *Proc. R. Soc.* **A309,** 53.
Collins, W. C. (1973*a*). *Phys. Stat. Sol.* (*b*) **56,** 291.
—— (1973*b*). *Ibid* (*b*) **57,** 63.
—— and Schneider, I. (1972). *Ibid.* (b) **51,** 769.

COMINS, J. D. and WEDEPOHL, P. T. (1966). *Solid State Commun.* **4,** 537.
COMPTON, W. D. and RABIN, H. (1964). *Solid state physics* (ed. F. Seitz and D. Turnbull) Academic Press Inc. New York, **16,** 121.
CRAWFORD, J. H. and SLIFKIN, L. M. (eds.) (1972). *Point defects in solids,* Vol. 1, Plenum Press, New York and London.
DELBECQ, C. J., HAYES, W., and YUSTER, P. H. (1961). *Phys. Rev.* **121,** 1043.
——, SMALLER, B., and YUSTER, P. H. (1958). *Ibid* **111,** 1235.
DEN HARTOG, H. W. (1969). Thesis, Groningen University.
—— (1970). *Phys. Stat. Sol.* **38,** 457.
—— and ARENDS, J. (1967*a*). *Ibid* **22,** 131.
—— ——(1967*b*). *Ibid* **23,** 713.
—— and FLIM, G. J. (1972). *Ibid* (b) **50,** K 53.
——, MOLLEMA, P., and SCHAAFSMA, A. (1973). *Ibid* (a) **55,** 721.
——, TINBERGEN, W., and PERDOK, W. G. (1970). *Ibid* (a) **2,** 347.
ELCOMBE, M. M. and PRYOR, A. W. (1970). *J. Phys. C, Sol. St. Phys.* **3,** 492.
ELLIOTT, R. J., HAYES, W., JONES, G. D., MACDONALD, H. F., and SENNETT, C. T. (1965). *Proc. R. Soc.* **A289,** 1.
FEDDER, R. C. (1970*a*). *Phys. Rev.* B, **2,** 32.
—— (1970*b*). *Ibid* B, **2,** 40.
FELTHAM, P. and ANDREWS, J. (1965). *Phys. Stat. Sol.* **10,** 203.
FLEMING, R. J. (1966). *Ibid* **18,** K 139.
FLYNN, C. P. (1971). *Communications in solid state Physics*, p. 150.
FITCHEN, D. B. (1968). *Physics of color centers* (ed. W. Beall Fowler), p. 294, Academic Press, New York.
FEOFILOV, P. P. (1953). *Dokl. Akad. Nauk* SSSR **92,** 545.
—— (1961). *The physical basis of polarized emission,* Consultants Bureau, New York.
—— and KAPLYANSKII, A. A. (1962). *Sov. Phys. Usp.* **5,** 79.
FOWLER, W. B. (ed.) (1968). *Physics of color centers,* Academic Press, New York.
GILBERT, T. L. and WAHL, A. C. (1971). *J. Chem. Phys.* **55,** 5247.
GÖRLICH, P. and KARRAS, H. (1962). *Phys. Stat. Sol.* **2,** 988.
——, ——, and KOCH, A. (1965). *Ibid* **12,** 203.
——, ——, and LEHMANN, R. (1961*a*). *Ibid* **1,** 389.
——, ——, —— (1961*b*). *Ibid* **1,** 525.
——, ——, —— (1963). *Ibid* **3,** 98.
——, ——, SYMANOWSKI, C., and ULLMANN, P. (1968). *Ibid* **25,** 93.
—— and ULLMANN, P. (1972). *Ibid* (b) **50,** 577.
GOURARY, R. S. and ADRIAN, F. J. (1960). *Solid state physics.* (ed. F. Seitz and D. Turnbull) Academic Press Inc., New York, **10,** 127.
HAARER, D. and PICK, H. (1967). *Z. Phys.* **200,** 213.
HAGSTON, W. E. (1970). *Phys. Stat. Sol.* **39,** 551.
HAKEN, H. (1963). *Polarons and excitons* (ed. C. G. Kuper and G. D. Whitfield), p. 295, Oliver and Boyd, Edinburgh and London.
HALL, J. L. and SCHUMACHER, R. T. (1962). *Phys. Rev.* **127,** 1892.
HALL, T. P. P., LEGGEAT, A., and TWIDELL, J. W. (1969). *J. Phys. C, Sol. St. Phys.* **2,** 1590.
——, ——, —— (1970). *Ibid* **3,** 2352.
HARKER, A. H. (1972). Unpublished data; also A.E.R.E. Report TP568 (1974).
HARMER, A. L. (1971). Thesis, Oxford.
HAYES, W. (1970). *Magnetic resonance* (ed. C. K. Coogan, N. S. Ham, S. N. Stuart, J. R. Pilbrow and G. V. H. Wilson) p. 271, Plenum Press, New York and London.

—— (1969). *Proc. int. conf. on science and technology of non-metallic crystals*, New Delhi (ed. S. C. Jain), p. 239.
—— and Hodby, J. W. (1966). *Proc. R. Soc.* **A294,** 359.
——, Kirk, D. L., and Summers, G. P. (1969). *Sol. St. Comm.* **7,** 1061.
—— and Lambourn, R. F. (1973*a*). *J. Phys. C, Solid State Phys.* **6,** 11.
—— —— (1973*b*). *Phys. Stat. Sol.* (*b*) **57,** 693.
——, ——, Rangarajan, G., and Ritchie, I. M. (1973). *J. Phys. C, Sol. St. Phys.* **6,** 27.
——, ——, and Smith, P. H. S. (1970). *Ibid* **3,** 1797.
——, ——, and Stott, J. P. (1974) *J. Phys. C, Solid State Phys.* (to be published).
—— and Stoneham, A. M. (1969). *Phys. Lett.* **29A,** 519.
—— and Stott, J. P. (1967). *Proc. R. Soc.* **A301,** 313.
—— and Twidell, J. W. (1962). *Proc. phys. Soc.* London **79,** 1295.
Henderson, B. and Wertz, J. E. (1968). *Adv. Phys.* **17,** 749.
Henry, C. H., Schnatterly, S. E., and Slichter, C. P. (1965). *Phys. Rev.* **137A,** 583.
Hersch, H. N. (1966). *Ibid* **148,** 928.
Hirschfelder, J., Eyring, H., and Rosen, N. (1936). *J. chem. Phys.* **4,** 121.
Hobbs, L. W., Hughes, A. E., and Pooley, D. (1973). *Proc. R. Soc.* **A332,** 167.
Hodby, J. W. (1969). *J. Phys. C, Sol. St. Phys.* **2,** 404.
Holstein, T. F. (1959). *Ann. Phys.* (New York) **8,** 325, 343.
Hughes, A. E. (1967). *J. Physique* **28,** C4, 55.
—— and Henderson, B. (1972). *Point defects in solids*, Vol. 1 (ed. J. H. Crawford and L. M. Slifkin) Plenum Publishing Corporation, New York, p. 381.
Ishii, T., Tomiki, T., and Ueta, M. (1958). *J. Phys. Soc. Japan* **13,** 1411.
——, ——, —— (1959). *Ibid* **14,** 1415.
Ivey, R. F. (1947). *Phys. Rev.* **72,** 341.
Jette, A. N. and Das, T. P. (1969). *Ibid* **186,** 919.
——, ——, and Gilbert, T. L. (1969). *Phys. Rev.* **184,** 884.
Kabler, M. N. and Patterson, D. A. (1967). *Phys. Rev. Lett.* **19,** 652.
Kalder, K. A. and Malysheva, A. F. (1971). *Opt. Spectrosc.* **31,** 135.
Kamikawa, T., Kazumata, Y., Kikuchi, A., and Ozawa, K. (1966). *Phys. Lett.* **21,** 126.
—— and Ozawa, K. (1968). *J. Phys. Soc. Japan*, **24,** 115.
Känzig, W. (1962). *J. Phys. Chem. Sol.* **23,** 473.
—— and Woodruff, T. O. (1958). *J. Phys. Chem. Solids* **9,** 70.
Kaplyanskii, A. A. (1964). *Opt. Spectrosc.* **16,** 329.
—— and Medvedev, V. N. (1967). *Ibid* **23,** 404.
Karras, H. (1961*a*). *Phys. Stat. Sol.* **1,** 68.
—— (1961*b*). *Ibid* **1,** 160.
Kazumata, Y. (1969). *Phys. Stat. Sol.* **34,** 377.
Keller, F. J. and Patten, F. W. (1969). *Solid State Commun.* **7,** 1603.
Kiss, Z. J. and Staebler, D. L. (1965). *Phys. Rev. Lett.* **14,** 691.
Kolopus, J. L., Finch, C. B., and Abraham, M. M. (1970). *Phys. Rev.* **2,** 2040.
Kubo, K. (1966). *J. Phys. Soc. Japan.* **21,** 1300.
Lambourn, R. F. (1962). Thesis, Oxford.
Lefrant, S., Jumeau, D., and Taurel, L. (1972). *Phys. Stat. Sol.* **50,** K101.
Loudon, R. (1964). *Proc. phys. Soc.* Lond. **84,** 379.
Lüty, F. (1953). *Z. Phys.* **134,** 596.
—— (1961). *Halbleiter Probleme* (Braunschweig: Friedrick Weiweg).
Macias, A. (1968). *J. chem. Phys.* **48,** 3464.
Maslen, V. W. (1967). *Proc. phys. Soc.* Lond. **91,** 466.

Markham, J. J. (1966). *F. Centers in alkali halides*, Academic Press, New York.
Marzke, R. F. and Mieher, R. L. (1969). *Phys. Rev.* **182,** 453.
Matei, L. (1971). *Solid State Commun.* **9,** 1281.
Mayerl, M. (1951). *S.B. Akad. Wiss. Wien* IIa, **160,** 31.
McCall, R. T. and Grossweiner, L. I. (1967). *J. appl. Phys.* **38,** 284.
Messner, D. and Smakula, A. (1960). *Phys. Rev.* **120,** 1162.
Messier, D. R. (1968). *J. Electrochem. Soc.* (U.S.A.) **115,** 397.
Merz, J. L. and Pershan, P. S. (1967). *Phys. Rev.* **162,** 217.
Mollwo, E. (1934). *Nachr. Gesell. Wiss. Göttingen* **6,** 79.
Mott, N. F. and Littleton, M. J. (1938). *Trans. Faraday Soc.* **34,** 485.
Mukerji, A., Tanton, G. A., and Williams, J. E. (1967). *Phys. Stat. Sol.* **22,** K19.
Murray, R. B. and Keller, F. J. (1967). *Phys. Rev.* **153,** 993.
Neeley, V. I., Gruber, J. B., and Gray, W. J. (1967). *Ibid* **158,** 809.
Norgett, M. J. and Stoneham, A. M. (1973*a*). *J. Phys. C, Solid State Phys.* **6,** 229.
—— —— (1973*b*). *Ibid* **6,** 238.
Okamoto, F. (1961). *Phys. Rev.* **124,** 1090.
Osiko, V. V. (1965). *Sov. Phys. Sol. State* **7,** 1047.
Patterson, D. A. and Fuller, R. G. (1967). *Phys. Rev. Lett.* **18,** 1123.
Pekar, S. I. (1951). *Researches in the electron theory of crystals* (AEC (1963) translation 5575).
—— and Deigen, M. F. (1948). *Z.E.T.P.* **18,** 481.
Phillips, W. and Duncan, R. C. Jr. (1971). *Metall. Trans.* **2,** 769.
Pierce, C. B. (1964). *Phys. Rev.* **135,** A83.
Pooley, D. (1966*a*). *Proc. Phys. Soc.* **87,** 245.
—— (1966*b*). *Ibid* **87,** 257.
Przibram, K. (1956). *Irradiation colours and luminescence*, Pergamon Press, London.
—— (1959). *Z. Phys.* **154,** 111.
Rodine, E. T. and Land, P. L. (1971). *Phys. Rev.* **B4,** 2701.
Rose, B. F. and Schneider, E. E. (1968). (1971) *Phys. Lett.,* **34A,** 27.
Runciman, W. A., Stager, C. V. and Crozier, M. H. (1963). *Phys. Rev. Lett.* **11,** 204.
Rzepka, E., Baltog, I., Lefrant, S., Yuste, M., and Taurel, L. (1973) *Phys. Stat. Sol.* (b) **57,** 383.
Schneider, I. (1966). *Ibid* **17,** 1009.
—— (1970). *Ibid* **24,** 1296.
Schulman, J. H. and Compton, W. D. (1962). *Color centers in solids*, Pergamon Press.
——, Ginther, R. J., and Kirk, R. D. (1952). *J. chem. Phys.* **20,** 1966.
Schlesinger, M. and Menon, A. K. (1969). *Can. J. Phys.* **47,** 1637.
Scouler, W. J. and Smakula, H. (1960). *Phys. Rev.* **120,** 1154.
Seager, C. H., Welch, D. O., and Royce, B. S. H. (1972). *Phys. Stat. Sol.* (b), **40,** 609.
Sierro, J. (1965). *Phys. Rev.* **138,** A648.
Simpson, J. H. (1949). *Proc. R. Soc.* **A197,** 269.
Singh, R. S., Galipeau, D. W., and Mitra, S. S. (1970). *J. Chem. Phys.* **52,** 2341.
Slichter, C. P. (1963). *Principles of magnetic resonance*, Harper & Row, p. 195.
Sligar, S. G. and Blum, H. (1971). *Phys. Rev.* **B3,** 3587.
Smakula, A. (1950). *Ibid* **77,** 408.
—— (1953). *Ibid* **91,** 1570.
Smith, D. Y. (1965). *Phys. Rev.* **137,** A154.

—— and DEXTER, D. L. (1972). *Progr. opt.* **10,** 167.
SMITH, W. V. (1957). *G.E.C. Knolls Atomic Power Laboratory Report* KAPL-1720.
SMOLUCHOWSKI, R., LAZARETH, O. W., HATCHER, R. D., and DIENES, G. J. (1971). *Phys. Rev. Lett.* **27,** 1288.
SONDER, E. and SIBLEY, W. A. (1972). *Point defects in solids,* Vol. 1, (ed. J. H. Crawford and L. M. Slifkin) Plenum Publishing Corp., New York, p. 201.
——, SIBLEY, W. A., ROWE, J. E., and NELSON, C. M. (1967). *Phys. Rev.* **153,** 1000.
SONG, K. (1970). *J. chem. Phys.* **31,** 1389.
—— (1971). *Solid State Commun.* **9,** 1263.
SPAETH, J. M. (1969). *Phys. Stat. Sol.* **34,** 171.
STONEHAM, A. M. (1967). Unpublished work.
—— (1972). *Phys. Stat. Sol.* (b) **52,** 9.
——, HAYES, W., SMITH, P. H. S., and STOTT, J. P. (1968). *Proc. R. Soc.* **A306,** 369.
SWANK, R. K. and BROWN, F. C. (1963). *Phys. Rev.* **130,** 34.
TANTON, G. A., STETTLER, J. D., SHATAS, R. A., WILLIAMS, J. E., and MUKERJI, A. (1968). *Bull. Am. Phys. Soc.* **13,** 903.
TORRENS, I. M. and CHADDERTON, L. T. (1967). *Phys. Rev.* **159,** 671.
TOWNSEND, P. D. (1973). *J. Phys. C. Sol. St. Phys.* **6,** 961.
TWIDELL, J. W. (1970). *J. Phys. Chem. Solids,* **31,** 299.
TZALMONA, A. and PERSHAN, P. S. (1969). *Phys. Rev.* **182,** 906.
UETA, M., KONDO, Y., HIRAI, M., and YOSHINARI, T. (1969). *J. Phys. Soc. Japan* **26,** 1000.
VAN DOORN, C. Z. (1961–62). Phillips Research Report Suppl.
WAGNER, G. R. and MURPHY, J. (1972). *Phys. Rev.* B, **6,** 1638.
WATSON, R. E. (1958). *Ibid* **111,** 1108.
WELBER, B. (1964). *Ibid* **136,** A1408.
—— (1965). *J. Chem. Phys.* **40,** 606.
WICKERSHEIM, A. K. and HANKING, B. M. (1959). *Physica* **25,** 569.
WOOD, R. F. (1970). *Phys. Stat. Sol.* **42,** 849.
YAMASHITA, J. and KUROSAWA, T. (1958). *J. Phys. Chem. Solids,* **5,** 34.

5

RARE-EARTH IONS IN CUBIC SITES

5.1. Introduction

EARLY investigations into the strong fluorescent properties of the natural mineral fluorite revealed the source to be traces of rare-earth metal ions which can efficiently change ultraviolet radiation into visible radiation. Thus when interest in the optical and paramagnetic properties of rare-earth ions in low concentrations in solids began to expand in the early 1950's, CaF_2 was chosen as one of the more convenient anhydrous hosts. This material is almost ideal for the purpose since it is relatively easy to grow in large single-crystal form and it readily accepts rare-earth dopants. In addition, the band gap for these materials is large so that the optical spectra can be investigated to wavelengths as short as 1200 Å.

Alkaline earth fluorides doped with rare-earth ions provide a rich field for spectroscopic and solid state studies. In this chapter we shall be concerned almost entirely with the spectroscopic aspects, concentrating on ions in sites with cubic symmetry. The material covered here will provide background for the discussion of the solid state aspects dealt with in Chapters 6 and 7.

With the discovery of the solid state laser, CaF_2 was one of the first materials to be intensively examined for possible new lasers because there then existed preliminary optical data on the rare-earth ions in this material (Feofilov, 1956) and also because of the ease with which it can be produced with excellent optical and mechanical properties. Thus, for example, $CaF_2{:}Sm^{2+}$ became the second successful solid-state laser material after ruby (Sorokin and Stevenson, 1961). Indeed the impetus to search for new and better laser materials produced an explosive growth in the study of rare-earth doped alkaline earth fluoride crystals after 1961. One of the most important new developments was the discovery of how to produce stable divalent rare-earth ions in these hosts, adding $CaF_2{:}Tm^{2+}$ and $CaF_2{:}Dy^{2+}$ to the list of successful laser materials (Kiss and Duncan, 1962*a*, *b*; see also Smith and Sorokin, 1966).

In time more efficient laser materials were found and CaF_2 became less important in this field. However, spectroscopic studies of fluorites doped with rare-earth ions have continued to be productive in the development of new techniques. One of the first demonstrations of an infrared up-converter using energy transfer was in $BaF_2:Yb^{3+}:Tm^{3+}$ (Ovsyankin and Feofilov, 1966*a*, *b*). The most efficient room-temperature photochromic materials ever discovered are reduced crystals of CaF_2 containing La, Ce, Gd, or Tb (see Chapter 7). In addition, crystals containing divalent rare-earth ions with strong absorption bands and large magnetic circular dichroism, turned out to be ideal materials for optical-paramagnetism experiments in solids in the tradition which Kastler first developed for gases. These latter experiments have led to the development of a solid-state optically-pumped microwave maser (Sabisky and Anderson, 1967) and a new acoustic phonon spectrometer (Anderson and Sabisky, 1971).

The separate discussion of rare-earth ions at cubic sites comes about because the RE^{2+} ions form a distinct spectroscopic class in these hosts, existing almost entirely with cubic symmetry. Taken together with the RE^{3+} ions at cubic sites, a unique model system is formed for investigating the interaction of the $4f$ electrons with the eight equivalent F^- ions located at the corners of a cube. Once this interaction is well understood then, in principle, the study of the ions at non-cubic sites should be treated as a perturbation or distortion of this higher symmetry case (Chapter 6). It should be emphasized however, that at the present time the data and theory even for the cubic sites are far from complete. A list of the rare-earth ions according to the number of $4f$ electrons and their ground electronic state at cubic sites in CaF_2 is given in Table 5.1.

After the crystals are grown with the rare-earth fluoride dopant the rare-earth ions are usually found in the trivalent state occupying the divalent cation site. The fluorite lattice is very adaptable and provides the necessary charge compensation in a variety of ways, leading to a number of different microscopic sites in a single crystal (Chapter 6). In crystals grown in an atmosphere of HF gas, sometimes called type II crystals, excess fluorine ions provide the compensation by occupying any of several different nearby interstitial sites. The binding energy of these F^- ions to the RE^{3+} ions is sufficiently weak that a few percent of the RE^{3+} ions appear to be located at sites of perfect cubic symmetry (see § 3.2.3). For example, Kiro and Low (1968) have shown using endor that in $CaF_2:Yb^{3+}$ cubic symmetry of some Yb^{3+} ions is maintained out to the third shell of fluorine lattice ions. The major spectroscopic problem for the trivalent rare-earth ions is the separation of the cubic spectrum from those of the other sites. A number of methods developed for doing this are discussed in § 5.2 (see also § 6.3).

TABLE 5.1
Electronic ground states of rare-earth ions at cubic sites in CaF_2

Electron number	Rare-earth ion		State	
1	La^{2+}		$5d$	^2E
		Ce^{3+}	$4f$	$^2F_{\frac{5}{2}}(\Gamma_8)$
2	Ce^{2+}		$4f5d$	Γ_5
		Pr^{3+}	$4f^2$	$^3H_4(\Gamma_5)$
3	Pr^{2+}		$4f^3$	$^4I_{\frac{9}{2}}(\Gamma_8)$
		Nd^{3+}		
4	Nd^{2+}		$4f^4$	$^5I_4(\Gamma_3)$
		Pm^{3+}		
5	Pm^{2+}		$4f^5$	$^6H_{\frac{5}{2}}(\Gamma_8)$
		Sm^{3+}		
6	Sm^{2+}		$4f^6$	$^7F_0(\Gamma_1)$
		Eu^{3+}		
7	Eu^{2+}		$4f^7$	$^8S_{\frac{7}{2}}$
		Gd^{3+}		
8	Gd^{2+}		$4f^75d(?)$	
		Tb^{3+}	$4f^8$	$^7F_6(\Gamma_3)$
9	Tb^{2+}		$4f^85d(?)$	
		Dy^{3+}	$4f^9$	$^6H_{\frac{15}{2}}(\Gamma_8)$
10	Dy^{2+}		$4f^{10}$	$^5I_8(\Gamma_3)$
		Ho^{3+}		$^5I_8(\Gamma_3?)$
11	Ho^{2+}		$4f^{11}$	$^4I_{\frac{15}{2}}(\Gamma_6)$
		Er^{3+}		$^4I_{\frac{15}{2}}(\Gamma_7)$
12	Er^{2+}		$4f^{12}$	$^3H_6(\Gamma_3)$
		Tm^{3+}		
13	Tm^{2+}		$4f^{13}$	$^2F_{\frac{7}{2}}(\Gamma_7)$
		Yb^{3+}		
14	Yb^{2+}		$4f^{14}$	$^1S_0(\Gamma_1)$
		Lu^{3+}		

Crystals doped with the fluorides of Eu, Sm and to a lesser degree Tm, are found to contain a large fraction of divalent rare-earth ions when grown in a reducing atmosphere of H_2 gas or excess rare-earth metal. The conversion of the other rare-earth ions to the divalent state was first achieved by using ionizing radiation, usually X-rays or gamma rays (Hayes and Twidell, 1961). Free electrons are found to be trapped by trivalent ions located at cubic sites to produce the divalent state (for further discussion see Chapter 7). This procedure readily converts only that few per cent of the ions initially located at cubic sites to the divalent state and these are only marginally stable at room temperature or under intense optical illumination. Stable and complete conversion to the divalent state can be achieved by removing the charge-compensating F^- interstitial ions through either solid state electrolytic reduction (Guggenheim and Kane, 1964; Fong, 1964) or even better, by reduction of the crystals in an alkaline metal atmosphere at approximately 700 °C (Kiss and Yocom, 1964; see also Phillips and Duncan, 1971). Free electrons find their way to the rare-earth ions during the process and convert the ions to the divalent state; more than 98% of the ions are found at sites of cubic symmetry for doping levels of less than 0·1%. The vapour reduction process has been successfully applied to all the rare-earth ions except La, Ce, Gd, and Tb, which form instead the more complex photochromic centre (Chapter 7).

The spectroscopic task of finding the $4f$ energy levels of the divalent ions is also much simpler than that for the trivalent ions because the $4f5d$ levels form a convenient series of pump bands in the visible spectrum for exciting the ions. This lowering of the $5d$ energy levels upon going to the divalent state makes a striking difference to the appearance of the crystals; those containing only RE^{3+} ions are as clear as the pure host while those containing RE^{2+} ions are strongly coloured. These bands also add to the richness of the spectroscopic problem since it becomes possible to study these mixed configuration states, especially in La^{2+} and Ce^{2+} where the $5d$ levels are lowered to the point where they form the ground state (§ 5.3).

5.2. $4f^n$ Electronic states

5.2.1. *Hamiltonian for $4f^n$ electrons*

The very narrow width, ~ 1 cm^{-1}, of the optical transitions within the $4f^n$ configuration of the rare-earth ions in solids demonstrates directly that the chemical binding with the surrounding ligand ions is not very sensitive to which state within this configuration the ion occupies. The coupling is weak because the $4f$ orbitals lie inside the $5s$ and $5p$ orbitals, which keep them fairly well isolated from the neighboring ions. The calculation of the energy levels of this configuration therefore starts with those of the free ion and the effects of the lattice are then added as a small perturbation. There are many general references on this problem, but, in presenting a brief introduction we shall primarily follow Dieke (1968) and Abragam and Bleaney (1970).

The method for obtaining the free-ion energy levels has been a complex mixture of theory and experiment, with the theory generally giving a guide to the organization of experimental data and the experiments providing measurements of the relevant model parameters

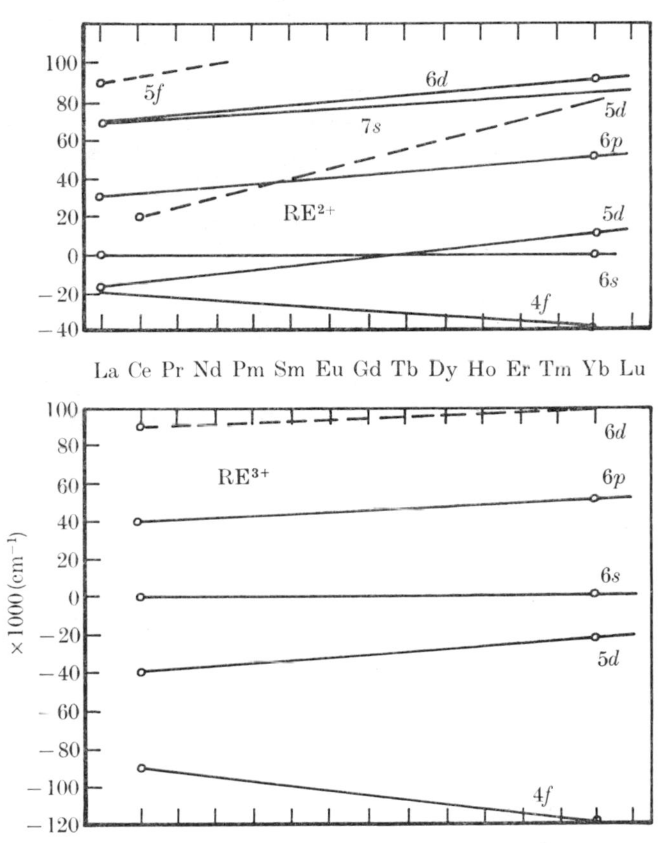

FIG. 5.1. Relative position of the centres of the principal configurations of the divalent and trivalent rare-earth ions [Dieke and Crosswhite, 1963].

(Dieke, 1968). The zeroth-order approximation begins with a Hartree-Fock calculation of the partially filled shells, with the electrons in the xenon core and the nuclear charge providing a central potential, $V_0(r_i)$:

$$H_0 = \sum_i (p_i^2/2m + V_0(r_i)). \tag{5.1}$$

This simple one-electron form of the energy leads to all the levels of a given configuration having the same eigenvalues. Dieke used a variety of experimental data to obtain approximate values for the configuration energy of each of the divalent and trivalent rare-earth ions (Fig. 5.1). The wide separation between the configurations in the trivalent ions is maintained through the addition of other terms in the Hamiltonian; the ground state of these ions therefore always arises from the $4f^n$ configuration and all the optical properties in the visible spectrum are due to the weak transitions within this configuration. The reduction of the coulomb term in the potential upon going to the divalent state brings the configurations much closer together so that the $4f^{n-1}5d$ configuration plays a more important role in these ions. The $4f^n$ configuration cannot mix with the nearby $4f^{n-1}6s$ and $4f^{n-1}5d$ configurations because of parity and hence it still remains fairly pure of any admixture.

The next major terms in the free-ion Hamiltonian are the spin-orbit coupling and the electron repulsion terms:

$$H_1 = \sum_i \zeta(l_i . s_i) + \sum_{ij} \frac{e^2}{r_{ij}}. \tag{5.2}$$

It is possible to show that within the $4f^n$ configuration this operator introduces only four independent parameters, the spin-orbit coupling constant ζ, and three Slater integrals of the type

$$F^k = \int_0^\infty \int_0^\infty \frac{r_<^k}{r_>^{k+1}} R_i^2(r_i) R_j^2(r_j) r_i^2 r_j^2 \, dr_i \, dr_j \qquad (k = 2, 4, 6). \tag{5.3}$$

If the wavefunctions are hydrogenic then these three integrals have constant ratios and most calculations use these ratios to reduce the number of independent parameters associated with the second term in eqn (5.2) to one.

Using modern computers and group theory it is possible to handle almost any combination of these parameters starting with any representation. However, because the electron repulsion term dominates and L and S are good quantum numbers of this operator, it has become conventional to label the levels with the set of quantum numbers $|4f^n JLS\rangle$. The values of L and S are taken to be those obtained in the limit of zero spin-orbit coupling with no energy level crossings being allowed in the limiting process. The diagonal matrix elements for the spin-orbit terms in this representation can be obtained from the operator, $\zeta L . S$, which gives rise to a series of spin-orbit multiplets. In most of the cases to be discussed L and S are quite good quantum numbers. When off-diagonal matrix elements of the spin-orbit coupling are incorporated the calculation is said to include intermediate coupling.

Dieke [1968] has given an overall picture of the lowest levels of the $4f^n$ configuration using data from the spectra of the trivalent rare-earth ions in $LaCl_3$ (Fig. 5.2). This figure contains a great amount of information about the free-ion levels of this configuration and frequent reference is made to it. The energy levels of the divalent ions are spaced approximately the same as those of the trivalent ions with about a 15% reduction in the energy scale. Most of the levels shown in Fig. 5.2 cannot be observed in the divalent state since the levels of the $4f^{n-1}5d$ configuration overlap them (§ 5.3).

The perturbation introduced by the host crystal is written as a sum of one-electron operators which are consistent with the local symmetry of the substitutional site. Under the assumption that the electrons in the unfilled shell are described by pure $4f$ electrons it is possible to reduce the Hamiltonian to the form

$$H_c = \sum_i B^4\left[\frac{4\pi}{9}\right]^{\frac{1}{2}}\{Y_4^0(\theta_i, \phi_i)+(\tfrac{5}{14})^{\frac{1}{2}}[Y_4^4(\theta_i, \phi_i)+Y_4^{-4}(\theta_i, \phi_i)]\}+$$

$$+B^6\left[\frac{4\pi}{13}\right]^{\frac{1}{2}}\{Y_6^0(\theta_i, \phi_i)-(\tfrac{7}{2})^{\frac{1}{2}}[Y_6^4(\theta_i, \phi_i)+Y_6^{-4}(\theta_i, \phi_i)]\} \quad (5.4)$$

for a site of cubic (O_h) symmetry. The $Y^k(\theta_i, \phi_i)$ are normalized spherical harmonics and B^4 and B^6 are unspecified parameters.

It is now well established that covalency effects make the major contribution to the crystal-field splittings. The effect of the electrostatic potential due to a lattice of point charges (§ 5.2.3) is relatively small and so the reason why eqn (5.4) works so well has to be examined in more detail. If we assume that the real wavefunctions contain small admixtures of other orbitals $|j\rangle$, centred either on the rare-earth ion or the ligand ions, we then have the crystal-field Hamiltonian and perturbed wavefunctions

$$H_c = H_c(4f)+H'(4f, j), \quad (5.5)$$

$$|\alpha\rangle = |4f, \alpha\rangle+\sum_j \frac{|j\rangle\langle j|\, H'|4f, \alpha\rangle}{E_j-E_\alpha}. \quad (5.6)$$

Here the perturbed wavefunctions are labeled with the same quantum numbers as the pure $4f$ states since the perturbation is small and no new levels are brought into the problem. The matrix elements of the crystal-field Hamiltonian to first order can be written as

$$\langle\alpha|\, H_c\,|\beta\rangle = \langle 4f, \alpha|\left\{H_c(4f, 4f)+\sum_j \frac{H'\,|j\rangle\langle j|\, H'}{E_j-E_\alpha}\right\}|4f, \beta\rangle \quad (5.7)$$

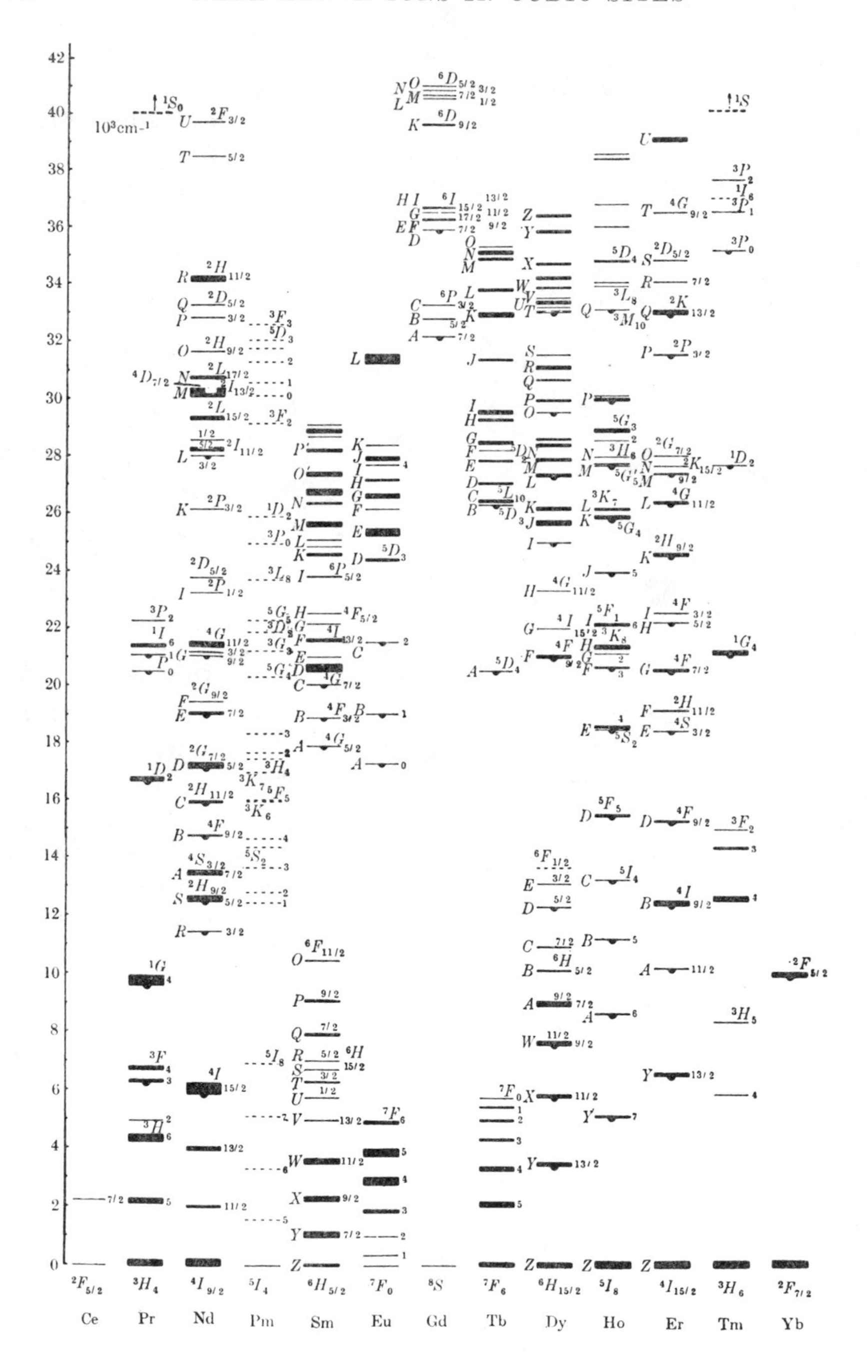

FIG. 5.2. Stark manifolds of triply-ionized rare-earths in anhydrous $LaCl_3$. The pendant half-circles indicate fluorescing levels [Dieke, 1968].

(The effects of wavefunction overlap can also be put into this form (§ 5.2.3).) The net factor within the large brackets is a one-electron operator for pure $4f$ states and it must reflect the cubic symmetry of the problem. Hence even when the second term makes the major contribution to the matrix elements of the crystal-field splittings, the problem still remains *formally* one of solving for the splittings within a pure $4f$ configuration. This means that eqn (5.4) can be used under rather general conditions, the effects of covalency and electrostatic shielding being taken up into the parameters B^4 and B^6.

This result has additional consequences because the $4f$ wavefunctions which result from fitting the observed energy splittings are those which represent the major part of the wavefunctions for the real states, eqn (5.6). Thus these $4f$ wavefunctions can be used to calculate magnetic splitting factors to within a few percent and also relative fluorescent intensities. Conversely, epr measurements on the magnetic ground states which are formally sensitive to the ratio of the parameters B^4 and B^6 can be expected to give a relevant measurement of this ratio. The $4f$ wavefunctions of course do not give information on properties such as the transferred hyperfine interaction with the neighbouring fluorine ions which depend directly on the admixed ligand parts of the wavefunctions.

In order not to add to the confusion that exists about the definition of the parameters B^4 and B^6 we shall generally use the traditional terms $A_4\langle r^4\rangle$ and $A_6\langle r^6\rangle$, which arise from the angular-dependent part of the simple electrostatic point-charge Hamiltonian.

$$A_4\langle r^4\rangle = \tfrac{1}{8}B^4 \quad \text{and} \quad A_6\langle r^6\rangle = \tfrac{1}{16}B^6 \tag{5.8}$$

5.2.2. *Divalent rare-earth $4f^n$ optical properties*

The determination of the crystal-field parameters is primarily done through optical studies, with epr providing some additional information in certain cases (§ 5.2.4). The $4f$–$4f$ optical transitions of ions at cubic sites are weak magnetic-dipole in character, satisfy the selection rule $\Delta J = 0, \pm 1$, and are most easily observed in fluorescence rather than by direct absorption. The occurrence of fluorescence however depends on the multi-phonon decay processes being slower than or comparable to the radiative decay rates. This generally means that the energy gap between the levels must be fairly large ($\geqslant 5000\ \text{cm}^{-1}$; see Weber, 1968 and Riseberg and Moos, 1968). This trend can be seen in Fig. 5.2. In most of the divalent rare-earth ions the $4f$–$5d$ levels begin at 10 000

cm^{-1} above the ground state and so only those $4f$ levels which are lower in energy can be observed.

Excellent optical data are available for about half of the ions in the series. Very strong $4f$–$4f$ fluorescence is observed for $Dy^{2+}(4f^{10})$ (Kiss, 1965), $Ho^{2+}(5f^{11})$ (Weakliem and Kiss, 1967) and $Tm^{2+}(4f^{13})$ (Kiss, 1962) in all three alkaline earth fluoride hosts. In CaF_2 sharp zero-phonon lines are observed in fluorescence from the lowest $4f5d$ level to the lower $4f^n$ levels in $Sm^{2+}(4f^6)$ (Feofilov, 1956, Wood and Kaiser, 1962, and Sorokin *et al.*, 1962) and in Eu^{2+} (Kaplyanskii and Feofilov, 1962). Most of the levels of the $4f^2$ configuration have been observed in Ce^{2+} as sharp absorption lines from the mixed $4f5d$ ground state (Alig *et al.*, 1969). Zeeman studies have been carried out on all the above ions and stress measurements have been made on all but Ce^{2+}.

The ions $Pr^{2+}(4f^3)$ and $Nd^{2+}(4f^4)$ have no wide gaps in their energy level structure and hence do not fluoresce. $Er^{2+}(4f^{11})$ does not fluoresce because the ${}^3H_4$ manifold lies lower than ${}^3H_5$ and optical transitions from ${}^3H_4$ to the ground ${}^3H_6$ are forbidden ($\Delta J > 1$). With sufficiently

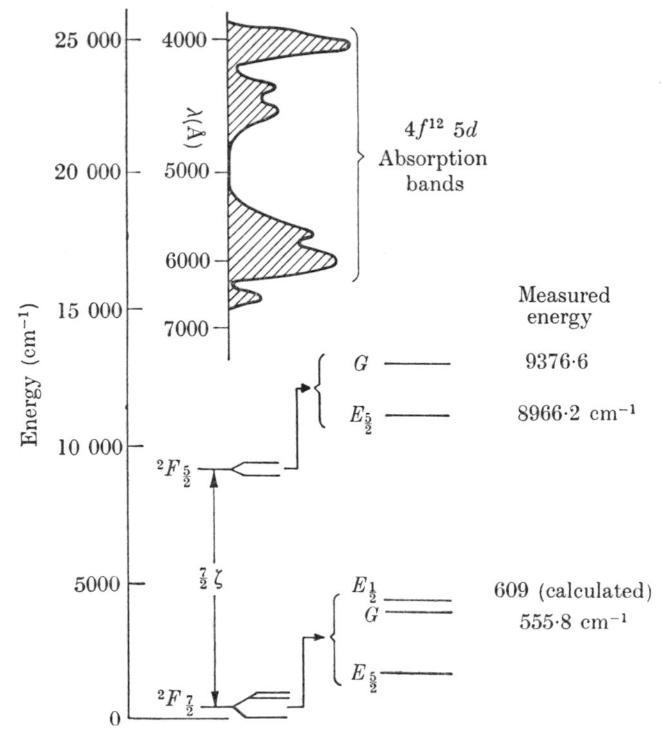

FIG. 5.3. Energy levels of the $4f^{13}$ configuration in CaF_2:Tm^{2+} and the $4f$–$5d$ absorption bands [Kiss 1962 as modified by Sabisky and Anderson 1966].

high optical resolution the $4f$–$4f$ transitions can be observed in direct absorption using samples $\sim$1 cm thick and ion concentrations of $\sim$0·1%. Some lines of Nd^{2+} have been observed in absorption by Arkhangel'skaya and Shraiber (1968). The allowed transitions in Pr^{2+} occur rather far in the infrared, which would make it very difficult to observe them in this fashion. However, the $^3H_6$—$^3H_5$ transitions in Er^{2+} at 7000 cm^{-1} might be accessible to direct absorption studies.

The optical properties of the three ions Dy^{2+}, Ho^{2+}, and Tm^{2+} are quite similar. In Fig. 5.3 the crystal-field-split components of the two lowest J manifolds of CaF_2:Tm^{2+} are shown along with the $4f$–$5d$ absorption bands. When excited in the $4f$–$5d$ bands the ions quickly decay through non-radiative processes to the first J manifold just above the ground state. Phonon processes generally maintain thermal equilibrium between the crystal-field levels of this excited manifold; fluorescence to the crystal-field components of the ground manifold takes place at a slower rate. The quantum efficiency for the fluorescence in CaF_2:Tm^{2+} is about 5%, and the fluorescent lifetime is 3 ms (Kiss, 1962).

The fluorescence spectrum of Ho^{2+} in CaF_2, SrF_2, BaF_2, and $SrCl_2$ as shown in Fig. 5.4 is typical of what is observed with these ions. Most of the energy comes from direct transitions between electronic levels; the two very intense lines in this case actually can be resolved with higher resolution into 5 pure electronic lines. Associated with each electronic line is a group of weaker vibronic sidebands, which move closer to the parent line and become sharper as one goes from CaF_2 to $SrCl_2$. Generally there are five peaks in the vibronic structure and it seems possible to assign symmetries to each peak through the observation that some are missing, depending on the symmetry of the emitting and terminal electronic states (Axe and Sorokin, 1963). The frequency and symmetry of the vibronic energies are given in Table 5.2.

The linewidth of the pure electronic transitions are narrow ($\sim$1 cm^{-1}), which also reflects the weak dynamic coupling of the $4f$ electrons to the lattice. The $^2F_{\frac{5}{2}}(\Gamma_7) \rightarrow {}^2F_{\frac{7}{2}}(\Gamma_7)$ transition in CaF_2:Tm^{2+} is unusually narrow (Fig. 5.5). This is due in part to the fact that only local strains of Γ_1 symmetry can directly shift this line. Also, because the emitting and terminal levels are the lowest crystal-field components of the two J manifolds, and the energy separations to the higher crystal-field levels are large ($\sim$500 cm^{-1}), most dynamic phonon processes are weak and are rapidly frozen out as the temperature is lowered. The solid line in Fig. 5.5 represents the temperature dependence of the linewidth

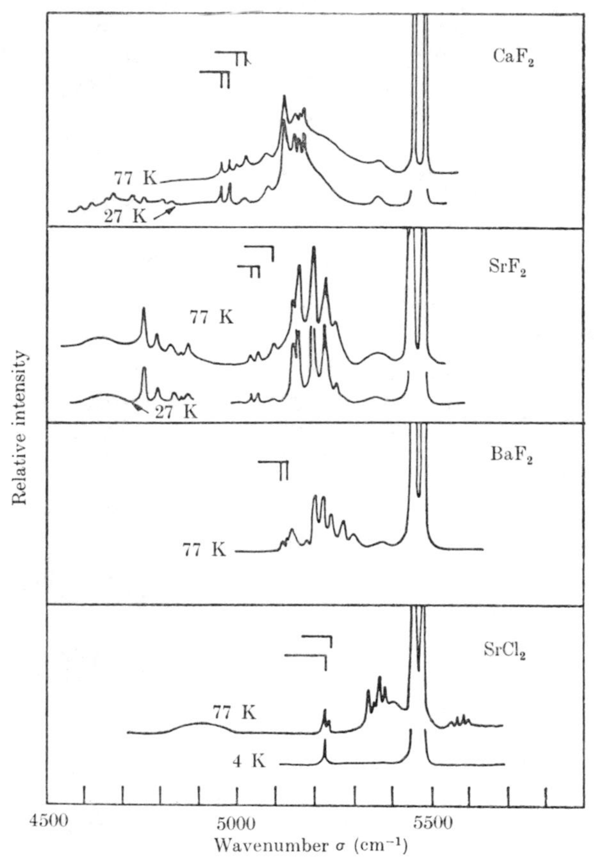

Fig. 5.4. Low-resolution emission spectra, $^4I_{\frac{13}{2}}$ to $^4I_{\frac{15}{2}}$, of $Ho^{2+}(4f^{11})$ in CaF_2, SrF_2, BaF_2, and $SrCl_2$ [Weakliem, and Kiss, 1967].

expected of the Raman-like phonon process which should be dominant in this case. The $^2F_{\frac{5}{2}}(\Gamma_7) \to {}^2F_{\frac{7}{2}}(\Gamma_8)$ transition has a limiting linewidth at low temperatures of 12 cm^{-1}, which is determined by the spontaneous multiphonon decay rate of the $^2F_{\frac{7}{2}}(\Gamma_8)$ level to the ground state.

Divalent thulium has the simplest energy level scheme of the divalent rare-earth ions which have been observed. The $4f^{13}$ configuration, a single hole in the $4f$ shell, consists of two J manifolds, $^2F_{\frac{5}{2}}$ and $^2F_{\frac{7}{2}}$, which are separated by $\sim$9000 cm^{-1}. The excited $^2F_{\frac{5}{2}}$ manifold is split in the cubic crystal field into two levels (Γ_7 and Γ_8) and the ground $^2F_{\frac{7}{2}}$ manifold into three (Γ_6, Γ_7, and Γ_8) (see Fig. 5.3). Since there are at most two levels of the same symmetry, the complete Hamiltonian can easily be diagonalized exactly. Actually, first-order perturbation theory

TABLE 5.2

Vibronic sidebands of 4f–4f transitions (cm^{-1})

(Tentative phonon symmetry assignments)

System	Γ_3	Γ_5	Γ_4	Γ_4	Γ_2	Reference
$CaF_2:Dy^{2+}$	100	180	270	—	400	Kiss (1965)
$CaF_2:Ho^{2+}$	—	160	232	310	370	Weakliem and Kiss (1967)
$CaF_2:Tm^{2+}$	88	180	257	348	390	Kiss (1962)
$SrF_2:Sm^{2+}$	90	140	216	282	349	Axe and Sorokin (1963); Kaplyanskii and Przhevaskii (1966); Cohen and Guggenheim (1968)
$SrF_2:Dy^{2+}$	85	155	218	295	360	Kiss (1965).
$BaF_2:Sm^{2+}$	—	—	186	244	—	Kaplyanskii and Przhevaskii (1966)
$BaF_2:Dy^{2+}$	79	140	182	255	305	Kiss (1965)
$CaF_2:Er^{3+}$	135, 228, 375					Aizenberg *et al.* (1971)
$CaF_2:Dy^{3+}$	136					Al'tshuler *et al.* (1970)
$SrF_2:Dy^{3+}$	143					Al'tshuler *et al.* (1970)
$BaF_2:Dy^{3+}$	127					Al'tshuler *et al.* (1970)
$CdF_2:Dy^{3+}$	83					Al'tshuler *et al.* (1970)

is completely adequate for this case. The simplicity of the $4f$ configuration for Tm^{2+} and the large amount of data available for this ion, gives it a position of special importance in the series.

Fluorescence has been observed from the ${}^2F_{\frac{5}{2}}(\Gamma_7)$ level to the ${}^2F_{\frac{7}{2}}(\Gamma_7)$ and ${}^2F_{\frac{7}{2}}(\Gamma_8)$ levels and absorption from the ground ${}^2F_{\frac{7}{2}}(\Gamma_7)$ level to the ${}^2F_{\frac{5}{2}}(\Gamma_7)$ and ${}^2F_{\frac{5}{2}}(\Gamma_8)$ levels in CaF_2 (Kiss, 1962), and in SrF_2 and BaF_2 (Weakliem, 1973). This gives the position of all the Tm^{2+} levels except that of the ${}^2F_{\frac{7}{2}}(\Gamma_6)$ level. This level has never been directly observed since fluorescence from the ${}^2F_{\frac{5}{2}}(\Gamma_7)$ level to this state is forbidden and that from the ${}^2F_{\frac{5}{2}}(\Gamma_8)$ level occurs at such a high temperature that it is obscured by the vibronic sidebands of the other lines. Therefore there are only three independent energy splittings known and these are just sufficient to determine the spin-orbit coupling constant and the two crystal-field parameters; they are given for the three alkaline earth fluoride hosts in Table 5.3. The spin-orbit coupling constant (2513 cm^{-1}) is identical in all three hosts and is slightly larger than the free-ion value of 2503 cm^{-1} given by Sugar (1970).

Divalent holmium ($4f^{11}$) has a more complicated free-ion energy level scheme (see Fig. 5.2) and only crystal-field levels in the two lowest manifolds (${}^4I_{\frac{15}{2}}$ and ${}^4I_{\frac{13}{2}}$) have been observed. Because these two arise from the same term ($L = 6$ and $S = 3$) there is no expectation that the data will reveal any precise information about the Slater integrals F_k. Even so Weakliem and Kiss (1967) used intermediate coupling in their calculations by including the levels ${}^2L_{\frac{15}{2}}$, ${}^2K_{\frac{15}{2}}$, ${}^2K_{\frac{13}{2}}$,

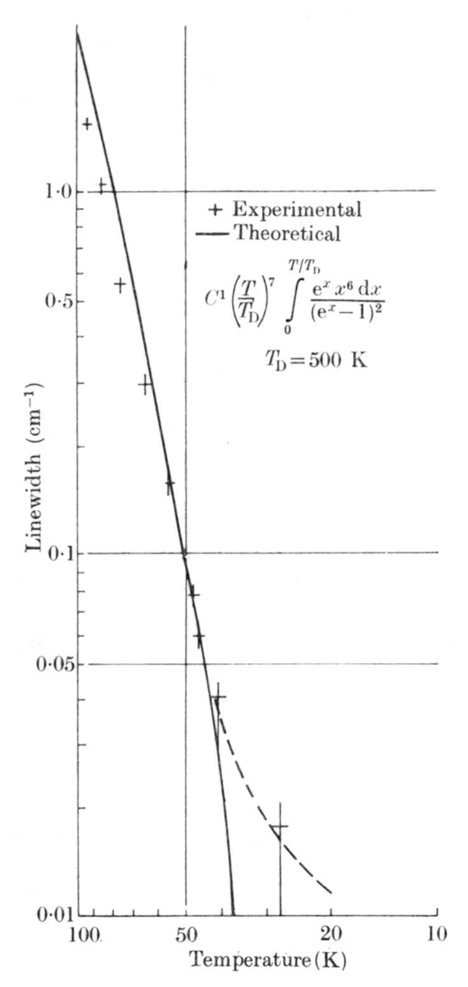

FIG. 5.5. Temperature dependence of the linewidth of the $^2F_{\frac{5}{2}}(\Gamma_7)$ to $^2F_{\frac{7}{2}}(\Gamma_7)$ transition in CaF_2:Tm^{2+} [Duncan and Kiss, 1973].

and $^2I_{\frac{13}{2}}$ in their basis set of wavefunctions and by using estimates for the term energies. The levels of different J which are admixed by the crystal field, $^4I_{\frac{11}{2}}$ and $^4I_{\frac{9}{2}}$, completed their basis set. All four of the components of the ground manifold were determined experimentally as well as the lowest two in the excited manifold. This gave six independent data points to be fitted with three parameters, which they were able to do extremely well. The parameters they obtained are given in Table 5.4.

Kiss (1965) has observed all the crystal-field-split components of the two lowest manifolds, $^5I_8$ and $^5I_7$, in divalent dysprosium ($4f^{10}$). This

TABLE 5.3
Crystal-field parameters for Tm^{2+} *and* Yb^{3+} $(4f^{13})$

	$A_4\langle r^4\rangle$ cm^{-1}	$A_6\langle r^6\rangle$ cm^{-1}	Ref.
$CaF_2:Tm^{2+}$	−189·2	27·7	Kiss (1962) Bleaney (1964)
$SrF_2:Tm^{2+}$	−165·0	22·7	Weakliem (unpublished but reported in Hayes and Smith (1971))
$BaF_2:Tm^{2+}$	−138·8	18·8	Weakliem (unpublished but reported in Hayes and Smith (1971))
$SrCl_2:Tm^{2+}$	−75(20)	16(16)	Alig *et al.* (1973)
$CaF_2:Yb^{3+}$	−213	34·3	Kiro and Low (1970)

TABLE 5.4
Crystal-field parameters for $Ho^{2+}(4f^{11})$

Host	$A_4\langle r^4\rangle$ cm^{-1}	$A_6\langle r^6\rangle$ cm^{-1}
CaF_2	−295	29·1
SrF_2	−249	24·1
BaF_2	−210	20·8
$SrCl_2$	−110	15·6

All measurements from Weakliem and Kiss (1967).

gives thirteen independent data points to fit. However he restricted his basis set to only these two free-ion levels plus $^5I_6$ and so his parameters are not as reliable as those for Ho^{2+}. The parameters he obtained are given in Table 5.5. The spin-orbit coupling constant he found is larger than that found for Ho^{2+}, although it should be smaller, since this constant is observed to be a monotonically increasing function with increasing Z in the trivalent ions (Dieke, 1968). This probably is a direct consequence of having used such a limited basis set.

Three J manifolds $^3H_4$, $^3H_5$, $^3H_6$ of $Ce^{2+}(4f^2)$ have been observed in absorption from the $4f5d$ ground state (Alig *et al.*, 1969). A total of

TABLE 5.5
Crystal-field parameters for $Dy^{2+}(4f^{10})$

Host	$A_4\langle r^4\rangle$ cm^{-1}	$A_6\langle r^6\rangle$ cm^{-1}
CaF_2	−235	31·5
SrF_2	−209	28·1
BaF_2	−185	24·6

All measurements from Kiss (1965)

seven energy levels are known, with some g-values and symmetry assignments given by Zeeman studies. The spin-orbit splittings are only 2 to 3 times larger than the crystal-field splittings, which makes it essential to use a full basis set and carry out a complete diagonalization of the Hamiltonian. This has been done, but the fit is not entirely satisfactory and the results, particularly for $A_6\langle r^6\rangle$, are questionable.

The excellent optical data on $SrF_2:Sm^{2+}(4f^6)$ includes detailed Zeeman studies (Wood and Kaiser, 1962 and Feofilov *et al.*, 1962), but the analysis to obtain crystal-field parameters apparently has never been carried out. A point-charge calculation has been performed and shown to give splittings which are much smaller than those observed (Vetri and Bassoni, 1968). The data obtained for $CaF_2:Sm^{2+}$ is too limited to obtain crystal-field parameters (see § 5.3.3.).

The only $4f$ level of $Eu^{2+}(4f^7)$ not covered by the $5d$ configuration consists of the $^8S_{\frac{7}{2}}$ manifold. The absence of orbital angular momentum in this manifold leads to very small crystal-field splittings ($\sim$1 cm^{-1}), which are best studied by magnetic resonance (§ 5.2.4).

5.2.3. *Trivalent rare-earth $4f^n$ optical properties*

The present state of the optical data and the number of crystal-field parameters known for trivalent ions located at cubic sites is comparable to that of the divalent ions. In principle almost all the optical data that have been obtained on these ions in $LaCl_3$, given in Fig. 5.2, could be found in the alkaline earth fluorides. But the very complicated spectroscopic problem of sorting out the many different charge-compensated sites has only begun, so the cubic spectrum and its analysis is known only for Yb^{3+}, Dy^{3+}, Er^{3+}, and Gd^{3+} in some of the hosts. The non-cubic sites generally lack inversion symmetry and therefore do not have the rigid selection rules for the fluorescence which apply to the cubic sites. In addition the vibronic spectra can be relatively more intense and occur at slightly different frequencies for each of the non-cubic sites. When these facts are considered and also the fact that the concentration of the cubic sites is much smaller than that of non-cubic sites it is easy to understand why the spectra are known for so few cases.

A number of experimental techniques have been used to help separate out the cubic spectra. Probably the most powerful is the observation that at liquid nitrogen temperature (78 K) ionizing radiation such as X-rays tends to stimulate selectively the fluorescence of the ions at cubic sites (Vakhidov *et al.*, 1970; also see Makovsky *et al.*, 1962, and Merz and Pershan, 1967*a,b*). This occurs because the trivalent ions at cubic sites form very efficient electron traps (i.e. can readily be converted to the divalent state) and so electron-hole recombination preferentially takes place at these sites (see Chapter 7). Voronko *et al.* (1971) has also noted that the fraction of rare-earth ions at the cubic sites can even increase under continual X-ray illumination at 78 K. Another useful spectroscopic feature is that the cubic-site fluorescent lifetimes tend to be somewhat longer than the corresponding distorted sites, and this can also be used to help separate the different spectra. Al'tshuler *et al.* (1970) in sorting out the cubic spectra of Dy^{3+} used one set of crystals doped in the usual fashion, with an excess of F^- ions

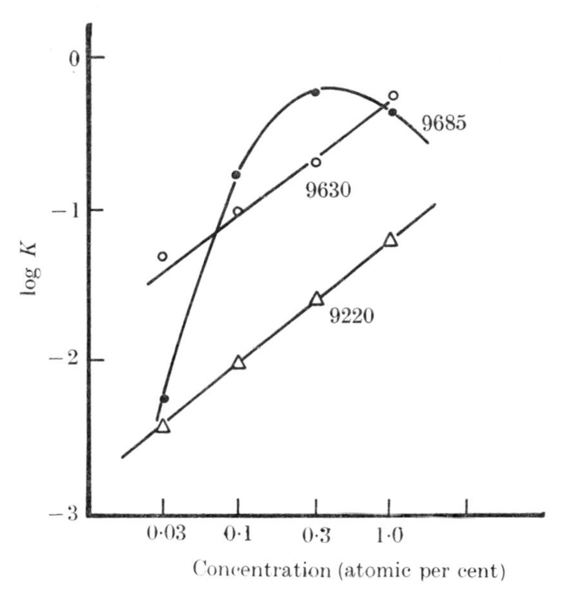

FIG. 5.6. Concentration dependence of the extinction coefficient for three optical transitions in $CaF_2:Yb^{3+}$ [Voronko *et al.*, 1969].

providing the charge compensation, and a second set where Na^+ provided the compensation (see § 6.4.2). The only spectrum common to both sets of crystals was the cubic one. Voronko *et al.* (1969) have correlated sets of optical lines in $CaF_2:Yb^{3+}$ as a function of Yb concentration (Fig. 5.6). In this case the fraction of ions at cubic sites is highest at the lowest doping level. The final proof of any site assignment rests on Zeeman or stress studies, which also enable level symmetries to be determined.

The history of the discovery of the cubic spectrum of $CaF_2:Yb^{3+}$ ($4f^{13}$ isoelectronic to Tm^{2+}) is illustrative of how difficult the problem can be even in this simple case. Kirton and White (1969) firmly established that the $^2F_{\frac{5}{2}}(\Gamma_7) \rightarrow {}^2F_{\frac{7}{2}}(\Gamma_7)$ transition occurs at 9630 Å using Zeeman spectroscopy. However detailed studies by this group (Kirton and McLaughlan, 1967 and Kirton and White, 1969) failed to identify any other optical transitions. The $^2F_{\frac{5}{2}}(\Gamma_7) \rightarrow {}^2F_{\frac{7}{2}}(\Gamma_8)$ transition was found in fluorescence to occur at 10 270 Å and the $^2F_{\frac{7}{2}}(\Gamma_7) \rightarrow {}^2F_{\frac{5}{2}}(\Gamma_8)$ transition was found in absorption at 9220 Å by two other groups (Voronko *et al.*, 1969 and Kiro and Low, 1970). Voronko *et al.* (1969) demonstrated that the 10 270 Å fluorescent life time was identical to that observed for the 9630 Å line at 77 K and 300 K in several crystals with a wide range of Yb doping levels, suggesting that both transitions originate on the same level. In addition the same group showed that the intensities of the 9220 Å and 9630 Å lines have the same dependence on Yb concentration (Fig. 5.6). It is interesting to see that both of these lines can be seen in the absorption spectrum published by Kiss (1962).

TABLE 5.6

Crystal-field parameters for $Dy^{3+}(4f^9)$

Host	$A_4\langle r^4\rangle$ cm^{-1}	$A_6\langle r^6\rangle$ cm^{-1}	Ref.
CaF_2	−274	48·5	Nara and Schlesinger (1971)
CaF_2	−257	41·9	Al'tshuler *et al.* (1970)
SrF_2	−232	43·8	Al'tshuler *et al.* (1970)
BaF_2	−207	46·4	Al'tshuler *et al.* (1970)
CdF_2	−275	41·0	Al'tshuler *et al.* (1970)

The calculation of crystal-field parameters from the data is identical to that for CaF_2:Tm^{2+}. Kiro and Low (1970) have correctly given these parameters, although they gave an incorrect position for the $^2F_{\frac{7}{2}}(\Gamma_6)$ level. It is somewhat puzzling that the magnetic resonance studies of Baker *et al.* (1968) gave a different position for the $^2F_{\frac{7}{2}}(\Gamma_8)$ level.

All the cubic crystal-field levels of $Dy^{3+}(4f^9)$ in the manifolds $^6H_{\frac{15}{2},\frac{13}{2},\frac{11}{2}}$ have been observed by Al'tshuler *et al.* (1970) using fluorescence originating from the $^4F_{\frac{9}{2}}$ manifold; these are known in CdF_2 as well as in the series CaF_2, SrF_2, and BaF_2. They fitted their data using a basis set expanded to include all levels of different J mixed by the crystal field and obtained the parameters given in Table 5.6. Nara and Schlesinger (1971) recalculated the parameters for CaF_2 including intermediate coupling and obtained somewhat different results also given in the table. Their numbers are probably somewhat better,

TABLE 5.7

Crystal-field parameters for $Gd^{3+}(4f^7)$

Host	$A_4\langle r^4\rangle$ cm^{-1}	$A_6\langle r^6\rangle$ cm^{-1}
CaF_2	−270	49·5
SrF_2	−240	41·4
BaF_2	−216	31·1

All measurements from O'Hare *et al.* (1970).

although the fit to the data did not improve as much as one would have expected.

The cubic spectrum of $Gd^{3+}(4f^7)$ first identified by Makovsky (1966, 1967) in CaF_2, SrF_2, and BaF_2 has been confirmed by detailed Zeeman studies by Detrio (1969) and Gilfanov *et al.* (1967). This is an interesting case since it is the only one where levels of two different terms, $^6P_{\frac{5}{2},\frac{7}{2}}$ and $^8S_{\frac{7}{2}}$, have been observed. Very good calculations for the crystal-field parameters have been carried out by O'Hare, Detrio and Donlan (1969) and O'Hare and Donlan (1969) who have clearly demonstrated an increase in the precision of the fit by expanding their basis set. They also included the spin-spin and spin-other-orbit interactions and used the "free-ion" parameters found by Judd and Crosswhite (1968) for Gd^{3+} in $LaCl_3$. The results of the calculation are given in Table 5.7. They also found that the electrostatic parameters, F_K, are reduced on going from $LaCl_3$ to the fluorides.

Using cathodoluminescence at 78 K, Aizenberg *et al.* (1971) have identified optical transitions which terminate on the crystal field components of the ground $^4I_{\frac{15}{2}}$ manifold of Er^{3+} at cubic sites in CaF_2, SrF_2, and BaF_2. They analysed their data using a basis set which included only the ground $^4I_{\frac{15}{2}}$ manifold and thus their crystal-field parameters given in Table 5.8 are not entirely suitable for making

TABLE 5.8

Crystal-field parameters† for $Er^{3+}(4f^{11})$

Host	$A_4\langle r^4\rangle$ cm^{-1}	$A_6\langle r^6\rangle$ cm^{-1}
CaF_2	−245	38·8
SrF_2	−223	34·4
BaF_2	−208	30·4

All measurements from Aizenberg *et al.* (1971).

† Only the $^4I_{\frac{15}{2}}$ term was used in the analysis of the data.

comparisons with the other measurements. For instance, their value of $A_4\langle r^4\rangle$ is smaller in magnitude than was found for the isoelectronic divalent ion Ho^{2+}, while the evidence in the other systems strongly favours the trivalent ions having larger crystal-field parameters. It should also be noted that Nara and Schlesinger (1971) found that their expanded basis set gave larger crystal-field parameters in the case of $CaF_2:Dy^{3+}$.

The information on the crystal-field parameters for the rest of the trivalent rare-earth ions at cubic sites is at the present time rather speculative. Values for various ions and their references are listed in Table 5.9. One approach used to obtain some of these values is to fit a

TABLE 5.9

Crystal-field parameters of rare-earth ions in CaF_2

System	$A_4\langle r^4\rangle$ cm^{-1}	$A_6\langle r^6\rangle$ cm^{-1}	Reference
$CaF_2:Tb^{3+}(4f^3)$	-420	38	Rabbiner (1967)
$CaF_2:Sm^{3+}(4f^5)$	-114	39	Schlesinger and Nirenberg (1969)
$CaF_2:Er^{3+}(4f^{11})$	-310	44	Weber and Bierig (1964)

set of lines using only the cubic Hamiltonian. This is not as unreasonable as it sounds since if ten or more lines are fitted with only three parameters it would be quite convincing. But it is necessary to prove that the optical transitions are purely electronic and belong to the same site symmetry. In addition, the calculation has to include a suitably large basis set so that the results are reliable. Rabbiner's (1967) results for $Tb^{3+}(4f^8)$ are suspect since he was limited to rather crude calculations. The more recent work of Nara and Schlesinger (1971b) using this approach for $CaF_2:Sm^{3+}$ is unreliable since their rms fit of 45 cm^{-1} is much too large; in this case the splittings of each J manifold are not very much greater than 45 cm^{-1}.

There also exist some bits of optical data which have not been developed to the point where crystal-field parameters can be determined. Merz and Pershan (1967a, b) using thermoluminescence and Vakhidov *et al.* (1970) using gamma-stimulated fluorescence have found interesting lines in $CaF_2:Ho^{3+}$ and $CaF_2:Er^{3+}$, which almost certainly come from ions at cubic sites. The transitions involved are ${}^5S_2 \rightarrow {}^5I_8$ for Ho^{3+} and ${}^4S_{\frac{3}{2}} \rightarrow {}^4I_{\frac{15}{2}}$ for Er^{3+}; both are spin allowed but highly forbidden magnetic-dipole transitions which require $\Delta L = 6$. Merz and

Pershan (1967*a*, *b*) have also reported a spectrum which they have tentatively identified as due to Sm^{3+} at cubic sites in CaF_2. Voron'ko *et al.* (1971*b*) have found a spectrum in $CaF_2{:}Nd^{3+}$ using X-ray-stimulated fluorescence ($^4F_{\frac{3}{2}} \rightarrow {}^4F_{\frac{9}{2}}$) which they feel arises from cubic sites. In all three cases there are more lines observed than the electronic level structure allows; thus there remains the problem of separating the vibronic and pure electronic transitions, assuming that the latter are even observed in these spectra. Finally, Zakharchenya and Rusanov

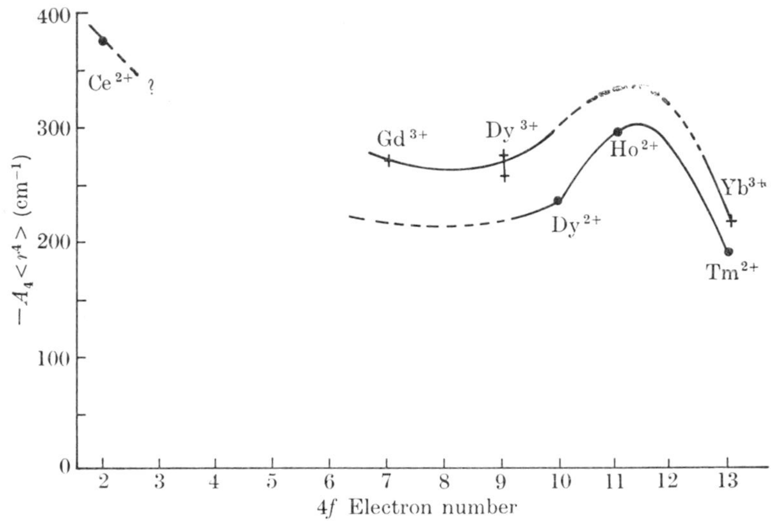

FIG. 5.7. Crystal-field parameter $A_4\langle r^4\rangle$ as a function of 4*f* electron number in CaF_2.

(1966) have identified, using Zeeman studies, the $^7F_0 \rightarrow {}^5D_1$ absorption line (5253·4 Å) and the $^5D_0 \rightarrow {}^7F_1$ emission line (5904·7 Å) Eu^{3+} at cubic sites in CaF_2.

The crystal-field parameters $A_4\langle r^4\rangle$ and $A_6\langle r^6\rangle$, which have been firmly established for the rare-earth ions located at cubic sites in CaF_2, are plotted against the 4*f* electron number in Figs. 5.7 and 5.8 respectively. One feature readily apparent from these plots is that the parameters for the trivalent ions are about 10–15% larger than those for the isoelectronic divalent ions. The value of $A_4\langle r^4\rangle$ for Ce^{2+} is probably fairly accurate, and thus there is a definite general trend for the fourth-order parameter to decrease as the 4*f* electron number increases.

The value of $A_4\langle r^4\rangle$ for Ho^{2+} stands out as an exception to this trend and it is difficult to decide whether this is a true effect or an artifact of

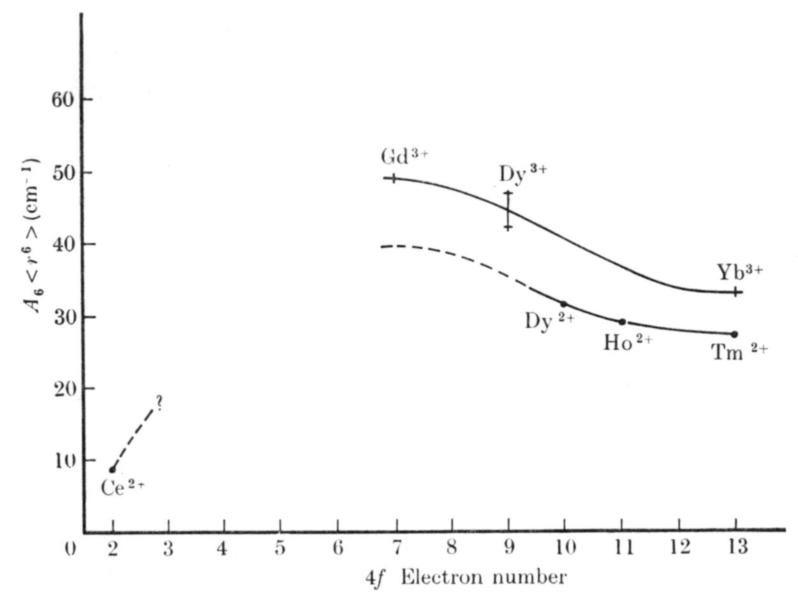

FIG. 5.8. Crystal-field parameter $A_6\langle r^6\rangle$ as a function of $4f$ electron number in CaF_2.

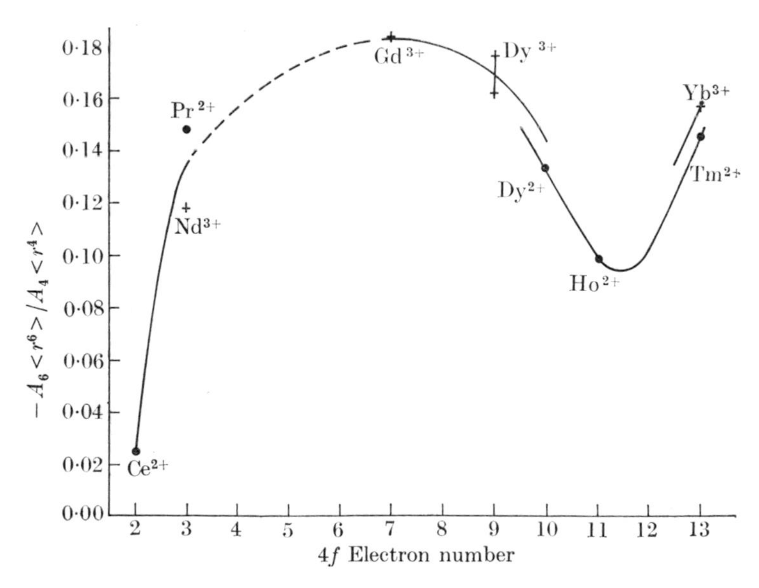

FIG. 5.9. The ratio of the crystal-field parameters (see Figs. 5.7 and 5.8) as a function of $4f$ electron number in CaF_2.

the more extensive calculation used by Weakliem and Kiss (1967). The very small value of $A_6\langle r^6\rangle$ for Ce^{2+} is most likely wrong since Alig *et al.* (1969) could not fit the data very satisfactorily. The remaining values of $A_6\langle r^6\rangle$ do show a very smooth decrease with increasing electron number. In Fig. 5.9 the ratio $A_6\langle r^6\rangle/A_4\langle r^4\rangle$ is plotted against the $4f$ electron number; here there is additional information from magnetic resonance experiments for Pr^{2+} and Nd^{3+} (see § 5.2.4). This plot shows that a small value of $A_6\langle r^6\rangle$ for Ce^{2+} is inconsistent with the other data.

These figures can be interpolated to provide reasonable values of the crystal-field parameters for those ions which have not been directly measured. By using the crystal-field splittings of each J manifold given by Lea, Leask, and Wolf (1962) it is possible to reconstruct the energy level schemes of these ions. The calculations do not include J mixing, which is relatively more important for those ions with low $4f$ electron number and hence small spin-orbit coupling, and so this procedure provides in these cases only rough first-order estimates. This approach is however useful for future studies of these energy levels and it very likely does give the proper ground states.

Figs. 5.10 and 5.11 give the $4f$ electronic energy level structure of the rare-earth ions located at cubic sites in CaF_2 for the first half ($n = 2$–7) and the second half ($n = 8$–13) of the series respectively. Each configuration is exhibited with either a divalent or trivalent ion using the best data available. To obtain the level structure of the isoelectronic ion, a good procedure is to scale the energy by 15% either up or down depending on whether one goes from the divalent to the trivalent ion or vice versa. Where the energy levels are closely spaced the order of the levels can change: for example, the ground state of Ho^{2+} has Γ_6 symmetry, while in Er^{3+} it is Γ_7. The energy levels for $Nd^{3+}(4f^3)$, $Nd^{2+}(4f^4)$, and $Tm^{3+}(4f^{12})$ were obtained using the interpolation procedure described above. The energy levels for $Sm^{2+}(4f^6)$ are those observed by optical studies in SrF_2 (Wood and Kaiser, 1962), since corresponding detailed data is lacking in CaF_2. The $Sm^{3+}(4f^5)$ levels are those given by Merz and Pershan (1967a, b). The rest of the data was obtained from references given in §§ 5.2.2 and 5.2.3.

5.2.4. *Magnetic resonance of $4f^n$ electrons*

The interaction of the $4f^n$ electrons with the fluoride hosts can be studied in several different ways using magnetic resonance. Measurements of g-values can be used directly to identify the site symmetry and

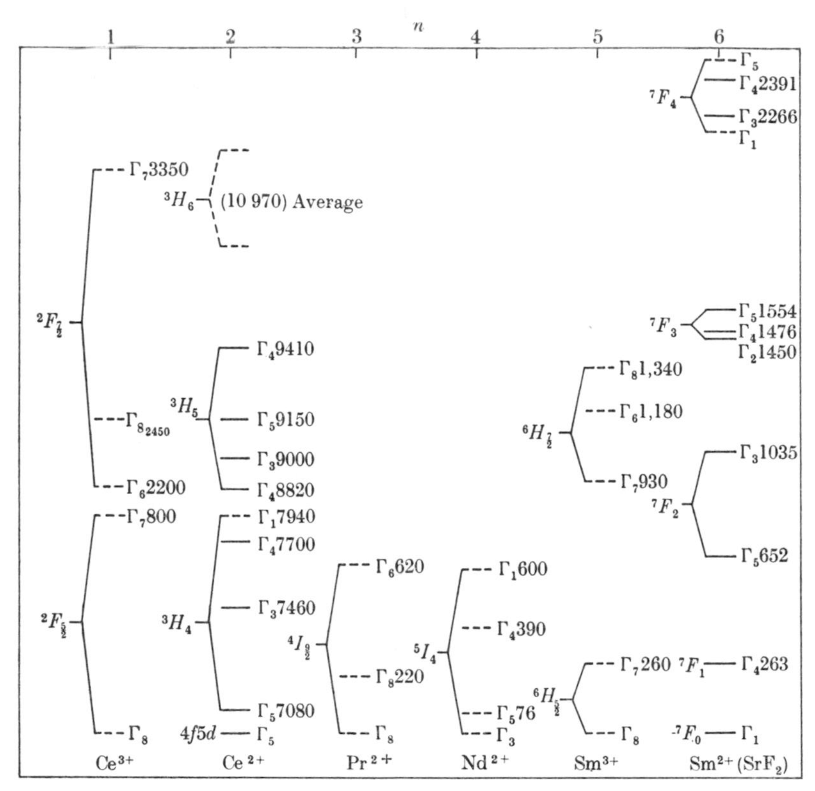

FIG. 5.10. $4f$ electronic level structures of rare-earth ions [$n = 1$ to 6] at cubic sites in CaF_2. Levels determined by experiment are indicated by solid lines, those by extrapolation and unconfirmed data are indicated by dashed lines. The spacing of the centres of the terms in Ce^{3+} and Ce^{2+} are at one half the scale of the crystal-field components. The energy of each level relative to the ground state is given in cm^{-1}.

the group representation of the ground level (Abragam and Bleaney, 1971). From this identification it is possible to obtain information about the crystal-field parameters. The early review papers on the rare-earth ions in the alkaline earth fluorides by Weber and Bierig (1964) and Ranon (1964) illustrate this very well. Endor measurements with the neighbouring fluorine ions give direct measurements of the admixture of the ligand orbitals with the metal-ion wavefunction. These measurements also provide the best information about the location of the neighbouring fluorine ions. Where it is possible to make detailed calculations of the $4f$ wavefunction (primarily the $4f^{13}$ configuration), deviations of the calculated g-values from the measured g-value can be used to make some inferences about the admixed orbitals. And finally, spin-lattice relaxation measurements, stress-induced shifts

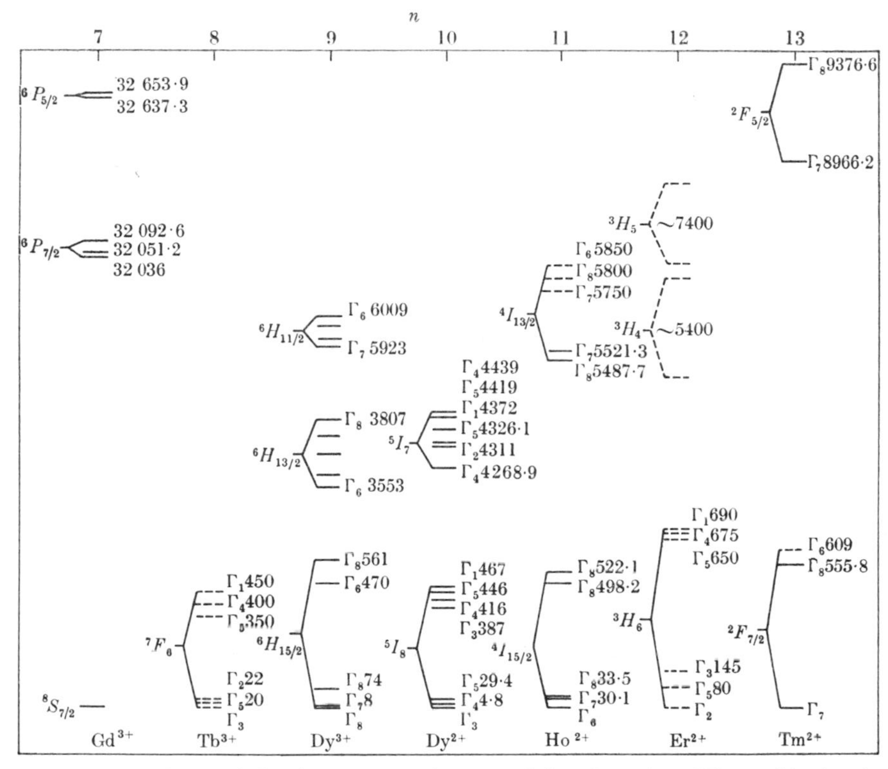

FIG. 5.11. 4f electronic level structures of rare-earth ions [n = 7 to 13] at cubic sites in CaF_2. Levels determined by experiment are indicated by solid lines, those by extrapolation and unconfirmed data are indicated by dashed lines. The spacing of the terms are at one tenth the scale of the crystal-field components, except for Gd^{3+} where the ^6P manifold is centered at $\sim\frac{1}{30}$ scale and all levels within this manifold drawn at full scale. The energy of each level relative to the ground state is given in cm^{-1}.

and temperature-induced shifts can provide information on the dynamic orbit-lattice interaction.

All the rare-earth ions with a $4f^n$ magnetic ground state have been observed by paramagnetic resonance in the cubic site in CaF_2 and in many of the other hosts; the g-values observed in CaF_2 are given in Table 5.10. (The measurements on Ce^{3+} (Dvir and Low, 1960) and on Nd^{3+} (Vincow and Low, 1960) are probably correct but have never been confirmed.) In addition, measurements have been made on a number of excited states which have been populated thermally, $Ho^{2+}(\Gamma_6, \Gamma_8)$ and $Dy^{3+}(\Gamma_7)$, or by optical excitation, $Tm^{2+}(^2F_{\frac{5}{2}}(\Gamma_7))$.

$Dy^{2+}(4f^{10})$ is a very unusual even-electron system since epr has been observed in four different crystal-field levels at the cubic site in CaF_2. The ground state is a non-magnetic Γ_3 doublet; 4·863 cm^{-1} above this

TABLE 5.10

Paramagnetic resonance g-values of rare-earth ions in cubic sites in CaF_2

Ion		Level [energy]†	g (observed)	g (calculated)	Reference
Ce^{3+}	$4f^1$	${}^2F_{\frac{5}{2}}(\Gamma_8)$	2·00; 3·1 ± 0·1	0·858; 2·000; 3·143	Dvir and Low (1960)
Pr^{2+}	$4f^3$	${}^4I_{\frac{9}{2}}(\Gamma_8)$	3·20; 0·53	—	Merritt *et al.* (1966)
Nd^{3+}	$4f^3$	${}^4I_{\frac{9}{2}}(\Gamma_8)$	1·10 ± 0·05; 2·26 ± 0·02	—	Vincow and Low (1960)
Sm^{3+}	$4f^5$	${}^6H_{\frac{5}{2}}(\Gamma_8)$	—	0·286; 0·667; 1·047	
Dy^{3+}	$4f^9$	${}^6H_{\frac{15}{2}}(\Gamma_8)$	2·63 ± 0·05; 5·48 ± 0·15; 13·7 ± 0·3	—	Bierig and Weber (1963)
					Low (1964)
		${}^6H_{\frac{15}{2}}(\Gamma_7)$[8 cm⁻¹]	7·468 ± 0·002	7·555	Bierig and Weber (1963)
Dy^{2+}	$4f^{10}$	${}^4I_8(\Gamma_3)$	see refs.	0	{Sabisky (1964)
		${}^4I_8(\Gamma_4)$[4·867 cm⁻¹]	4·82 ± 0·02	−4·89	{Mergerian *et al.* (1967)
					{Vinogradov *et al.* (1969)
		${}^4I_8(\Gamma_5)$[29·4 cm⁻¹]	4·92 ± 0·03	−5·00	Sabisky and Anderson (1964)
		${}^4I_7(\Gamma_4)$[4268·9 cm⁻¹]	4·69 ± 0·03	4·66	Sabisky and Anderson (1964)
Ho^{2+}	$4f^{11}$	${}^4I_{\frac{15}{2}}(\Gamma_6)$	5·912 ± 0·003	−5·974	Sabisky (1966)
					Lewis and Sabisky (1963)
					Hayes *et al.* (1963)
		${}^4I_{\frac{15}{2}}(\Gamma_7)$[30·1 cm⁻¹]	6·76 ± 0·03	6·80	Sabisky (1966)
Er^{3+}	$4f^{11}$	${}^4I_{\frac{15}{2}}(\Gamma_7)$	6·785	6·80	Baker *et al.* (1959)
					Dvir and Low (1960)
					Ranon and Low (1963)
Tm^{2+}	$4f^{13}$	${}^2F_{\frac{7}{2}}(\Gamma_7)$	3·451 ± 0·001	3·477	Hayes and Twidell (1961)
					Bessent and Hayes (1965)
					Anderson and Sabisky (1971)
		${}^2F_{\frac{5}{2}}(\Gamma_7)$[8966·2 cm⁻¹]	1·453 ± 0·002	−1·478	Sabisky and Anderson (1966)
Yb^{3+}	$4f^{13}$	${}^2F_{\frac{7}{2}}(\Gamma_7)$	3·438 ± 0·002		Hayes and Twidell (1961)
					Baker *et al.* (1969)

† [Energy above the ground state].

is a Γ_4 triplet (Mergerian *et al.*, 1965 and Vinogradov *et al.*, 1969); and at 28·6 cm^{-1} there is a second triplet, Γ_5 (Kiss *et al.*, 1965). These levels are strongly mixed and split by a magnetic field. Resonance absorption has been observed between the split components of the ground state and first excited state by Sabisky (1964) and the second excited state by Sabisky and Anderson (1964). In addition there is a metastable level, $^5I_7(\Gamma_5)$, 4268 cm^{-1} above the ground state, which is readily populated by optical pumping into the $4f$–$5d$ bands and directly observed by magnetic resonance absorption techniques (Sabisky and Anderson, 1964). This fascinating system has not been exploited for the detailed information it could yield on the dynamic coupling of the $4f$ electrons with the CaF_2 lattice; it is an even-electron system which responds to strains in first order and there are many levels available for measurement.

The $4f^7$ configuration (Eu^{2+} and Gd^{3+}) with its $^8S_{\frac{7}{2}}$ ground state presents a special case since the crystal-field splittings of the S term are very small ($\sim$1 cm^{-1}) and their origin has not been clearly elucidated. The crystal-field parameters for this state are defined through the effective spin-Hamiltonian

$$H_c = \frac{b_4}{60}[O_4^0+5O_4^4]+\frac{b_6}{1260}[O_6^0-21O_6^4], \tag{5.9}$$

where the O_n^m are spin operators of order 4 and 6 which exhibit Γ_1 symmetry; these are tabulated by Abragam and Bleaney (1970). The experimental values are given in Tables 5.11 and 5.12. Detailed crystal-field calculations by O'Hare *et al.* (1969) for $CaF_2:Gd^{3+}$ indicate that as much as one half of the value of the parameters can be accounted for through admixture of excited $4f$ levels into the ground manifold, but the rest must be due to covalency admixtures with the ligand orbitals (see also Wybourne, 1966).

The importance of electron paramagnetic resonance for determining crystal-field parameters in these systems has largely been replaced by the more direct and precise optical measurements. The $4f^3$ configuration (Pr^{2+} and Nd^{3+}) is the only important case where no useful optical data exist; here epr measurements of the ground Γ_8 state provide the ratio of the crystal-field parameters. This is possible because there are two Γ_8 levels within the ground $^4I_{\frac{9}{2}}$ manifold and so the g-values are sensitive to this ratio (see Lea, Leask, and Wolf, 1962). The measurements by Merritt *et al.* (1966) on Pr^{2+} give 0·15 for $A_6\langle r^6\rangle/A_4\langle r^4\rangle$. Ranon (1964) has reinterpreted the measurements of Vincow and Low

TABLE 5.11

Spin-Hamiltonian parameters for the $^8S_{\frac{7}{2}}$ level in Eu^{2+} *and* $Gd^{3+}(4f^7)$

System	$b_4(10^{-4}\ cm^{-1})$	$b_6(10^{-4}\ cm^{-1})$	g	References
$CaF_2:Eu^{2+}$	−55·50 (5)	+0·235 (1)	1·9910 (5)	Baker and Williams (1962) Hurren *et al.* (1969)
$CaF_2:Ge^{3+}$	−46·6 (3)	0·07 (30)	1·9918 (10)	Low (1958) Ryter (1957)
$SrF_2:Eu^{2+}$	−44·67 (5)	0·178 (1)	1·9916 (5)	Hurren *et al.* (1969)
$SrF_2:Gd^{3+}$	−41·0 (5)	0·6 (4)	1·9923 (10)	Sierro (1963)
$BaF_2:Eu^{2+}$	−35·22 (1)	0·12 (2)	1·9918 (3)	Hurren *et al.* (1969)
$BaF_2:Gd^{3+}$	−37·8 (1)	0·14 (3)	1·9916 (5)	Boatner and Reynolds (1970) Sierro (1963)
$SrCl_2:Eu^{2+}$	−16·10 (3)	0·02 (3)	1·9928 (5)	Low and Rosenburger (1959) Reynolds and Boatner (1970)
$SrCl_2:Gd^{3+}$	−10·0 (5)	0·07 (10)	1·9906 (10)	Low and Rosenburger (1959)
$CdF_2:Gd^{3+}$	−47·4 (3)	0(0·3)	1·992 (2)	Baker and Williams (1962)

(1961) on Nd^{3+} to obtain a ratio of 0·12. Neither of these results includes J mixing, which could be significant in this case. It is worth noting that the epr measurements on Dy^{3+} by Low (1964) gave $A_6\langle r^6\rangle/A_4\langle r^4\rangle = 0{\cdot}17$, which agrees quite well with the values obtained from the optical measurements of 0·16 by Nara and Schlesinger and 0·18 by Al'tschuler *et al.* (1970).

As can be seen in Table 5.10 most of the remaining measured g-values agree very well with the calculated values. In most cases the small discrepancies are due to inaccuracies in the $4f$ wavefunctions used in the

TABLE 5.12

Central ion (A) *and first-shell fluorine* (A_s, A_p) *hyperfine constants for the $^8S_{\frac{7}{2}}$ level in* $^{151}Eu^{2+}$ *and* Gd^{3+}

System	$A(^{151}Eu)$ (MHz)	A_s (MHz)	A_p (MHz)	References
$CdF_2:Eu^{2+}$	—	−1·94	+4·10	Valentin (1969)
$CaF_2:Eu^{2+}$	−102·907 (1)[a] −98·3[b]	−2·23	4·01	Baker and Hurrell (1963)
$SrF_2:Eu^{2+}$	−97·0[b]	−1·93	3·80	Valentin (1969)
$BaF_2:Eu^{2+}$	−95·8[b]	−1·65	3·52	Baberschke (1970)
$CaF_2:Gd^{3+}$	—	−1·84 (2)	5·04 (2)	Bill (1969)

[a] Baker and Williams (1962); [b] Hurren *et al.* (1969).

calculations. The wavefunctions for Ho^{2+} are quite good and so in this case one could argue that there is evidence for the presence of an orbital reduction factor. However, the Tm^{2+} data give a much more direct measurement of this factor.

The g-values of the ground ${}^2F_{\frac{7}{2}}(\Gamma_7)$ level and excited ${}^2F_{\frac{5}{2}}(\Gamma_7)$ level of Tm^{2+} can be calculated with precision because of the simplicity of the $4f^{13}$ configuration. The discrepancy between the calculated and observed g-values therefore can be directly interpreted in terms of a reduction in the matrix elements of the orbital angular momenta, either due to covalency effects or dynamic coupling to the lattice (Inoue, 1963). Hayes and Smith (1971) have shown that the ground and excited state measurements give the same orbital reduction factor as defined by Axe and Burns (1966) in both CaF_2 and SrF_2 (Table 5.13). Baker (1968) has shown that in CaF_2 eighty per cent of this reduction factor can be attributed to covalency. Anderson, Call, Stott, and Hayes (1974) argue that the covalency contribution should decrease on going from CaF_2 to BaF_2 as do all the other covalency-dependent factors (see § 5.2.5), but the measured values of the orbital reduction factor increases slightly instead. This would imply that the contribution due to the Inoue effect is increasing, which is consistent with the measured increase of the dynamic orbit-lattice coupling on going from CaF_2 to BaF_2 (Sabisky and Anderson, 1970).

The hyperfine interactions with the rare-earth nuclei have been measured for most of the ions, either directly from the epr spectra or in detail using endor. The endor measurements by Baker *et al.* (1969) on ${}^{171}Yb^{3+}$ and ${}^{173}Yb^{3+}$ provided a precise measurement of the pseudo-nuclear moment, which is due to admixture of excited states into the ground state by the magnetic field. From this they could infer the position of the nearby ${}^2F_{\frac{7}{2}}(\Gamma_8)$ level which, curiously, disagrees with the position of this level obtained from the optical measurements. These measurements on Yb, however, were unable to provide a measure of the hyperfine anomaly as Baker and Williams (1962) also found for ${}^{151}Eu$ and ${}^{153}Eu$. The hyperfine anomaly allows the contribution of the core polarization to be determined (see Abragam and Bleaney, 1970); this can also be obtained if the hyperfine interaction constants are known in levels in two different J manifolds. Sabisky and Anderson (1966) measured the hyperfine splitting in the excited state, ${}^2F_{\frac{5}{2}}(\Gamma_7)$, of $CaF_2:Tm^{2+}$ and found a core polarization which was three times larger than Bleaney (1964) obtained by interpolating from the Eu measurements. Hayes and Smith (1971) made the same measurement in

TABLE 5.13

Spin-Hamiltonian parameters for Tm^{2+} *and* Yb^{3+} *in cubic sites in the alkaline earth fluorides*

System		A(MHz)	g	$1-k$[a]	References
$CaF_2:Tm^{2+}$	$^2F_{\frac{7}{2}}(\Gamma_7)$	(−)1101·374 (4)	(+)3·4510 (10)[b]	0·0115 (3)	Bessent and Hayes (1965)
	$^2F_{\frac{5}{2}}(\Gamma_7)$	(+)1160 (6)	(−)1·449 (1)	0·0110 (10)	Sabisky and Anderson (1966)
$SrF_2:Tm^{2+}$	$^2F_{\frac{7}{2}}(\Gamma_7)$	(−)1103·5 (10)	(+)3·4448 (10)[b]	0·0114 (3)	Hayes and Smith (1971)
	$^2F_{\frac{5}{2}}(\Gamma_7)$	(+)1163 (6)	(−)1·449 (1)	0·0101 (10)	Hayes and Smith (1971)
$BaF_2:Tm^{2+}$	$^2F_{\frac{7}{2}}(\Gamma_7)$	(−)1104·2 (4)	(+)3·4360 (10)[b]	0·0124 (3)	Anderson *et al.* (1973)
$CaF_2:Yb^{3+}$	$^2F_{\frac{7}{2}}(\Gamma_7)$	(^{171}Yb)(+)2638·70 (5)	(+)3.438 (2)	0·018 (2)	Baker, Blake, and Copland (1969)
		(^{173}Yb)(−) 727·094 (60)			
$SrF_2:Yb^{3+}$	$^2F_{\frac{7}{2}}(\Gamma_7)$	—	3·438 (5)	—	Ranon and Yariv (1964)
$BaF_2:Yb^{3+}$	$^2F_{\frac{7}{2}}(\Gamma_7)$	—	3·430 (5)	—	Antipin *et al.* (1967)

[a] See Hayes and Smith (1971) and Baker (1968).
[b] Measured by E. S. Sabisky and reported in Hayes and Smith (1971).

$SrF_2:Tm^{2+}$, applied a slightly different analysis of the data for both cases, and arrived at essentially the same conclusion.

The hyperfine interactions with the neighbouring fluorine ions is more directly related to the interaction of the 4*f* electrons with the host lattice. Detailed endor measurements have been made on CaF_2 doped with Eu^{2+} (Baker and Hurrell, 1963), Yb^{3+} (Baker *et al.*, 1968, Kiro and Low, 1968) and Gd^{3+} (Bill, 1970). Tm^{2+} has been investigated in CaF_2 (Bessent and Hayes, 1966), SrF_2 (Hayes and Smith, 1971) and BaF_2 (Anderson *et al.*, 1974). The Tm^{2+} and Yb^{3+} results are given in Table 5.14. In all the cases studied the symmetry of the fluorine ions in each shell was found to be cubic. Unusual structure was observed in the first-shell endor resonances in $SrF_2:Tm^{2+}$ which was suggestive of local dynamic distortions, but the later work in $BaF_2:Tm^{2+}$ demonstrated that this structure is an instrumental effect associated with frequency mixing by the resonant-ion system. The tensor hyperfine coupling with the second, third and fourth shells of fluorine ions is in all cases, except possibly Yb^{3+}, identical to that calculated using a pure magnetic-dipolar interaction with the fluorine ions located at the normal host-lattice positions. For Yb^{3+} there is a slight indication of a collapse of the second shell of fluorine ions. The evidence for a non-zero scalar hyperfine interaction with the second shell of fluorine ions is of the order of the experimental uncertainty in all cases. The data for the first shell of fluorine ions contains appreciable covalent contributions; in the series of measurements with Tm^{2+} and Eu^{2+} it is possible to extract plausible radial positions for these fluorines.

The radial position of the first-shell fluorine ions is a very difficult and yet important question to answer. There simply is no direct method available to measure the radial location of neighbouring ions in this or any other impurity-ion problem, except in those rare cases (as above) where there is only a pure magnetic-dipolar superhyperfine interaction. Baberschke (1971) proposed that a reasonable method for obtaining the covalent contribution to the tensor hyperfine constant for the nearest neighbours was to assume that it scaled linearly with the scalar hyperfine component. He applied this concept to Eu^{2+} in CdF_2, CaF_2, SrF_2, and BaF_2. Anderson *et al.* (1974) have carried out the same analysis for Tm^{2+} in CaF_2, SrF_2, and BaF_2. In Fig. 5.12 the difference between the measured tensor superhyperfine constant, A_p, and that calculated from the magnetic dipolar interaction, A_d, using two model locations of the F^- ions, is plotted as a function of the measured component, A_s, for all three hosts. The radial positions used for curve (a) are those of

TABLE 5.14

Fluorine endor parameters (MHz) *for the ground state of* Tm^{2+} *and* Yb^{3+}

System	A_s (1st shell)	A_p (1st shell)	A_s (2nd shell)	A_p (2nd shell)	References
$CaF_2:Tm^{2+}$	(+)2·584 (10)	12·283 (10)	0·010 (10)	1·386 (5)	Bessent and Hayes (1965)
$SrF_2:Tm^{2+}$	2·023 (20)	11·112 (20)	0·000 (10)	1·165 (10)	Hayes and Smith (1970)
$BaF_2:Tm^{2+}$	1·560 (20)	10·020 (20)	0·000 (15)	0·958 (15)	Anderson *et al.* (1974)
$CaF_2:Yb^{3+}$	1·67 (1)	17·57 (10)	0·015 (15)	1·391 (10)	Baker *et al.* (1968) Kiro and Low (1968).

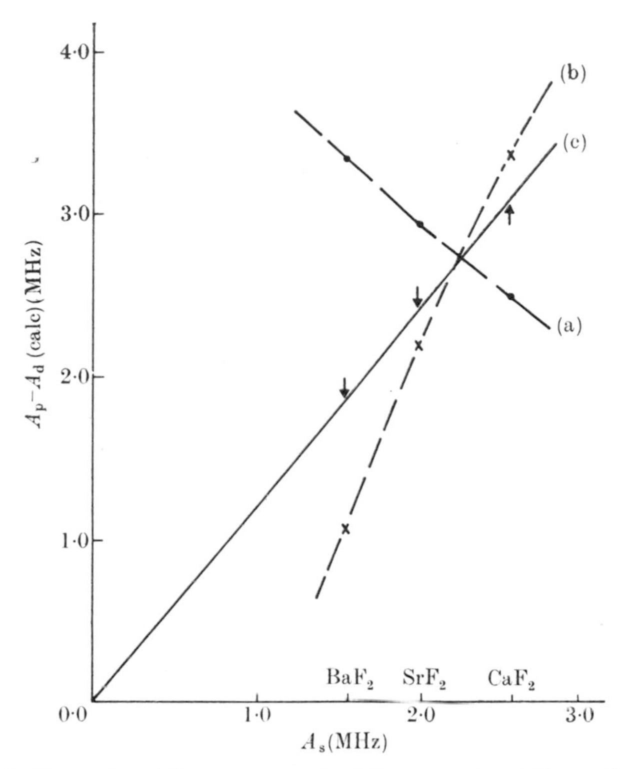

FIG. 5.12. Tensor hyperfine component of the nearest-neighbour fluoride ions A_p minus a dipolar component A_d calculated for three models discussed in the text, plotted against the scalar components A_s as measured for Tm^{2+} in CaF_2, SrF_2, and BaF_2 [Anderson *et al.*, 1974].

the normal host lattice. In curve (b) a fixed radial distance equal to the sum of the estimated Tm^{2+} ionic radius of 1·08 Å (Axe and Burns, 1966) and the F^- ionic radius of 1·36 Å was used. The true position should lie between these two extremes. The second constraint is that the covalent contribution to A_s and A_p should scale linearly with each other (see § 5.2.5). Thus the points in Fig. 5.12 for the ions at their true location should lie on a line drawn through the intersections of curves (a) and (b) and passing through the origin. Using these points it is possible to calculate the position of the first shell (Table 5.15). The results show a small expansion from the normal host position in CaF_2 of 0·06 Å, a slight contraction of −0·04 Å in SrF_2 and a much larger contraction of −0·17 Å in BaF_2. In general the fluorine ions tend to remain closer to the Tm^{2+}–F^- spacing of 2·44 Å than the normal host-lattice positions. Baberschke's results for Eu^{2+}, also in the table, are in good agreement with those of Tm^{2+}.

TABLE 5.15
First shell cation-anion spacings (Å)

System	Host lattice[a]	Distorted position	Difference	Reference
$CaF_2:Tm^{2+}$	2·36	2·42	+0·06	Anderson *et al.* (1974)
$SrF_2:Tm^{2+}$	2·50	2·46	−0·04	
$BaF_2:Tm^{2+}$	2·68	2·51	−0·17	
$CdF_2:Eu^{2+}$	2·333	2·453	+0·12	Baberschke (1970) and
$CaF_2:Eu^{2+}$	2·366	2·450	+0·08	Hurren *et al.* (1969)
$SrF_2:Eu^{2+}$	2·511	2·511	—	
$BaF_2:Eu^{2+}$	2·685	2·583	−0·10	
EuF_2	2·51			
TmF_2	2·44 as estimated by (Axe and Burns (1966))			

[a] The lattice spacings given for Tm^{2+} are corrected for thermal contraction to ~20 K; the Eu^{2+} spacings are room temperature values.

Hurren *et al.* (1969) used hydrostatic pressure-induced changes of the crystal-field parameters of the $^8S_{\frac{7}{2}}$ ground state of Eu^{2+} in CaF_2, SrF_2, and BaF_2 to obtain the location of the first-shell fluorine ions. Their data for $\log_{10}(b_4)$ are plotted against $\log_{10}\delta$ in Fig. 5.13, where δ is the cation-anion separation in the host crystal. Assuming the local compressibility is the same as the bulk crystal and noting that the Sr^{2+} ionic radius (1·20 Å) and the Eu^{2+} ionic radius (1·18 Å) are essentially

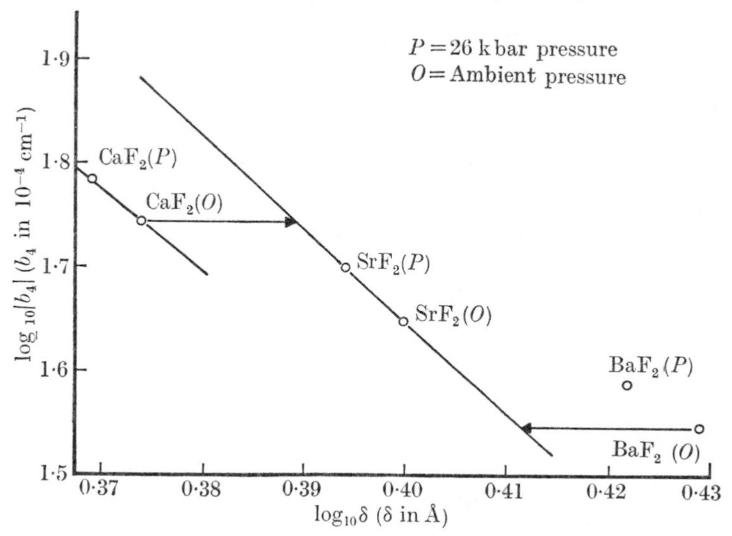

FIG. 5.13. $\log_{10}|b_4|$ as determined for Eu^{2+} in CaF_2, SrF_2, and BaF_2 at ambient and 26 kbar hydrostatic pressure plotted as a function of the logarithm of the normal host metal-ion to nearest-neighbour F^- ion spacing, δ. Translation of the ambient pressure data points for CaF_2 and BaF_2, as indicated by arrows, provides a measure of the Eu^{2+}–F^- spacing in these hosts [Hurren *et al.*, 1969].

the same, they shifted horizontally the CaF_2 and BaF_2 data as shown in the second diagram in the figure to obtain the relaxed local-ion positions. These positions are identical to those obtained from the endor measurements. The authors also found that the data for b_6 in such a plot lay on a straight line using these ion positions, lending support to their procedure. In fact, it should be noted that there is a degree of ambiguity in this procedure since the local strain does not have to be the same as that of the bulk. The agreement with the endor results suggests that the local hydrostatic compressibility is in fact much the same as in the bulk.

The dynamic coupling of the $4f$ electrons with the host lattice has been investigated using spin-lattice relaxation and stress-induced shifts of g-values. A general review of this subject has recently been published by Orbach and Stapleton (1972), which contains most of the data available on the rare-earth ions in the alkaline earth fluorides. There is therefore little reason to duplicate all these date here. An additional reason for not covering this field in detail is that the theory of the dynamic interaction is in a rather primitive stage. For example, it is claimed that the stress-induced g-shifts in $CaF_2{:}Ho^{2+}$ and $ThO_2{:}Yb^{3+}$ can be adequately explained by the electrostatic point-charge model (Sroubek *et al.*, 1968) and indeed the same model gave quite accurate direct one-phonon spin-lattice relaxation rates for $CaF_2{:}Tm^{2+}$ (Sabisky and Anderson, 1970). But the dynamic interaction constants were found to increase upon going from $CaF_2{:}Tm^{2+}$ to $BaF_2{:}Tm^{2+}$, while the electrostatic model predicts just the opposite. Since the static crystal-field parameters are dominantly covalent in nature, a theory was recently developed to take this into account with a suitably distributed electronic charge which moves rigidly with the host lattice (Ivanenko and Malkin, 1970, 1972; Malkin *et al.*, 1971). These authors have claimed that their theory explains the Tm^{2+} results. This must be fortuitous, since Abragam *et al.* (1972) have recently shown that the dynamic spin-lattice coupling due to local rotations of the host lattice by the phonons, accounts almost completely for the coupling previously attributed to local strains of Γ_5 symmetry in $CaF_2{:}Tm^{2+}$. Baker and van Ormondt (1973) have directly confirmed that the Γ_5 coupling is very small through stress induced g-shifts in $CaF_2{:}Tm^{2+}$.

5.2.5. *Microscopic origin of 4f crystal-field parameters*

The crystal-field parameters of rare-earth ions have generally been interpreted in terms of the electrostatic point-charge model, but recent work has shown that overlap and covalency effects have a more dominant role. Divalent thulium in the alkaline earth fluorides has played an important part in this discussion, because it is the simplest rare-earth system known and there exists a large quantity of excellent data for it. The crystal-field parameters, first-shell fluorine super-hyperfine constants and orbital reduction factors are known for Tm^{2+} in CaF_2, SrF_2, and BaF_2, and in addition for Yb^{3+} in CaF_2 (see § 5.2.2–4).

The electrostatic point-ion contributions to the crystal-field parameters in CaF_2:Tm^{2+} have been calculated out to the fourth shell of neighbouring ions by Bleaney (1964). If one modifies the calculation to

TABLE 5.16

Experimental and calculated point-charge crystal-field parameters (cm^{-1})

Host	b_4 (exp)	b_4 (calc)	b_6 (exp)	b_6 (calc)
CaF_2	45·8	16·6 14·4	5·16	1·07[a] 0·83[b]
SrF_2	40·0	12·3 10·7	4·23	0·70[a] 0·55[b]
BaF_2	33·6	8·8 7·7	3·50	0·44[a] 0·34[b]

[a] Values of $\langle r^4 \rangle$ and $\langle r^6 \rangle$ Given by Watson and Freeman (1962).
[b] Values of $\langle r^4 \rangle$ and $\langle r^6 \rangle$ Given by Wakim *et al.* (1972).

[*Note.* $b_4 = \frac{8}{33}A_4\langle r^4 \rangle = 0{\cdot}2424A_4\langle r^4 \rangle$

$$b_6 = \frac{16}{2{\cdot}6 \times 33}A_6\langle r^6 \rangle = 0{\cdot}1865A_6\langle r^6 \rangle].$$

take the relaxation of the first shell of fluorines into account (§ 5.2.4) the crystal-field parameters given in Table 5.16 are obtained. It is seen that the calculation gives values which are approximately one quarter of the observed ones. This discrepancy has remained in spite of the various attempts to modify the theory. The fourth and sixth moments of the $4f$ electron radius in Tm^{2+} as calculated by Watson and Freeman (1962) are essentially the same as those calculated recently by Wakim *et al.* (1972). It is very difficult to invoke strong polarization effects of the ligand ions since Tm^{2+} has the same charge and approximately the same size as the cations it replaces. Anti-shielding effects due to the $5p$ shell simply do not occur in the $4f^{13}$ case (Watson and Freeman, 1967).

Jorgenson *et al.* (1963) made a very early attempt to fit the $CaF_2:Tm^{2+}$ data using a one-parameter weak anti-bonding model with rather inconclusive results in this particular case, but good qualitative results for a variety of other systems. Axe and Burns (1966) carried out a more detailed calculation and found that rough estimates of the covalency parameters give much better agreement with the observed splittings than the best point-ion electrostatic calculations. Watson and Freeman (1967) in examining this result suggested that it was the admixture arising from the non-orthogonality of the $4f$ electron wavefunction to the F^- ion wavefunctions which makes the major contribution. Detailed calculations by Anisomov, Rakauskas, and Dagys (1969), Anisomov and Dagys (1971), and Anisomov, Vala, and Dagys (1971) confirm that overlap effects dominate pure covalency and other indirect interactions. Stedman and Newman (1971) have analysed the experimental data on the rare-earth ions in the alkaline earth fluoride hosts in terms of the covalency picture developed by Newman (1971). They make the observation that configuration admixture effects must be small, since going from the divalent to the isoelectronic trivalent ion makes a minor change in the crystal-field parameters (see Figs. 5.7 and 5.8). This however is a difficult comparison to make without more detailed analysis since the $4f$ electron wavefunction is more compact in the trivalent ion and the neighbouring fluorine ions are pulled in by the extra positive charge. Baker (1968) has handled the Axe and Burns model in a phenomenological fashion to fit the superhyperfine interaction constants, the crystal field parameters and the orbital reduction factors in $CaF_2:Tm^{2+}$ and $CaF_2:Yb^{3+}$.

In the Axe and Burns (1966) covalency model the natural energy parameters are the relative splittings of the $4f$ orbital in the absence of spin-orbit coupling. In a cubic field the $4f$ orbitals split into three levels of symmetry a_{2u}, t_{1u}, and t_{2g}, leading to two energy parameters which have been defined as

$$\begin{aligned} \theta &= E(t_{1u})-E(t_{2g}), \\ \Delta &= E(t_{2g})-E(a_{2u}). \end{aligned} \tag{5.10}$$

The parameters θ and Δ are analogous to $10Dq$ in the more familiar d orbital case. These parameters can also be directly related to the crystal-field parameters, as discussed further on.

The metal-ion orbitals only interact with those sets of ligand orbitals

TABLE 5.17

Linear combinations of ligand orbitals in cubic MX_8 complexes which can bond with f orbitals on the central ion. [Taken from Axe and Burns (1966)]

Representation (bond type)	Designation	Central ion orbital	Ligand orbitals	Group overlap ($S_{M\nu}$)
$a_{2u}(s \text{ and } p\sigma)$	$\vert\beta\rangle$	$(105)^{\frac{1}{2}}xyz$	$(8)^{-\frac{1}{2}}[\sigma_1+\sigma_2+\sigma_3+\sigma_4-\sigma_5-\sigma_6-\sigma_7-\sigma_8]$	$(40/9)^{\frac{1}{2}}\langle p\sigma \vert f\rangle$
$t_{2u}(\pi)$	$\vert\epsilon_1\rangle$	$\frac{1}{2}(105)^{\frac{1}{2}}z(x^2-y^2)$	$(32)^{-\frac{1}{2}}[(\eta_2+\eta_3+\eta_5+\eta_8-\eta_1-\eta_4-\eta_6-\eta_7)$ $+\sqrt{3}(\xi_1+\xi_3+\xi_6+\xi_8-\xi_2-\xi_4-\xi_5-\xi_7]$	
	$\vert\epsilon_2\rangle$	$\frac{1}{2}(105)^{\frac{1}{2}}x(y^2-z^2)$	$(32)^{-\frac{1}{2}}[(\eta_2+\eta_4+\eta_5+\eta_7-\eta_1-\eta_3-\eta_6-\eta_8)$ $+\sqrt{3}(\xi_1+\xi_3+\xi_6+\xi_8-\xi_2-\xi_4-\xi_5-\xi_7)]$	$(40/9)^{\frac{1}{2}}\langle p\pi \vert f\rangle$
	$\vert\epsilon_3\rangle$	$\frac{1}{2}(105)^{\frac{1}{2}}y(z^2-x^2)$	$(8)^{-\frac{1}{2}}(\eta_1+\eta_2+\eta_7+\eta_8-\eta_3-\eta_4-\eta_5-\eta_6)$	
$t_{1u}(s \text{ and } p\sigma)$	$\vert\delta_1\rangle$	$\frac{1}{2}(7)^{\frac{1}{2}}z(5z^2-3r^2)$	$(8)^{-\frac{1}{2}}(\sigma_1+\sigma_4+\sigma_6+\sigma_7-\sigma_2-\sigma_3-\sigma_5-\sigma_8)$	
	$\vert\delta_2\rangle$	$\frac{1}{2}(7)^{\frac{1}{2}}x(5x^2-3r^2)$	$(8)^{-\frac{1}{2}}(\sigma_1+\sigma_3+\sigma_6+\sigma_8-\sigma_5-\sigma_7-\sigma_2-\sigma_4)$	$-(32/27)^{\frac{1}{2}}\langle p\sigma \vert f\rangle$
	$\vert\delta_3\rangle$	$\frac{1}{2}(7)^{\frac{1}{2}}y(5y^2-3r^2)$	$(8)^{-\frac{1}{2}}(\sigma_1+\sigma_2+\sigma_7+\sigma_8-\sigma_3-\sigma_4-\sigma_5-\sigma_6)$	
$t_{1u}(\pi)$	$\vert\delta_1\rangle$	$\frac{1}{2}(7)^{\frac{1}{2}}z(5z^2-3r^2)$	$(32)^{-\frac{1}{2}}[(\xi_2+\xi_3+\xi_5+\xi_8-\xi_1-\xi_4-\xi_6-\xi_7)$ $+\sqrt{3}(\eta_1+\eta_4+\eta_6+\eta_7-\eta_2-\eta_3-\eta_5-\eta_8)]$	
	$\vert\delta_2\rangle$	$\frac{1}{2}(7)^{\frac{1}{2}}x(5x^2-3r^2)$	$(32)^{-\frac{1}{2}}[(\xi_2+\xi_4+\xi_5+\xi_7-\xi_1-\xi_3-\xi_6-\xi_8)$ $+\sqrt{3}(\eta_2+\eta_4+\eta_5+\eta_7-\eta_1-\eta_3-\eta_6-\eta_8)$	$(1/2)^{\frac{1}{2}}\langle p\pi \vert f\rangle$
	$\vert\delta_3\rangle$	$\frac{1}{2}(7)^{\frac{1}{2}}y(5y^2-3r^2)$	$(8)^{-\frac{1}{2}}(\xi_1+\xi_2+\xi_7+\xi_8-\xi_3-\xi_4-\xi_5-\xi_6)$	

σ_i is a sigma orbital (along the internuclear distance) centered on the X ion and η_i and ξ_i are π orbitals (perpendicular to the internuclear distance). As can be seen $4f$ orbitals that transform as the a_{2u} representation can form sigma but not π bonds with the ligands, etc. The coefficients relating the group overlaps and the ion pair overlaps are given in the last column and are the same for $2s$ and $2p\sigma$ overlap integrals. The positions of atoms F referred to the nucleus M at the centre of the cube, as origin, are (1) $(a/\sqrt{3})(1, 1, 1)$, (2) $(a/\sqrt{3})(\bar{1}, 1, \bar{1})$, (3) $(a/\sqrt{3})(1, \bar{1}, \bar{1})$, (4) $(a/\sqrt{3})(\bar{1}, \bar{1}, 1)$, where a is the length of the cube edge. The Miller indices for the vectors to describe the ligand $2p$ orbitals are

$$\begin{array}{lll} \sigma_1(\bar{1}, \bar{1}, \bar{1}) & \xi_1(1, \bar{2}, 1) & \eta_1(1, 0, \bar{1}) \\ \sigma_2(1, \bar{1}, 1) & \xi_2(\bar{1}, \bar{2}, \bar{1}) & \eta_2(\bar{1}, 0, 1) \\ \sigma_3(\bar{1}, 1, 1) & \xi_3(1, 2, \bar{1}) & \eta_3(1, 0, 1) \\ \sigma_4(1, 1, \bar{1}) & \xi_4(\bar{1}, 2, 1) & \eta_4(\bar{1}, 0, \bar{1}) \end{array}$$

The positions and associated vectors for atoms (5), (6), (7), and (8) are obtained from those for (1), (2), (3), and (4) by inversion.

located on the eight neighbouring fluorine ions which transform according to the same representation. A suitable representation of the basis functions for the $4f$ metal-ion and $2s$, $2p\sigma$, and $2p\pi$ ligand orbitals is reproduced in Table 5.17. If the covalency parameters are to be treated in a phenomenological way there is no need to consider explicitly higher-order metal-ion orbitals ($6s$, $6p$, etc), nor ligand-ligand overlap. (In fact these make a small contribution compared to the direct overlap between the metal ion and ligands.)

The wavefunction for the antibonding hole, which determines the optical and magnetic properties of the Tm^{2+}: $8F^-$ complex, has the form

$$\psi_A(\Gamma_i) = \phi_m(\Gamma_i) - \sum_\nu \lambda_{\Gamma_i\nu}\chi_L(\Gamma_i\nu), \tag{5.11}$$

where ν indicates the bond type (i.e. $\nu = 2s$, $2p\sigma$, or $2p\pi$), ϕ_m is the metal-ion wavefunction and χ_L are the ligand-ion wavefunctions. The admixture, or covalency parameters $\lambda_{\Gamma\nu}$ are given by expressions of the form

$$\lambda_{\Gamma\nu} = \frac{\langle\phi|\, h\, |\chi\rangle - S\langle\phi|\, h\, |\phi\rangle}{\langle\chi|\, h\, |\chi\rangle - \langle\phi|\, h\, |\phi\rangle}, \tag{5.12}$$

where the detailed subscripts have been dropped to emphasize the essential form. The factor S represents the general overlap parameters, $S(\Gamma\nu) = \langle\phi_m(\Gamma) \mid \chi_L(\Gamma\nu)\rangle$, and h represents the one-electron Hamiltonian of the complex. Axe and Burns obtained approximate values for the $S(\Gamma\nu)$ using Hartree-Fock wavefunctions for Eu^{2+} and F^-.

Using perturbation theory the energy of the complex can be written as

$$E(\Gamma) = \langle\phi(\Gamma)|\, h\, |\phi(\Gamma)\rangle + \Delta E(\Gamma), \tag{5.13}$$

where

$$\Delta E(\Gamma) = \sum_\nu \lambda^2_{\Gamma\nu}\{\langle\phi(\Gamma)|\, h\, |\phi(\Gamma)\rangle - \langle\chi(\Gamma_\nu)|\, h\, |\chi(\Gamma\nu)\rangle\}. \tag{5.14}$$

This can be rewritten in the form

$$\Delta E(\Gamma) = \langle\phi|\left\{\sum_\nu \frac{[h - \langle\phi|\, h\, |\phi\rangle]\, |\chi(\nu)\rangle\langle\chi(\nu)|\, [h - \langle\phi|\, h\, |\phi\rangle]}{\langle\chi|\, h\, |\chi\rangle - \langle\phi|\, h\, |\phi\rangle}\right\} |\phi\rangle, \tag{5.15}$$

where the factor within the large brackets is a pure $4f$ one-electron operator with the cubic symmetry of the complex. This is a more explicit form of eqn (5.7) which includes the effects of overlap. This is the equation which allows the parameters θ and Δ to be directly related to the crystal-field parameters:

$$\Delta = 10b_4 + 84b_6, \qquad \theta = 8[b_4 - 7b_6]. \tag{5.16}$$

It should also be noted that the denominator contains the metal-ion energy, $\langle\phi_m|\,h\,|\phi_m\rangle$, which must be assumed to be a constant for the above correspondence to be made. However, in those ions with a large spin-orbit coupling constant (Tm^{2+} has one of the largest), it could turn out that the crystal-field parameters may depend to a small extent on the J manifold from which they are derived.

The Wolfsberg-Helmholtz approximation is made in order to be able to carry the calculation of the energies further. This amounts to assuming that the off-diagonal matrix elements of the Hamiltonian can be written as

$$\langle\phi|\,h\,|\chi\rangle = [\langle\phi|\,h\,|\phi\rangle + \langle\chi|\,h\,|\chi\rangle]S, \tag{5.17}$$

which puts the admixture coefficients in the form

$$\lambda = \frac{S\langle\chi|\,h\,|\chi\rangle}{\langle\chi|\,h\,|\chi\rangle - \langle\phi|\,h\,|\phi\rangle}. \tag{5.18}$$

Using results obtained for NiF_6^{-4} and VF_6^{-4} as guides, Axe and Burns estimate that

$$\langle 2p|\,h\,|2p\rangle - \langle\phi|\,h\,|\phi\rangle = -100\times10^3\ \text{cm}^{-1}$$

and

$$\langle 2s|\,h\,|2s\rangle - \langle\phi|\,h\,|\phi\rangle = -200\times10^3\ \text{cm}^{-1} \tag{5.19}$$

Thus, by letting $\langle\phi|\,h\,|\phi\rangle$ act as the only remaining free variable they found that the best fit to the optical data was obtained with $\langle\phi|\,h\,|\phi\rangle \sim -70\times10^3\ \text{cm}^{-1}$. The resulting admixture coefficients then gave superhyperfine coupling coefficients which were in reasonable agreement with those measured.

Baker (1968) took a different approach to this problem by treating a limited set of the admixture coefficients as variable parameters. Only factors $\lambda_{a_{2u}s}$, $\lambda_{t_{2u}p\sigma}$, and $\lambda_{t_{2u}p\pi}$ enter into most of the theoretical expressions relating to the experimental data. Relabelling these as λ_s, λ_σ, and λ_π for convenience, the fluorine superhyperfine constants given by Baker (1968) are:

$$A_p = \frac{P}{140}\{(9+2\cdot5\,\kappa)\lambda_\pi^2 + 4\lambda_\sigma^2 + 25\surd(\tfrac{2}{3})\lambda_\sigma\lambda_\pi\}$$

and

$$A_s = \frac{-P}{140}\{-2\cdot5\,\kappa\lambda_\pi^2 + 10\kappa\lambda_\sigma^2 + 14\surd(\tfrac{2}{3})\lambda_\sigma\lambda_\pi\} + \frac{\lambda_s^2}{14}A_{2s}, \tag{5.20}$$

where for the fluorine atom $\kappa \simeq -0\cdot1$, $P = 6\cdot3$ GHz, and $A_{2s} = 46$GHz. Baker also estimated, using the fluorine hyperfine data for $CaF_2{:}Eu^{2+}$, that the indirect coupling through the $5p$ shell would make an additional contribution to A_s and A_p of the order of $-0\cdot9$ MHz and $-0\cdot5$ MHz respectively.

The Axe and Burns model contains two orbital reduction factors,

$$k = \frac{\langle a_{2u}| L_z |t_{2u}\rangle}{\langle fa_{2u}| L_z |ft_{2u}\rangle}$$

and

$$k' = \frac{\langle t_{2u}| L_z |t_{2u}\rangle}{\langle ft_{2u}| L_z |ft_{2u}\rangle}, \tag{5.21}$$

which are defined to give unity in the limit of no ligand admixture. They find that the calculated second factor, k', is unity to first order and the first one can be calculated from

$$1-k = \lambda_s^2+\lambda_\pi^2+\lambda_\sigma^2+\sqrt{(\tfrac{2}{3})}\lambda_\sigma\lambda_\pi. \tag{5.22}$$

A small contribution must be added to both reduction factors to take into account the dynamic interaction with the phonons (Inoue, 1963). The excited and ground state epr in $CaF_2:Tm^{2+}$ and $SrF_2:Tm^{2+}$ clearly show that k' is closely equal to 1, in excellent agreement with this model. However, no one has calculated the Inoue contributions to k'.

Using the energy values given in eqn (5.10), the covalency contribution to the energy parameter Δ is given by

$$\Delta = [3\lambda_s^2+(\lambda_\sigma^2-\lambda_\pi^2)]\times 10^5 \text{ cm}^{-1}. \tag{5.23}$$

The other energy parameter θ involves more of the admixture coefficients:

$$\theta = [3\lambda_{t_{1u}s}^2+\lambda_{t_{1u}p\pi}^2+\lambda_{t_{1u}p\sigma}^2-\lambda_{t_{2u}p\pi}^2]\times 10^5 \text{ cm}^{-1}. \tag{5.24}$$

The electrostatic point-charge contribution must be added to (5.23) and (5.24) before a direct comparison is made with the data.

Baker (1968) found he could obtain satisfactory fluorine hyperfine constants for $CaF_2:Tm^{2+}$ by using a value of λ_s which was slightly larger than that of Axe and Burns and requiring the fluorine ions to move slightly outward from their normal lattice position. (This expansion is consistent with the discussion of the positions of the first-shell fluorine ions given in § 5.2.4.) The values of the parameters are given in Table 5.18. The parameters also gave a value for $1-k$ of 0·008, which is reasonably close to the experimental value of 0·011. (Axe and Burns (1966) apparently made a numerical error in calculating this factor.) And finally, the results for the energy splittings are reasonably close to the observed values (Table 5.18). Baker could also find reasonable fits to the data for $CaF_2:Yb^{3+}$ by keeping the ratio $\lambda_\pi/\lambda_\sigma$ the same as that found in $CaF_2:Tm^{2+}$ and scaling the remaining two factors to match the fluorine hyperfine constants (Table 5.18). These parameters

TABLE 5.18

Orbital admixture parameters and energies for Tm^{2+} *and* Yb^{3+} *in* CaF_2

System	$\lambda_{2p\sigma}$	$\lambda_{2p\pi}$	λ_{2s}	Δ (cm^{-1})	θ (cm^{-1})	$1-k$	Reference
CaF_2:Tm^{2+}	0·0619	0·0393	0·0288	442	32	0·0081	Axe and Burns (1966)
CaF_2:Tm^{2+}	0·0625	0·0400	0·0374	580	78	0·0089	Baker (1968)
CaF_2:Yb^{3+}	0·086	0·062	0·038	891	−9	0·017	Baker (1968)

give an orbital reduction factor of 0·017, which compares favourably with the experimental value of 0·018. Baker also calculated the energy parameters for this system using eqn (5.23), since he thought the crude estimates used to obtain this equation for the divalent ion were at least as uncertain as the additional binding energy for the trivalent ion.

Anderson *et al.* (1974) argued that in the series CaF_2:Tm^{2+}, SrF_2:Tm^{2+}, and BaF_2:Tm^{2+} it is reasonable to suppose that all the admixture parameters would be maintained in the same ratio. This would make the covalency contributions to all the experimental factors scale linearly with one another. The observed crystal-field parameters, b_4 and b_6,† minus the electrostatic point-ion contribution, do scale linearly with the experimental scalar fluorine hyperfine constants A_s (Fig. 5.14). The extrapolated intercept of a straight line drawn through the data does not pass through the origin, but is close to the value of −0·9 MHz estimated by Baker for the $5p$ contribution to A_s in CaF_2.

Using the first-shell fluorine-ion positions given in Table 5.15, it is found that the covalent contributions to b_4 and b_6 scale as R^{-12}. This is a higher power-law than the value of 8·2 obtained by Axe and Burns (1966), but it is consistent with that found by Stedman and Newman (1971) for the rare-earth ions in oxide compounds. This high power-law is also in reasonable agreement with the s and $p\pi$ overlap calculations by Burns and Axe (1967) for Eu^{2+}–O^{2-} and Eu^{2+}–F^-.

A curious feature worth noting is that in all three alkaline earth fluoride hosts the covalency contribution to θ is essentially zero. There is no obvious reason for this fact, especially since one orbital type does not dominate the others in this problem. Even with the electrostatic point-ion contribution included, θ is at least one order of magnitude smaller than Δ. Through the relationship between the crystal-field parameters and θ given by eqn (5.16), it is immediately seen that if $\theta \approx 0$ then $b_6/b_4 = \frac{1}{7}$ or $A_6\langle r^6\rangle/A_4\langle r^4\rangle = 0{\cdot}185$, which is the approximate value observed for the rare-earth ions in CaF_2 (see Fig. 5.10).

† See the footnote to Table 5.16 for a definition in terms of parameters of eqns (5.8).

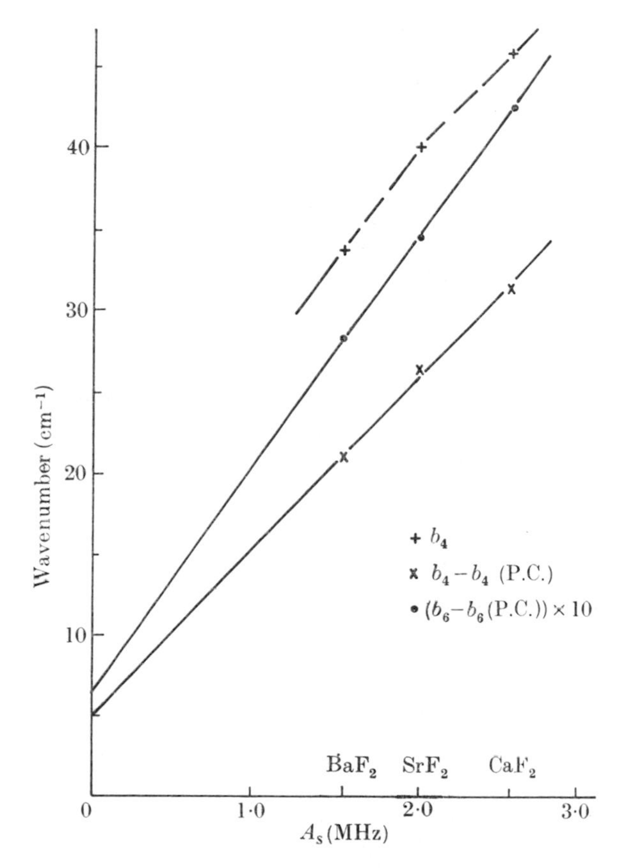

FIG. 5.14. Crystal-field parameters b_4 and b_6 minus the calculated electrostatic component as a function of the first-shell F^- scalar hyperfine constant A_S for Tm^{2+} in CaF_2, SrF_2, and BaF_2 [Anderson *et al.*, 1974].

This particular ratio ensures that the $^2F_{\frac{7}{2}}(\Gamma_8)$ and $^2F_{\frac{7}{2}}(\Gamma_6)$ levels will be almost degenerate in Tm^{2+} and by implication it may be this same feature, acting as a hidden symmetry, which leads to the crystal-field levels in the maximum J manifolds of the other rare-earth ions occurring in two widely-spaced groups. This at least is a better explanation for this occurrence in Tm^{2+} than the semi-classical reasons given by Trammell (1963).

The covalency model also predicts monotonically decreasing crystal-field parameters as a function of increasing $4f$ electron number (the overlap integrals given by Burns and Axe (1967) exhibit this behaviour). This general trend does seem to be confirmed by the limited data available (Figs. 5.8 and 5.9) except for the $A_4\langle r^4\rangle$ value of Ho^{2+} and the $A_6\langle r^6\rangle$ value of Ce^{2+}.

5.3. $4f^{n-1}5d$ Electronic states

5.3.1. *The $4f^{n-1}5d$ configuration*

Strong optical transitions are allowed between the $4f^n$ and $4f^{n-1}5d$ configuration, but these are not as rich in zero-phonon lines as transitions within the $4f^n$ configuration. The scarcity of zero-phonon lines is due to the fact that the different configurations are associated with different equilibrium positions of neighbouring ions in the lattice. This results in large Franck–Condon shifts, giving rise to the emission of many phonons in the optical transitions and hence to broad bands. There are, of course, notable exceptions where sharp optical lines are observed.

Dieke and Crosswhite (1963) in a general survey of the divalent and trivalent rare-earth ions have given the relative energies of the various configurations in the free ions (Fig. 5.1). When the ions are placed into the alkaline earth fluorides at cubic sites the $5d$ configuration is split into two levels, t_{2g} and e_g, the latter lying lower in energy by approximately 10^4 cm^{-1}. This makes the $4f^n$ and $4f^{n-1}5d$ configurations lie very close together in the divalent ions, while in Ce^{3+} the $4f \rightarrow 5d$ transitions occur in the ultraviolet ($\sim$3000 Å) and at even shorter wavelengths in the other trivalent ions. In $CaF_2:Ce^{3+}$ only one optical transition has been assigned to the cubic site (Manthey, 1973) and so it is impossible to obtain the crystal-field splitting of the d electron in this case. This discussion of the $4f^{n-1}5d$ configuration, therefore, is limited to the divalent ions, which are found mostly in the alkaline earth fluorides and $SrCl_2$.

The coupling of the $4f$ electrons with the $5d$ electron is most easily visualized in a representation where the two d orbitals, t_{2g} and e_g, couple to the $4f^{n-1}$ core; this is split by the $4f$–$4f$ electrostatic interaction and spin-orbit coupling, as illustrated in Fig. 5.15 for $CaF_2:Tm^{2+}$ (Alig *et al.*, 1973). In this case the $4f^{12}6s$ configuration also has to be considered and in this illustration is assumed to be degenerate with the $4f^{12}5d$ configuration. The remaining interactions: $H_{SO}(d)$, d spin-orbit coupling; $H_{CF}(f)$, $4f$ crystal field; $H_{EE}(fd)$, f–d electrostatic; $H_{EE}(fs)$, f–s electrostatic and $H_{CI}(f^2d, f^2s)$, configuration interactions, are all only slightly smaller than the first three interaction terms (2000–5000 cm^{-1}), so that perturbation theory is inadequate for use in detailed calculations. Loh (1968) has surveyed the $4f^n \rightarrow 4f^{n-1}5d$ spectra of a number of the divalent ions, and has found in a very qualitative sense that when the $4f^{n-1}$ core states are widely spaced, the bands do appear to show structure corresponding to the $4f^{n-1}$ core energies; this is repeated at a

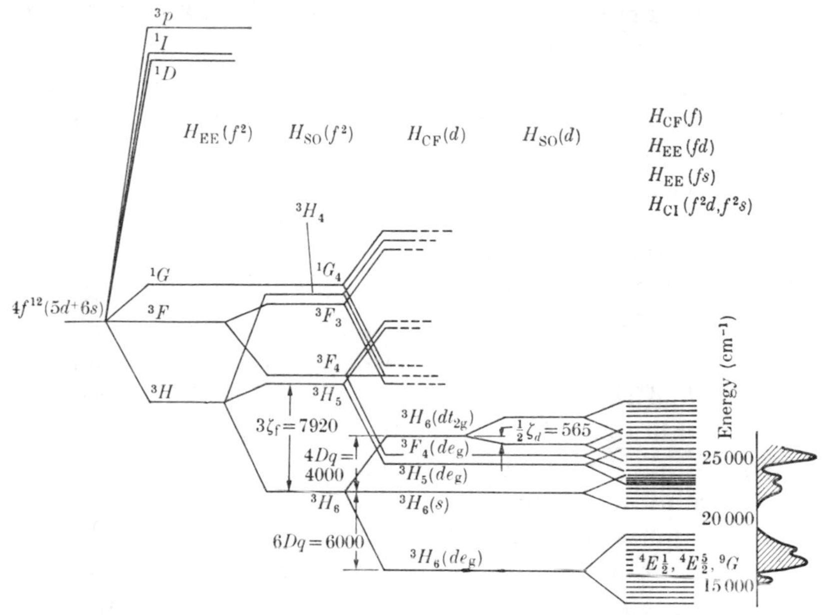

FIG. 5.15. The energies of selected levels of the $4f^{12}(5d+6s)$ configuration (see text) of Tm^{2+} in a strong cubic crystal field. The subscripts EE, SO, CF, and CI refer respectively to the electrostatic, spin–orbit, crystal field, and configuration interactions; the letters s, d, and f refer to the electrons undergoing the interaction. The broad absorption bands in CaF_2:Tm^{2+} are shown on the right. The spin–orbit parameters of the f and d electrons are denoted ζ_f and ζ_d respectively. The crystal-field parameter of the d electron is denoted Dq; the crystal field levels of the d electron are denoted e_g and t_{2g} [Alig *et al.*, 1973].

separation of $\sim 10^4$ cm^{-1}, the crystal-field splitting of the d orbital. The theory of this configuration in cubic crystal fields is also discussed by Starostin (1967) and Eremin (1969).

McClure and Kiss (1963) have calculated the change in the lowest $4f^n \rightarrow 4f^{n-1}5d$ excitation energy as one progresses through the divalent rare-earth series. In any calculation of this type the host lattice modifies the Coulomb interaction with the core nucleus in such a manner that the configuration energies shown in Fig. 5.1 are slightly shifted with respect to one another by an amount which cannot be calculated. Using the lowest observed $4f^n \rightarrow 4f^{n-1}5d$ excitation in CaF_2:Sm^{2+} as a reference point they could successfully predict where the bands are observed to begin in all the remaining ions except for La^{2+}, Ce^{2+}, Gd^{2+}, and Tb^{2+}. In these four cases they were not aware that their data came primarily from the photochromic centre (see Chapter 7), but they surmised that the ground states of these ions in CaF_2 arise from the $4f^{n-1}5d$ configuration. This has been verified for

$La^{2+}(5d)$ (Hayes and Twidell, 1963) and $Ce^{2+}(4f5d)$ (Alig *et al.*, 1969), but not so far for $Gd^{2+}(4f^7 5d)$ and $Tb^{2+}(4f^8 5d)$.

The most recent detailed calculations by Weakliem (1972) on CaF_2:Eu^{2+} found that the $4f$–$5d$ electrostatic (Coulomb) interaction parameters have to be reduced by almost a factor of 2 from the free-ion values. This large reduction had been suggested earlier by Yanase and Kasuya (1970) in their study of EuF_2 and by Weakliem *et al.* (1970) in their survey of the magnetic circular dichroism of the $4f^{n-1} \rightarrow 4f^{n-1}5d$ bands. This result has also been confirmed by Alig *et al.* (1973), who found that the spread of the levels arising from $^3H_6(e_g)$ in CaF_2:Tm^{2+} and $SrCl_2$:Tm^{2+} (see Fig. 5.15) is primarily due to this electrostatic interaction, and the free-ion parameters give a much wider energy spread than is observed in the lowest group of levels in these crystals. This result is not completely unexpected since a similar reduction is known for the $5d$–$5d$ Coulomb parameters and has been given the name nephelauxetic effect by Jorgenson (1962), meaning 'cloud increasing'. The occurrence of such a large reduction in this term could easily be the source of the poor fit achieved to the sharp optical lines observed in CaF_2:Ce^{2+} (Alig *et al.*, 1969) and to the electronic Raman scattering lines observed by Kiel and Scott (1970) in this crystal. It is also probable that the detailed study of $SrCl_2$:Yb^{2+} (Piper *et al.*, 1967) will have to be reexamined. The values of $10Dq$ for the crystal-field splitting of the d electron obtained in these few cases where detailed comparisons with data were attempted give the following reasonable values: 17×10^3 cm^{-1} for CaF_2:Ce^{2+} (Alig *et al.*, 1969); 15×10^3 cm^{-1} for CaF_2:Eu^{2+} (Weakliem, 1972); and 8×10^3 cm^{-1} for $SrCl_2$:Yb^{2+} (Piper *et al.*, 1967). Eremin (1970) has estimated from optical absorption data that $10Dq$ for Yb^{2+} has the values 16×10^3 cm^{-1} in CaF_2; 14×10^3 cm^{-1} in SrF_2 and 12×10^3 cm^{-1} in BaF_2. Loh (1973) has recently given qualitative arguments that $10Dq$ for $SrCl_2$:Yb^{2+} is nearer to 13×10^3 cm^{-1}.

5.3.2. *Properties of the* $4f^n \rightarrow 4f^{n-1}5d$ *absorption bands*

The earliest survey of the $4f^n \rightarrow 4f^{n-1}5d$ bands in CaF_2 by McClure and Kiss (1963) were made on gamma-irradiated crystals which not only contained the photochromic centres (Chapter 7), but also have other radiation-induced centres which produce spurious absorption bands in the near ultraviolet region of the spectrum. The surveys by Loh (1968) and by Weakliem *et al.* (1970) give better pictures of the divalent absorption bands since they were carried out on chemically reduced samples. One characteristic is that the lowest band observed for each ion moves to higher energies as one goes from CaF_2 to BaF_2, reflecting the weakening of the interaction of the *d* electron with the host lattice.

The $4f^n$–$4f^{n-1}5d$ transitions are electric dipole in nature. However the oscillator strength is spread over as many as 10^3 levels so that the strength associated with each prominent peak in the band has a typical value of 10^{-2}–10^{-3}. Absolute absorption coefficients at the peaks of some of the bands have been given by Arkhangel'skaya *et al.* (1967), and these are very useful for measuring the concentration of the divalent ions in these crystals (Table 5.19). The results were obtained by

TABLE 5.19

Optical absorption coefficients of selected peaks of the $4f^n \rightarrow 4f^{n-1}5d$ bands of divalent rare-earth ions

System	Optical wavelength (μm)	k [cm^{-1}] at 0·01 mol %	σ[10^{-18} cm^2]	References
$CaF_2:Nd^{2+}$	1475	8·1	3·3	Arkhangel'skaya (1967)
$CaF_2:Sm^{2+}$	620	9·3	3·8	
$CaF_2:Eu^{2+}$	338	17·9	7·3	
$CaF_2:Dy^{2+}$	715	6·2	2·5	
$CaF_2:Ho^{2+}$	897	4·2	1·7	
$CaF_2:Er^{2+}$	910	6·8	2·5	
$CaF_2:Tm^{2+}$	590	8·6	3·5	
$CaF_2:Tm^{2+}$	590	5·3	2·1	Anderson and Sabisky (1971)
$SrF_2:Tm^{2+}$	575	4·3	2·1	
$BaF_2:Tm^{2+}$	560	3·1	1·9	

carefully measuring the absolute absorption coefficient of $CaF_2:Eu^{2+}$ in which almost all the europium added to the melt during crystal growth is incorporated into the crystal in the divalent state. This measurement was then transferred to the other rare-earth ions by using crystals containing both Eu^{2+} and one of the rare earths produced in the divalent state through the photochemical process

$$Eu^{2+}+Re^{3+} \xrightarrow{h\nu} Eu^{3+}+Re^{2+},$$

where the frequency of excitation coincides with the Eu^{2+} absorption bands. In this process a certain fraction of the Eu^{2+} ions are converted to the trivalent state and the coexisting rare-earth ions in the trivalent state capture the electron and are converted to the divalent state. Under the assumption that all the electrons are captured in this way, the reduction of the Eu^{2+} concentration is equal to the concentration of Re^{2+} produced. The absolute absorption coefficients obtained by this method for CaF_2:Tm^{2+} are smaller than those reported by Anderson and Sabisky (1971), which were obtained from examining a large sample of chemically reduced crystals. One check on these latter values is that the absorption cross-section of the first large peak is the same in CaF_2:Tm^{2+} and BaF_2:Tm^{2+}.

The photochemical process described above occurs because most of the $4f^{n-1}5d$ levels are degenerate with the conduction band of the host crystals, as demonstrated by the photoconductivity measurements of

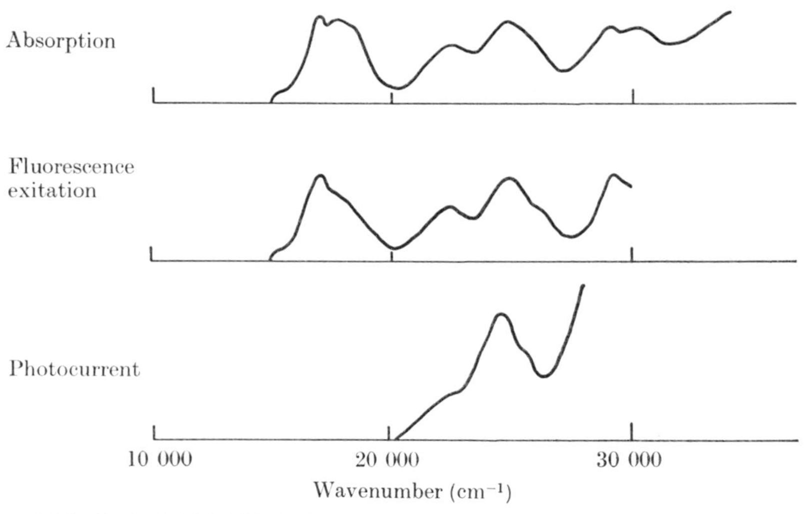

FIG. 5.16. Optical $4f^{13}$–$4f^{12}5d$ absorption bands $(^2F_{\frac{5}{2}}(\Gamma_7) \to {}^2F'_{\frac{7}{2}}(\Gamma_7))$, fluorescence excitation spectra, and photocurrent per incident photon for CaF_2:Tm^{2+} [Anderson and Kiss, 1964].

Anderson and Kiss (1964) on CaF_2:Tm^{2+} and Heyman (1969) on CaF_2:Ho^{2+}. This is seen in Fig. 5.16 where the photocurrent per incident photon in CaF_2:Tm^{2+} is shown to have the same modulated structure as the absorption and fluorescence excitation spectrum of the $^2F_{\frac{5}{2}}(\Gamma_7) \to {}^2F_{\frac{7}{2}}(\Gamma_7)$ transition in this crystal. This photocurrent is independent of temperature to the extent that thermally assisted processes can be ruled out. A simple perturbation calculation of the autoionization of a $5d$ electron into a conduction band with a parabolic density of states, predicts a threshold behaviour of $(\nu-\nu_0)^{\frac{7}{2}}$ for the photocurrent per absorbed photon. The CaF_2:Tm^{2+} data can be fitted to this form with $\nu_0 = 13\ 500\ \text{cm}^{-1}$, which puts the ground state about 1 eV below the conduction band.

When the $4f^n$ ground state of the divalent ion is paramagnetic (see Table 5.10) these absorption bands show very rich magneto-optical properties. This is illustrated in Fig. 5.17 by the magnetic circular dichroism (MCD) and magnetic linear dichroism (MLD) of $CaF_2:Eu^{2+}$, taken from the magneto-optical survey of all of the rare-earth ions by Weakliem *et al.* (1970). The large MLD of this system was first noted by Zakharchenya and Ryskin (1963) and the Faraday rotation (MOR) and MCD of the sharp lines in $CaF_2:Eu^{2+}$ and $SrF_2:Eu^{2+}$ have been

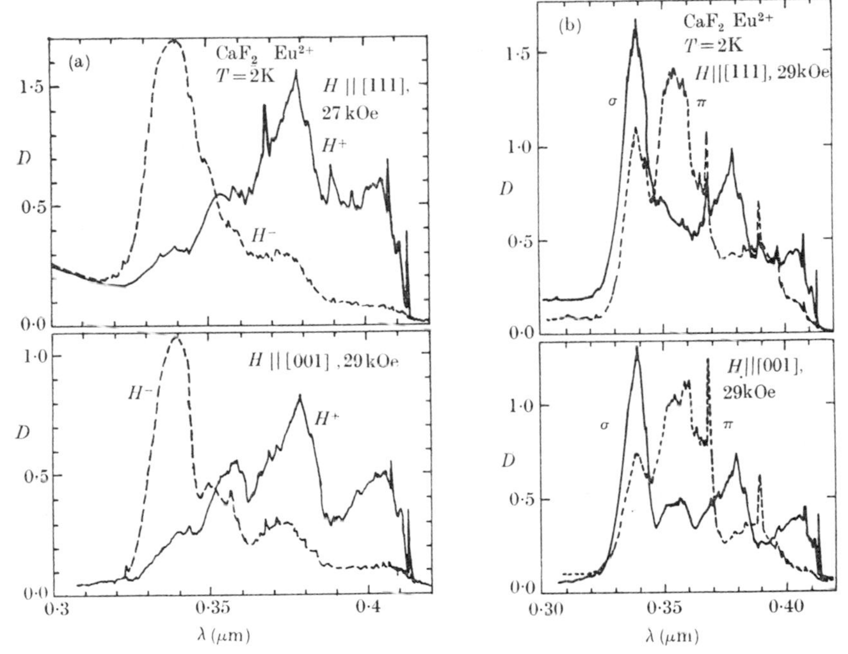

FIG. 5.17. Magnetic circular dichroism (a) and magnetic linear dichroism (b) of CaF_2: Eu^{2+}; top figures $H \parallel$ [111], bottom $H \parallel$ [001] for both (a) and (b). π denotes polarization with the electric vector of the light parallel with the external magnetic field and σ denotes perpendicular polarization [Weakliem *et al.*, 1970].

investigated in more detail by Shen (1964). Brief notes have also been reported on the MOR (Alekseyeva *et al.*, 1967) and MCD (Inari, 1968) of $CaF_2:Ho^{2+}$ and the MOR of $CaF_2:Dy^{2+}$ (Alekseyeva *et al.*, 1968). Similar brief notes were published on the MOR (Shen, 1964) and MCD (Anderson *et al.*, 1966) of $CaF_2:Tm^{2+}$. Finally Margerie (1967) has studied the MCD of several sharp lines in $CaF_2:Sm^{2+}$.

The very large magneto-optical properties arise in these systems because of the large spin-orbit coupling of the $4f$ electrons, which allows the electric dipole operator to couple strongly with the magnetic sublevels. This large effect would be washed out if the phonons accompanying the optical transitions have any symmetry other than Γ_1; this apparently is not the case in these systems. It follows that the symmetry of the electronic states which contribute to the bands can be determined through the magneto-optical studies, and this has been exploited in the calculations by Weakliem (1972) on $CaF_2:Eu^{2+}$ and Alig (1973) on $CaF_2:Tm^{2+}$.

The magneto-optical properties of the bands make these crystals ideal systems for study by optical-rf double resonance techniques in the tradition of Kastler (1951). $CaF_2:Tm^{2+}$ with its simple magnetic ground state has been the most exploited system in this respect. It has been used as the basis of an optically-pumped microwave maser (Sabisky and Anderson, 1967) and for nuclear polarization studies through optical pumping (Mollenauer *et al.*, 1968; Grant *et al.*, 1971). It also forms the basis of an acoustic-phonon spectrometer (Anderson and Sabisky, 1971) which has been used for measurements of the van der Waals force (Sabisky and Anderson, 1973). A number of double resonance experiments have also been carried out on the ground state and on an excited state of $CaF_2:Eu^{2+}$ (Chase, 1970).

5.3.3. *Sharp line spectra of* $4f^n-4f^{n-1}5d$ *transitions*

Sharp red fluorescence lines (~7100 Å) were first observed in CaF_2 by Feofilov (1956) and later were used in an optically-pumped laser by Sorokin and Stevenson (1961), and Kaiser *et al.* (1961) (See also Sorokin *et al.*, 1962). These lines have been extensively studied in all of the alkaline earth fluorides and $SrCl_2$; the primary sources of information are Wood and Kaiser (1962) and Feofilov and Kaplyanskii (1962). The upper levels have clearly been identified as belonging to the $4f^55d$ configuration, the primary fluorescence arising from a Γ_1 state at 14 360 cm^{-1} in CaF_2 and terminating on the $^7F_1(\Gamma_4)$ level of the $4f^6$ configuration. A sharp line in absorption is also observed from the ground $^7F_0(\Gamma_1)$ state to a Γ_4 state lying at 14 489 cm^{-1}. Kaplyanskii and Przhevaskii (1966) have found that fluorescence is induced under stress from a Γ_3 level which lies 28 cm^{-1} above the Γ_1 state. It is a reasonable approximation to assume these levels primarily arise from the $^6H_{\frac{5}{2}}(e_g)$ level of the configuration; the e_g identification is confirmed by the fact that stress measurements along the (111) axis produce no splittings. Also, rough calculations of the g-values for the Γ_4 state have given 0·38 (Zvereva and Makarov, 1967), which agrees reasonably well with the Zeeman measurements of 0·31 by Margerie (1967) and of 0·36 by Zakharchenya and Ryskin (1962).

The measurements by Kaplyanskii and Przhevaskii (1966) show, as one expects, that the $5d$ levels shift about 10 times more than the $4f$ levels under stress. This strong coupling of the $5d$ electron to the host lattice results in the vibronic sidebands being comparable in intensity to the electronic lines. Axe and Sorokin (1963) have studied the vibronic sidebands of Sm^{2+} in some detail and find the main frequency is at 335 cm^{-1} in CaF_2 and 292 cm^{-1} in SrF_2 and is Γ_3 in character (see also Runciman and Stager, 1962, and Cohen and Guggenheim, 1968). They also discovered that the $4f$–$5d$ absorption bands in $SrCl_2$:Sm^{2+} are very rich in vibronic structure and this has been discovered to be true for all the other rare-earth ions in $SrCl_2$.

The $5d$ levels markedly shift to higher energy upon going from CaF_2 to one of the other hosts. This has the unusual feature of exposing the $^5D_0$ level of the $4f^6$ configuration at 14 600 cm^{-1} and fluorescence is therefore observed to originate from this level in all hosts except CaF_2. Huang and Moos (1968) have also observed absorption from the $^5D_0$ level into the $4f^55d$ bands in SrF_2 and $SrCl_2$ and made rough comparisons to the free-ion levels.

Divalent europium is the only other case where sharp line fluorescence from the $4f^{n-1}5d$ configuration to the $4f$ ground state is observed in the alkaline earth fluorides; the fluorescence line at 4130 Å in CaF_2 is also observed in absorption (Kaplyanskii and Feofilov, 1962). Stress measurements (Kaplyanskii and Przevskii, 1965) and Zeeman measurements (Zacharchenya *et al.*, 1965, Kisliuk *et al.*, 1968, and Eremin *et al.*, 1971) have clearly shown that the $4f^65d$ level associated with the 4130 Å has Γ_8 symmetry; in a rough sense it can be approximated as $^7F_0(^2e_g)$. A number of calculations have been made to explain the structure in the band associated with this line (Freiser *et al.*, 1968, Yanase and Kasuya, 1970 and Eremin, 1969) as well as its magnetic properties. The most recent work by Weakliem (1972), who performed a rather full calculation on the $^7F_{0,1,\ldots,6}(^2e_g)$ levels, is the best reference for this ion. His calculation qualitatively explains the magneto-optical properties of the absorption bands.

A detailed study of the excited Γ_8 level of Eu^{2+} was carried out by Chase (1970) in CaF_2, SrF_2, and BaF_2 using optical detection of paramagnetic resonance. His g-values for this state (H along the 100 axis) are 4·08(2) and 3·28(2) and the corresponding hyperfine constants are 2360(30) MHz and 420(8) MHz. For comparison, the calculated

${}^7F_0({}^2e_g)$ level g-values are 2·2 and 2·0, while Weakliem's (1972) calculation is able to give g-values of 4·4 and 3·0 only if the electrostatic coupling between $4f$ and $5d$ electrons is reduced from the free-ion values. Chase also discussed evidence for the presence of a dynamic Jahn-Teller effect in this state (see § 5.3.4).

Finally, Chase observed essentially the same magnetic parameters in all the alkaline earth fluorides. This was particularly puzzling in the case of BaF_2 since the fluorescence spectrum of this crystal does not resemble that observed in the other two hosts.

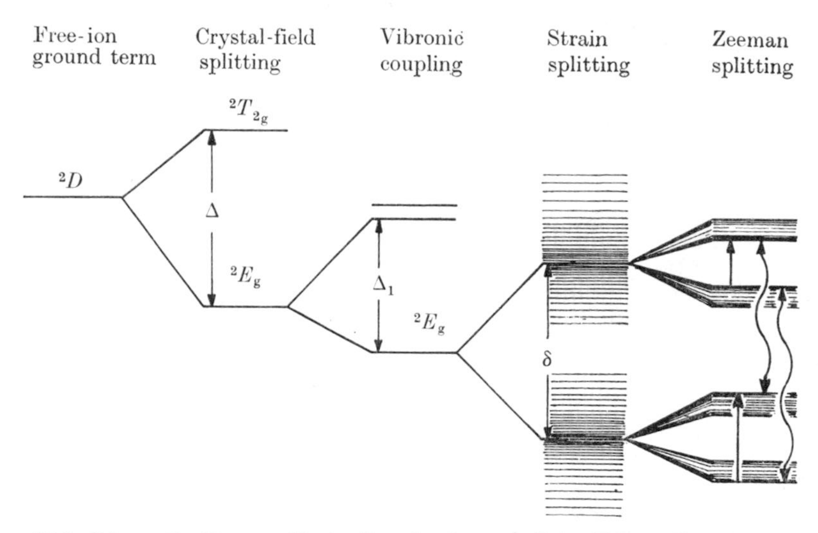

FIG. 5.18. Schematic diagram illustrating the dynamic Jahn-Teller effect observed for d^1 orbitals in the alkaline earth fluorides and $SrCl_2$. The ${}^2D$ free-ion term is split by the cubic crystal field into ${}^2E_g$ and ${}^2T_{2g}$ states, with the ${}^2E_g$ lowest. A weak to moderate vibronic interaction results in a ${}^2E_g$ ground state and ${}^2A_1$ and ${}^2A_2$ levels at an energy denoted as Δ_1. The ground ${}^2E_g$ level is split by random strains into two Kramers doublets with a mean spacing indicated by δ. The Kramers degeneracy is removed by an applied magnetic field and epr transitions (represented by straight arrows) are induced by the microwave magnetic field. The wavy arrows represent transitions resulting from vibronic relaxation [Herrington *et al.*, 1973].

5.3.4. *Jahn-Teller effect*

Paramagnetic resonance studies of the ground ${}^2e_g$ level of the d orbital of divalent ions at cubic sites in the alkaline earth fluoride hosts and $SrCl_2$ have all revealed the Jahn-Teller effect. In these systems the Jahn-Teller distortion energy is smaller than the zero-point energy of the nuclei; this is called the weak-to-moderate Jahn-Teller regime and has been extensively analysed by Ham (1972). A qualitative picture of this system can be obtained as follows, using Fig. 5.18 which is taken from Herrington *et al.* (1973). The lower ^2e_g crystal-field component of the ^2D term couples to a local vibrational mode of Γ_3 symmetry to produce a set of three vibronic levels of symmetry Γ_1, Γ_2, and Γ_3. (A vibronic level is by definition one which is described by a linear combination of products of electronic and vibrational wave functions). The lowest vibronic state has Γ_3 symmetry and in a certain limit [a small warping potential term] the Γ_1 and Γ_2 levels are nearly degenerate at an energy Δ_1 (sometimes designated as 3Γ) above the ground state. Random strains split the ground vibronic level and mix the vibronic levels. Each of the resulting states are Kramers doublets and paramagnetic resonance is observed between the magnetic splittings of these levels. The applied magnetic field in this case will also mix the vibronic states.

The observed paramagnetic resonance lineshapes are determined by summing over the sets of ions experiencing different local strains. When the strain splittings of the ground vibronic state are smaller than the vibronic energy Δ_1 (the dynamic Jahn-Teller situation) the envelope of the absorption line has maxima at the approximate locations given by

$$h\nu(m) = g_1\beta H + A_1 m \pm q(g_2\beta H + A_2 m)[1-3(\zeta^2\eta^2+\xi^2\zeta^2+\eta^2\xi^2)], \quad (5.25)$$

where m refers to the nuclear-spin magnetic quantum number and ξ, η, and ζ are the direction cosines of the magnetic field relative to the crystal axis [see eqn (21.76) in Abragam and Bleaney, 1970]. The parameter q depends on the strength of the Jahn-Teller coupling, being 1 in the limit of no coupling and $\frac{1}{2}$ in the limit of strong coupling.

When the random strains are large compared to the vibronic splitting Δ_1 (the static Jahn-Teller situation) the paramagnetic resonance absorption spectra have maxima which follow the pattern associated with tetragonal symmetry. The spectra are therefore described with resonance parameters $g_{\parallel}$, $g_{\perp}$, $A_{\parallel}$, and $A_{\perp}$.

As the temperature is raised the acoustic phonons induce transitions between the vibronic and strain-split components of each ion. When this transition rate becomes faster than the difference in the Larmor frequencies of each component, paramagnetic resonance absorption is observed only at the average frequency. Thus in both the static and dynamic cases described above, as the temperature of the sample is raised, a single central isotropic resonance line is observed to grow at the expense of the rest of the resonances. In the dynamic case it is also possible that one of the singlet vibronic levels will become thermally

TABLE 5.20

Spectroscopic parameters for the dynamic Jahn-Teller systems

System	g_1	qg_2	$A_1(10^{-4}\ cm^{-1})$	$qA_2(10^{-4}\ cm^{-1})$	$qQ(10^{-4}\ cm^{-1})$	q	References
$CaF_2:{}^{45}Sc^{2+}(3d^1)$	1·9719 (5)	−0·0211 (5)	−65·8 (5)	−24·1 (5)	−0·12 (5)	0·69	Hochli and Estle (1967) Hochli (1967)
$SrF_2:{}^{45}Sc^{2+}(3d^1)$	1·964 (1)	−0·028 (1)	−67·1 (5)	−24·1 (5)	−0·12 (5)	0·73	Herrington *et al.* (1972)
$BaF_2:{}^{45}Sc^{2+}(3d^1)$	1·9555 (5)	−0·0309 (5)	−68·3 (5)	−24·5 (5)	−0·12 (8)	0·66	Herrington *et al.* (1972)
$SrCl_2:{}^{45}Sc^{2+}(3d^1)$	1·9531 (10)	−0·0442 (8)	−62·7 (8)	−31·8 (8)	−0·20 (5)	0·86	Herrington *et al.* (1973)
$SrCl_2:Y^{2+}(4d^1)$	1·9289 (10)	−0·0478 (8)	+24·3 (8)	+5 (8)	—	0·65	Herrington *et al.* (1973)
$SrCl_2:{}^{139}La^{2+}(5d^1)$	1·8808 (10)	−0·0687 (8)	−119·5 (8)	−18·8 (8)	0·15 (5)	0·57	Herrington *et al.* (1970, 1971, 1972)

populated as the temperature is raised. This too would have an isotropic resonance line, which would appear in the same place as the motionally narrowed line, and it is very difficult to know which is being observed.

The first Jahn-Teller system observed in CaF_2 was $La^{2+}(5d^1)$, investigated by Hayes and Twidell (1963). At temperatures below 10 K they observed a tetragonally distorted spectrum with the spectroscopic parameters $g_{\parallel} = 2{\cdot}00(1)$, $g_{\perp} = 1{\cdot}904(2)$, $A_{\parallel} = 37(9)\times10^{-4}\ \text{cm}^{-1}$, and $A_{\perp} = 62{\cdot}6(1{\cdot}0)\times10^{-4}\ \text{cm}^{-1}$. This spectrum merged into an isotropic one at 10–12 K and at 20 K they observed

$$g = 1{\cdot}937(2) \quad \text{and} \quad A = 52(1)\times10^{-4}\ \text{cm}^{-1},$$

which are equal to the averages of the anisotropic parameters given by

$$\begin{aligned} g_{\text{av}} &= \tfrac{1}{3}[g_{\parallel}+2g_{\perp}], \\ A_{\text{av}} &= \tfrac{1}{3}[A_{\parallel}+2A_{\perp}]. \end{aligned} \tag{5.26}$$

This fits the description of the static Jahn-Teller case, where the static strains are larger than the vibronic splittings and hence freeze the ions into predominantly tetragonally distorted sites. This type of situation has also been reported in $CaF_2{:}Y^{2+}(4d^1)$ by O'Conner and Chen (1964), but has not been confirmed by other authors (see also § 7.3.2).

The spectrum of $Sc^{2+}(3d^1)$ (Hochli and Estle, 1967 and Ham, 1968) clearly shows that the $3d^1$ electron experiences a dynamic Jahn-Teller effect in CaF_2, in contrast to the $4d^1$ and $5d^1$ orbitals. As the temperature is lowered below 4 K a tetragonal spectrum is never observed; instead an anisotropic spectrum appears which can be fitted by eqn (5.25). The quadrupole coupling of this system has also been measured (Herrington *et al.*, 1972); if the ion were sitting at a site of pure cubic symmetry this would be zero. The spectroscopic parameters are given in Table 5.20.

The ion $Sc^{2+}(3d^1)$ has also been found to exhibit the same dynamic Jahn-Teller effect in SrF_2, BaF_2, and $SrCl_2$, with the spectroscopic parameters given in Table 5.20. Also, $Y^{2+}(4d^1)$ and $La^{2+}(5d^1)$ have been shown to exhibit dynamic rather than static Jahn-Teller effects in $SrCl_2$.

The value of q (eqn 5.25) is estimated using a pure crystal-field perturbation calculation for g_1 and g_2, which leads to the equation (Ham, 1972)

$$q = \frac{qg_2}{g_1-2{\cdot}0023}. \tag{5.27}$$

It is seen in Table 5.25 that q is largest in $SrCl_2{:}Sc^{2+}$ and smallest in

BaF_2, and this indicates that the Jahn-Teller interaction is weakest in $SrCl_2$ and strongest in BaF_2. The values of q also show that the Jahn-Teller coupling is largest for the $5d^1$ orbital and smallest for the $3d^1$ orbital in $SrCl_2$, as would be expected. Herrington *et al.* (1973) have estimated that in $SrCl_2$ the Jahn-Teller coupling ranges from 2000 to 7000 cm^{-1}, which puts the first excited vibronic state roughly 100 cm^{-1} above the ground state. The random strain splittings are also estimated to be of the order of 1 cm^{-1}. However, there are no direct measurements of these quantities available at this time.

Chase (1970) using combined optical and magnetic resonance techniques found the lowest level of the $4f^6 5d$ configuration in $CaF_2:Eu^{2+}$ (approximately $^7F_0(^2e_g)$ in character) to exhibit a dynamic Jahn-Teller effect. Because of the uncertainties in the coupling of the $4f$ and $5d$ electrons with each other (see § 5.3.3) it is impossible to make a precise calculation of the g-values of this state in the absence of the Jahn-Teller effect and hence to obtain a value for q. However, Chase did make the reasonable suggestion that the levels which are observed nearby (within $\sim$10 cm^{-1}) are the excited vibronic components and thus it is possible to obtain a direct measurement of the position of these levels; it is not possible to do this in the other systems discussed in this section. It would seem that a fruitful line of investigation would be to follow up Chase's suggestion and analyse the stress measurements by Kaplyanskii and Przhevaskii (1966) not only on these levels in $CaF_2:Eu^{2+}$ but also in $CaF_2:Sm^{2+}$. It is also clear that calculations in the future on the electronic structure of the mixed $4f^{n-1}5d$ configuration will have to include a discussion of the Jahn-Teller effect, especially if estimates of Zeeman splittings are required.

References

ABRAGAM, A. and BLEANEY, B. (1970). *Electron paramagnetic resonance of transition ions*, Clarendon Press, Oxford.

ABRAGAM, J., JACQUINOT, J. F., CHAPELLIER, M., and GOLDMAN, M. (1972). *J. Phys.* **C5**, 2629.

AIZENBERG, I. B., MALKIN, B. Z., and STOLOV, A. L. (1971). *Fiz. Tver. Tela* **13,** 2566 [*Sov. Phys.–Sol. State* **13,** 2155].

ALEKSEEVA, L. A., and FEOFILOV, P. P. (1967). *Optika Spetrosk.* **22,** 996 (*Opt. Spectrosc.* **22,** 545).

——, STAROSTIN, N. V., and FEOFILOV, P. P. (1967). *Optika Spektrosk.* **23,** 259 (*Opt. Spectrosc.* **23,** 140).

—— —— (1968). *Opt. Spektrosk.* **24,** 145 (*Opt. Spectrosc.* **24,** 72).

ALIG, R. C., KISS, Z. J., BROWN, J. P., and MCCLURE, D. S. (1969). *Phys. Rev.* **186,** 276.

——, DUNCAN, R. C. Jr., and MOKROSS, B. K. (1973). *J. chem. Phys.* **59,** 5837.

AL'TSHULER, N. S., EREMIN, M. V., LUKS, R. K., and STOLOV, A. L. (1970). *Fiz. Tver. Tela* **11**, 3483 (*Sov. Phys.–Sol. State* **11**, 2921).
ANDERSON, C. H., and KISS, Z. J. (1964). *Bull. Am. Phys. Soc.* **9**, 87.
——, WEAKLIEM, H. A., and SABISKY, E. S. (1966). *Phys. Rev.* **143**, 223.
—— and SABISKY, E. S. (1969). *Ibid* **178**, 547.
—— and SABISKY, E. S. (1971). *Physical acoustics*, Vol. 8, eds. Mason and Thurston, Academic Press, Inc., New York and London.
——, CALL, P., STOTT, J. P., and HAYES, W. (1974). (To be published).
ANISIMOV, F., RAKAUSKAS, R., and DAGYS, R. (1969). *Phys. Stat. Sol.* **35**, K75.
—— and DAGYS, R. (1971). *Ibid* **B44**, 821.
——, VALA, A., and DAGYS, R. (1971). *Ibid* **A6**, K15.
ANTIPIN, A. A., KATYSHEV, A. N., KURKIN, I. N., and SHEKUN, L. Ya. (1968). *Sov. Phys. Sol. St.* **9**, 2684–9.
ARKHANGEL'SKAYA, V. A., KISELYEVA, M. N., and SHRAIBER, V. M. (1967). *Opt. Spektrosk.* **23**, 509 (*Opt. Spectrosc.* **23**, 275).
—— and SHRAIBER, V. M. (1968). *Opt. Spektrosk.* **24**, 635 [*Opt. Spectrosc.* **24**, 338].
AXE, J. D. and SOROKIN, P. P. (1963). *Phys. Rev.* **130**, 945.
—— and BURNS, G. (1966). *Ibid* **152**, 331.
BABERSCHKE, K. (1971). *Phys. Lett.* **34A**, 41.
BAKER, J. M. (1968). *J. Phys.* **C1**, 1670.
——, BLAKE, W. B. J., and COPLAND, G. M. (1969). *Proc. R. Soc.* **A309**, 119.
——, DAVIES, E. R., and HURRELL, J. P. (1968). *Ibid* **A308**, 403.
——, HAYES, W., and JONES, D. A. (1959). *Proc. phys. Soc.* Lond. **73**, 942.
—— and HURRELL, J. P. (1963). *Ibid* **82**, 742.
—— and WILLIAMS, F. I. B. (1962). *Proc. R. Soc.* **A267**, 283.
—— and VAN ORMONDT, D. (1973). (To be published).
BESSENT, R. G. and HAYES, W. (1965). *Proc. R. Soc.* **A285**, 430.
BIERIG, R. W. and WEBER, M. J. (1963). *Phys. Rev.* **132**, 164.
BILL, H. (1969). *Phys. Lett* **29A**, 593.
BLEANEY, B. (1964). *Proc. R. Soc.* **A277**, 289.
BLEANEY, B. (1967). *La structure hyperfine magnetique des atomes et des molecules*, p. 13. Colloques Internationaux du C.N.R.S., Paris.
BURNS, G. and AXE, J. D. (1966). *Optical properties of ions in crystals*, ed. H. M. Crosswhite and H. W. Moos, Interscience, New York.
BOATNER, L. A., REYNOLDS, R. W., and ABRAHAM, M. M. (1970). *J. Chem. Phys.* **52**, 1248.
CHASE, L. L. (1970). *Phys. Rev.* **B2**, 2308.
COHEN, E., and GUGGENHEIM, H. J. (1968). *Ibid* **175**, 354.
DETRIO, J. A. (1969). *Ibid* **185**, 494.
DIEKE, G. H. (1968). *Spectra and energy levels of rare earth ions in crystals*, Wiley, New York.
—— and CROSSWHITE, H. M. (1963). *Appl. Optics* **2**, 675.
DUNCAN, R. C. Jr. and KISS, Z. J. (1973). Private communication.
DVIR, M. and LOW, W. (1960). *Proc. Phys. Soc.* **75**, 136.
EREMIN, M. V. (1969). *Opt. Spektrosk.* **26**, 578 (*Opt. Spectrosc.* **26**, 317).
—— (1970). *Opt. Spektrosk.* **29**, 100 [*Opt. Spectrosc.* **29**, 53).
——, ZAKHARCHENYA, B. P., RYSKIN, A. Yu, and STEPANOV, Yu. A. (1970). *Fiz. Tver. Tela* **13**, 1128 [*Sov. Phys.–Sol. State* **13**, 934].
FEOFILOV, P. P. (1956). *Opt. Spektrosk.* **1**, 992.
—— and KAPLYANSKII, A. A. (1962). *Ibid* **12**, 493 (*Opt. Spectrosc.* **12**, 272).
FONG, F. (1964). *J. chem. Phys.* **41**, 2291.
FREISER, M. J., METHFESSEL, S., and HOLTZBERG, F. (1968). *J. appl. Phys.* **39**, 900.

GILFANOV, F. Z., LEUSHIN, A. M., and STOLOV, A. L. (1967). *Fiz. Tver. Tela* **9,** 1357 (*Sov. Phys.–Sol. State* **9,** 1061].
GRANT, W. B., MOLLENBAUER, L. F., and JEFFRIES, C. D. (1971). *Phys. Rev.* **4B,** 1428.
GUGGENHEIM, H. and KANE, J. V. (1964). *App. Phys. Lett* **4,** 172.
HAM, F. S. (1968). *Phys. Rev.* **166,** 307.
—— (1972). *Electron paramagnetic resonance*, ed. S. Geschwind, Plenum, New York.
HAYES, W., JONES, G. D., and TWIDELL, J. W. (1963). *Proc. phys. Soc.* Lond. **81,** 371.
—— and SMITH, P. H. S. (1971). *J. Phys.* **C4,** 840.
—— and TWIDELL, J. W. (1961). *J. chem. Phys.* **35,** 1521.
—— —— (1963). *Proc. phys. Soc.* Lond. **82,** 330.
HERRINGTON, J. R., DISCHLER, B., ESTLE, T. L., and BOATNER, L. A. (1970). *Phys. Rev. Lett.* **24,** 984.
——, ESTLE, T. L., and BOATNER, L. A. (1971). *Phys. Rev.* **B3,** 2933.
——, ——, and BOATNER, L. A. (1972). *Ibid* **B5,** 2500.
——, ——, ——. (1973). *Ibid* **B7,** 3003.
HEYMAN, P. M. (1969). *App. Phys. Lett.* **14,** 81.
HOCHLI, U. T. (1967). *Phys. Rev.* **162,** 262.
HOCHLI, U. T. and ESTLE, T. L. (1967). *Phys. Rev. Lett.* **18,** 128.
HUANG, J. W. and MOOS, H. W. (1968). *J. chem. Phys.* **49,** 2431.
HURREN, W. R., NELSON, H. M., LARSON, E. G., and GARDNER, J. H. (1969). *Phys. Rev.* **185,** 624.
INARI, T. (1968). *J. Phys. Soc. Japan* **25,** 639.
INOUE, M. (1963). *Phys. Rev. Lett.* **11,** 196.
IVANENKO, Z. I. and MALKIN, B. Z. (1970). *Fiz. Tverd. Tela* **11,** 1859 (*Sov. Phys.—Sol. State* **11,** 1498).
—— —— (1972). *Fiz. Tverd. Tela* **14,** 153 (*Sov. Phys.–Sol. State* **14,** 122).
JØRGENSEN, C. K. (1962). *Progress in inorganic chemistry*, ed. F. A. Cotten, Interscience, New York.
——, PAPPALARDO, R., and SCHMIDTKE, H. H. (1963). *J. chem. Phys.* **39,** 1422.
JUDD, B. R. and CROSSWHITE, H. M. (1968). *Phys. Rev.* **174,** 89.
KAISER, W., GARRETT, C. G. B., and WOOD, D. L. (1961). *Phys. Rev.* **123,** 766.
KAPLYANSKII, A. A. FEOFILOV, P. P. (1962). *Opt. Spektrosk.* **13,** 235 (*Opt. Spectrosc.* **13,** 1291).
—— and PRZHEVASKII, A. K. (1962). *Opt. Spektrosk.* **13,** 882 (*Opt. Spectrosc.* **13,** 508).
—— —— (1966). *Opt. Spektrosk.* **20,** 1045 (*Opt. Spectrosc.* **20,** 577).
KASTLER, A. (1951). *C.R. Libd. scare. Acad. Sci, Paris* **232,** 953.
KIEL, A. and SCOTT, J. F. (1970). *Phys. Rev.* **B2,** 2033.
KIRO, D., and LOW, W. (1968). *Phys. Rev. Lett.* **20,** 1010.
—— —— (1970). *Magnetic resonance*, eds. Coogan, C. K., Ham, F. S., Stuart, S. N., Pilbrow, J. R., and Wilson, G. V. N., Plenum Press, New York.
KIRTON, J. and MCLAUGHLAN, S. D. (1967). *Phys. Rev.* **155,** 279.
—— and WHITE, A. M. (1969). *Ibid* **178,** 543.
KISLIUK, P., TIPPINS, H. H., MOORE, C. A., and POLLACK, S. A. (1968). *Ibid* **171,** 336.
KISS, Z. J. (1962). *Ibid.* **127,** 718.
—— (1965). *Ibid* **137,** A1749.
——, ANDERSON, C. H., and ORBACH, R. (1965). *Ibid* **137,** A1761.
—— and DUNCAN, R. C. (1961*a*). *Proc. IRE* **50,** 1531.
—— (1961*b*). *Ibid* **50,** 1532.
—— and YOCOM, P. N. (1964). *J. chem. Phys.* **41,** 1511.

LEA, K. R., LEASK, M. J. M., and WOLF, W. P. (1962). *J. Phys. Chem. Solids* **23,** 1381.
LEWIS, H. R. and SABISKY, E. S. (1963). *Phys. Rev.* **130,** 1370.
LOH, E. (1968). *Ibid* **175,** 533.
—— (1973). *Ibid* **B7,** 1846.
LOW, W. (1958). *Ibid* **109,** 265.
—— (1964). *Ibid* **134,** A1479.
—— and ROSENBURGER, U. (1959). *Ibid* **116,** 621.
MANTHEY, W. J. (1973). *Ibid* **B8,** 4086.
MAKOVSKY, J. (1966). *Phys. Lett.* **19,** 647.
—— (1967). *J. chem. Phys.* **46,** 390.
——, LOW, W., and YATSIV, S. (1962). *Phys. Lett.* **2,** 186.
MALKIN, B. Z., IVANENKO, Z. I., and AIZENBERG, I. B. (1971). *Fiz. Tver. Tela* **12,** 1873 (*Sov. Phys.–Sol. State* **12,** 1491).
MARGERIE, J. (1967). *Physica's Grav.* **33,** 238.
MCCLURE, D. S. and KISS, Z. J. (1963). *J. chem. Phys.* **39,** 3251.
MERGERIAN, D., HARROP, I. H., STOMBLER, M. D., and KRIKORIAN, K. C. (1967) *Phys Rev.* **153,** 349.
MERRITT, F. R., GUGGENHEIM, H., and GARRETT, C. G. B. (1966). *Ibid* **145,** 188.
MERZ, J. L. and PERSHAN, P. S. (1967*a*). *Ibid* **162,** 217.
—— —— (1967*b*). *Ibid* **162,** 235.
MOLLENAUER, L. F., GRANT, W. B., and JEFFRIES, C. D. (1968). *Phys. Rev. Lett.* **20,** 488.
NARA, H. and SCHLESINGER, M. (1971*a*). *Solid State Commun.* **9,** 1247.
—— —— (1971*b*). *Phys. Rev.* **B3,** 58.
NEWMAN, D. J. (1971). *Adv. Phys.* **20,** 197.
O'CONNOR, J. R. and CHEN, J. H. (1964). *Appl. Phys. Lett* **5,** 100.
O'HARE, J. M., DETRIO, J. A., and DONLAN, V. L. (1969). *J. chem. Phys.* **51,** 3937.
—— and DONLAN, V. L. (1969). *Phys. Rev.* **185,** 416.
ORBACH, R. and STAPLETON, H. J. (1972). *Electron paramagnetic resonance*, ed. S. Geschwind, Plenum, New York.
OVSYANKIN, V. V. and FEOFILOV, P. P. (1966*a*). *Zh. eksper. teor. Fiz.* **3,** 494 (*Sov. Phys. JETP Letters* **3,** 322).
—— —— (1966*b*) *Zh. eksper. teor. Fiz.* **4,** 471 [*Sov. Phys. JETP Letters* **4,** 317].
PHILLIPS, W. and DUNCAN, R. C. (1971). *Metal. Trans.* **2,** 769.
PIPER, J. S., BROWN, J. P., and MCCLURE, D. S. (1967). *J. chem. Phys.* **46,** 1353.
RABBINER, N. (1967). *J. opt. Soc. Am.* **57,** 217.
RANON, U. (1964). *J. Phys. Chem. Solids* **25,** 1208.
—— and HYDE, J. S. (1966). *Phys. Rev.* **141,** 259.
—— and LOW, W. (1963). *Ibid.* **132,** 1609.
—— and YARIV, A. (1964). *Phys. Lett.* **9,** 17–19.
RECTOR, C. W., PANDEY, B. C., and MOOS, H. W. (1966). *J. chem. Phys.* **45,** 171.
REYNOLDS, R. W. and BOATNER, L. A. (1970). *Ibid* **52,** 3851.
RISEBERG, L. A. and MOOS, H. W. (1968). *Phys. Rev.* **174,** 429.
RYTER, Ch. (1957). *Helv. Phys. Acta* **30,** 353.
RUNCIMAN, W. A. and STAGER, C. V. (1962). *J. chem. Phys.* **37,** 196.
SABISKY, E. S. (1964). *Ibid* **41,** 892.
— (1966). *Phys. Rev.* **141,** 351.
—— and ANDERSON, C. H. (1964). *Phys. Rev. Lett.* **13,** 754.
—— —— (1966). *Phys. Rev.* **148,** 194.
—— —— (1967). *IEEE Trans.* **QE-3,** 287.
—— —— (1970). *Phys. Rev.* **B1,** 2028.

SABISKY, E. S. and ANDERSON, C. H. (1973) *Ibid* **A7,** 790.
SCHLESINGER M. and NERENBERG, M. (1969). *Phys. Rev.* **178,** 568–71.
SHEN, Y. R. (1964). *Ibid* **134,** A661.
SIERRO, J. (1963). *Helv. Phys. Acta* **36,** 505.
SMITH, W. V. and SOROKIN, P. P. (1966). *The laser*, McGraw-Hill, New York.
SOROKIN, P. P. and STEVENSON, M. J. (1961). *IBM J. Res. Develop.* **5,** 56.
—— —— LANKARD, J. R., and PETTIT, G. D. (1962). *Phys. Rev.* **127,** 503.
SROUBEK, Z., TACHIKI, M., ZIMMERMAN, P. H. and ORBACH, R. (1968). *Phys. Rev.* **165,** 435.
STAROSTIN, N. V. (1967). *Opt. Spektrosk.* **23,** 486 [*Opt. Spectrosc.* **23,** 260].
STEDMAN, G. E. and NEWMAN, D. J. (1971). *J. Phys. Chem. Solids* **32,** 2001.
SUGAR, J. (1970). *J. opt. Soc. Am.* **60,** 454.
TITLE, R. S. (1963). *Phys. Lett.* **6,** 13.
TRAMMELL, G. T. (1963). *Phys. Rev.* **131,** 932.
VAKHIDOV, SH. A., KAIPOV, B., and TAVSHUNSKII, G. A. (1970). *Opt. Spektrosk.* **28,** 949 (*Opt. Spectrosc.* **28,** 515).
VALENTIN, R. (1969). *Phys. Lett.* **30A,** 344.
VETRI, G. and BASSONI, F. (1968). *Nuovo Cimento* **55B,** 504.
VINCOW, G. and LOW, W. (1960). *Phys. Rev.* **122,** 1390.
VINOGRADOV, E. A., ZVEREVA, G. A., IRISOVA, N. A., MANDEL'SHTAM, T. S., PROKHOROV, A. M., and SHMONOV, T. A. (1969). *Fiz. Tver. Tela* **11,** 335 (*Sov. Phys.–Sol. State* **11,** 268).
VORONKO, YU. K., DENKER, B. I., and OSIKO, V. V. (1971). *Fiz. Tver. Tela* **13,** 178 [*Sov. Phys.–Sol. State* **13,** 141].
——, DMITRUK, M. V., OSIKO, V. V., and SHCHERBAKOV, I. A. (1971). *Fiz. Tver. Tela* **13,** 1611 (*Sov. Phys.–Sol. State* **13,** 1348).
——, OSIKO, V. V., and SHCHERBAKOV, I. A. (1969). *Zh. eksp. teor. Fiz.* **56,** 151 (*JETP Sov. Phys.* **29,** 86).
WAKIM, F. G., SYNEK, M., GROSSGUT, R., and DAMOMMIO, A. (1972). ***Phys. Rev.*** **A5,** 1121.
WATSON, R. E. and FREEMAN, A. J. (1962). *Ibid* **127,** 2058.
—— —— (1967). *Ibid* **156,** 251.
WEAKLIEM, H. A. (1972). *Ibid* **B6,** 2743.
—— (1973). (see Hayes W. and Smith, P. H. S. (1971)).
——, ANDERSON, C. H., and SABISKY, E. S. (1970). *Phys. Rev.* **B1,** 4354.
—— and KISS, Z. J. (1967). *Ibid* **157,** 277.
WEBER, M. J. (1968). *Ibid* **171,** 283.
—— and BIERIG, R. W. (1964). *Ibid* **A134,** 1492.
WOOD, D. L. and KAISER, W. (1962). *Ibid* **126,** 2079.
WYBOURNE, B. G. (1966). *Ibid* **148,** 317.
YANASE, A. and KASUYA, T. (1970). *Supp. Prog. theor. Phys.* **46,** 388.
ZAKHARCHENYA, B. P., MAKAROV, V. P., and RYSKIN, A. YU. (1964). *Opt. Spektrosk.* **17,** 219 (Opt. Spectrosc. **17,** 116).
——, RUSANOV, I. B., and RYSKIN, A. Y. (1965). *Opt. Spektrosk.* **18,** 999 (*Opt. Spectrosc.* **18,** 563).
—— —— (1966). *Fiz. Tver. Tela* **8,** 41 (*Sov. Phys.–Sol. State* **8,** 31).
—— and RYSKIN, A. YU. (1962). *Opt. Spektrosk.* **13,** 875 (*Opt. Spectrosc.* **13,** 501).
—— —— (1963). *Opt. Spektrosk.* **14,** 309 (*Opt. Spectrosc.* **14,** 163).
ZVEREVA, G. A. and MAKAROV, V. P. (1967). *Fis. Tver. Tela* **9,** 2994 (*Sov. Phys.–Sol. State* **9,** 2356].

6

RARE-EARTH AND OTHER IMPURITY IONS IN NON-CUBIC SITES

6.1. Introduction

WHEN a lattice ion is replaced by an impurity ion of the same charge the lattice symmetry at the impurity site may not be disturbed. In CaF_2, for example, the Ca^{2+} site has O_h symmetry, so that simple substitution of an impurity ion at the Ca^{2+} site leads to an environment of cubic symmetry. Situations of this sort were discussed for paramagnetic impurity ions in the previous chapter, and particular emphasis was placed in that chapter on crystal-field effects. There are a number of reasons why the symmetry of an impurity site may be lower than that of the lattice ion replaced, and in this chapter we discuss these lower symmetry situations. We shall be concerned here more with general solid state aspects of the impurity-lattice system rather than with crystal-field effects, though a study of low-symmetry-site structures does involve a study of crystal-field effects. We shall be concerned entirely with magnetic ions, primarily rare-earth ions.

The substitution for Ca^{2+} of an ion of different charge requires the incorporation somewhere in the crystal of an additional compensating charge to preserve overall electrical neutrality. This additional charge may be remote from the impurity ion, in which case it does not perturb the cubic symmetry of the site of the impurity. However, if the charge compensation is local the symmetry of the site is lowered.

Even divalent substitutional ions occur in sites with non-cubic symmetry. This may be either because the ion is associated with some other type of centre or defect in the solid, or because the disparity in size between the ion and the available site is so great as to allow low-symmetry distortion of the structure or movement of the impurity ion away from the position of high symmetry.

Such charge compensation and distortion effects occur generally when impurity ions are incorporated in a solid, but they have probably been studied more intensively in alkaline earth fluorides than in any other material, particularly for rare-earth-impurity ions. A large amount of experimental evidence has been collected on the nature of these distorted sites, mainly by epr and endor and to a lesser extent by optical spectroscopy.

The lowering of the symmetry of the site from cubic introduces additional terms into the crystal field. A considerable variety of related crystal-field parameters have been used in the literature. We shall standardize this variety and quote values of $A_n^m\langle r^n\rangle$ for reasons given in § 5.2.1. The relations between the various parameters which are used in the literature are discussed by Hutchings (1964). The cubic crystal field (§ 5.2.1) involves only two parameters $A_4\langle r^4\rangle$ and $A_6\langle r^6\rangle$ for *f* electrons; and as we shall be mainly concerned with rare-earth ions we shall discuss the problem in terms of *f*-electrons: for *d*-electrons one may set the sixth-order terms to zero. One commonly finds sites of tetragonal symmetry about $\langle 001\rangle$ requiring five parameters, $A_2^0\langle r^2\rangle$, $A_4^0\langle r^4\rangle$, $A_4^4\langle r^4\rangle$, $A_6^0\langle r^6\rangle$, and $A_6^4\langle r^6\rangle$, and sites of trigonal symmetry about $\langle 111\rangle$ requiring six parameters, $A_2^0\langle r^2\rangle$, $A_4^0\langle r^4\rangle$, $A_4^3\langle r^4\rangle$, $A_6^0\langle r^6\rangle$, $A_6^3\langle r^6\rangle$, and $A_6^6\langle r^6\rangle$. Numerous cases of lower symmetry have also been observed requiring even more crystal-field parameters.

Crystal-field parameters are usually deduced from the energy levels of the ground configuration determined by optical absorption and fluorescence. In cubic symmetry it is possible to evaluate the two crystal-field parameters from a knowledge of the energies of the lowest J manifold alone, except for Ce^{3+} and Sm^{3+} where the ground $J = \frac{5}{2}$ state is split into only two levels. To calculate the larger number of parameters for the lower-symmetry sites requires more information. The lower-symmetry crystal field raises the degeneracy of Γ_8 quartets and Γ_4 and Γ_5 triplets, so that there are more energy levels; and also the lowering of the symmetry relaxes the selection rules, so that generally there are more optical lines for lower-symmetry sites than for cubic sites. For most ions even a complete determination of the energy levels of the lowest J manifold does not give sufficient information to determine the crystal-field parameters; it is necessary to have, in addition, information about energy levels of excited J manifolds, or information about the eigenfunctions which describe states, which can be obtained from optical-Zeeman spectroscopy or from epr. For trigonal and lower symmetries, for example, it is not possible to evaluate the crystal-field parameters even from a complete knowledge of the energies of all the states of the configuration f^1 (Ce^{3+}) and f^{13} (Yb^{3+} and Tm^{2+}).

In contrast to optical spectroscopy, normal epr gives information about only the ground state, and it is often difficult to deduce more than the symmetry of the site from this information; nevertheless § 6.5 on epr measurements is a long one simply because a very large number of sites have been investigated in this way. For some sites endor has been used to construct detailed models of the local surroundings of an impurity ion and the charge-compensation mechanism.

In § 6.2 we discuss some of the problems of determining the nature of impurity sites which are common to all of the methods. In the remaining sections of the chapter we discuss the details of non-cubic sites which have been elicited by different measurement techniques: in § 6.3 by endor, where only a few sites have been studied but very detailed information is obtained; in § 6.4 by optical measurement, for which only a few ions have been investigated thoroughly, mainly in tetragonal sites; and in § 6.5 by epr and spin-lattice relaxation, by means of which most of the non-cubic sites have been investigated. Section 6.6 discusses some epr measurements of interactions between pairs of neighbouring paramagnetic ions.

The experimental techniques discussed in this chapter are the same as those covered in Chapter 5, and the paramagnetic ions considered are largely the same. The main difference between this chapter and the previous one is simply in the lower symmetry of the site of the paramagnetic ion. Although the perturbation of the cubic site on going to lower symmetry is often a large one, the properties of the cubic site form a valuable starting point for discussion of the lower symmetry sites. Much of the basic theory for this chapter has been discussed in Chapter 5, and the reader would benefit from a previous reading of that chapter.

6.2. The structure of impurity sites

6.2.1. *Problems of determining structures*

Whatever the mechanism which produces a low symmetry site, and whatever the symmetry of the individual sites, the crystal preserves its overall cubic symmetry, so that exactly equivalent but differently oriented sites will occur related by the rotations of the cubic space group of the crystal. For example there are three similar tetragonal sites with principal axes [001], [010], and [100]. For trigonal sites there are four equivalent $\langle 111\rangle$ directions of the threefold axis. For orthorhombic sites with axes along twofold axes of the cubic space group such as [001], [110], and $[\bar{1}10]$ there are six equivalent groups, and for those with one axis along $\langle 110\rangle$ but the other two not along symmetry directions in the $\{110\}$ plane there are twelve equivalent groups.

In the absence of an applied perturbation (uniaxial stress, electric or magnetic field) any set of equivalent but differently oriented sites, such as the three tetragonal sites, have identical energy levels so that the optical spectra are superimposed. The superposition of the lines from the differently oriented sites in zero applied field does not allow one to use the characteristic polarization of the optical spectra for low-symmetry sites to aid the interpretation of the spectra. Therefore it is not possible to recognize the symmetry of the site from zero-field measurements, and if sites of different symmetry occur simultaneously it is not possible to differentiate the optical lines belonging to different types of site. Hence the analysis of optical spectra is very difficult.

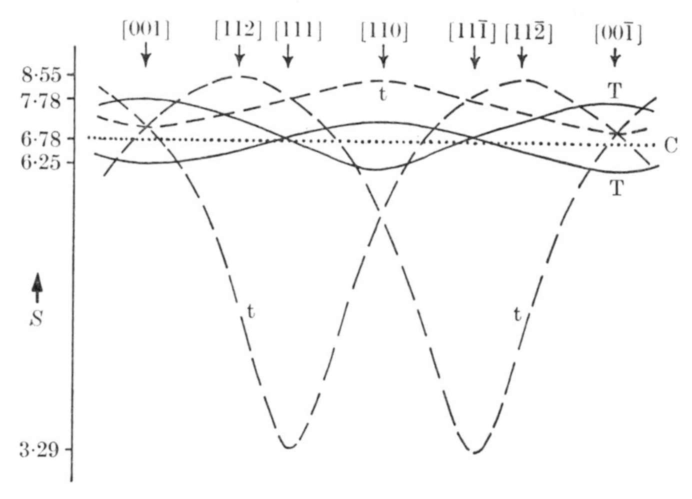

Fig. 6.1. Angular variation of the epr spectrum of Er^{3+} in CaF_2 in the [110] plane, taken from Rector, Pandey, and Moos (1966). The lines are labelled with their symmetry: C = cubic, T = tetragonal, and t = trigonal.

At low-symmetry sites the epr spin-Hamiltonian, involving Zeeman and hyperfine interactions, and residual Stark splittings for S-state ions, has the same symmetry as that of the site. Hence the angular variation of epr lines as a function of the direction of the external field has the symmetry of the site. If this angular variation can be followed as the magnetic field is rotated the assignment of site symmetries is fairly easy. Fig. 6.1 shows an example of the angular variation of cubic, tetragonal and trigonal sites in the same crystal. However, when sites of different symmetry, or several different sites with the same point symmetry but different spin-Hamiltonian parameters, occur in the same specimen the spectrum can be very complex and difficult to interpret. If the anisotropy is not very great the contributions from differently oriented sites of the same type may not be resolved from one another; in this case one obtains an isotropic epr line with perhaps an angular variation of linewidth, which is indistinguishable from that due to a cubic site. It is probable, at least for rare-earth ions, that sites where the charge-compensating ion is beyond $2a_0$ have too small an anisotropy for them to be resolved. Hence the "remote" charge compensation needed for an apparently cubic site does not in fact have to be very remote (see also the end of § 6.2.2).

Samples prepared in different ways contain different sites in differing concentrations. These differing concentrations are reflected in different intensities of both epr and optical lines. If, upon varying some external parameter such as the concentration of impurities, heat treatment or irradiation, the relative intensities of some optical transitions remain unchanged even though the absolute intensity changes considerably, there is a strong presumption that these lines correspond to the same centre. A similar constancy of the relative intensity of epr and optical lines also suggests that the lines originate in the same centre. However, such correlation can be misleading as the concentration of two different centres may be caused to change in the same way by the variation of external parameters.

The symmetry of the sites responsible for optical lines may be found by studying the optical-Zeeman spectrum, as this shows the same symmetry as the epr spectrum. The spectra are more complicated than epr spectra because the Zeeman spectrum of an optical line involves the Zeeman splitting of both levels between which the optical transition occurs. However, such measurements do form an unambiguous way of assigning site symmetries to spectral lines. A further method is to reduce the site symmetry by applying external uniaxial stress or an electric field and observing the effect upon the optical lines. Beaumont, Harmer, and Hayes (1972) have shown how the use of both uniaxial stress and electric field may enable one to determine the symmetry of the site of a defect (see § 4.3.3).

The identification of optical lines corresponding to a particular site may enable one to calculate the energies of the crystal-field levels of some of the J manifolds, though the common occurrence of vibronic sidebands require care in the interpretation of the spectral lines. The difficulty is compounded by the fact that vibrational energies and crystal-field energies for rare-earth ions are of comparable magnitude.

Information about the energy Δ of the first excited state is sometimes available from the observation of an Orbach relaxation process giving a spin-lattice relaxation rate proportional to $\exp(-\Delta/kT)$ for the epr transitions (Finn, Orbach, and Wolf, 1961). However, the interpretation of relaxation rates must be made very cautiously when there are ions with different site symmetries present, as cross relaxation could mean that the relaxation rate of one site was in fact determined by the spin-lattice relaxation rate for a site of different symmetry, or even by that of an unrelated impurity.

The g-values in the Zeeman spin-Hamiltonian both for ground and excited states depend upon the admixtures of $|J, M_J\rangle$ states, which in turn depend upon the crystal-field parameters. Hence the g-values determined by epr and optical-Zeeman spectroscopy may be used together with the optically determined energies of the excited states to evaluate the crystal-field parameters.

The great difficulty in obtaining enough information to make a complete analysis of the crystal field is indicated by the fact that there are very few examples where it has been done.

Identification of the symmetry of a site, and even complete evaluation of the crystal-field parameters, does not indicate the cause of the reduced symmetry. In much of the literature the evidence used to propose models for sites of certain symmetries is highly circumstantial. In many cases the presence or absence of certain ions is inferred from the way in which the samples were prepared. Such assignments are little more than inspired guesses, and they have sometimes led to incorrect models.

There are some other pieces of evidence which may be used in constructing a model for a site. If for axial symmetry the $A_2^0\langle r^2\rangle$ term is dominant the sign of the g-value anisotropy, $(g_{\parallel}-g_{\perp})$, depends upon the sign of $A_2^0\langle r^2\rangle$. This has often been used as an indication of the sign of the effective compensating charge, but such arguments must be used very cautiously (see § 6.3).

Further information can sometimes be obtained from the hyperfine structure on the epr line. A consequence of the lowering of the symmetry of the site of the paramagnetic ion is that the site symmetries of neighbouring diagmagnetic ions are also lowered, affecting transferred-hyperfine-interaction tensors. Sometimes resolved transferred hyperfine structures have helped in the identification of centres by suggesting the presence of particular nuclei, or groups of equivalent nuclei. However, in many cases the transferred hyperfine structure is not resolved at all, or not sufficiently resolved for it to be interpreted.

The only reliable way to obtain precise information about the surroundings of the paramagnetic ion is to use endor. This not only identifies the nature of any nearby nuclei with non-zero nuclear spin, but also measures the magnitude of the magnetic field which the paramagnetic ion exerts at the nuclear site. Hence one can find the positions of the nuclei in the crystal structure and measure the amount of covalent bonding between the diamagnetic and the paramagnetic ions. A fuller discussion of the use of endor and some of the sites it has been used to identify is given in § 6.3.

In the following sections we shall discuss the large mass of data which has been collected on non-cubic sites. Unfortunately the quantity of data is large because of the very large number of different types of

TABLE 6.1

Some possible mechanisms for producing non-cubic sites with various symmetries

	Tetragonal
1.	Displacement of paramagnetic ion along (100)
2.	Interstitial at $(00\frac{1}{2})$
3.	Interstitials at both $(00\frac{1}{2})$ and $(00-\frac{1}{2})$
4.	Substitution or vacancy at (001)
5.	Interstitial at $(00\frac{3}{2})$
	Trigonal
1.	Displacement of paramagnetic ion along (111)
2.	Substitution or vacancy at $(\frac{1}{2}\frac{1}{2}\frac{1}{2})$
3.	Simultaneous substitution or vacancies at $(\frac{1}{2}\frac{1}{2}-\frac{1}{2})$, $(\frac{1}{2}-\frac{1}{2}\frac{1}{2})$, and $(-\frac{1}{2}\frac{1}{2}\frac{1}{2})$
4.	Interstitial at (111)
5.	Substitution at $(\frac{3}{2}\,\frac{3}{2}\,\frac{3}{2})$
	Rhombic (simple)
1.	Displacement of paramagnetic ion along (011)
2.	Substitution at (011)
	Rhombic (complex)
1.	Interstitial at (221) (rather far away)
2.	Substitution at $(\frac{1}{2}\frac{1}{2}\frac{3}{2})$

site, but there are very few sites about which sufficient is known for a complete characterization to be made. Many of the sites have been observed by only one set of workers, and many have not been studied completely.

6.2.2. *Some typical structures*

It is perhaps valuable, in advance of the discussion of individual examples to consider some of the possibilities which have been suggested for producing sites of the observed symmetries. Table 6.1 lists some of these, but it does not attempt to be exhaustive. One could invent much more complicated arrangements with the appropriate symmetry, and there certainly are cases observed where all of the nearest eight fluorine ions are replaced by other ions (see § 6.3).

There is no standard notation in the literature for the sites we wish to discuss. We therefore propose to label tetragonal sites Tg, trigonal sites Tr and rhombic sites Rh. Where more than one site of a particular symmetry occurs for a particular ion we shall use a numerical subscript to distinguish them, but this does not imply any relation between similarly numbered sites for different ions. Any evidence which indicates that sites for different ions may be similar will be discussed in the text. If sites are characterized by more details than their symmetry alone this will be indicated by information given in a bracket about the nature of the site (e.g. the Tg(F_i^-) site discussed in the next paragraph).

Two simple sites, one of which occurs for most of the rare-earth ions and the other of which probably occurs for many of them, have been positively identified by endor (see § 6.3). In the first which we call Tg(F_i^-) an interstitial F^- ion occupies the nearest interstitial site at ($00\frac{1}{2}$), forming a tetragonal site. In the other a trigonal site Tr(O_s^{2-}) is formed by the substitution of an O_s^{2-} ion for a *nn* F^- ion.

The concentrations of magnetic impurity ions of various symmetries depend upon the details of the sample preparation and subsequent heat treatment, and also upon the total concentration of the magnetic impurity and other impurities in the melt. The incorporation of impurities has been discussed in § 3.2. A number of workers have studied the relative concentration of sites of different symmetry produced under various conditions and have discussed mechanisms to account for their observations (Voron'ko, Osiko, and Shcherbakov, 1969; Brown, Roots, Williams, Shand, Groter, and Kay, 1969; Gil'fanov, Livanova, Orlov, and Stolov, 1970; Miner, Graham, and Johnston, 1972). There is some evidence cited by Miner *et al.* that cubic sites may not be produced only by distant charge compensation, but also that small regions of a separate phase may be formed in which substitutional trivalent impurity ions and interstitial F_i^- ions are incorporated in a new cubic crystal structure (see also O'Hare, 1972).

The stability of a site may depend, among other factors, upon the radii of the substitutional ion and the ion it replaces. There is a steady decrease in radius through the rare-earth ions (the lanthanide contraction) from 1·02 Å for Ce^{3+} to 0·94 Å for Yb^{3+} (Evans 1966); the same source gives the radii of the alkaline earth ions to be 0·99 Å for Ca^{2+}, 1·13 Å for Sr^{2+} and 1·35 Å for Ba^{2+}. The site Tg(F_i^-) is known to occur for Yb^{3+} and Ce^{3+} and for several intermediate ions in CaF_2, and does not appear to occur at all in BaF_2. Brown, Roots, Williams, Shand, Groter, and Kay (1969) have shown, by studying SrF_2 crystals doped with rare-earth ions under conditions where oxygen is excluded, that the non-cubic sites are tetragonal for the large rare-earth ions and are trigonal for the smaller ones, with a change over around Dy^{3+}. They account for this using a model in which the surrounding eight F^- ions contract onto the rare-earth ion by an amount which depends upon Δr, the difference in ionic radius between the alkaline earth and the rare-earth ions. Whether the site $(00\frac{1}{2})$ or $(\frac{1}{2}\frac{1}{2}\frac{1}{2})$ is energetically favourable for an interstitial F^- ion depends upon this contraction and so upon the radius of the rare-earth ion. An electrostatic model predicts the change over from tetragonal to trigonal at Pr^{3+}/Nd^{3+} for BaF_2, at Dy^{3+}/Ho^{3+} for SrF_2 and not at all for CaF_2. This model does seem to be in rough accord with the observed occurrence of tetragonal and trigonal sites, though no positive identification has been made of a trigonal site with an interstitial F_i^- ion at $(\frac{1}{2}\frac{1}{2}\frac{1}{2})$.

6.3. Identification of sites using endor

It is possible to discuss low-symmetry sites in terms of the magnitudes of the parameters of the crystal field, without any knowledge of the atomic arrangement around the paramagnetic ion. However, the atomic arrangement is of interest and importance, and most references speculate about the atomic arrangement even when there is virtually no evidence except for the symmetry of the site. In view of this we shall first discuss those sites which have been definitely identified, in order to be able to compare them with less well characterized sites in later discussion.

Endor is an ideal method for obtaining information about the atomic arrangement around the paramagnetic ion (Baker, Davies, and Reddy, 1972) as it investigates the nmr of those nuclei which are close to the paramagnetic ion. The terms in the spin-Hamiltonian which involve the nuclear spin operators are given in the equation

$$\mathscr{H} = g_N\beta_N \mathbf{H} \cdot \mathbf{I} + \mathbf{S} \cdot \mathbf{A} \cdot \mathbf{I}. \tag{6.1}$$

We have here restricted ourselves to those terms which are present when $S = I = \frac{1}{2}$; they are also the major terms when S or I is greater than $\frac{1}{2}$. In general there are $2I$ transitions with selection rules $\Delta I_z = \pm 1$, $\Delta S_z = 0$ for each value of S_z; for $S = I = \frac{1}{2}$ there are two transitions which give sufficient information to evaluate g_N and $\mathbf{A}$. If $g_N\beta_N H \gg \mathbf{A}$

these lines are equally spaced about $\nu_N = g_N \beta_N H/h$, the nmr frequency of the free nucleus, and are separated by **A**; for smaller values of H the off-diagonal elements of **S . A . I** may produce asymmetric shifts. Fig. 6.2 shows the endor spectrum of the tetragonal site for Ce^{3+} in CaF_2 around the free ^{19}F nmr frequency; this shows clearly the symmetrical disposition of the endor of second and third shell F^- ions where **A** is relatively small, but considerable asymmetry for *nn* F^- ions where **A** is relatively large. A study of the angular dependence of the endor transitions upon the direction of the external field H enables one to

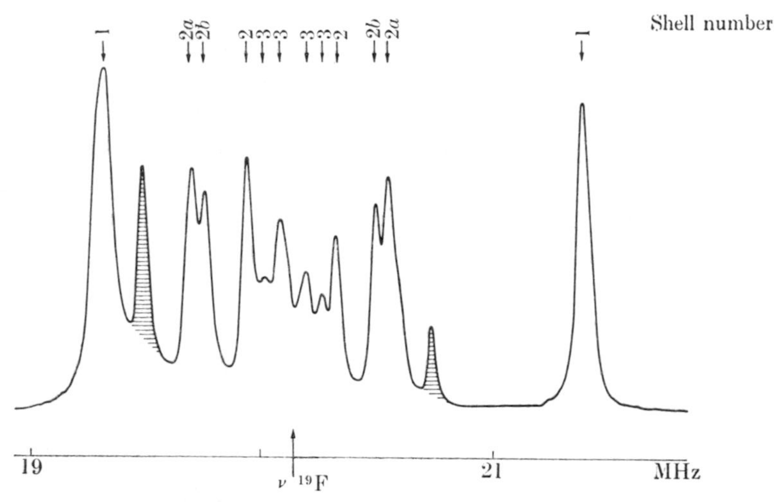

FIG. 6.2. The endor spectrum of ^{19}F nuclei for $Tg(F_i^-)$ sites of Ce^{3+} in CaF_2, taken from Baker, Davies, and Reddy (1972). The lines originating from different shells of nuclei around the Ce^{3+} ion are labelled, and those originating from the interstitial ion are shaded. The free ^{19}F nmr frequency is indicated at the bottom of the diagram.

evaluate all of the components of the **A** tensor. The symmetry of the **A** tensor reflects the symmetry of the bond between the paramagnetic ion and the nucleus, and so may depend both upon the symmetry of the site of the paramagnetic ion and the direction of the line joining the ion to the nucleus.

The measurement of g_N uniquely identifies the nucleus, and the measurement of **A** gives information from which its position may be deduced. Endor lines tend to be very narrow, so that each nucleus gives a separate, well resolved spectrum. In high-symmetry directions such as that shown in Fig. 6.2, the lines from several nuclei may be coincident, but small rotations of the direction of **H** cause these lines to split up. By a careful measurement of the angular variation of the endor lines it is often possible to construct a complete map of the surroundings of the paramagnetic ion. The interpretation is complicated by the fact that there are two contributions to **A**. One arises from purely dipolar interaction between the known magnetic moment of the paramagnetic ion and that of the nucleus, and may be calculated precisely in terms of the position of the nucleus. The other, due to covalent bonding, is much more difficult to interpret.

For ions with well-localized unpaired electrons, such as rare-earth ions, the covalent interaction is appreciable only for ligand ions. For iron-group ions the effects of covalency extend further from the paramagnetic ion, and for loosely bound electrons such as those in *F* centres the covalent contributions may be appreciable out to quite distant nuclei (see § 4.2).

For rare-earth nuclei the components of **A** for nuclei beyond the ligand shell are almost entirely due to dipolar interaction, so that it is possible to deduce the position of the nucleus from the measured **A** values to quite high precision. The precision is limited both by experimental inaccuracy and uncertainty in the covalent contribution. The order of magnitude of the latter may be estimated by finding the sum of the values of A_{qq}/g_q for the three principal directions q of the g-tensor of the paramagnetic ion. For purely dipolar interaction this sum is zero; if not, the sum divided by $\sum_q g_q^{-1}$ gives an indication of the covalent contribution to **A**. For ligand nuclei the covalent interaction generally enhances the dipolar contribution to **A**, so that the endor transitions due to ligand nuclei can be recognized by **A** values somewhat larger than one would expect for purely dipolar interaction. Fig. 6.2 shows an extreme example where the ligand endor lines would be coincident with the free ^{19}F nmr frequency ν^{19}F for purely dipolar interaction, and those of the interstitial F^- ion would be considerably closer to ν^{19}F. Further, the absence of any **A** values larger than those expected for ligand nuclei with dipolar interaction indicates the absence of ligand nuclei with nuclear spin. In this way, for example, van Gorkom (1970) showed that there are no *nn* F^- ions for type II cubic centres of Mn^{2+} in CaF_2, and Reddy, Davies, Baker, Chambers, Newman, and Ozbay (1971) showed that there are no *nn* F^- ions for one of the trigonal centres, and only one *nn* F^- ion for another trigonal centre, of Yb^{3+} in CaF_2 (see § 6.3.1).

6.3.1. *Examples of sites measured by endor*

The site which we have called $Tg(F_i^-)$ has been unambiguously identified by endor. ^{19}F endor indicates that there are ^{19}F nuclei close to all of the first, second, and third-shell F^- ion sites around the rare-earth-impurity ion, and in addition shows the two endor lines from an interstitial F^- ion. The presence of the interstitial F^- ion (and H_i^- and D_i^- ions at the same site) in the common $Tg(F_i^-)$ site was recognized by the relatively large **A** value, indicating that although it is not in the *nn* shell, it cannot be further away than the nearest interstitial site along ⟨100⟩. The symmetry of its angular variation shows that it is situated along the ⟨100⟩ axis. The presence of the interstitial F^- ions has been confirmed by endor for $Tg(F_i^-)$ sites of Ce^{3+}, Nd^{3+}, Yb^{3+}, and U^{3+} (see e.g. Kiro and Low, 1970).

It is also possible to prepare samples in which the charge compensation is achieved by an H_i^- ion instead of F_i^- at the nearest interstitial site ($\frac{1}{2}00$). Endor measurements on this site for Ce^{3+} are reported by Baker, Davies, and Reddy (1969).

Recently a considerable amount of work has been done on the infra-red spectra due to local modes of vibration of impurity ions in alkaline-earth halides. This type of investigation is discussed in § 2.5.5 where it is shown that the local modes of H_i^- and D_i^- ions confirm the model of the $Tg(H_i^-)$ site for all rare-earth ions in CaF_2.

The power of the endor method of investigating low-symmetry centres is very well illustrated by measurements on trigonal centres of Yb^{3+} in CaF_2. One centre, referred to as T_2 by McLaughlan and Newman (1965), shows only seven F^- ions in the first shell; further fluorine shells are complete. Oxygen endor in crystals prepared with ^{17}O substituted for natural oxygen confirms that the missing *nn* site is occupied by an O_s^{2-} ion. Its position in the first shell is indicated by sizeable covalent contribution to both the **A** tensor and the quadrupole interaction tensor (Reddy, Davies, Baker, Chambers, Newman, and Ozbay, 1971). As it is well characterized we shall refer to the T_2 centre as $Tr(O_s^{2-})$.

Another centre, referred to as T_1 by McLaughlan and Newman, shows complete outer F^- shells but only one F^- ion, along the trigonal axis, in the first shell. Endor in ^{17}O substituted specimens shows that there is an O_s^{2-} ion on the trigonal axis on the opposite side of the Yb^{3+} ion in the *nn* shell, and three other off-axis O_s^{2-} ions in the first shell. Fig. 6.3 shows a possible model of this site, though exact atomic positions cannot be determined. The site produces complete charge compensation, but is rather unexpected and certainly could not be predicted from epr measurements alone. As this T_1 centre involves four substitutional oxygen ions we shall label it $Tr(O_{4s}^{2-})$.

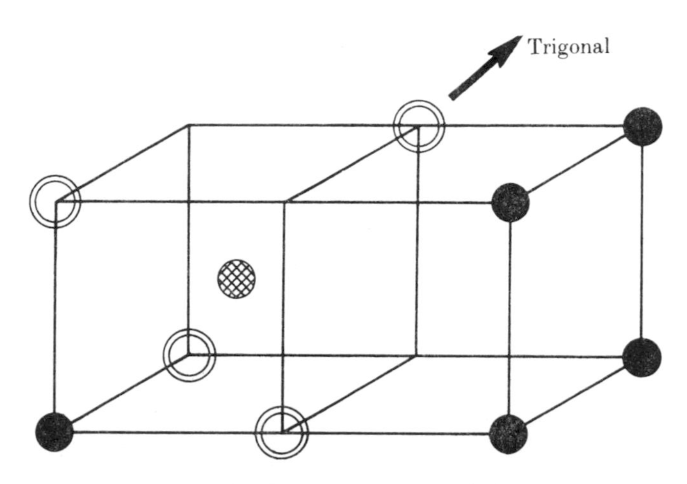

FIG. 6.3. A possible model for the T_1 centre of Yb^{3+} in CaF_2, with only one F^- nearest neighbour along the trigonal axis indicated by the heavily shaded circle. The double circles represent O^{2-} ions, one of which lies along the trigonal axis opposite to the F^- ion, and the other three form an equilateral triangle whose plane is perpendicular to the trigonal axis. Quite large displacements of the various ions could occur while still preserving the trigonal symmetry.

The site called T_{3H} by McLaughlan and Newman was shown by endor (Reddy *et al.*, 1971) to be similar to $Tr(O^{2-}_{4s})$ except that the remaining F^- ion in the *nn* shell is replaced by H^-, so that the eight *nn* F^- ions are replaced by $O^{2-}_4H^-$, and we label this site $Tr(O^{2-}_{4s}, H^-_s)$. The optical absorption due to the local mode of the H^-_s ion has been observed for this centre (see e.g. Newman, 1969). Such local modes have also been observed for RE^{2+}–H^- trigonal centres, in which one F^- ion in the *nn* shell is replaced by H^-, but endor has not yet been reported in this type of centre (see § 6.5.9 for a discussion of epr in this centre).

Trigonal sites have also been investigated in ThO_2 and CeO_2. Here Th^{4+} ions are replaced by trivalent rare-earth ions so the sign of the compensating charge required is opposite to that in CaF_2. Two types of trigonal sites are formed by Dy^{3+}, Er^{3+}, and Yb^{3+} substitution; these depend upon the method of sample preparation, the details of which are listed later (p. 382) in Table 6.9 (Zonn, Katyshev, Mitrofanov, and Pol'skii, 1969). The sites labelled Tr_1 exhibit a two-line transferred-hyperfine-structure indicating some sort of ligand substitution, as ThO_2 and CeO_2 have only a very low abundance of isotopes with nuclear spin $I \neq 0$. Endor (Er^{3+} in CeO_2 by Zonn *et al.*, 1969, and Yb^{3+} in ThO_2 by Reddy, unpublished) confirms that one of the O^{2-} ligands is replaced by an F^-

ion making the site Tr(F^-). The sites labelled Tr_2 show no transferred-hyperfine-structure and no endor, showing that the charge-compensation mechanism is either remote or involves ions without nuclear spin. The trigonal Er^{3+} site in ThO_2 was attributed by Michoulier and Wasiela (1971) to a divalent ion (possibly Pb^{2+} from the flux) in an interstitial ($\frac{1}{2}\frac{1}{2}\frac{1}{2}$) site, but the close similarity of the g-values to those of T_1 in CeO_2 makes it likely that it corresponds to a Tr(F^-) site and that the transferred-hyperfine-structure is unresolved in the rather large line width.

These examples show that a great deal of information about the surroundings of the paramagnetic ion can be obtained from endor measurements, but these measurements are laborious and time consuming, and this probably accounts for the limited use which has so far been made of the technique. As an example of the amount of detail which can be obtained by endor we return to Tg(F_i^-), the tetragonal sites produced by an F^- ion in the nearest interstitial site along $\langle 100 \rangle$. A careful study of the endor of the other F^- ions around the paramagnetic ion, both for Ce^{3+} and Yb^{3+} in CaF_2, has shown that the large interstitial F^- ion pushes its neighbouring F^- ions radially outwards by about 9% and attracts the rare-earth ion towards it by about 6% of the undisturbed distances (Baker, Davies, and Hurrell, 1968). The distance of the interstitial ion from its nearest neighbours, 2·58 Å, is somewhat smaller than twice the ionic radius of the F^- ion (2·72 Å) given by Evans (1966), which is the ionic separation one expects for a 'hard sphere' model. Kiro and Low (1970) have also measured the atomic spacings for an H^- ion in the interstitial position, obtaining 2·66 Å compared with the sum of the ionic radii of F^- and H^- equal to 2·90 Å. In each case the outward shift of the surrounding F^- ions is about 60% of that predicted by a 'hard sphere' model. For both F^- and H^- interstitial ions the distortion is much smaller in the second shell of F^- ions around the interstitial, and may be assumed to be damped out altogether at the third shell.

6.3.2. *Endor data for the* Tg(F_i^-) *site for* Yb^{3+} *and crystal-field estimates*

The very detailed information about the ionic displacements caused by the incorporation of the interstitial ion, discussed in the last section, may be used in order to estimate the crystal field at the paramagnetic ion. Although such an estimate is rendered very uncertain by imprecisely known effects of covalent bonding (see § 5.2.5), it does illustrate one very important point.

In much of the literature it is assumed that the effect of the charge-compensating ion upon the crystal field may be represented to a first approximation simply by the superposition on the cubic crystal field of additional terms arising from a point charge. Such a charge leads to a term in $A_2^0\langle r^2\rangle$ for an axial site whose sign depends upon the sign of the effective charge-compensating ion. The A_2^0 term is often assumed to be dominant, and making this assumption one may often deduce the sign of A_2^0 from the g-values of the ground state determined by epr (§ 6.5) Hence, *if these assumptions were valid* one could find the effective sign of the compensating charge.

A simple point-charge calculation of the contributions to the crystal field for the tetragonal site shows that the ionic displacements of the other ions make major contributions to the crystal-field parameters.

TABLE 6.2

Contributions to the value of $A_2^0\langle r^2\rangle$ *in* cm^{-1} *for* $Tg(F_i^-)$ *and* $Tg(H_i^-)$ *sites*

Site	Interstitial ion charge	Interstitial ion dipole	Displaced near F^- ions	Polarization	Distant ions	Total
$Tg(F_i^-)$†	+1200	+220	−880	−290	−110	+140
$Tg(H_i^-)$‡	+1290§	+800	−270	−190	—	+1630

† Baker and Blake (1970) estimated for Yb^{3+}

‡ Ždánský and Edgar (1971) estimated for Gd^{3+}.

§ This includes a contribution due to the displacement of the Gd^{3+} ion.

For example a calculation by Baker, Davies, and Hurrell (1968) took account of the changes in the crystal field caused by (i) the interstitial F_i^- ion, (ii) the movement of the rare-earth ion towards the interstitial, (iii) the outward movement of the eight F^- ions around the interstitial, and (iv) the polarization of these eight F^- ions due to the charge on the interstitial. The calculation was performed for Yb^{3+}, but similar results would obtain for Ce^{3+}. The numerical values of these various contributions are tabulated by Baker and Blake (1970) and are given in the first line of Table 6.2, except that we have added a contribution from the electrostatic polarization of the interstitial F^- ion. This shows that the direct contribution of the interstitial charge of about $+1200\ cm^{-1}$ is almost cancelled by the other effects. Hence, bearing in mind the uncertainties in the calculation due to the effects of covalency and shielding which are discussed by Baker and Blake, the calculation could yield either sign for the resultant $A_2^0\langle r^2\rangle$ term, and is certainly consistent with the experimental value of about $+200\ cm^{-1}$.

Clearly one cannot with confidence use the measured sign of $A_2^0\langle r^2\rangle$ to make deductions about the sign of the effective compensating charge. Hence some of the simple direct deductions which appear in the literature must be treated with caution.

It is interesting to compare this calculation with a similar one performed by Žďánský and Edgar (1971) for sites of Gd^{3+} associated with H_i^- interstitial ions. The second line of Table 6.2 lists the various contributions. The most notable difference is that the large polarizability of H^- compared with F^- introduces a large contribution from the electric dipole moment on the H^- ion. In fact the dipole moment used was not that calculated using the free ion polarizability, but was deduced from a theory of the frequencies of the local modes of vibration of the H^- ion discussed in Chapter 2. The effective point charge on the H^- ion was also found in the same calculation to be $0{\cdot}85e$, so that when a 28% increase in the value of $\langle r^2\rangle$ between Yb^{3+} and Gd^{3+} is taken into account the contribution to $A_2^0\langle r^2\rangle$ from the charge on the interstitial ion is about the same in both cases. Žďánský and Edgar used a nominal value for the outward displacement of the surrounding F^- ions, which is much smaller than the displacements found in the endor experiments mentioned earlier; hence they greatly underestimate the contribution produced by these displacements. Scaling up the value in the first line of column 3 to allow for the increase in the value of $\langle r^2\rangle$ and the 30% greater displacement changes the figure of $-270\ \mathrm{cm^{-1}}$ to $-1450\ \mathrm{cm^{-1}}$. If one also includes the effect of the displacement of distant ions the resultant $A_2^0\langle r^2\rangle$ term is about $+350\ \mathrm{cm^{-1}}$. The equivalent figure for Yb^{3+} in a $Tg(H_i^-)$ site would be about $+280\ \mathrm{cm^{-1}}$, about twice the value for $Tg(F_i^-)$. Yb^{3+} is not observed in $Tg(H_i^-)$ sites by epr, but the change in $(g_{\parallel}-g_{\perp})$ between $Tg(F_i^-)$ and $Tg(H_i^-)$ sites for Ce^{3+} and Nd^{3+}, given in Table 6.8 (see p. 380), suggests an increase in $A_2^0\langle r^2\rangle$ by 10 and 12% respectively. Žďánský and Edgar suggest that it may be about 25% for Gd^{3+}. There is some confusion about the actual value of $A_2^0\langle r^2\rangle$ for $Tg(F_i^-)$ and $Tg(H_i^-)$ sites in Gd^{3+}, but Table 6.4 (see p. 361) indicates a value of about $300\ \mathrm{cm^{-1}}$, which is in good accord with the above modification of Žďánský and Edgar's estimate.

It is of interest to consider another centre reported by Jones, Peled, Rosenwaks, and Yatsiv (1969) comprising a neutral H atom in the $(00\tfrac{1}{2})$ interstitial site next to a Gd^{3+} ion. No epr has been observed for this centre. Its optical absorption spectrum indicates that the ${}^6P_{7/2}$ levels are similar to those of the $Tg(F_i^-)$ and $Tg(H_i^-)$ centres, with a crystal-field splitting which is intermediate between these two centres.

This is a large crystal field for a neutral atom. One may suppose that it is largely caused by the dipole moment induced on the H atom by the extra positive charge on the Gd^{3+} ion. The atomic polarizability of H gives a calculated value of $A_2^0\langle r^2\rangle$ of about 185 cm^{-1}.

These predictions of simple electrostatic models are remarkably close to the measured values. Lest one should be too carried away by the apparent success of these calculations two points should be emphasized. First, electrostatic screening has been ignored, and this is usually very important for the $A_2^0\langle r^2\rangle$ term, reducing its magnitude by up to 70%. Secondly, covalent effects have been ignored; this is often legitimate for calculations of $A_2^0\langle r^2\rangle$ because the contribution from charges at a distance does not converge rapidly as the distance increases, so that only a modest fraction is contributed by the ligands, and this fraction alone is affected by covalency. However, because of the cubic symmetry of the CaF_2 lattice, the contribution of distant ions in the undistorted part of the lattice averages to zero, so that in $Tg(F_i^-)$ sites the ligands make the major contribution to $A_2^0\langle r^2\rangle$, and covalency is therefore important. Baker and Blake (1970) demonstrate the importance of covalent effects by a comparison of the magnitude of $A_2^0\langle r^2\rangle$ deduced from the quadrupole interaction for ^{173}Yb in $Tg(F_i^-)$ sites and the magnitude deduced from the crystal field.

In addition to calculating $A_2^0\langle r^2\rangle$ Baker *et al.* (1969) also use a point-charge model to calculate changes in the other crystal-field parameters, and then use two models to attempt to calculate scaling factors to allow for covalency. That such scaling factors are quite large, about four for fourth order and ten for sixth-order terms, is clear from comparison with the measured crystal-field parameters for cubic symmetry. The uncertainties in the scaling factors makes the estimate fairly inaccurate, though it comes remarkably close to the experimental values; however it does indicate that the interstitial ion causes major changes in all of the crystal-field parameters. One aspect of the model which probably enables it to predict reasonable results for the changes in A_4^0, A_4^4, and A_6^4 is that all of the contributions (i) to (iv) cause changes in the same direction, so that there are not competing effects as there are for A_2^0.

As the calculation has been performed explicitly for Yb^{3+}, it is desirable to compare it with experimentally determined parameters for this ion. To determine the parameters one requires knowledge of the energies of all of the states of both the $J = \frac{7}{2}$ and $J = \frac{5}{2}$ manifolds. One optical transition only has been observed from the lowest state of the $J = \frac{5}{2}$ manifold to the ground state (Kirton and McLaughlan, 1967).

In addition, optical-Zeeman studies (Kirton and White, 1969) and epr studies give the g-values of the lowest states of each manifold. To supplement this information and obtain the crystal-field parameters Baker and Blake used an endor method involving the Yb nuclei, as opposed to the ligand endor discussed earlier in this section.

The method is similar to that used by Baker, Blake, and Copland (1969) for Yb^{3+} in cubic sites (see also § 5.2.4). It is based upon the use of the high accuracy of endor to measure second-order effects of matrix elements of the Zeeman and hyperfine Hamiltonians which couple the ground state to excited states of the lowest J manifold. The magnitudes of the matrix elements are known, so that the measured second-order effect may be used to calculate the energy denominators. For example, in cubic symmetry the second-order effects of matrix elements $\langle\Gamma_7|\, a\mathbf{I}\,.\,\mathbf{J}\,|\Gamma_8\rangle$ and $\langle\Gamma_7|\, g_J\beta\mathbf{H}\,.\,\mathbf{J}\,|\Gamma_8\rangle$ between the ground state $|\Gamma_7\rangle$ and the excited state $|\Gamma_8\rangle$ lead to a term in $\mathbf{H}\,.\,\mathbf{I}$ in the spin-Hamiltonian for $|\Gamma_7\rangle$. This term has the same form as the nuclear Zeeman interaction $g_n\beta_n\mathbf{H}\,.\,\mathbf{I}$ and so leads to an apparent nuclear g-factor which is different from the true nuclear g-factor. The measured pseudo-nuclear g-factor was used to calculate the energy of the Γ_8 level. Comparison with the optically determined energy of this excited state shows a 5% discrepancy which is outside the claimed experimental errors. A reanalysis of similar data for Tm^{2+} shows a similar discrepancy, of the same sign but somewhat smaller, and here the experimental errors of the two determinations just overlap. These discrepancies indicate that some unknown factor disturbs the second-order calculations. Similar factors presumably occur in the lower-symmetry site and render calculations based on second-order effects inaccurate to about 5%. However, even within such limitations a determination of the crystal-field parameters is valuable: indeed, often the accurately determined positions of excited states cannot be fitted to a crystal-field Hamiltonian to better than $\sim$10 cm^{-1}.

In tetragonal symmetry all of the spin-Hamiltonian parameters have two components, parallel and perpendicular to the tetragonal axis (see Baker and Blake, 1970). Measurement of these parameters gives twice as much information about the excited states as one obtains from similar measurements in cubic symmetry.

The g-values of the lowest state of the $J = \frac{5}{2}$ manifold enable one to calculate all of the states of that manifold to quite high accuracy. The g-values of the ground state enable one to calculate the two states of the $J = \frac{7}{2}$ manifold to similar accuracy. The energy of the excited Γ_7

state in this manifold is determined to be 450(20) and 430(20) cm^{-1} respectively from two independent parameters in the endor spin-Hamiltonian (see Baker and Blake, 1970). The energy of the lowest Γ_7 state of $J = \frac{5}{2}$ is found to be 10 325(1) cm^{-1} from optical spectroscopy. Four additional pieces of data are used in a least squares analysis leading to the final assignment of energy levels in Fig. 6.4, two of them

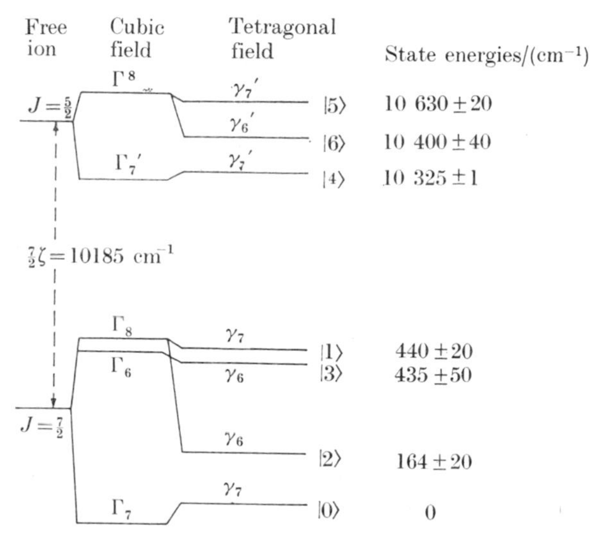

FIG. 6.4. The energy levels of Yb^{3+} in CaF_2 in sites of cubic and tetragonal, $Tg(F_{\bar{1}})$, symmetry, taken from Baker and Blake (1970).

involving second-order effects of the crystal-field coupling to the $J = \frac{5}{2}$ manifold. The uncertainties in the energy values arise from uncertainties in the epr and endor data, and within these uncertainties the nine pieces of experimental data are internally consistent.

There is just sufficient information in the energy levels of the $J = \frac{5}{2}$ and $\frac{7}{2}$ manifolds to obtain the five crystal-field parameters and the spin-orbit coupling parameter, though it is probably reasonable to assume that the spin-orbit coupling parameter is the same as for Yb^{3+} in cubic symmetry so that the available information slightly over-determines the crystal-field parameters. The parameters are given in Table 6.3. Clearly the departures from cubic symmetry are large, particularly for $A_4^0\langle r^4\rangle$.

These crystal-field parameters will be compared with those derived for other ions from optical spectroscopy in § 6.4.

TABLE 6.3
Crystal-field parameters (cm^{-1}) *for* Yb^{3+} *in cubic and tetragonal* $Tg(F_i^-)$ *sites*

Symmetry	$A_2^0\langle r^2\rangle$	$A_4^0\langle r^4\rangle$	$A_4^4\langle r^4\rangle$	$A_6^0\langle r^6\rangle$	$A_6^4\langle r^6\rangle$
cubic	0	−186	−930	34	−714
tetragonal	189	−47	−693	47	−644

6.3.3. *Endor measurement of transferred-hyperfine-structure parameters*

There have been a number of endor measurements of the values of the elements of transferred-hyperfine-interaction tensors **A** for ligand ions. We have mentioned earlier that there are contributions to **A** from both magnetic dipole-dipole interaction and from overlap and covalency. Chapter 5 shows that one can make some progress in interpreting the overlap and covalency in sites of cubic symmetry. For lower-symmetry sites this is difficult for two reasons; (a) the metal-ion wavefunctions are different and not always as well known as they are for cubic symmetry, (b) ionic displacements change the overlap of the ligand and metal-ion wavefunctions. Precise knowledge of the metal-ion wavefunctions and the ionic displacements would allow one to calculate the overlap parameters and to estimate the covalent contribution to **A**. However, the precision of knowledge is not great enough to make such calculations generally fruitful.

Such calculations might be of value, however, when the symmetry of the ligand bond is axial (as indeed it is for the ligand bonds in cubic sites). Two types of this sort of bond have been studied; (a) that to the interstitial ion in $Tg(F_i^-)$, $Tg(H_i^-)$, and $Tg(D_i^-)$ sites, and (b) that to the on-axis ligands in the trigonal Yb^{3+} centres $Tr(O_{4s}^{2-})$ (one F^- and one O^{2-} ligand on the trigonal axis), $Tr(O_s^{2-})$ (F^- and O^{2-}) and $Tr(O_{4s}^{2-}, H_s^-)$ (H^- and O^{2-}); there are also three similar examples in ThO_2 and CeO_2 for $Tr(F_s^-)$ sites.

Table 6.4 gives the values of $A_{||}$ and $A_{\perp}$ measured for these examples. Each component has contributions from magnetic dipole-dipole interaction, A_{dip}, and from s and p electrons on the ligand which have become unpaired by interaction with the metal ion so that

$$A_{||} = A_{||dip} + A_{||s} + A_{||p},$$

and there is a similar expression for $A_{\perp}$. A_{dip} can be calculated from the known g-values and crystal structure, and allowance can be made for the displacements of the ions indicated by endor measurements of

TABLE 6.4

Transferred hyperfine-coupling parameters measured by endor for ligand ions in CaF_2

Tetragonal Tg(X_1^-) sites

Ion pair	$A_{\parallel}$	$A_{\parallel(dip)}$	$A'_{\parallel}$	$A_{\perp}$	$A_{\perp(dip)}$	$A'_{\perp}$	Ref.
Ce^{3+}–F^-	$+19{\cdot}36(1)$	$+13{\cdot}57$	$+5{\cdot}79$	$-1{\cdot}20(1)$	$-3{\cdot}08$	$+1{\cdot}88$	a,b
Ce^{3+}–H^-	$+23{\cdot}04(1)$	$+15{\cdot}43$	$+7{\cdot}59$	$-3{\cdot}65(2)$	$-3{\cdot}26$	$-0{\cdot}39$	a,b
Ce^{3+}–D^-	$+3{\cdot}550(15)$	$+2{\cdot}26$	$+1{\cdot}29$	$-0{\cdot}562(2)$	$-0{\cdot}47$	$-0{\cdot}09$	a
Nd^{3+}–F^-	$+47{\cdot}57(2)$	$+18{\cdot}19$	$+29{\cdot}38$	$-8{\cdot}17(2)$	$-2{\cdot}68$	$-5{\cdot}49$	b
Nd^{3+}–H^-	$+43{\cdot}3(2)$	$+21{\cdot}5$	$+23{\cdot}8$	$-2{\cdot}49(1)$	$-2{\cdot}21$	$-0{\cdot}28$	b
U^{3+}–F^-	$+57{\cdot}62(2)$	$+14{\cdot}79$	$+42{\cdot}83$	$-16{\cdot}80(2)$	$-3{\cdot}93$	$-12{\cdot}87$	b
Yb^{3+}–F^-	$-8{\cdot}70(4)$	$-7{\cdot}1$	$-1{\cdot}6$	$+12{\cdot}39(1)$	$+8{\cdot}9$	$+3{\cdot}5$	c

Trigonal Yb^{3+} centres

Centre	Neigh-bour	$A_{\parallel}$	$A_{\parallel(dip)}$	$A'_{\parallel}$	$A_{\perp}$	$A_{\perp(dip)}$	$A'_{\perp}$	Ref.
Tr(O^{2-}_{4s})	F^-	$+35{\cdot}526(3)$	$+7{\cdot}53$	$+28{\cdot}00$	$-14{\cdot}677(2)$	$-12{\cdot}5$	$-2{\cdot}2$	d
Tr(O^{2-}_{s})	F^-	$+32{\cdot}110(9)$	$+8{\cdot}1$	$+24{\cdot}0$	$-14{\cdot}144(8)$	$-12{\cdot}5$	$-1{\cdot}6$	d
Tr(O^{2-}_{4s} H^-_s)	H^-	$+15{\cdot}548(5)$	$+9{\cdot}2$	$+6{\cdot}3$	$-11{\cdot}136(4)$	$-13{\cdot}0$	$+1{\cdot}9$	d
Tr(O^{2-}_{4s})	O^{2-}	$+4{\cdot}118(5)$	$+1{\cdot}09$	$+3{\cdot}03$ +[21]		$-1{\cdot}81$		d
Tr(O^{2-}_{s})	O^{2-}	$+9{\cdot}81(1)$	$+1{\cdot}17$	$+8{\cdot}64$ +[60]		$-1{\cdot}81$		d
Tr_1	F^-†	$+29{\cdot}87(1)$	$+24{\cdot}4$	$+5{\cdot}5$	$-3{\cdot}96(1)$	$-7{\cdot}8$	$+3{\cdot}8$	d
Tr_1	F^-‡	$+38(3)$	$+24{\cdot}5$	$+13$		$-9{\cdot}0$		e
Tr_1	F^-§	$+58.6(2)$	$+10{\cdot}8$	$+39{\cdot}8$	$+27{\cdot}4(2)$	$-5{\cdot}4$	$+32{\cdot}8$	f

† In ThO_2. ‡ In CeO_2. § $Pb^{3+}(5d^{10}6s)$ in ThO_2.

[a] Baker, Davies, and Reddy, 1969. [b] Kiro and Low, 1970. [c] Baker, Davies and Hurrell, 1968. [d] Reddy, unpublished. [e] Zonn, Katyshev, Mitrofanov, and Pol'skii, 1969. [f] Rohrig and Schneider, 1969.

The bond symmetry is axial so that the spin-Hamiltonian of the thfs interaction is $\mathscr{H} = A_{\parallel}S_zI_z + A_{\perp}(S_xI_x + S_yI_y)$. The parameters are given in MHz; $A_{\parallel}$ and $A_{\perp}$ are the measured values, $A_{\parallel}$ (dip) and $A_{\perp}$ (dip) are the calculated dipolar contribution, and $A'_{\parallel}$ and $A'_{\perp}$ are the residue which must be attributed to covalent bonding. See p. 363 for a discussion of the figures in square brackets.

more distant nuclei; hence the non-dipolar contribution, labelled A' in Table 6.4, may be calculated. s and p-wavefunctions of the ligand may be admixed in two ways, either through direct overlap and covalency between them and the $4f^n$-wavefunctions, or through overlap and covalent interaction with the outer s and p-shells of the metal ion; it should be remembered that the latter are subject to core polarization due to exchange interaction with the $4f^n$-wavefunctions. In general it is the uncertain contribution from the core-polarization mechanism which makes the parameters difficult to estimate. So far, no detailed calculations have been made of the non-dipolar contributions. However, some general comments can be made.

Firstly, the configuration of H^- is $1s^2$ so that as a first approximation there should be no p-electron contribution. However, electrostatic polarization admixes p_z-orbitals, allowing a p-electron contribution to

σ-bonding only. Although the p-electron admixture may be small the greater overlap of $2p_z$-functions, with greater radial extent than the $1s$-functions, may make the two contributions comparable. If only σ-bonding occurs the calculation of A parameters is considerably simplified, and $|A_{\|s}| = |A_{\perp s}|$ and $|A_{\|p}| = 2\,|A_{\perp p}|$, the relative sign depending upon the sign of the s and p-electron admixtures into the metal-ion wavefunction. A notable feature of both of the results for $Tg(H_i^-)$ sites is that $A'_{\perp} \approx 0$, indicating that $A_{\perp s} \approx -A_{\perp p}$, so that $A_{\|p} \approx 2A_{\|s}$. That the latter parameters should have the same sign is to be expected on the above electrostatic polarization model, but their similar magnitude is perhaps unexpected.

Secondly, for Ce^{3+} and Yb^{3+} in $Tg(X_i^-)$ sites the nature of the ground state wavefunction is such that σ-bonding cannot occur between $4f^n$ and an on-axis ligand. Hence for $Ce^{3+}(H_i^-)$ and $Ce^{3+}(D_i^-)$ one expects a purely dipolar interaction. There is indeed a much smaller A' than in $Nd^{3+}(H_i^-)$, where σ-bonding is allowed, but the observed non-dipolar interaction for Ce^{3+} indicates that there must be σ-bonding, either with core-polarized states of the metal ion, or with states of different parity which are admixed by components of the crystal field which do not have a centre of symmetry. Secemski *et al.* (1970) suggest that the similarity between the values of A' for $Re^{3+}(F_i^-)$ and $Re^{3+}(H_i^-)$ shows that the core polarization of the closed shells dominates the direct overlap of the $4f^n$ configuration, but this argument is not supported by detailed calculation. The ratio of the A parameters for $Ce^{3+}(H_i^-)$ and $Ce^{3+}(D_i^-)$ are nearly equal to the ratio of the nuclear g-factors multiplied by the ratio of the electronic g-factors, the slight difference presumably being due to the effect of different zero-point vibrational amplitudes.

As mentioned above, σ-bonding can occur for Nd^{3+} and U^{3+} in $Tg(X_i^-)$ sites. The configurations $4f^3$ and $5f^3$ have very similar states. The greater spread of both the $5f$-wavefunctions and of the $6s^2 6p^6$ outer shells on the U^{3+} ion accounts for the greater magnitude of its A parameters.

For Yb^{3+} in trigonal sites σ-bonding is possible for on-axis ligands. $Tr(O_{4s}^{2-})$ and $Tr(O_{4s}^{2-}H_s^-)$ are identical except for the difference between F^- and H^- and a small difference in the trigonal crystal field; the much smaller bonding contribution for H^- is presumably due to the absence of first-order p-electron on H^-, though this contrasts with the situation in $Tg(H_i^-)$ sites. Note, however, that in the absence of detailed information we have calculated A_{dip} on the assumption that the Yb^{3+}–F^- distance is the same as the Yb^{3+}–H^- distance; if the larger size of the

H^- ion is assumed to produce a 6% larger spacing, as it does in the interstitial position in $Tg(H_i^-)$ sites, the contribution to A' would be close to zero.

It is interesting to compare O_s^{2-} and F^- in $Tr(O_{4s}^{2-})$ and $Tr(O_s^{2-})$ centres as both ions have the same configuration, though O^{2-} is generally considered to be more covalent. The figures in square brackets in Table 6.4 are A' scaled up in the ratio $g_n(^{19}F)/g_n(^{17}O)$ for comparison with the ^{19}F parameters. This indicates that the covalency for O^{2-} in $Tr(O_s^{2-})$ sites is much greater than that of F^-, which is consistent with a displacement of the Yb^{3+} ion by about 9% towards the O_s^{2-} ion deduced from endor of the more distant F^- ions. For $Tr(O_{4s}^{2-})$ sites the O^{2-} covalency is smaller than that of F^-, which is again consistent with the displacement of 5% towards F^- suggested by the endor of more distant ions (Reddy, private communication).

Table 6.4 also contains two measurements of interactions with F_s^- ions in ThO_2 and CeO_2. The lattice dimensions are similar to those of CaF_2, but the distortion of the ionic positions around the Yb^{3+} ion are unknown. These two examples make an interesting comparison with those in CaF_2 because the sign of the $A_2^0\langle r^2\rangle$ term in the crystal field is opposite, due to the effective positive charge on the F_s^- ion site in ThO_2 and CeO_2. The consequent change in the wavefunction is reflected in a much smaller contribution to A', though some of the reduction may be caused by electrostatic repulsion of the Yb^{3+} ion from the effective positive charge on the F_s^- ion. Table 6.4 also contains the result of a measurement on Pb^{3+} associated with a single unpaired $6s$ electron in a $Tr(F_s^-)$ site in ThO_2.

The considerable detail with which hyperfine-interaction parameters are known for these several examples of ions in known positions with quite well known wavefunctions makes this a promising situation for future careful calculations of the hyperfine interactions from first principles.

6.4. Optical measurements

We discuss optical measurements next, not because there is a large amount of interpreted information, but because in principle it is possible to deduce all of the crystal-field parameters for a particular site if sufficient information is available. In fact, although the literature contains a great deal of information about optical spectra of non-cubic sites, there is very little interpretation in terms of firm crystal-field parameters. A number of problems make interpretation difficult:

(a) uncertainty in identification of optical lines with particular sites, (b) effects of vibronic side bands, (c) effects of intermediate coupling in the free ion, (d) J mixing due to the crystal field, (e) insufficient information about the levels in any one J manifold. Although in principle it is possible to calculate the values of n crystal-field parameters from n energy levels, the solution is not unique, but depends upon the identification of the eigenfunctions associated with the levels. Hence it is necessary either to identify the eigenfunctions, or to make inspired guesses about them. For example it may be possible to obtain approximate crystal-field parameters from other methods and then make small adjustments to the parameters to fit the optical data. Alternatively if a lot of optical information is available one may use a trial and error method to obtain the best fit to all of the optical data.

If, as a first approximation, one ignores effects of intermediate coupling and J mixing one can sometimes deduce the crystal-field parameters by stages. Any manifold whose J is 1 or $\frac{3}{2}$ is split only by A_2^n terms, so for an axial field the A_2^0 term may be determined uniquely. The validity of ignoring the admixtures of wavefunctions may sometimes be checked by comparing the value of $A_2^0\langle r^2\rangle$ calculated for two such manifolds. Then any manifold whose J is 2 or $\frac{5}{2}$ is split only by A_2^n and A_4^n terms, so that if $A_2^0\langle r^2\rangle$ has already been found one can, for axial symmetry, calculate $A_4^0\langle r^4\rangle$ and $A_4^4\langle r^4\rangle$. Finally by using manifolds with greater multiplicity one may calculate the sixth-order terms.

We shall discuss the few positive identifications of sites and calculations of crystal-field parameters, make some comments on other uninterpreted data, and give a bibliography of the uninterpreted data. The majority of interpreted optical data for non-cubic sites relate to tetragonal sites, probably $Tg(F_i^-)$. This will be discussed first in § 6.4.1. We wish to compare crystal-field parameters deduced from optical data with those for Yb^{3+} given in § 6.3.2, and as there is more data for ions in the second half of the rare-earth group, we shall discuss the ions in the order of decreasing Z. The optical data for sites of lower symmetry will be discussed in § 6.4.2.

6.4.1. *Optical measurements for tetragonal sites*

In § 6.3.2 crystal-field parameters for Yb^{3+} in the $Tg(F_i^-)$ site in CaF_2 were derived from endor measurements. Only one optical transition is observed for this site, and indeed for any site of Yb^{3+} in CaF_2 (Kirton and McLaughlan, 1967) so that there is no optical confirmation of the crystal field parameters given in Table 6.3. For comparison with this data we discuss the optical data for other ions which are thought to be in this sort of tetragonal site.

The most complete analysis has been made for Er^{3+} by Smirnov (1970). Fig. 6.5 shows Smirnov's interpretation of the data from the references listed in his paper using crystal-field parameters given in Table 6.5. These parameters were calculated using certain selected bits of information, but Fig. 6.5 shows the quality of the general fit. $A_2^0\langle r^2\rangle$ was calculated from the splitting of the ${}^4S_{\frac{3}{2}}$ term. The other parameters were calculated using the g-values of the ground and first-excited state, both of which are determined by epr (§ 6.5), and the centre of gravity of the ground term which was determined optically.

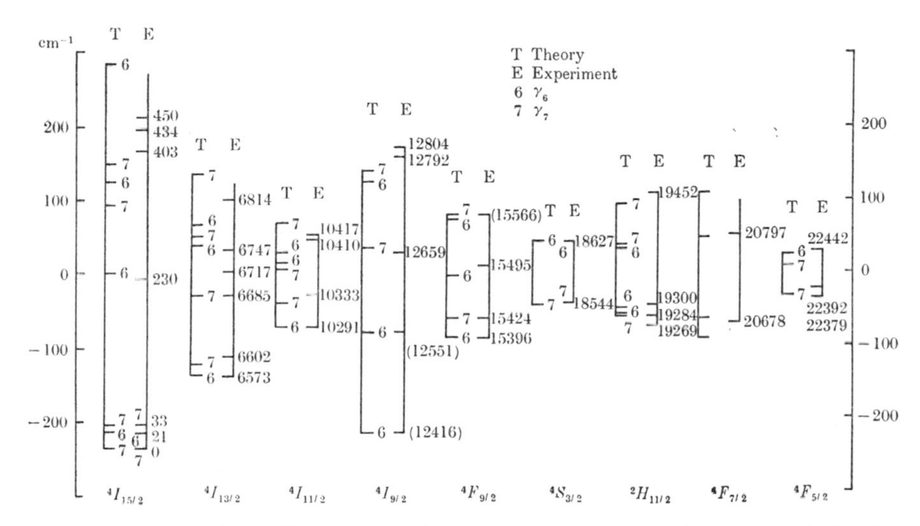

FIG. 6.5. Crystal-field splittings of various energy levels of Er^{3+} in tetragonal sites in CaF_2 taken from Smirnov (1970). For each J manifold the experimentally determined energies are shown on the left and Smirnov's theoretical fit on the right. The labels 6 and 7 indicate whether the states have γ_6 or γ_7 symmetry. The energies are given in wavenumbers above the ground state, those in brackets being somewhat uncertain.

The crystal-field parameters are comparable to those found for Yb^{3+} from endor. The agreement between them is remarkably good when one considers the large uncertainties in both determinations. At first sight it appears that the distortion from cubic symmetry is larger for Er^{3+}, because $A_2^0\langle r^2\rangle$ is larger and the ratios A_4^0/A_4^4 and A_6^0/A_6^4 are further from the values appropriate to cubic symmetry. It is possible however that taking J mixing into account in Er^{3+} would reduce the value of $A_2^0\langle r^2\rangle$ and this would alter the values of the other parameters. There is no direct confirmation that the tetragonal site for Er^{3+} is

TABLE 6.5

Crystal-field parameters in cm^{-1} *for* $Tg(F_i^-)$ *sites*

Ion	Host	$A_2^0\langle r^2\rangle$	$A_4^0\langle r^4\rangle$	$A_4^4\langle r^4\rangle$	$A_6^0\langle r^6\rangle$	$A_6^4\langle r^6\rangle$	Ref.
Yb^{3+}	CaF_2	+189	−47	−693	+47	−644	a
Er^{3+}	CaF_2	+355	−36	−921	+54	−505	b
Dy^{3+}	CaF_2	+149	−280	−940	+50	−228	c
Tb^{3+}	SrF_2	+115	−210	−1000	+30	−600	d
Gd^{3+}	CaF_2	+339	−47	−3655	+23	−641	
Gd^{3+}	SrF_2	+205	−34	−3023	+33	−586	e
Gd^{3+}	SrF_2	+103	−214	−910	+117	−250	f
Eu^{3+}	CdF_2	+150	400	2200			g
Eu^{3+}	CaF_2	+250	−176	−1240			h
Sm^{3+}	CaF_2	+98	+20	+1850	+220	−1200	i
Sm^{3+}	CdF_2	+39	0	+1650	+136	−1550	i
Nd^{3+}	CaF_2	430					
Pr^{3+}	CaF_2	+250	−44	+150	+37	+273	
Ce^{3+}	CaF_2	+170	−292	−2150	+62	+455	k

[a] Baker and Blake, 1970. [b] Smirnov, 1970. [c] Nara and Schlesinger, 1972. [d] see § 6.4: Antipin, Livanova and Shekun, 1968. [e] Ivoilova and Leushin, 1972. [f] O'Hare, 1971. [g] Based on Kingsley and Prener, 1962. [h] Bokii, Gaigerova, Gaiduk, Dudnik, and Murav'ev, 1968. [i] Rabbiner, 1967. [j] Hargreaves, 1972. [k] Manthey, 1972.

$Tg(F_i^-)$, but the similarity of the crystal-field parameters makes it highly likely that the sites for Er^{3+} and Yb^{3+} are the same.

Schlesinger and Kwan (1971) have observed luminescence of Dy^{3+} in CaF_2, and Nara and Schlesinger (1972) have interpreted some of the spectral lines in terms of a site of tetragonal symmetry. A comparison of the crystal-field splittings of the ground ${}^6H_{\frac{15}{2}}$ manifold and the first-excited ${}^6H_{\frac{13}{2}}$ manifold for tetragonal and cubic symmetries indicates that the non-cubic part of the tetragonal crystal field is small. The energies of these two manifolds are fitted to an rms deviation of 7·8 cm^{-1} by the parameters listed in Table 6.5. It is suggested that the tetragonal site is $Tg(F_i^-)$ but this is unconfirmed. Comparison of the crystal-field parameters with those of other tetragonal sites shows some discrepancies, particularly in the values of $A_4^0\langle r^4\rangle$ and $A_6^4\langle r^6\rangle$, but the magnitude of $A_2^0\langle r^2\rangle$ is in reasonable accord with other ions.

The excited terms of Gd^{3+} in a tetragonal site thought to be $Tg(F_i^-)$ have been identified in CaF_2 by several workers including Schlesinger and Nerenberg (1969), Gil'fanov, Dobkina, Stolov, and Livanova (1966) and Gil'fanov, Livanova, Stolov, and Khodyrev (1967), and in SrF_2 by Gil'fanov *et al.* and by Detrio, Yanney, Ferralli, Ware, and Donlan (1970). A considerable number of authors have attempted to interpret this data; the extensive literature is summarized in the references of Schlesinger and Nerenberg (1969), O'Hare (1971), and

Ivoilova and Leushin (1972). Difficulties arise because the Russell-Saunders approximation for the f^7 configuration produces zero for all first-order matrix elements of the spin-orbit and crystal-field interactions, so the fine structure and Stark splittings are due to effects of higher order than those in other rare-earth ions. The difficulties of fitting experimental data to crystal-field parameters are well illustrated by the extensive work of O'Hare (1971) and Ivoilova and Leushin (1972). The parameters obtained by these authors for SrF_2 are given in Table 6.5; they are seen to differ considerably, but by nothing like so large a margin as some of the earlier calculations. It is not easy to see how the two calculations differ. O'Hare states that he obtained good intermediate-coupled wavefunctions for the free ion before considering the crystal field; Ivoilova and Leushin imply that they have also accounted for the breakdown of Russell-Saunders coupling and configurational interaction, which admixes other configurations of the same parity as $4f^7$. Both authors have diagonalized the complete matrix including the ^8S, ^6P, ^6D, and ^6I terms. This was necessary as earlier interpretations showed that if parameters were chosen separately to fit only ^6P or ^6I terms one obtained different parameters for each term. The large variation between the parameters listed in Table 6.5 and those found by other authors may well derive from the methods used to take into account the off-diagonal matrix elements of the crystal field between these terms. Both O'Hare and Ivoilova and Leushin claim to fit the energy levels for SrF_2 with an rms deviation of about 7 cm^{-1}, though the parameters used to do this are quite different. Their methods of calculation appear to be similar so the differences are likely to arise either from the starting free-ion wavefunctions, or from the identification of the Stark components of the various terms. Comparison with the parameters for other ions in Table 6.5 shows up discrepancies for both sets of parameters, though those of Ivoilova and Leushin are discrepant only in the rather large value of $A_4^4\langle r^4\rangle$, and these authors do suggest some possible mechanisms for this discrepancy. However, these differences leave a sense of unease about the whole theoretical interpretation, and it would clearly be valuable to have them resolved.

The situation for Gd^{3+} is further complicated by Jones, Peled, Rosenwaks, and Yatsiv (1969) and Ždánský and Edgar (1971) who report and interpret spectra in the $Tg(H_i^-)$ site obtaining a value of $A_2^0\langle r^2\rangle = +1900$ cm^{-1}. The electrostatic model of Ždánský and Edgar is discussed in § 6.3.2 and summarized in Table 6.2, and gives a value of $A_2^0\langle r^2\rangle$ of $+1630$ cm^{-1}, but we have explained why this is a very considerable overestimate. It is not clear why the value of $A_2^0\langle r^2\rangle$ deduced from substantially the same optical spectrum as that for the Tg(F_i^-) site is so much larger than the values quoted in Table 6.5.

Eu^{3+} has been studied in tetragonal centres in CaF_2 by Bokii, Gaigerova, Gaiduk, Dudnik, and Murav'ev (1968), and in centres thought to be tetragonal in CdF_2 (which has almost identical lattice parameters to CaF_2) by Kingsley and Prener (1962). Fig. 6.6 shows the

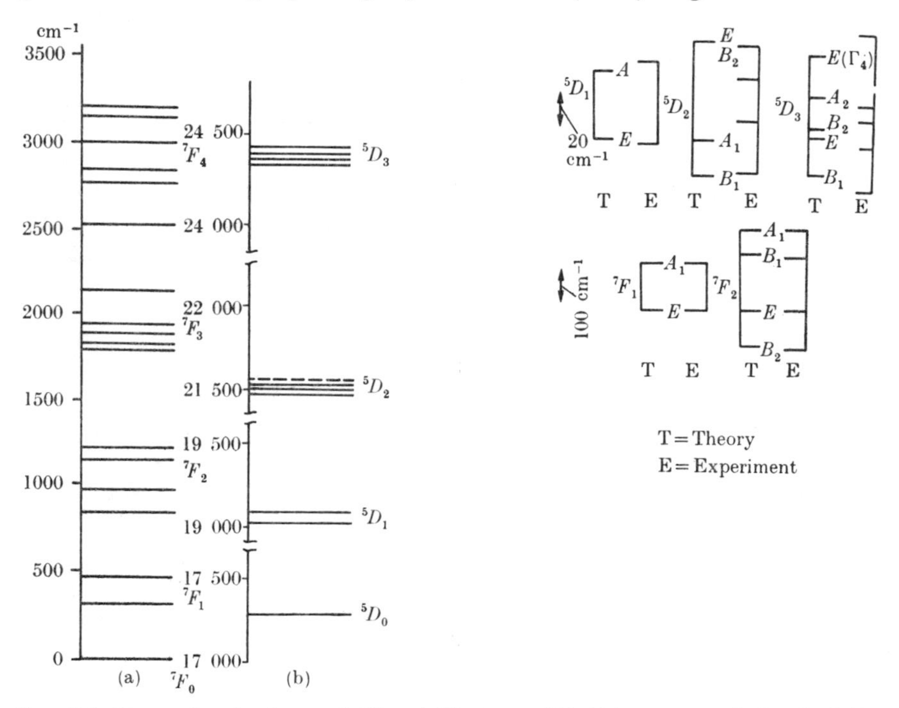

FIG. 6.6. Energy levels of several ^7F and ^5D states of Eu^{3+} in tetragonal sites in CaF_2, shown on left (a) and (b); and a comparison of the crystal-field splittings for several manifolds with those calculated using the parameters in Table 5 (Bokii, Gaigerova, Gaiduk, Dudnik, and Munav'ev, 1968).

energy levels of the lowest ^7F_J manifolds and some excited ^5D_J manifolds in CaF_2, and also indicates the quality of the fit produced by the crystal-field parameters listed in Table 6.5. The CdF_2 data were not interpreted in terms of crystal-field parameters, but a rough calculation using $J = 1$ and $J = 2$ manifolds and ignoring J mixing leads to the parameters listed in Table 6.5; this is a considerable oversimplification since it is apparent that the crystal-field splittings in Fig. 6.6 are not small compared with the separation of the J manifolds.

Optical-fluorescence spectra have been observed for Sm^{3+} in what are taken to be tetragonal sites in both CaF_2 and CdF_2 indicating that the energy level schemes are very similar in these two crystals. The fluorescence occurs between the $^4G_{\frac{5}{2}}$ level at around 17 650 cm^{-1} and the lowest three manifolds of ^6H_J. Rabbiner (1967) interpreted the crystal-field splittings of these ^6H_J manifolds on the assumption of LS coupling and diagonalized only the matrix for the crystal field within each J manifold. The fact that twelve energy levels could be fitted to five crystal-field parameters with an rms deviation of about 30 cm^{-1} was taken as evidence that the site symmetry is indeed tetragonal. As the overall crystal-field splittings, 200–300 cm^{-1}, are appreciable compared with the separation, $\sim$1200 cm^{-1}, between the different J manifolds, there will be appreciable corrections to which Rabbiner attributes the quite large deviation between the experimentally observed levels and those predicted on the simple model.

Comparison of the crystal-field parameters, listed in Table 6.5, with those of other ions shows that the magnitudes are reasonable, but that the signs of the fourth-order terms are puzzling. A small positive $A_4^0\langle r^4\rangle$ may be reasonable as all of the values of $A_4^0\langle r^4\rangle$ are very much smaller than for cubic symmetry, so that a somewhat larger distortion for Sm^{3+} might change the sign of this parameter. However, a positive sign for the larger parameter $A_4^4\langle r^4\rangle$ is difficult to account for.

In the approximation that Rabbiner used the ground manifold $^6H_{\frac{5}{2}}$ is not affected by sixth-order terms. It is described by a matrix of the following form:

$$\begin{array}{c|ccc|} \pm\frac{5}{2} & a & 0 & b \\ \pm\frac{1}{2} & 0 & c & 0 \\ \mp\frac{3}{2} & b & 0 & d \end{array}, \qquad (6.2)$$

where

$$\begin{aligned} a &= 10\alpha_J A_2^0\langle r^2\rangle + 60\beta_J A_4^0\langle r^4\rangle, \\ b &= 12\sqrt{5}\beta_J A_4^4\langle r^4\rangle, \\ c &= -8\alpha_J A_2^0\langle r^2\rangle + 120\beta_J A_4^0\langle r^4\rangle, \\ d &= -2\alpha_J A_2^0\langle r^2\rangle - 180\beta_J A_4^0\langle r^4\rangle, \end{aligned}$$

$$\alpha_J = \frac{13}{315}, \qquad \beta_J = \frac{26}{27\times 385}.$$

There is not enough information to solve for three crystal-field parameters, but in hindsight the large energy separation between the extreme states of the manifold is due to the large off-diagonal element b,

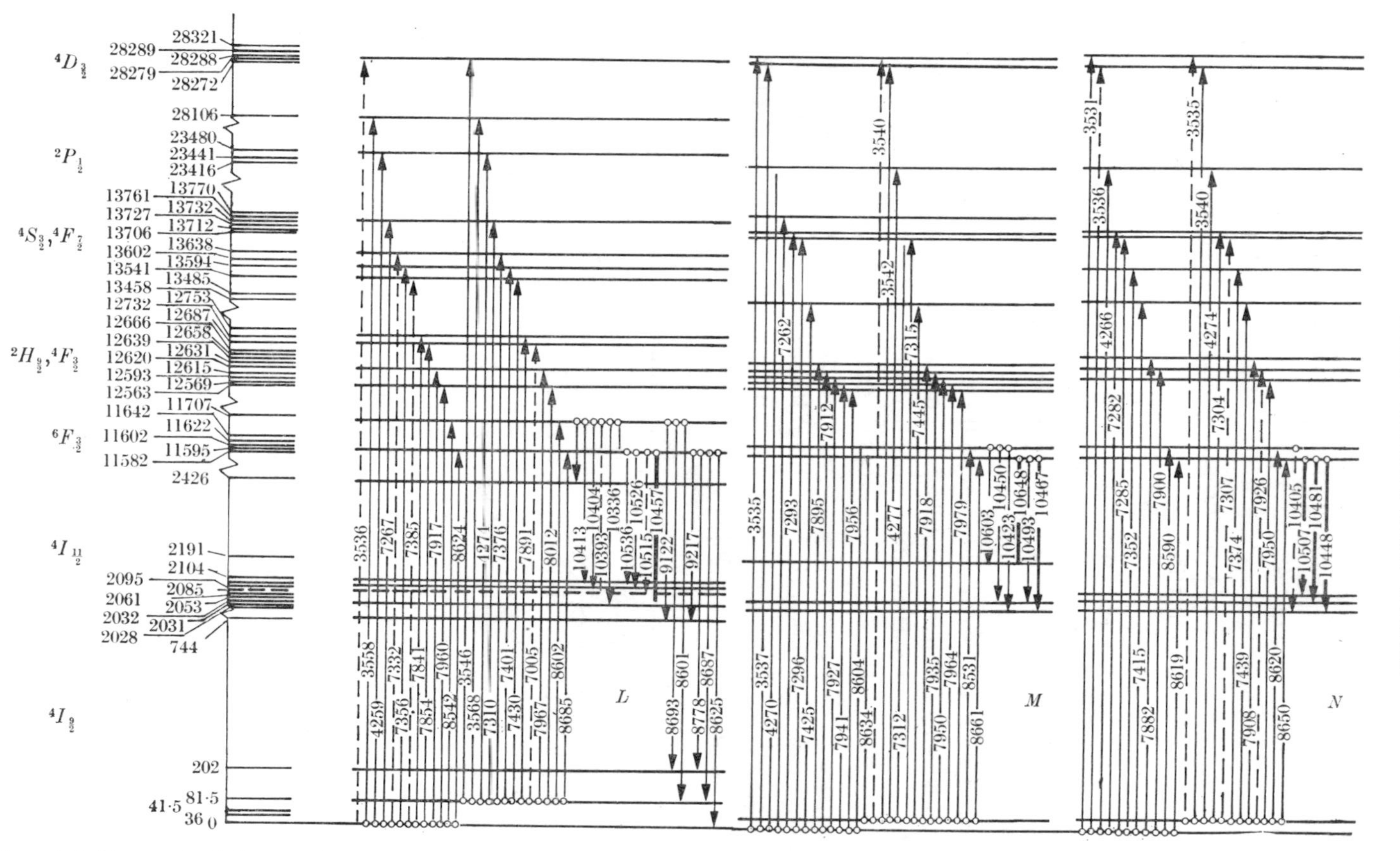

FIG. 6.7. Energy-level diagram for Nd^{3+} ions in three sites in CaF_2. *L* corresponds to the $Tg(F_i^-)$ site, and *M* and *N* are both rhombic sites. This diagram, taken from Voron'ko, Keminskii, and Osiko (1966*b*) shows energies in cm^{-1} on the left and the wavelengths of the observed optical transitions in Å.

so that the magnitude of b is well determined; however, the fitting of the levels does not determine its sign. The same is true of the off-diagonal elements in the other manifolds, though the relative signs of $A_4^4\langle r^4\rangle$ and $A_6^4\langle r^6\rangle$ should be correctly determined. However, using Rabbiner's parameters it appears that for ${}^6H_{\frac{7}{2}}$ most of the off-diagonal elements are dominated by $A_6^4\langle r^6\rangle$, and one wonders how much difference would be made by changing the sign of $A_4^4\langle r^4\rangle$. Such a change would bring the parameters into good accord with those of other ions. However, in this context it is interesting to consider the ground state of Nd^{3+} in $Tg(F_i^-)$ sites. In § 6.4 it is shown that the g-values of the ground state indicate that it is an admixture of $|\pm\frac{9}{2}\rangle$ and $|\pm\frac{1}{2}\rangle$ with a very small amplitude of $|\mp\frac{7}{2}\rangle$. This small amplitude requires a very small crystal-field matrix element $\langle\frac{7}{2}|\,V_{cf}\,|\frac{1}{2}\rangle$, which implies that $A_4^4\langle r^4\rangle \approx -0{\cdot}8\,A_6^4\langle r^6\rangle$. This relative sign for $A_4^4\langle r^4\rangle$ and $A_6^4\langle r^6\rangle$ does correlate with Rabbiner's parameters for Sm^{3+}. It seems very unlikely that there should be a change of sign of such large parameters between the first half and the second half of the rare-earth group, so these signs remain a mystery.

Rabbiner calculates the eigenstates appropriate to his model. The ground state is $\{\cos\theta\,|\pm\frac{5}{2}\rangle - \sin\theta\,|\mp\frac{3}{2}\rangle\}$ with $\cos\theta = 0{\cdot}61$, and a change of sign of $A_4^4\langle r^4\rangle$ would merely change the sign of $\sin\theta$. This state has $g_{\parallel} \approx 0$ and $g_{\perp} \approx 0{\cdot}62$. These g-values are close to those measured for a tetragonal site for Sm^{3+} in CaF_2 by Antipin, Kurkin, Livanova, Potvorova, and Shekun (1965) for which $g_{\parallel} = 0{\cdot}00(6)$, and $g_{\perp} = 0{\cdot}823(3)$. This site has also been observed by Evans and McLaughlan (1966), and Ashburner, Newman, and McLaughlan have confirmed that it is $Tg(F_i^-)$ by noting the close similarity of its g-values with those of sites which were shown to be $Tg(H_i^-)$ and $Tg(D_i^-)$. The g-values for these sites are given in Table 6.9 (p. 382), and Newman's method of observing the local modes of vibration of the hydride ion is discussed in § 2.6.5. The inexact match between the g-values calculated using Rabbiner's wavefunctions and those measured by epr could well be caused by J mixing.

There is considerable optical data for Nd^{3+} in tetragonal sites, and here endor studies have confirmed that the site is $Tg(F_i^-)$. Fig. 6.7 shows the energy-level diagram deduced from the optical data by Voronko, Keminskii, and Osiko (1966*b*). If J mixing is ignored the splitting of the ${}^4F_{\frac{3}{2}}$ state corresponds to $A_2^0\langle r^2\rangle \approx 430\ \text{cm}^{-1}$. This cannot be confirmed using the ${}^4S_{\frac{3}{2}}$ state as that state overlaps ${}^4F_{\frac{7}{2}}$ making identification of the levels uncertain. Many, but not all, of the levels of the two lowest manifolds $^4I_{\frac{9}{2}}$ and $^4I_{\frac{11}{2}}$ are known, but no interpretation has been made in terms of crystal-field parameters. The correlation of this spectrum with the site $Tg(F_i^-)$ has been made by a measurement of the Orbach process in the spin-lattice relaxation. This gives the position of the first-excited state at 74(5) cm^{-1} in reasonable agreement with the optically determined value of 81 cm^{-1}.

There have been some recent measurements on Pr^{3+} in tetragonal sites (Hargreaves, 1972), and also on Ce^{3+} where all but one of the energy levels of $4f^1$ were determined from fluorescence of the lowest $5d$ level (Manthey, 1972). Here again there appear to be some discrepancies between the signs of $A_4^4\langle r^4\rangle$ and $A_6^4\langle r^6\rangle$ and those of other ions in the table.

Table 6.5 (p. 366) lists the values of the crystal-field parameters for the ions we have discussed. As we have indicated, there are sufficient optical data available to make it possible to calculate more precise parameters. However, even these rough values do show some consistency.

Hargreaves (1967) has measured optical transitions for U^{3+} in the $Tg(F_i^-)$ site, which has also been identified by endor. He gives all of the energy levels for the lowest manifolds $^4I_{\frac{9}{2}}$ and $^4I_{\frac{11}{2}}$ for both CaF_2 and SrF_2, but these levels are not interpreted in terms of crystal-field parameters.

Stacy, Edelstein, McLaughlin, and Conway (1972) have observed tetragonal centres by optical-Zeeman spectroscopy for Am^{2+}, Cm^{3+}, and Nd^{3+} in CaF_2.

6.4.2. *Optical data for sites of trigonal and lower symmetry*

Kirton and McLaughlan (1967) have observed absorption lines for Yb^{3+} in many different sites in CaF_2. These were correlated with epr data and hence the optical lines were assigned to particular sites. Kirton and White (1969) have measured the Zeeman effect for all of the optical lines and have confirmed all but one of the assignments of Kirton and McLaughlan. A feature of Yb^{3+} is that it appears to exhibit only one absorption line for each type of site, presumably between the lowest states of the $J = \frac{5}{2}$ and $J = \frac{7}{2}$ manifolds. Hence one obtains very little information about the crystal field from the optical data.

Voron'ko, Osiko, and Shcherbakov (1969) have studied absorption and luminescence of crystals containing a wide range of concentrations of Yb^{3+} ions. They find several lines which they attribute to tetragonal centres, but they do not deduce an energy-level scheme. They do deduce an almost complete energy-level scheme for the $J = \frac{5}{2}$ and

$J = \frac{7}{2}$ manifolds for a centre of unknown symmetry. This centre has an overall splitting of only 146 cm^{-1} in the $J = \frac{7}{2}$ manifold, compared with 650 cm^{-1} in cubic symmetry, which is very difficult to understand in terms of any likely distortion.

Rabbiner (1969) has also observed fluorescence for Sm^{3+} in a trigonal site in CaF_2 in the spectral region 0·72 to 0·55 μ from excited states to several ^6H_J manifolds. Using the model discussed above for tetragonal sites he fits 18 energy levels with six crystal-field parameters with an rms deviation of 40 cm^{-1}; the parameters are listed in Table 6.6.

TABLE 6.6

Crystal-field parameters for trigonal sites in CaF_2 *in* cm^{-1}

Ion	Ref.	$A_2^0\langle r^2\rangle$	$A_4^0\langle r^4\rangle$	$A_4^3\langle r^4\rangle$	$A_6^0\langle r^6\rangle$	$A_6^3\langle r^6\rangle$	$A_6^6\langle r^6\rangle$
Sm^{3+}	a	+84	0	−8500	0	−1000	−550
Sm^{3+}	b	+59	−34	−440	+2·4	−41	+740
Sm^{3+}	c	0	+75	−2130	+69	+850	+660
Gd^{3+}	d	+712	−225	0			
Dy^{3+}	e(i)	+103	+165	4902	+54	458	+874
		(+93)	(+157)	(4573)	(+45)	(373)	(+823)
Dy^{3+}	e(ii)	+701	+188	3067	+20	841	+796
		(+713)	(+184)	(2938)	(+17)	(691)	(+771)
Dy^{3+}	e(iii)	−104	+229	5199	+95	1713	+218

Where signs are not given they have not been determined.
[a] Rabbiner, 1969. [b] Nara and Schlesinger, 1971. [c] Parameters for cubic symmetry for comparison. [d] Schlesinger and Nerenberg, 1969. [e] Eremin, Luks, and Stolov 1971; (i) is attributed to the fluorine centre, (ii) to type I(O^{2-}) centre, and (iii) to type II (? OH^-) centre; bracketed numbers refer to SrF_2.

Nara and Schlesinger (1971) have interpreted the absorption spectrum of Sm^{3+} in trigonal sites in CaF_2 in the spectral region 0·9 to 1·7 μ. This absorption occurs between the ground state and the ^6F_J manifolds and $^6H_{\frac{15}{2}}$, which lie between 6000 and 11 000 cm^{-1}; hence they give information about different levels from the fluorescence measurements. This data has been interpreted using a complete intermediate-coupling scheme, starting with accurate free-ion wavefunctions and taking into account off-diagonal crystal-field matrix elements between all states up to 11 000 cm^{-1}. Above this there is a natural gap in the level diagram, the next excited state being $^4G_{\frac{5}{2}}$ at around 17 650 cm^{-1}, so that one can legitimately neglect other off-diagonal elements. The energies of 14 levels are fitted with an rms deviation of 40 cm^{-1} with the parameters given in Table 6.6. Considering the sophistication of the model used in this calculation it is surprising that the fit is not better.

It is not clear whether the two spectra correspond to the same trigonal site. As more than one trigonal site has been observed for many ions it is possible that they refer to different sites. Nara and Schlesinger conclude after using the intermediate-coupling model that the effect of J mixing is not very serious, so one would expect Rabbiner's assumptions to give reasonable results. The predicted energy levels for the ${}^6H_J$ manifolds obtained using Nara and Schlesinger's parameters give an rms deviation of 55 cm^{-1}, which is not significantly worse than the fit obtained by Rabbiner.

No epr has been observed for Sm^{3+} in trigonal sites. This might be expected if the ground state were $|\pm\frac{3}{2}\rangle$, for which $g_\perp \approx 0$, and the epr might be too weak to observe ($|\pm\frac{3}{2}\rangle$ is the ground state for both sets of parameters in Table 6.6). These analyses of Sm^{3+} spectra show that it is very difficult to obtain agreement between data from different sources, and indicates that there are so many difficulties in the interpretation of optical data that one must treat information about crystal fields obtained from this data with great caution.

The transitions ${}^6P_{\frac{5}{2}}$, ${}^6P_{\frac{7}{2}}$ to ${}^8S_{\frac{7}{2}}$ for Gd^{3+} in trigonal sites in CaF_2 have been observed by Schlesinger and Nerenberg (1969). The crystal-field parameters obtained from a good fit of the ${}^6P_J$ data are given in Table 6.6. There is clearly little relation between these parameters and either set for Sm^{3+}. Again it is not known whether these data refer to the same trigonal site, nor by how much the parameters for Gd^{3+} would be changed by using an intermediate coupling model.

Ensign and Byer (1972) have studied optical-emission and absorption spectra of rhombic sites of Er^{3+} in CdF_2 with monovalent charge compensators (Na^+, Li^+, K^+, and Ag^+), and they have correlated these spectra with the epr spectra whose g-values are given in Table 6.8 (p. 380). The overall crystal-field splitting of the lowest multiplet, ${}^4I_{\frac{15}{2}}$, was accounted for by a predominantly cubic crystal field with $A_4\langle r^4\rangle = -245$ cm^{-1} and $A_6\langle r^6\rangle = 40$ cm^{-1}; and the g-values of the ground doublet and the splitting of the $\Gamma_8^{(1)}$ states of ${}^4I_{\frac{15}{2}}$ were accounted for in terms of simple addition of axial terms along the $\langle 110\rangle$ direction. The charge-compensating ion was assumed to be substituting for Cd^{2+} at $(\frac{1}{2}\frac{1}{2}0)$.

Ensign and Byer (1973) have correlated epr and optical measurements on a trigonal centre of Er^{3+} in CdF_2 (Tr_2 in Table 6.9; p. 382). The energies of all of the levels of the ${}^4I_{\frac{15}{2}}$ ground state were determined. Arguments are advanced for supposing that the site Tr_1 is probably $Tr(O_{4s}^{2-})$ and that Tr_2 is probably $Tr(O_{4s}^{2-})$.

Luks, Saitkulov, and Stolov (1969) and Luks, Livanova, and Stolov (1970) have studied Dy^{3+} ions in a number of site symmetries. For crystals of CaF_2 grown in fluoridizing atmosphere, spectra of four centres were observed and partial energy-level diagrams for the manifolds $^6H_{\frac{13}{2}}$ and $^6H_{\frac{15}{2}}$ were constructed for three of the centres labelled B, C, and D. However, only the cubic centre C was identified. Centre D was associated with Na^+ doping and probably corresponds to the rhombic centre observed using epr. Crystals of CaF_2 and SrF_2 grown in a fluoridizing atmosphere, but annealed for several hours in air, showed two spectra. In each the Zeeman effect for one particular line showed that they have trigonal symmetry. It seems likely that they correspond to $Tr(O_{4s}^{2-})$ and $Tr(O_s^{\,-})$ centres of Yb^{3+}. Only some of the levels for $^6H_{\frac{15}{2}}$ and $^6H_{\frac{13}{2}}$ were found for each centre, but for both centres all levels of $^6H_{\frac{11}{2}}$ were found with overall separation of 449 cm^{-1} and 466 cm^{-1} for so called type I centres in CaF_2. Three properties suggest that type I centres correspond to $Tr(O_{4s}^{2-})$: (a) the crystal-field splittings are larger than for type II centres, (b) the somewhat smaller splittings for CaF_2 in spite of the smaller lattice constant, which was also observed for such centres formed by Gd^{3+} and (c) they correspond to stronger oxidizing conditions of formation. A more complete energy-level diagram for the rhombic centres formed by Na^+ doping is given by Al'tshuler, Eremin, Luks, and Stolov (1970) who observed their luminescence in CaF_2, SrF_2, and BaF_2. References to earlier work on Dy^{3+} is given in the three papers we have cited.

Eremin, Luks, and Stolov (1971) have analysed the crystal-field splittings of trigonal Dy^{3+} centres, both those discussed above and a third attributed to F_i^- ions in $(\frac{1}{2}\frac{1}{2}\frac{1}{2})$ sites. The crystal-field parameters are listed in Table 6.6. The calculations were performed by taking the known cubic crystal field and superimposing a point change appropriate to the model of charge compensation in the centre. An iterative procedure was then used to refine these parameters to fit the $^6H_{\frac{11}{2}}$, $^6H_{\frac{13}{2}}$, and $^6H_{\frac{15}{2}}$ levels of all three centres. Using these parameters the eigenstates were calculated and also the g-values of the ground states. For the fluorine centre they found $g_{\parallel} = 16{\cdot}2$, $g_{\perp} = 0$ for SrF_2 and $g_{\parallel} = 16{\cdot}3$, $g_{\perp} = 0$ for BaF_2; for the type I centre they found similar values, $g_{\parallel} = 19{\cdot}7$, $g_{\perp} = 0$ for SrF_2 and $g_{\parallel} = 19{\cdot}8$, $g_{\perp} = 0$ for BaF_2. These g-values are similar to those measured by Weber and Bierig (1964) in CaF_2, $g_{\parallel} = 16(1)$, $g_{\perp} \approx 1$. Eremin *et al.* identify this centre in CaF_2 with type I sites on the ground that the larger crystal field causes greater admixture of high J levels into the ground state and so causes departures from $g_{\perp} = 0$. For type II sites the fact that $A_0^2\langle r^2\rangle < 0$ leads to very different g-values, $g_{\parallel} = 1{\cdot}43$, $g_{\perp} = 10{\cdot}3$, which are close to those measured by Kask, Kornienko, and Rybaltovskii (1966) for a γ-irradiated sample prepared under oxygenating conditions. In support of their model Eremin *et al.* point out the similarities in the ratio A_4^3/A_4^0 for cubic (28) and fluorine (29) sites compared with that for type I sites (16), which indicates a much greater distortion from cubic symmetry for the latter site.

A number of levels in the energy-level diagram have been found by Voron'ko, Keminskii, and Osiko (1966*b*) for Nd^{3+} ions in two different

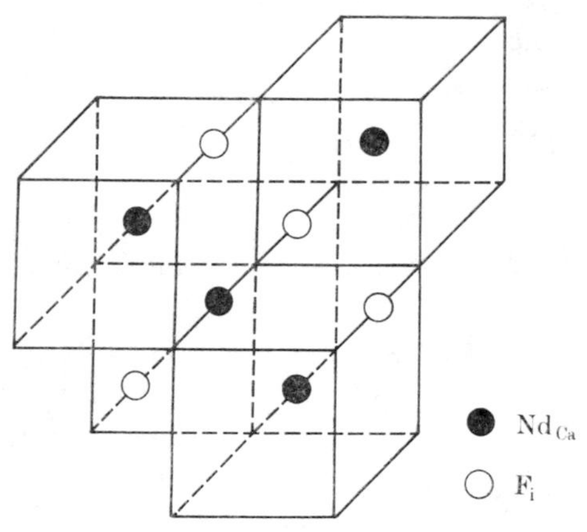

FIG. 6.8. Model of an aggregate of four Nd^{3+}–F^- pairs in CaF_2 proposed by Osiko and Shcherbakov (1971). Four cubes of the basic simple cubic F^- ion sublattice are shown with Nd^{3+} ions at each body centre in a substitutional site. The additional F^- ions are at the body centres of the adjacent four cubes which are not drawn. The whole complex occupies one complete unit cell of CaF_2 with atoms at each body-centre position.

rhombic sites in CaF_2 crystals grown in a fluoridizing atmosphere, in addition to the tetragonal centres discussed earlier. These spectra were correlated with rhombic epr spectra by measurement of an Orbach process in the spin-lattice relaxation, giving the energy of the first-excited state.

Osiko and Shcherbakov (1971) have calculated equilibrium concentration of aggregates of Nd^{3+}–F^- pairs where the F^- is assumed to occupy the nearest ($\frac{1}{2}00$) interstitial position. They have proposed that the rhombic centres observed by Voron'ko *et al.* are due to: (i) two Nd^{3+}–F^- pairs in adjacent lattice sites, (ii) four such pairs as shown in Fig. 6.8, but such an arrangement would give a trigonal spectrum for each Nd^{3+} ion. In either case one would expect sizeable interactions between the Nd^{3+} ions which would be observable in the epr spectrum, and Voron'ko *et al.* make no mention of such effects (but see § 6.6). Similar spectra have been observed by Voron'ko, Keminskii, and Osiko (1966*a*) for Er^{3+} in CaF_2. For both Er^{3+} and Nd^{3+} the effect of adding Y^{3+} has been studied. This tended to decrease the concentration of tetragonal centres and increase the concentration of more complex centres: however, for Er^{3+} no new spectra appeared, whereas for Nd^{3+} new spectra did appear. Analysis of the effects on Nd^{3+} spectra when associated with a range of rare-earth ions showed that this is an ion-size effect (Voron'ko, Mikaelyan, and Osiko, 1968). The shift of lines in both rhombic spectra are shown to be linearly dependent upon the difference in ionic radii of Nd^{3+} and the rare-earth ion codoped into crystals, except for a different trend between Er^{3+} and Lu^{3+}. The lack of effect on Er^{3+} when codoped with Y^{3+} may be due to the similar sizes of the two ions.

Toledano (1972) reports optical-fluorescence spectra of Nd^{3+} (mainly from ${}^4F_{\frac{3}{2}}$ to ${}^4I_{\frac{11}{2}}$) of four different sites in crystals of CaF_2 grown in oxidizing conditions. One of these sites only is produced when the crystals are doped with NdOF, and so it is presumed that the site involves only one O^{2-} ion. This site is shown to have trigonal symmetry by circular-dichroism measurements, and a table of a number of terms for the trigonal spectrum are quoted from Boccara (1971); the trigonal field is shown to be unusually large by the overlap of ${}^4F_{\frac{5}{2}}$ and ${}^2H_{\frac{9}{2}}$ which are clearly resolved for most other Nd^{3+} sites. The lack of epr is also cited as evidence for C_{3v} symmetry, as for positive $A_2^0\langle r^2\rangle$ one expects the lowest state to have a $g_{\perp} \approx 0$, which might lead to unobservably weak epr lines (see § 6.5.3). This site might be $Tr(O_s^{2-})$ since for Yb^{3+} the corresponding site has positive $A_2^0\langle r^2\rangle$. The other three sites which are produced by Nd_2O_3 doping are presumed to be rhombic sites. Toledano (1972) discusses possible models involving two O_s^{2-} ions and one F^- vacancy in the nn shell. He also discusses the optical evidence for covalent bonding between the Nd^{3+} ion and these neighbouring O_s^{2-} ions. Table 6.9 (p. 382) shows that a great variety of epr spectra have been observed in CaF_2 crystals doped with oxide, but no correlation has been made between optical and epr spectra.

Manthey and McClure (private communication; Manthey, 1972) have made an extensive study of the optical spectrum of Ce^{3+} in CaF_2. Spectra were observed for cubic sites, and for a variety of F^- and Na^+ compensated sites. The work on the $Tg(F_i^-)$ site was mentioned in

§ 6.4.1 and is included in Table 6.5 (p. 366). The symmetry of some of the sites was confirmed by uniaxial-stress measurements. Models of charge-compensation mechanisms were inferred from these symmetries and from comparison of point-charge calculations of crystal-field parameters with measured crystal-field splittings. This work is at an early stage, but promises to give a great deal of information about Ce^{3+} in a large variety of sites. Some of this information is discussed in § 6.5.1.

Amster and Wiggins (1969) report optical spectra of Eu^{3+} associated with oxygen ions, some of which may be due to $(Eu_2O_2)^{2+}$ centres (cf. Tm_2O_2 centres in § 6.5.9).

6.4.3. *Optical data: conclusion*

This discussion of optical measurements has shown that there is very little undisputable information about the crystal fields so far deduced from optical measurements. We have outlined the very great difficulties in this sort of work both in assigning the optical spectrum to a particular site so that enough information is available about its energy levels, and also in the theoretical interpretation which in many cases requires the use of intermediate-coupled wavefunctions and consideration of J mixing effects caused by off-diagonal matrix elements of the crystal field. For an unambiguous interpretation in terms of crystal-field parameters it is necessary to study the Zeeman effect of the optical transitions; a measurement of the g-values for the various states both checks the assignment to sites of a particular symmetry and helps in assigning wavefunctions appropriate to the levels. Clearly to obtain meaningful information from the optical data requires a lot of pains-taking work on the optical spectrum itself and as much correlation as possible with other experimental data.

6.5. Results of epr measurements

Table 6.7 lists the site symmetries for which epr has been reported for various trivalent ions, together with the symmetry of the ground state and any nearby excited states observable by epr for non-cubic sites. A casual glance at this table does not indicate any systematic trend in the occurrence of particular sites, because it does not indicate the relative probabilities of occurrence or any special conditions of crystal growth, heat treatment or irradiation that may be necessary to produce particular sites. In fact the observation of the appropriate trends requires very clean, careful chemistry in the crystal preparation and a study of the properties of the crystal as a function of impurity

TABLE 6.7
Summary of symmetries of non-cubic sites for ions with the configuration f^n.

Ion	f^n	Cubic	Tetragonal	Trigonal	Rhombic
Ce^{3+}	f^1	Ca?	Ca^2 Sr Ba	Ca^2	Ca^2
Pr^{3+}	f^2			Ca^2	
U^{4+}	f^2			Ca^3 Sr Ba	
Nd^{3+}	f^3	Ca	Ca^2 Sr		Ca^7Sr
U^{3+}	f^3		Ca Sr Ba		Ca
Sm^{3+}	f^5		Ca^2		
Pu^{3+}	f^5			Ba	
Gd^{3+}	f^7	Ca Ba Sr Cd	Ca Sr Cd	Ca^2 Sr^2 Ba Cd	
Tb^{4+}	f^7			Ca^2	
Tb^{3+}	f^8		Ca Sr^2	Ca	
Dy^{3+}	f^9	Ca	Ca	Ca^2	Ca^2
Ho^{3+}	f^{10}		Ca	Sr	
Er^{3+}	f^{11}	Ca Sr Ba	Ca Sr Ba	Ca^5 Sr Ba	
Tm^{2+}	f^{13}	Ca Sr Ba		Ca Sr Ba^2	Ba
Yb^{3+}	f^{13}	Ca Sr	Ca	Ca^3 Sr^3 Ba^3 Pb	Ca^5

The alkaline earth fluoride is indicated by the symbol of the alkaline earth element, and the index is the number of sites observed with a particular symmetry (sites which are closely similar have not been counted as separate). Some of the sites listed occur in very low concentration.

concentration. For example, many charge-compensation sites involve oxygen ions, so to eliminate these sites requires careful crystal growth in a reducing atmosphere. Centres involving oxygen may then be produced by heat treatment in air or water vapour. A considerable amount of recent Russian literature relates to the relative abundance of different sites and shows that there are definite trends. We have also already mentioned at the end of § 6.2.2 the work of Brown *et al.* (1969) on the relative stability of tetragonal and trigonal F_i^- charge-compensation sites.

Some systematic studies have been made of the effects of producing particular types of charge compensation. For example McLaughlan (1967) codoped crystals of CaF_2 with rare-earth and alkali ions, and Ensign and Byer (1972) have done the same thing for CdF_2. For all of the ions used this produced a simple (in the sense of Table 6.1) rhombic spectrum whose intensity was proportional to the alkali-ion concentration. For Yb^{3+} and Er^{3+} dopings crystals were grown with Li^+, Na^+, and K^+ also added giving the results shown in Table 6.8. It is thus clear that the association between the rare-earth ion and the alkali ion is a close one: it is likely that the alkali ion replaces the *nn* Ca^{2+} or Cd^{2+} ion.

TABLE 6.8

Rhombic spectra observed for Yb^{3+} *in* CaF_2 (*McLaughlan*, 1967) *and for* Er^{3+} *in* CdF_2 (*Ensign and Byer*, 1972) *when the rare-earth ion is codoped into the crystal with a monovalent metal ion.*

Rare-earth ion	Compensating ion	Ionic† radius	g_x	g_y	g_z
Yb^{3+}	Li^+	0·70	3·123(8)	3·008(8)	4·146(8)
Yb^{3+}	Na^+	1·00	3·289(5)	3·102(5)	3·926(5)
Yb^{3+}	K^+	1·37	3·180(5)	3·304(5)	3·817(5)
Er^{3+}	Li^+	0·70	5·412(5)	5·397(5)	9·071(5)
Er^{3+}	Na^+	1·00	5·898(5)	5·882(5)	8·274(5)
Er^{3+}	Ag^+	1·30	5·315(5)	6·473(5)	8·402(5)
Er^{3+}	K^+	1·37	4·999(5)	6·564(5)	8·333(5)

† The ionic radii, given in Å, are tabulated for eight-fold coordination and are related to those for six-fold coordination (see *Handbook of Chemistry and Physics*, 47th ed., edited by R. C. Weast and S. M. Selby, Chemical Rubber Publishing Co., Cleveland, Ohio, 1966) by the factor 1·03 (Evans, 1964, p. 44).

A similar rhombic spectrum was observed for Gd^{3+} by Smith and Cole (1970).

Another systematic study of this sort is the effect of doping the crystal with rare-earth oxide rather than fluoride. Most of the rhombic spectra (nearly all *complex* in the sense of Table 6.1) have been produced in this way, there being a number of such sites for some ions.

The g-values measured for the low-symmetry sites are listed in Table 6.9 (p. 382). We have indicated (a) where sites are well characterized and (b) where more than one researcher has observed the same site.

One of the problems is that of identifying the paramagnetic ion giving rise to the observed resonance, and small quantities of unexpected impurity can be misleading. Some of the rare-earth ions comprise 100% odd isotope with a characteristic hyperfine structure, but these are the non-Kramers ions (Pr^{3+}, Tb^{3+}, Ho^{3+}, Tm^{3+}) which are difficult to observe by epr. The other ions, except for Ce^{3+} which has only even isotopes, are mainly even isotopes with small percentages of odd isotopes. The epr consists of a strong central line flanked by a weak hyperfine structure. The hyperfine structure is characteristic, but the intensity of the spectrum is often so low that one would not expect to observe the hyperfine structure. In some cases the nature of the ion may be inferred from the g-values themselves; for example, a low symmetry distortion of a cubic site does not greatly change the sum of the three principal g-values of a doublet state, so that for an axial site $(g_{\parallel}+2g_{\perp}) \approx 3g_{\text{cubic}}$.

Many of the epr papers which describe the data listed in Table 6.9 contain discussions of possible mechanisms for producing sites of the particular symmetries observed. We shall not discuss each case, but we shall illustrate some of the arguments which may be used to handle the data by taking particular examples.

Most of the epr data refer to rare-earth ions in alkaline earth fluorides, and these results will be discussed first. Each configuration f^n is discussed separately, and as the data for the few actinide ions which have been measured indicate that they are similar to the rare-earth ions, configurations $4f^n$ and $5f^n$ will be discussed in the same section. For the first half of the group the f^n configurations are discussed in order of increasing n (except that Yb^{3+} $(4f^{13})$ is linked with $4f^1$). The second half of the shell is much less well understood and the order of sections, f^6 and f^8, f^{11}, and f^9 follows roughly an order of decreasing understanding. There follows a section on $Tm^{2+}(4f^{13})$, the only example of a non-cubic site for a divalent ion observed by epr. In these sections the sites for various ions are referred to by the label which has been assigned to them in Table 6.9, so these sections should be read in connection with reference to Table 6.9. The S-state configuration f^7 is discussed after all of the others because its spin-Hamiltonian is different and therefore requires special treatment. Section 6.5.10 discusses some non-cubic sites in ThO_2 and CeO_2 and §6.5.11 deals with $3d$ transition-group ions. Section 6.5.12 discusses a different type of information available from epr measurements of the Orbach-relaxation processes in the spin-lattice relaxation time, and § 6.5.13 discusses some recent measurements of the effect of external electric fields on the epr spectrum.

6.5.1. f^1 *and* f^{13}

We shall start by examining $Ce^{3+}(4f^1)$ and $Yb^{3+}(4f^{13})$. The single-electron or single-hole configuration is particularly simple as there are only two J manifolds, and the J values are small. Small J values lead to a small number of coefficients in the expansion of the states in $|J, M_J\rangle$ substates, and so make numerical analysis of the data easier. For Yb^{3+} the excited $J = \frac{5}{2}$ manifold at about 10 000 cm^{-1} is little admixed by the crystal field into the $J = \frac{7}{2}$ ground manifold, but for Ce^{3+} the excited $J = \frac{7}{2}$ manifold is only at about 2000 cm^{-1} so that there

TABLE 6.9

Details of sites observed for various ions with the configuration f^n

Ion	Host	Symmetry	g_1	g_2	g_3, Δ	mean g	$(g_{\|}-g_{\perp})$ δ(deg)	Reference	Confidence
Ce^{3+}	CaF_2	$Tg(F_i^-)$	3·038(3)	1·396(2)		1·943	+1·642	B59	a, b
		$Tg(H_i^-)$	3·150(3)	1·330(1)		1·937	+1·820	A68, K68	a, b
		$Tg(D_i^-)$	3·160(3)	1·327(1)		1·938	+1·833	A68, K68	a
		Tg_2	0·725(1)	2·402(2)		1·843	−1·677	M67	
		Tr_1	2·38(3)	<0·1		0·8	+2·38	W64	
		Tr_2	3·673(2)	<0·3		1·2	+3·6	M66	
		Rh_1	2·469(2)	1·288(2)	0·340(2)	1·366	0°	M67	
		Rh_2	0·844(1)	0·22(5)	3·286(1)	1·443	13·8°	M66	
		Rh_3	3·533(2)	<0·2	0·68(5)	1·47	40·7°	C72	
	SrF_2	$Tg(F_i^-)$	2·854(3)	1·472(2)		1·933	+1·382	A65, Br69	b
	BaF_2	$Tg(F_i^-)$	2·601(3)	1·555(2)		1·904	+1·046	A65	
Pr^{3+}	CaF_2	Tr_1	5·65(1)	<0·1	Δ = 0·05(1) cm^{-1}			M66b	c
		Tr_2	5·83(1)	<0·1	Δ = 0·05(1) cm^{-1}			M66b	c
Nd^{3+}	CaF_2	$Tg(F_i^-)$	4·412(8)	1·301(2)		2·338	+3·111	B56	a, b, c
		$Tg(H_i^-)$	4·470(3)	0·980(3)		2·143	+3·490	A68, K68	b
		$Tg(D_i^-)$	4·80(1)	0·967(1)		2·24	+3·83	A68	
		Tg_2	1·780(2)	2·798(3)		2·459	−1·018	M67	
		Rh_1	3·099(5)	1·039(2)	0·405(5)	1·514	0°	M67	
		Rh_2	1·17(1)	5·1(1)	0·58(1)	2·28	30·2(5)°	K66	
		Rh_3	2·05(1)	0·25(1)	4·3(1)	2·20	11·5(5)°	K66	
		Rh_4	1·94(1)	0·65(1)	4·4(1)	2·33	2·5(5)°	K66	
		Rh_5	1·83(3)	0·74(1)	4·65(5)	2·41	0°	K66	
		Rh_6	3·35(10)	2·08(5)	2·00(15)	2·48	0°	K67	
		Rh_7	1·2(1)	3·88(5)	2·14(5)	2·4	16(1)°	K65	
	SrF_2	$Tg(F^-)$	4·289(8)	1·505(2)		2·433	+2·784	B56	c
		Rh_6	3·62(2)	2.26(2)	1·95(2)	2·61	0°	K68	

Sm^{3+}	CaF_2	Tg(F_i^-)	0·000(6)	0·823(3)		0·548	−0·823	A65b, N	b, c
		Tg(H_i^-)	<0·1	0·868(1)		0·579	−0·868	A68	
		Tg(D_i^-)	<0·1	0·870(1)		0·580	−0·870	A68	
		Tg_2	0·907(10)	0·544(15)		0·665	+0·363	L64, W64	b, c
Tb^{3+}	CaF_2	Tg	17·768(20)		$\Delta = 0{\cdot}1711(2)\ cm^{-1}$			F62, K72	b, c
		Tr	17·28(1)	<0·25	$\Delta = 1{\cdot}056(1)\ cm^{-1}$			F62	c
	SrF_2	Tg_1	17·95(5)		$\Delta = 0{\cdot}48(1)\ cm^{-1}$			An68	c
		Tg_2	17·85(5)		$\Delta = 0{\cdot}69(1)\ cm^{-1}$			An68	c
Dy^{3+}	CaF_2	Tg	1·7(1)	2·82(5)		2·78	−1·1	B63	
		Tr_1	16(1)	<1		5·3	+16	B63	
		Tr_2	4·93(5)	1·50(5)		2·98	+3·43	B63	
		Rh_1	7·605(20)	11·79(5)	2·000(2)	7·13	0°	M67	
		Rh_2	1·55(3)	0·3(1)	15·8(5)	5·9	17·5(5)°	K66	
	ThO_2	Tr_2	1·625(2)	9·95(5)		7·18	−8·32	Ab67	
	CeO_2	Tr_2	1·632(2)	9·98(5)		7·20	−8·35	Ab67	
		Tr_1	2·24(1)	10·00(5)		7·41	−7·76	Z68	c
Ho^{3+}	CaF_2	Tg	14·8(1)		$\Delta = 0{\cdot}033(1)\ cm^{-1}$			Ko72b	c
	SrF_2	Tr	18·2(1)		$\Delta = 0{\cdot}66(1)\ cm^{-1}$			R69	c
Er^{3+}	CaF_2	Tg	7·78(2)	6·254(5)		6·76	+1·43	B59, R63	b
		Tg*	1·746(2)	9·16(1)		6·69	−7·41	B59, R63	b
		Tr_1	3·30(1)	8·54(2)		6·79	−5·24	R66, R63	b
		Tr_2	2·206(7)	8·843(10)		6·63	−6·637	R63, W64	b, c
		Tr_3	6·31(5)	2·14(4)		3·53	+4·17	W64	
		Tr_4	10·29(3)	1·475(5)		4·41	+8·81	Bo67	c
		Tr_5	—	2·51(5)				Bo67	
	SrF_2	Tg	10·04(5)	4·632(5)		6·63	+5·41	Br69	
		Tr_3	6·163(5)	7·038(7)		6·75	−0·875	Br69, An67a	b, c
	BaF_2	Tg	5·908(10)	7·411(10)		6·91	−1·503	Z64	
		Tr_3	5·94(2)	7·13(2)		6·73	−1·19	An67a	c
	CdF_2	Tr_2	2·875	8·334		6·514	−5·459	B72, E73	
		Tr_1	3·231	8·334		6·633	−5·103	E73	

TABLE 6.9 (*contd.*)

Ion	Host	Symmetry	g_1	g_2	g_3	mean g	$(g_{\|\|}-g_{\perp})$ δ	Reference	Confidence
Er^{3+}	ThO_2	Tg	3·462(3)	7·624(5)		6·24	−4·162	Ab65	
(cont.)		Tr	11·087(10)	4·337(5)		6·587	+6·750	Mi71	
	CeO_2	$Tr(F_s^-)$	10·30(5)	4·84(2)		6·66	+5·46	An67b, Ab66	a, b, c
		Tr_2	4·539(5)	7·399(7)		6·446	−2·860	Z69, Ab66	b
Yb^{3+}	CaF_2	$Tg(F_i^-)$	2·423(1)	3·878(1)		3·393	−1·455	R64, W64	a, b, c
		$Tr_1(O_{4s}^{2-})$	1·323(1)	4·389(4)		3·367	−3·066	M65, L63	a, b, c
		$Tr_2(O_s^{2-})$	1·421(1)	4·389(4)		3·400	−2·968	M65, L63	a, b, c
		$Tr_{3H}(O_{4s}^{2-}, H_s^-)$	1·516(2)	4·291(5)		3·270	−2·775	M65, R71	a, b, c
		$Tr_{3D}(O_{4s}^{2-}, D_s^-)$	1·512(2)	4·291(5)		3·269	−2·779	M65, R71	a, b, c
		Rh_1	3·289(5)	3·102(5)	3·926(5)	3·429	0°	M67, Ab68	b
		Rh_2	6·99(1)	1·355(2)	1·094(2)	3·15	3·2(2)	M66c	
		Rh_3	7·24(1)	0·992(2)	0·957(2)	3·06	26·3(2)°	M66c	
		Rh_4	—	1·241(2)	1·096(2)		8·8(5)°	M66c	
		Rh_5	6·45(2)	2·175(2)	1·667(2)	3·43	0°	M66c	
	SrF_2	Tr_1	1·345(2)	4·420(4)		3·395	−3·085	R64	
		Tr_2	2·804(3)	3·743(9)		3·430	−0·939	R64, An68b	b
	BaF_2	Tr_1	1·334(2)	4·405(5)		3·381	−3·071	R64	
		Tr_2	2·763(3)	3·768(3)		3·433	−1·005	R64, An68b	b
	PbF_2	Tr	0·971(5)	3·96(1)		2·96	−2·99	An68c	
	ThO_2	Tr_2	4·772(2)	2·724(1)		3·407	+2·048	Ab65	
		$Tr(F_s^-)$	4·495(2)	2·872(1)		3·413	+1·623	Rup	a
	CeO_2	Tr_2	4·733(4)	2·744(2)		3·407	+1·989	Ab66	
		Tr_1	4·17(2)	3·07(2)		3·44	+1·10	An67b, Z69	b
Tm^{2+}	CaF_2	Tr_1	4·74	2·74		3·41	+2·00	N69	
	SrF_2	Tr_1	4·790(3)	2·760(2)		3·437	+2·030	M72	
	BaF_2	Tr_1	4·713(3)	2·795(2)		3·438	+1·918	M72	
		Tr_2	2·23(1)	3·98(2)		3·40	−1·75	S67	
		Rh	5·71(1)	2·51(1)	0·36(1)	2·86	0°	F65	

U^{4+}	CaF_2	Tr_1	4·02(1)	<0·1	$\Delta = 0{\cdot}03(1)\ cm^{-1}$			M66b	
		Tr_2	5·66(2)	<0·1	$\Delta = 0{\cdot}03(1)\ cm^{-1}$			M66b	
		Tr_3	3·238(5)	<0·1	$\Delta = 0{\cdot}1$			Y62	
	SrF_2	Tr_3	2·85(5)					Y62	
	BaF_2	Tr_3	3					Y62	
U^{3+}	CaF_2	$Tg(F_i^-)$	3·535(3)	1·880(3)		2·332	+1·655	B56, T63	b, a
	SrF_2	Tg	3·433(8)	1·971(2)		2·358	+1·462	B56, T63	b
	BaF_2	Tg	3·337(2)	2·115(1)		2·522	+1·222		
	CaF_2	Rh	1·38(1)	2·85(2)	2·94(1)	2·39	19(1)°	M63	
Pu^{3+}	BaF_2	Tr	0·80(3)	1·300(5)		1·13	−0·50	E69	
Pb^{3+} ($3d^{10}6s$)	ThO_2	Tr	1·9706(7)	1·9642(7)		1·9663	−0·0064	Ro69	c

For axial sites $g_1 = g_{\parallel}$ and $g_2 = g_{\perp}$. For rhombic sites g_3 is always for the (110) direction; for those which are *simple* in the context of Table 6.1 the angle $\delta = 0°$ and g_1 corresponds to (001); for those which are *complex* δ is the angle between the principal direction of the g-tensor and (001). Δ is a zero-field splitting. The column labelled 'confidence' gives an indication of the reliability of the information about the site

[a] endor measurements (see § 6.3)

[b] reported by more than one reference

[c] characteristic hyperfine structure observed.

The sites are labelled Tg—tetragonal, Tr—trigonal, and Rh—rhombic. The numerical subscripts serve only to distinguish different sites of the same symmetry for any particular ion; sites with the same subscript for different ions are not necessarily correlated. When sites are well characterized they are labelled with a symbol in brackets which indicates the type of site, e.g. $Tg(F_i^-)$ corresponds to the tetragonal site with an interstitial F^- ion. For a complete discussion of the type of site indicated by these symbols see § 6.3.

References for Tables 6.9 and 6.16

A68 ASHBURNER, I. J., NEWMAN, R. C., and MCLAUGHLAN, S. D. (1968). *Phys. Lett.* **27A**, 212–4.

Ab65 ABRAHAM, M. M., WEEKS, R. A., CLARK, G. W., and FINCH, C. B. (1965). *Phys. Rev.* **137**, A138.

Ab66 ——, ——, ——, —— (1966). *Ibid* **148**, 350–2.

Ab67 ——, FINCH, C. B., RANBENHEIMER, L. J., SAFFAR, Z. M. EL. and WEEKS, R. A. (1967). *International conference on magnetic resonance and relaxation, Ljubljana* (1966) (Amsterdam: North-Holland Publishing Company).

Ab68 ABBRUSCATO, V. J., BANKS, E., and MCGARVEY, B. R. (1968). *J. Chem. Phys.* **49**, 903–11.

An65a ANTIPIN, A. A., KURKIN, I. N., CHIRKIN, G. K., and SHEKUN, L. YA. (1965). *Sov. Phys. Sol. St.* **6**, 1590–1.

An65b ——, ——, LIVANOVA, L. D., POTVOROVA, L. Z., and SHEKUN, L. YA. (1965). *Ibid* **7**, 1271–2.

An67a ——, ——, ——, ——, —— (1967). *Ibid* **8**, 2130–2.

An67b ——, ZONN, Z. N., IOFFE, V. A., KATYSHEV, A. N., and SHEKUN, L. YA. (1967). *Ibid* **9**, 521–2.

An67c ——, KATYSHEV, A. N., KURKIN, I. N., and SHEKUN, L. YA. (1967). *Ibid* **9**, 1070–3.

An68a ——, LIVANOVA, L. D., and SHEKUN, L. YA. (1968). *Ibid* **10**, 1025–9.

An68b ——, KATYSHEV, A. N., KURKIN, I. N., and SHEKUN, L. YA. (1968). *Ibid* **9**, 2684–9.

An68c —— and KURKIN, I. N. (1968). *Ibid* **10**, 994–5.

An68d ——, ZONN, Z. N., KATYSHEV, A. N., KURKIN, I. N., and SHEKUN, L. YA. (1968). *Ibid* **9**, 2080–3.

An71 —— and SKREBNEV, V. A. (1971). *Ibid* **12**, 1728–9.

B56 BLEANEY, B., LLEWELLYN, P. M., and JONES, D. A. (1956). *Proc. Phys. Soc.* **69B**, 858–60.

B59 BAKER, J. M., HAYES, W., and JONES, D. A. (1959). *Ibid* **73**, 942–5.

B63 BIERIG R. W., and WEBER M. J. (1963). *Phys. Rev.* **132**, 164–7.

B64 BIERIG R. W., WEBER, M. J., and WARSHAW, S. I. (1964). *Ibid* **134**, A1504–16.

B72 BYER, N. E., ENSIGN, T. C., and MULARIE, W. M. (1972). *Bull. Am. Phys. Soc.* **17**, 310 (DG15).

Bo67 BOBROVNIKOV, YU. A, ZVEREV, G. M., and SMIRNOV, A. I. (1967). *Sov. Phys. Sol. St.* **8**, 1750–6.

Br69 BROWN, M. R., ROOTS, K. G., WILLIAMS, J. M., SHAND, W. A., GROTER, C., and KAY, H. F. (1969). *J. Chem. Phys.* **50**, 891–9.

C72 CHAMBERS, D. N. and NEWMAN, R. C., 1972. *J. Phys. C, Sol. St.* 997–1007.

E66 EVANS, H. W., and MCLAUGHLAN, S. D. (1966). *Phys. Lett.* **23**, 638–9.

E69 EDELSTEIN, N., MOLLET, H. F., EASLEY, W. C., and MEHLHORN, R. J. (1969). *J. Chem. Phys.* **51**, 3281–5.

E73 ENSIGN T. C. and BYER N. E. (1973). *Phys. Rev.* **B7**, 907–12.

F62 FORRESTER P. A. and HEMPSTEAD C. F. (1962). *Ibid* **126**, 923–30.

F65 —— and MCLAUGHLAN S. D. (1965). *Ibid* **138**, A1682–8.

K64 KASK N. E. KORNIENKO L. S., and FAKIR, M. (1964). *Sov. Phys. Sol. St.* **6**, 430–3.

K65 ——, KORNIENKO, L. S., and RYBALTOVSKII, A. O. (1965). *Ibid* **7**, 532–3.

K66 ——, ——, —— (1966). *Ibid* **7**, 2614–9.

K67 ——, ——, and LARIONTSEV, E. G. (1967). *Ibid* **8**, 2058–62.

K68 —— —— (1968). *Ibid* **9**, 1795–7.

Ka68 KAFRI, A., KIRO, D., YATSIV, S., and LOW, W. (1968). *Solid St. Commun.* **6**, 573–4.

Ka69 KARRA, J. S. and WALDMAN, H. (1969). *Phys. Rev.* **183**, 441–52.

Ko72a KORNIENKO, L. S. and RYBALTOVSKII, A. O. (1972). *Sov. Phys. Sol. St.* **13**, 1609–15.

Ko72b KORNIENKO, L. S. and RYBALTOVSKII, A. O. (1972). *Ibid* **13**, 1785–6.
L63 LOW, W. and RANON, U. (1963). In *Paramagnetic resonance*, ed. W. Low, Academic Press, New York, 1963, p. 167–77.
L64 —— (1964). *Phys. Rev.* **134**, A1479–82.
M63 MAHLAB, E., VOLTERRA, V., LOW, W., and YARIV, A. (1963). *Ibid* **131**, 920–2.
M65 McLAUGHLAN, S. D. and NEWMAN, R. C. (1965). *Phys. Lett.* **19**, 552–4.
M66a —— and FORRESTER, P. A. (1965). *Phys. Rev.* **151**, 311–4.
M66b —— (1966). *Ibid* **150**, 118–20.
M66c ——, FORRESTER, P. A., and FRAY, A. F. (1966). *Ibid* **146**, 344–9.
M67 —— (1967). *Ibid* **160**, 287–9.
M72 MARSH, D. (1972). *J. Phys. C, Solid State* **5**, 863–70.
Mi71 MICHOULIER, J. and WASIELA, A. (1971). *Comptes Rendus Acad. Sci. Paris* **271**, B, 1002.
N69 NEWMAN, R. C. (1969). *Adv. in Phys.* **18**, 545–663
R69 RANON, U. and LEE, K. (1969). *Phys. Rev.* **188**, 539–45.
R63 —— and LOW, W. (1963). *Phys. Rev.* **132**, 1609–11.
R64 —— and YARIV, A. (1964). *Phys. Lett.* **9**, 17–19.
R66 RECTOR, C. W., PANDEY, B. C., and MOOS, H. W. (1966). *J. Chem. Phys.* **45**, 171–9.
R71 REDDY, T. RS., DAVIES, E. R., BAKER, J. M., CHAMBERS, D. N., NEWMAN, R. C., and OZBAY, B. (1971). *Phys. Lett.* **36A**, 231–2.
R up REDDY, T. RS. (unpublished).
Ro69 ROHRIG, R. and SCHNEIDER, J. (1969). *Phys. Lett.* **30A**, 371–2.
S67 SABISKY, E. S. and ANDERSON, C. H. (1967). *Phys. Rev.* **159**, 234–8.
T63 TITLE, R. S. (1963). In *Paramagnetic resonance*, ed. W. Low, Academic Press, New York, 1963, p. 178–88.
V65 VORON'KO, YU. K., ZVEREV, G. M., MESHKOV, B. B., and SMIRNOV, A. I. (1965). *Sov. Phys. Sol. St.* **6**, 2225–32.
W64 WEBER, M. J. and BIERIG, R. W. (1964). *Phys. Rev.* **134**, A1492–503.
Y62 YARIV, A. (1962). *Ibid* **128**, 1588–92.
Z64 ZVEREV, G. M. and SMIRNOV, A. I. (1964). *Sov. Phys. Sol. St.* **6**, 76–9.
Z68 ZONN, Z. N., KATYSHEV, A. N., KURKIN, I. N., and SHEKUN, L. YA. (1968). *Ibid* **9**, 1682–3.
Z69 ——, ——, MITROFANOV, YU. V., and POL'SKII, YU. E. (1969). *Ibid.* **11**, 284–7.

are larger admixtures into the ground $J = \frac{5}{2}$ manifold. However, in both ions to quite good accuracy one may ignore the excited manifolds.

Both ions exhibit tetragonal, trigonal and rhombic spectra, and Yb^{3+} exhibits a cubic spectrum; Dvir and Low (1960) have observed a spectrum which may be due to Ce^{3+} in cubic sites. The ground state of Ce^{3+} in a cubic site is a Γ_8 quartet described by the states listed in Table 6.10.

For a large axial distortion one may often assume that the $A_2^0\langle r^2\rangle$ term dominates, though for some sites other crystal-field terms may make comparable contributions. However, one may make a first approximation by considering the axial sites as cubic in zero order with a perturbation in $A_2^0\langle r^2\rangle$. The effects of such a perturbation on a Γ_8 state such as those given in Table 6.10 are (a) to raise the degeneracy by separating the two Kramers' doublets (which of the two is lower depends upon the sign of A_2^0) and (b) to change the value of $\cos\theta$ by admixture from the excited state in the ground manifold.

TABLE 6.10

States of the Γ_8 quartet for Ce^{3+} in a cubic site referred to $\langle 001\rangle$ and $\langle 111\rangle$ axes. The first-order g-values of the states regarded as doublets are given for arbitrary θ and for the value of θ appropriate to cubic symmetry

	$\langle 001\rangle$ axis	
State	$g_{\parallel}$	$g_{\perp}$
$\lvert\pm\frac{1}{2}\rangle$	0·86	2·58
$\cos\theta\,\lvert\pm\frac{5}{2}\rangle+\sin\theta\,\lvert\mp\frac{3}{2}\rangle$	$\frac{48}{7}\cos^2\theta-\frac{18}{7}$	$\frac{12\sqrt{5}}{7}\cos\theta\sin\theta$
$\cos\theta=\sqrt{\frac{5}{6}}$	3·15	1·43
	$\langle 111\rangle$ axis	
State	$g_{\parallel}$	$g_{\perp}$
$\lvert\pm\frac{3}{2}\rangle$	2·57	0
$\cos\theta\,\lvert\pm\frac{5}{2}\rangle+\sin\theta\,\lvert\mp\frac{1}{2}\rangle$	$\frac{36}{7}\cos^2\theta-\frac{5}{7}$	$\frac{18}{7}\sin^2\theta$
$\cos\theta=\sqrt{\frac{5}{3}}$	2·00	1·14

Table 6.10 also lists the g-values which would be observed for the states if they were separated doublets; when they are degenerate in cubic symmetry the matrix elements of the Zeeman interaction between the two doublets invalidate the expressions for $g_{\perp}$.

Consider first the sites of tetragonal symmetry. The single tetragonal site observed for Yb^{3+} has been analysed in considerable detail using endor, and was discussed in §§ 6.3.1 and 6.3.2. It was shown to correspond to positive A_2^0. Fluorine-endor measurements of this site and one of the tetragonal sites for Ce^{3+} show that the site is $Tg(F_i^-)$. Site Tg_2 of Ce^{3+} has g-values very close to those for $\lvert\pm\frac{1}{2}\rangle$, whereas for the other state the maximum value of $g_{\perp}=1{\cdot}92$ corresponding to $\cos\theta=1/\sqrt{2}$ for which $g_{\parallel}=0{\cdot}86$. Hence it is clear that Tg_2 corresponds to $\lvert\pm\frac{1}{2}\rangle$ which is the ground state for negative A_2^0. Site $Tg(F_i^-)$ has g-values very close to those given in Table 6.10 for $\cos\theta=\frac{5}{6}$ and so corresponds to positive A_2^0 (the expected sign, cf. Yb^{3+} above). Exact agreement is not expected between the experimental g-values and those calculated from the above states because there are small g-shifts due to covalency, virtual-phonon effects (Inoue, 1963) and admixtures from the nearby $J=\frac{7}{2}$ manifold.

Spin-lattice relaxation measurements on Ce^{3+} in $Tg(F_i^-)$ sites show no Orbach relaxation process indicating that the closest excited state is at an energy higher than 150 cm^{-1}. This level was estimated to be at 167 cm^{-1} by Manthey and McClure (private communication; Manthey, 1972), and their value of $A_2^0\langle r^2\rangle$ is given in Table 6.5 (p. 366).

The origin of the Tg_2 site for Ce^{3+} is uncertain, though it is associated with Na^+ ions, which may replace Ca^{2+} at the *nn* site at (100). The effective single additional negative charge at the Na_s^+ site would contribute a small positive $A_2^0\langle r^2\rangle$, but this could be overwhelmed by negative contributions due to displacements of the Ce^{3+} and its neighbouring F^- ions produced by the additional charge. Probably $A_2^0\langle r^2\rangle$ is small, but as the $|\pm\frac{1}{2}\rangle$ state is unadmixed one would need to know the energies of the excited states to obtain a value of $A_2^0\langle r^2\rangle$. No firm evidence of the nature of this site is available. Some evidence about the crystal fields at sites compensated with Na^+, both the tetragonal site and those of lower symmetry, has been obtained from optical measurements by Manthey and McClure (see § 6.4.2).

Similar arguments to those used above for the tetragonal centres may be used for the trigonal centres. For Ce^{3+} the centre Tr_1 has g-values close to those of $|\pm\frac{3}{2}\rangle$, whereas to obtain $g_\perp < 0{\cdot}1$ for the other state in Table 6.10 requires $\cos\theta \approx 1$ for which $g_\parallel = 4{\cdot}29$. Hence Tr_1 clearly corresponds to $|\pm\frac{3}{2}\rangle$ which is the ground state for negative A_2^0, provided that $A_2^0\langle r^2\rangle$ is the dominant term. Site Tr_2, however, fits much more closely with the other state in Table 6.10 for $\cos\theta = 0{\cdot}94$ ($g_\parallel = 3{\cdot}675$, $g_\perp = 0{\cdot}3$), and hence corresponds to positive A_2^0.

The site Tr_2 is produced by doping with oxide and hence may be associated with O^{2-} ions. Manthey (1972) has measured an optical spectrum from sites of trigonal symmetry in crystals which exhibit the Tr_2 epr spectrum. The fact that it shows the largest crystal field of all of the Ce^{3+} sites in CaF_2 investigated by Manthey suggests that the site may be $Tr(O_s^{2-})$, though this is unconfirmed. A study of the intensity of absorption bands which are observed when the crystals are warmed up above 4 K indicates that the first excited state (presumably $|\pm\frac{3}{2}\rangle$) lies at $60{\cdot}2(7)\ cm^{-1}$. If the splitting of Γ_8 into these two lowest levels were due entirely to an additional $A_2^0\langle r^2\rangle$ term it would require $A_2^0\langle r^2\rangle = +264\ cm^{-1}$.

Manthey observes the optical spectrum of another trigonal site for which the ground state is $\{\cos\theta\,|\pm\frac{5}{2}\rangle + \sin\theta\,|\mp\frac{1}{2}\rangle\}$, and the crystal field is smaller than that of the centre described above. It is suggested that this may be the $Tr(F_i^-)$ centre, and a possible association with the epr spectrum for Tr_1 is suggested, but neither of these suggestions is confirmed.

In contrast the trigonal centres for Yb^{3+} have been identified using ligand endor (§ 6.3.1). $Tr(O_s^{2-})$ has an O_s^{2-} ion replacing one of the *nn* F^- ions, $Tr(O_{4s}^{2-})$ has the arrangement shown in Fig. 6.3 (p. 353), and

$Tr(O^{2-}_{4s}, H^-_s)$ is the same as $Tr(O^{2-}_{4s})$ except that the single F^- ion in the first shell is replaced by H^- or D^-. All of these centres have an additional effective charge at the corner of the *nn* cube, and all correspond to positive A^0_2, as may be deduced from the sign of $(g_\parallel - g_\perp)$. For the Tr_2 centres of Ce^{3+} the positive sign of A^0_2 and the association with O^{2-} make it likely that the charge compensation in these centres is the same as either $Tr(O^{2-}_s)$ or $Tr(O^{2-}_{4s})$ for Yb^{3+}, but it is not known which.

For Yb^{3+} in SrF_2 and BaF_2 there are trigonal centres with g-values close to those for $Tr(O^{2-}_s)$ and $Tr(O^{2-}_{4s})$ centres in CaF_2, but it is not known to which of the two they correspond. In addition, for Yb^{3+} in SrF_2 and BaF_2, there are trigonal centres with positive A^0_2 of much smaller magnitude than in the $Tr(O^{2-}_s)$ and $Tr(O^{2-}_{4s})$ centres. Ranon and Yariv (1964) attribute these sites to F^-_i charge compensation in the interstitial site $(\frac{1}{2}\frac{1}{2}\frac{1}{2})$, but there is no endor evidence yet to confirm the suggestion. This is in accord with Brown *et al.* (1969) who suggest that this is the more stable of the F^-_i compensated centres for SrF_2 and BaF_2, and it is notable that $Tg(F^-_i)$ centres do not occur in these crystals.

Analysis of the rhombic sites is much more difficult because the very low symmetry makes all of the states for Ce^{3+} admixtures of the sort $\{a\,|\pm\frac{5}{2}\rangle + b\,|\pm\frac{1}{2}\rangle + c\,|\mp\frac{3}{2}\rangle\}$ and, although one can find values of a, b, and c to fit the g-values it is not possible to interpret these parameters in terms of crystal-field parameters without additional information. For Yb^{3+} the admixtures of $|\pm M_J\rangle$ states involve four coefficients which cannot be deduced from the g-values, so that even less information is available than for Ce^{3+}.

Site Rh_1 of Ce^{3+} is associated with Na^+ compensation, and it seems likely that it is produced by the replacement of a *nn* Ca^{2+} ion along $\langle 110\rangle$ by Na^+_s, but this possibility is not yet proved. A similar spectrum is observed for Yb^{3+} ions in crystals which contain Na^+. Further information about sites of Ce^{3+} ions of low symmetry produced by Na^+ compensation is provided by the optical measurements of Manthey and McClure (§ 6.4.2). This work is capable of giving an indication of the magnitude of the crystal field at each site. It also indicates the presence of a larger variety of sites than have been observed by epr. This work is at a preliminary stage, but it indicates that a lot of information might be obtainable from a thorough study, both by optical and epr spectroscopy, of the Na^+ compensated sites.

Various models have been suggested for Rh_2 of Ce^{3+} but they are purely speculative. The sites are produced when crystals are doped with

Ce_2O_3 suggesting some form of oxygen compensation. Rhombic sites are also produced by oxide doping for other rare-earth ions, there being four such sites for Yb^{3+} and five for Nd^{3+}. Most of these sites are *complex*, and only one of the principal axes of the g-tensor coincides with a symmetry axis of the crystal, i.e. $\langle 110 \rangle$. It follows from this that the site has reflection symmetry in the $\{110\}$ plane, but the angle δ through which the other principal axes are tilted away from crystal-symmetry directions is not simply related to the position of the charge-compensating ion. Hence little can be done to interpret *complex* rhombic spectra.

A third rhombic site Rh_3 has been found by Chambers and Newman in samples of CaF_2 containing Ce^{3+} and a low oxygen concentration into which hydrogen has been diffused. The epr spectrum was correlated with infra-red absorption due to a local mode of Ce^{3+}–H^- pairs. Such local modes for Re^{3+}–H^- (or D^-) pairs were found for all of the trivalent rare-earth ions using samples prepared in the same way as the Ce^{3+} sample. It is suggested that these centres may be sites in which two *nn* F^- ions have been replaced by O_s^{2-} and H_s^- (no axially symmetric sites are observed suggesting that these two ions do not replace F^- ions on either side of Re^{3+} along the same $\langle 111 \rangle$ direction). This proposed model has not yet been confirmed by endor.

6.5.2. f^2

In contrast to Ce^{3+}, Nd^{3+}, and Yb^{3+} oxide doping for $Pr^{3+}(4f^2)$ and $U^{4+}(5f^2)$ produces trigonal sites. No Pr^{3+} centres have been observed in crystals grown in fluoridizing atmospheres, but U^{4+} centres with trigonal symmetry have been observed by Yariv (1962). The assignment of the spectra to U^{4+} rather than U^{3+} was based upon the conditions necessary to prepare the specimens and upon the shape of the epr lines and the requirement of an rf field parallel to the trigonal axis, both of which are characteristic of systems with an even number of unpaired electrons. Title, Sorokin, Stevenson, Pettit, Scardefield, and Lankard (1962) independently assigned the epr spectrum to U^{4+} after studying the optical spectrum of the centres. Crystals grown with oxide doping exhibit two other U^{4+} trigonal centres and two Pr^{3+} trigonal centres. The ground state for cubic symmetry is Γ_5, whose components are listed in Table 6.11. The fact that the doublet is the ground state in the axial sites indicates that A_2^0 is positive in all of the sites if it is the dominant term. A positive term in $A_2^0\langle r^2 \rangle$ also increases the amplitude of $|\pm 4\rangle$ in the ground state and so increases the value of $g_{\parallel}$. To obtain the

TABLE 6.11

States of Γ_5 triplet ground state of f^2 in a cubic field referred to a $\langle 111\rangle$ axis. The g-values are appropriate to the separated doublet state

State	$g_{\parallel}$	$g_{\perp}$
$\sqrt{\frac{28}{54}}\,\lvert\pm4\rangle-\frac{5}{\sqrt{54}}\,\lvert\pm1\rangle+\frac{1}{\sqrt{54}}\,\lvert+2\rangle$	4·0	0
$\lvert+3\rangle \quad \lvert 0\rangle \quad \lvert-3\rangle$	0	0

observed g-values for Pr^{3+} requires sizable admixtures of excited states by $A_2^0\langle r^2\rangle \approx +300\ \text{cm}^{-1}$, and the admixtures are similar for the two sites. This strongly suggests the same sort of sites as $\text{Tr}(O_s^{2-})$ and $\text{Tr}(O_{4s}^{2-})$ for Yb^{3+} but there is no supporting evidence for this. The U^{4+} results are more difficult to discuss because of the unknown effects of intermediate coupling, though McLaughlan (1966*b*), suggests that he may have observed resonances for cubic U^{4+} at around $g = 2$, regarding the Γ_5 states as $S = 1$, which is what one expects for LS coupling. This suggests that for U^{4+} the site Tr_1 has very little distortion as its g-value is very close to that given in Table 6.11; ultrasonic paramagnetic resonance (upr) transitions have also been observed for this centre by Wetsel and Donoho (1965) giving further proof that it is U^{4+}; such transitions would not be expected for an ion with an odd number of electrons, and indeed were not observed by Wetsel and Donoho for the U^{3+} centre in CaF_2. Bowden, Meyer, McDonald, and Stettler (1969) have discussed the transferred hyperfine structure (thfs) in the epr and upr of this centre. They suggest that the small distortion of the centre is produced by Jahn-Teller self distortion and quote a private communication from Wilkinson and Hartmen indicating that the spin-lattice relaxation of the doublet shows that the singlet lies only about 6 cm^{-1} above the doublet. The thfs is interpreted in terms of interaction with eight *nn* F^- ions in a centre with inversion symmetry. The Tr_2 site for U^{4+} is similar to the two Pr^{3+} sites and presumably arises from the same mechanism. Tr_3 does not seem to be explicable in terms of a dominant $A_2^0\langle r^4\rangle$ term as the g-value implies a reduced amplitude of $\lvert\pm4\rangle$ in the ground state corresponding to negative $A_2^0\langle r^2\rangle$, which would make the singlet the ground state. Certainly the various trigonal sites for U^{4+} appear to require considerably different crystal fields, unlike those for Pr^{3+}.

6.5.3. f^3

The tetragonal site $Tg(F_i^-)$ for $Nd^{3+}(4f^3)$ and the equivalent actinide ion $U^{3+}(5f^3)$ was the first site of this sort to be observed for rare-earth ions in alkaline earth fluorides. It was deduced, primarily from the thfs of the surrounding fluorine nuclei, that the charge compensation was due to an interstitial F_i^- ion along ⟨100⟩ (Bleaney *et al.* (1956)). This has been confirmed by endor but the arguments from thfs are a nice illustration of what can be deduced in this way.

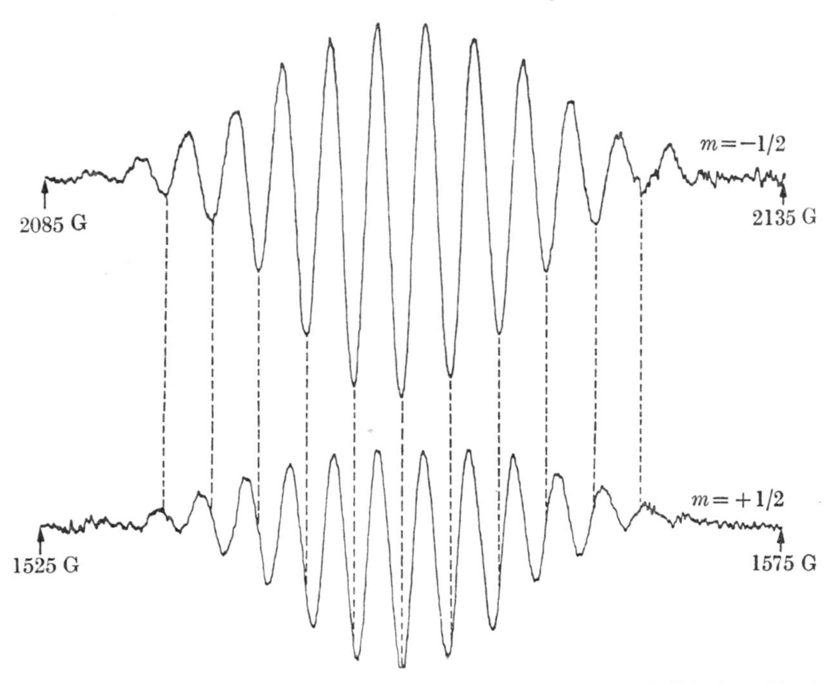

Fig. 6.9. Transferred fluorine-hyperfine structure on the epr line of Yb^{3+} in cubic sites in CaF_2 observed by Ranon and Hyde (1966). The plotted parameter is the second derivative of χ'' as a function of applied magnetic field. The symmetrical structure with a clear central line indicates that there are an even number of ^{19}F nuclei.

The arguments are based upon the details of the thfs pattern. If the paramagnetic ion interacts with n equivalent nuclei with nuclear spin I these will give $2nI$ equally spaced lines if nI is half integral and $(2nI+1)$ if nI is integral. The intensity of these lines increases as a binomial series towards the centre of the pattern; Fig. 6.9 shows such a pattern for Yb^{3+} in a cubic site where there are eight equivalent F nuclei. Counting the number of lines may be difficult because the weak outer lines may be lost in noise, but it is immediately obvious whether the number of lines is odd, when there is a single central line, or even when there is no central line. For integral I the pattern always has a central line; but for half-integral I, which is more relevant for CaF_2, the existence of a central line indicates that n is odd. When several groups of equivalent nuclei are present each thfs line from one group is further split by the thfs of the other groups, more distant groups giving an unresolved structure which merely contributes to the linewidth. The absence of a central line indicates that one of the groups with resolvable thfs comprises an odd number of equivalent nuclei.

For trigonal symmetry with H along the axis the nuclei must be equivalent in groups which are multiples of three if the nuclei lie off the axis, and nuclei on the axis may be equivalent in pairs or unique depending upon whether there is a centre of inversion symmetry or a plane of reflection symmetry perpendicular to the axis. Hence one nearly always expects to have no central line and one can only obtain information from the actual details of the thfs pattern. For tetragonal symmetry however, when H is along the axis, the nuclei off the axis must be equivalent in multiples of four. Hence the absence of a central line indicates that there is a nucleus of half-integral spin on the axis.

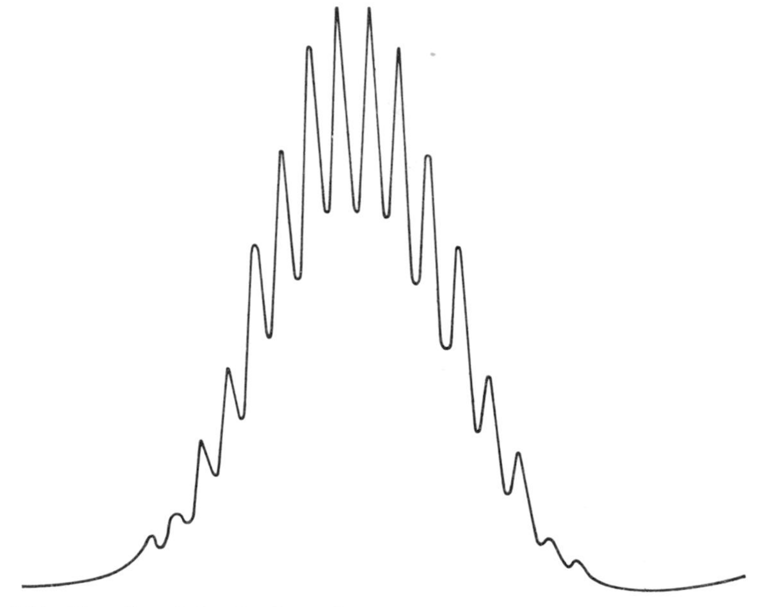

FIG. 6.10. Transferred fluorine-hyperfine structure on the epr line of U^{3+} in $Tg(F_i^-)$ sites in CaF_2 observed by Llewellyn (1956). χ'' is plotted directly as a function of applied magnetic field. The absence of a central line indicates that there is an odd number of ^{19}F nuclei, giving an indication of the presence of the charge-compensating interstitial fluorine ion.

For U^{3+} in $Tg(F_i^-)$ sites in CaF_2 the pattern shown in Fig. 6.10 has no central line indicating the presence of the charge-compensating interstitial F^- ion on the $\langle 100\rangle$ axis.

The ground state of Nd^{3+} in the cubic site is Γ_8 whose components are listed in Table 6.12. The values of a, b, and c and $\cos\theta$ are listed by Lea, Leask, and Wolf (1962) as a function of their parameter x (see § 5.2.4), and both optical and epr data indicate that $x \approx -0{\cdot}6$. The g-values for tetragonal sites cannot be made to fit the second state for any choice of $\cos\theta$, but correspond well with the first state for $a = 0{\cdot}8$, $b = 0{\cdot}6$, and $c \approx 0$. The splitting of the Γ_8 state for Nd^{3+} is not very sensitive to $A_2^0\langle r^2\rangle$ and the observed splitting is probably dominated by the change in $A_4^n\langle r^4\rangle$. Whether the splitting is produced by $A_2^0\langle r^2\rangle$ or by $A_4^n\langle r^4\rangle$, the observed ground state corresponds to a positive sign for

TABLE 6.12

States of Γ_8 quartet for Nd^{3+} in a cubic crystal field referred to a $\langle 001\rangle$ axis. The values of a, b, and c and of $\cos\theta$ correspond to $x \approx -0{\cdot}6$ taken from Lea, Leask, and Wolf (1962) and the g-values are calculated for isolated doublets

State	$g_{\parallel}$	$g_{\perp}$
$a\,\lvert\pm\frac{9}{2}\rangle+b\,\lvert\pm\frac{1}{2}\rangle+c\,\lvert\mp\frac{7}{2}\rangle$	$\frac{8}{11}(9a^2+b^2-7c^2)$	$\frac{8}{11}(6ac+5b^2)$
$a = 0{\cdot}463$, $b = -0{\cdot}556$		
$c = 0{\cdot}691$	$-0{\cdot}8$	$+2{\cdot}51$
$\cos\theta\,\lvert\pm\frac{5}{2}\rangle+\sin\theta\,\lvert\mp\frac{3}{2}\rangle$	$\frac{8}{11}(8\cos^2\theta-3)$	$\frac{16}{11}21\cos\theta\sin\theta$
$\cos\theta = 0{\cdot}999$, $\sin\theta = 0{\cdot}046$	$+3{\cdot}57$	$+0{\cdot}31$

the effective crystal-field parameter. The magnitude of the parameter is clearly large as there are certainly sizable components of $\lvert\pm\frac{7}{2}\rangle$ in the cubic state which are absent in the tetragonal state because of admixtures of excited states produced by the additional crystal field.

The tetragonal site observed for $U^{3+}(5f^3)$ has been shown by endor to be $Tg(F_i^-)$, and the similarity of the g-values to those of Nd^{3+} indicates a similar ground state with slightly different admixtures. The wavefunction defined by these g-values could be combined with the complete optical data for the two lowest J manifolds discussed in § 6.3 to obtain crystal-field parameters for the $5f^3$ ion which would make an interesting comparison with $4f^3$. U^{3+} also occurs in trigonal symmetry, which is not observed for Nd^{3+}, and $g_{\perp} \approx 0$ indicates that the ground state involves admixtures of $\lvert\frac{9}{2}\rangle$, $\lvert\frac{3}{2}\rangle$, $\lvert-\frac{3}{2}\rangle$, and $\lvert-\frac{9}{2}\rangle$; this corresponds to a negative value of $A_2^0\langle r^2\rangle$, if this is the dominant term. The epr is weak because of the very small value of $g_{\perp}$: if Nd^{3+}

occurs in similar sites the small value of $g_\perp$ may account for its not having been detected by epr.

The nature of the other tetragonal state for Nd^{3+}, produced by Na^+ codoping, is uncertain. For the second state in Table 6.12, $\cos\theta = 0{\cdot}85$ gives the closest approach to the observed g-values with $g_\parallel = 2{\cdot}03$ and $g_\perp = 3{\cdot}01$. However, a precise fit may be made for the other state with $a = 0{\cdot}61$, $b = 0{\cdot}66$, and $c = 0{\cdot}43$. The latter seems more likely, corresponding to a smaller positive $A_2^0\langle r^2\rangle$ than for the $Tg(F_i^-)$ site. However, Na^+ doping for Ce^{3+} gave a site with negative $A_2^0\langle r^2\rangle$, which would correspond to the other ground state for Nd^{3+}. There is clearly not enough evidence to decide between these two possibilities. The tetragonal site Tg_2 observed for Sm^{3+} by Weber and Bierig (1964), which is discussed below, may correspond to a negative value of $A_2^0\langle r^2\rangle$, and so may be a site of this type, though no specific Na codoping was used in the crystal preparation.

No trigonal sites have been found for Nd^{3+}, but there is a large number of rhombic sites. One of the rhombic sites Rh_6 corresponds to a situation in which the nn Ca^{2+} ion is replaced by a Nd^{3+} ion so forming a closely associated pair of Nd^{3+} ions. The charge compensation is supposed to be achieved by two F^- ions in the interstitial positions

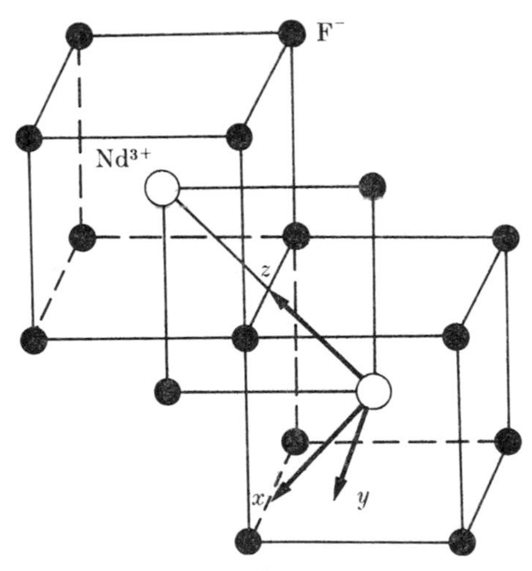

FIG. 6.11. Aggregate of two closely associated Nd^{3+}–F^- pairs, giving similar sites of rhombic symmetry for both Nd^{3+} ions. Note, however, that Catlow (1973) has performed a calculation on the stability of such a cluster for Y^{3+} ions and finds that a more stable situation is produced if the two F^- ions which are neighbours of both trivalent ions are allowed to relax towards interstitial sites. Such a distortion renders the rhombic symmetry complex. However, for La^{3+} ions rather than Y^{3+} Catlow found only a small binding energy, suggesting that such a distortion may only occur for the smaller rare-earth ions.

shown in Fig. 6.11; the interaction between these two ions is further discussed in § 6.6.

According to Kask, Kornienko, and Rybaltovskii (1965) γ-irradiation of crystals containing Nd^{3+} pairs converts one of the Nd^{3+} ions to Nd^{2+} so reducing the intensity of the Rh_6 spectrum and producing a new rhombic spectrum Rh_7. The presence of Nd^{2+} is confirmed by optical absorption spectroscopy. As the new rhombic spectrum is complex it cannot arise from a simple conversion of one of the Nd^{3+} to Nd^{2+}, as that would preserve the simple rhombic symmetry. It is suggested that an electron is removed from one of the F^- ions at a lattice site adjacent to both Nd^{3+} ions and is transferred to one of the Nd^{3+} ions. The new site is converted back to Rh_6 by heating to 150–200 °C.

6.5.4. f^5

The epr data for Sm^{3+} in $Tg(F_i^-)$ sites was discussed in comparison with the optical data in § 6.4.1. The g-values of the other site, Tg_2, correspond to $\cos\theta = 0{\cdot}88$ (the sign of $\sin\theta$ is undetermined as the relative sign of the g-values is not known). Ignoring J mixing gives a ground manifold specified by the matrix of eqn (6.2) for which the measured value of $\cos\theta$ gives a definite value to $|\tan 2\theta| = |2b/(a-d)|$. For the state observed to be the ground state, rather than $|\pm\frac{1}{2}\rangle$, requires $A_2^0\langle r^2\rangle$ either to be negative, or if it is positive to be less than 18 cm^{-1}.

Epr measurements on $Pu^{3+}(5f^5)$ are not directly comparable with those on Sm^{3+} because intermediate-coupling effects are so large in Pu^{3+} that even in a cubic crystal field the Γ_7 state is the ground state; for a pure ${}^6H_{\frac{5}{2}}$ state Γ_8 is the ground state as in Sm^{3+}.

6.5.5. f^8 *and* f^{10}

For the second half of the 4f-shell the large J values make the sort of analysis we have discussed so far very difficult to perform. Tb^{3+} and Ho^{3+}, which are only observed in axial sites, are easier to analyse than the others because the very large g-values indicate that for Tb^{3+} the ground state is almost pure $|\pm 6\rangle$ and for Ho^{3+} almost pure $|\pm 8\rangle$. Both of these ground states correspond to crystal fields with large positive $A_2^0\langle r^2\rangle$. For Tb^{3+} there is a splitting of the ground doublet due to the axial crystal field. The larger $A_2^0\langle r^2\rangle$ becomes relative to other crystal-field components the closer is the g-value to 18.

Antipin, Livanova, and Shekun (1968) have measured the Orbach spin-lattice relaxation rate for Tg_1 in SrF_2, and have found the first-excited state to lie at $\epsilon = 123$ cm^{-1}. Using this parameter, the g-value and the zero-field splitting $\Delta = 0{\cdot}48$ cm^{-1} of the ground doublet, and starting with the crystal-field parameters for cubic sites, they deduced the crystal-field parameters listed in Table 6.5 (p. 366), which strongly suggest that the site is $Tg(F_i^-)$. These parameters predict $\epsilon = 120$ cm^{-1}, $\Delta = 0{\cdot}29$ cm^{-1}, and $g_{\parallel} = 17{\cdot}97$; the disagreement in the value of Δ is not too serious as there are quite large contributions from excited J manifolds which were not taken into account. The magnitude of the Orbach relaxation rate, as opposed to its temperature dependence, was also predicted to within 25%. Considering the small amount of information used the crystal-field parameters deduced are in remarkable accord with those of other ions.

6.5.6. $4f^{11}$

Erbium is an example where there is some confusion about interpretations suggested by various authors. One cubic epr spectrum has been observed by several authors showing that Γ_7 is the ground state; then the optically determined positions of excited states show that $x \approx -0{\cdot}4$ with Γ_6 lying at 38 cm^{-1} and $\Gamma_8^{(1)}$ at 52 cm^{-1}. No resonance has been observed at $g = 6{\cdot}36$ appropriate to Γ_6, though Baker, Hayes, and Jones (1959) claim that the weak tetragonal resonance Tg_2 with $g_{\perp} = 9{\cdot}16$ corresponds to an excited state at 24 cm^{-1}; this was measured by the change in epr intensity between 20 and 14 K relative to the cubic Γ_7 resonance. This tetragonal resonance has also been observed by Ranon and Low (1963), who correctly determined $g_{\parallel} - 1{\cdot}746$, but they associated this with the ground state in tetragonal symmetry produced by the $Tg(F_i^-)$ site. This contrasts with Baker *et al.* who attributed the other tetragonal resonance Tg_1 to this ground state because it showed no temperature dependence of its intensity relative to the cubic Γ_7 resonance. Ranon and Low also observed Tg_1 in greater abundance than Tg_2, but the relative intensities of the two resonances appeared to depend a little upon the heat treatment of the sample, suggesting that they arose from completely independent sites. Rector, Pandey, and Moos (1966) show from optical-Zeeman studies that the only tetragonal site which they observe has a ground state with the g-values of Tg_1 and a first-excited state at 21 cm^{-1} with the g-values of Tg_2. The next-excited states are at 33 and 231 cm^{-1}. Further, their selection rules show that the ground state is derived from Γ_7, the first-excited state from Γ_6 and the second-excited state from Γ_8.

A model with simple addition of positive $A_2^0\langle r^2\rangle$ is not adequate to explain this situation because, as pointed out by Ranon and Low it causes an admixture of Γ_8 into Γ_7 which produces negative $(g_{\parallel}-g_{\perp})$. However, if the changes in $A_4^0\langle r^4\rangle$ and $A_6^0\langle r^6\rangle$ from the cubic values are correctly given by Tables 6.3 (p. 360) and 6.5 (p. 366) their effect is somewhat larger than that of $A_2^0\langle r^2\rangle$, and the decrease in the value of $A_4^0\langle r^4\rangle$ causes a positive value of $(g_{\parallel}-g_{\perp})$. The diagonal effect of $A_4^0\langle r^4\rangle$ on Γ_6 and the component of $\Gamma_8^{(1)}$ with which it is admixed are similar, and there is a large off-diagonal element pushing an antisymmetric admixture of these states downwards and the symmetric admixture upwards, presumably to 231 cm^{-1}. The g-values for the antisymmetric admixture are $g_{\parallel} = 1{\cdot}86$, $g_{\perp} = 9{\cdot}0$; these are very close to those of Tg_2, the first-excited state. The first-order effects of $A_n^0\langle r^n\rangle$ on Γ_7 and the other component of Γ_8 are small and the admixture of these two states is small, in agreement with the fact that the g-values of the ground state correspond to $\{\Gamma_7+0{\cdot}067\Gamma_8\}$. On the assumption that only these two close states are admixed the other state becomes $\{\Gamma_8-0{\cdot}067\Gamma_7\}$ which has a g-value in the $\langle 111\rangle$ direction of 3·8; this is close to the optically measured value of 3. Smirnov (1970) has calculated the g-values of the low-lying states using his crystal-field parameters (Table 6.5) and obtains excellent agreement. Although there is no endor evidence that the site is $Tg(F_i^-)$ the correspondence between Smirnov's parameters and other parameters in Table 6.5 leave little doubt that the site is $Tg(F_i^-)$.

The $Tg(F_i^-)$ site also appears to occur in SrF_2 with considerably larger $(g_{\parallel}-g_{\perp})$ though the complexities of calculating $(g_{\parallel}-g_{\perp})$ mentioned above make it difficult to compare this with the difference in crystal field parameters between CaF_2 and SrF_2 indicated by Table 6.5. The tetragonal site for BaF_2 has the opposite sign for $(g_{\parallel}-g_{\perp})$. This may indicate that the site is not $Tg(F_i^-)$, or that the changes in crystal-field parameters are sufficiently different from CaF_2 and SrF_2 to give negative $(g_{\parallel}-g_{\perp})$, or it may be that the state derived from Γ_6 has been pushed into the lowest position.

Brown *et al.* (1969) do not expect $Tg(F_i^-)$ to occur for BaF_2 as $Tr(F_i^-)$ is shown to be more stable, and for SrF_2 both sites are expected to occur as they have approximately equal probability. Several types of trigonal centres have been observed (see Table 6.9, p. 382). Tr_1 and Tr_2 in CaF_2 are similar and correspond to relatively large distortions; also Tr_2 is strongly associated with oxygen treatment suggesting that Tr_1 and Tr_2 may be the same sites as $Tr(O_s^{2-})$ and $Tr(O_{4s}^{2-})$ for Yb^{3+}. The much smaller distortion observed for the trigonal sites Tr_3 in SrF and BaF_2 may be associated with the site $Tr(F_i^-)$ expected by Brown *et al.*

The g-values of the other trigonal sites observed in CaF_2 are so far from the cubic g-values that it is difficult to construct a plausible model for them.

6.5.7. f^9

The situation for Dy^{3+} is rather more difficult to analyse than that for Er^{3+}. In cubic symmetry $\Gamma_8^{(1)}$ is the ground state, but epr is also observed for Γ_7 at 8 cm^{-1}, and there is another Γ_8 state at about 80 cm^{-1}. The low-symmetry crystal field is likely to admix all of these states considerably. As for Er^{3+}, the $A_2^0\langle r^2\rangle$ and $A_4^0\langle r^4\rangle$ terms are comparable and tend to oppose each other in the relevant matrix elements so that there are no firm starting points for the calculation of g-values. As Γ_7 and $\Gamma_8^{(1)}$ are so close in cubic symmetry it is likely that they will be severely admixed by the low-symmetry components, but a state $\{\sin\theta\,\Gamma_7+\cos\theta\,\Gamma_8^{(1)}\}$ has a very wide range of possible g-values. A state fairly close to a symmetric admixture of Γ_7 and $\Gamma_8^{(1)}$ has relatively low g-values, and so seems likely to form the basis of the only observed tetragonal site, but for an exact fit of the g-values one requires appreciable admixtures of $\Gamma_8^{(2)}$. A similar state with somewhat different admixtures would account for the trigonal site Tr_2. Tr_1 is more difficult to account for as a $g_{\parallel} \approx 16$ requires a state with a high amplitude of $|\pm\frac{15}{2}\rangle$ or $|\pm\frac{13}{2}\rangle$ though a large positive $A_2^0\langle r^2\rangle$ would give a ground state with a large amplitude of $|\pm\frac{15}{2}\rangle$.

Byer, Ensign, and Mularie (1972) report a trigonal centre in oxygen-fired CdF_2 containing Dy^{3+}, and the same group (Byer and Mularie, 1972) find that Na^+ compensation produces a *simple* rhombic site for Dy^{3+} in CdF_2.

6.5.8. $4f^{13}Tm^{2+}$

It is rather surprising to find a variety of non-cubic sites for Tm^{2+}, as charge compensation is unnecessary and a stable cubic site is formed in all alkaline-earth halides. Sabisky and Anderson (1967) report non-cubic sites which are present in low abundance in crystals of CaF_2, SrF_2, and BaF_2 doped with TmF_3 and reduced by vapour baking; for 0·02% Tm concentration about 1% of the Tm^{2+} sites are non-cubic. However, for some γ-irradiated BaF_2:Tm crystals it was found that about 30% of the Tm^{2+} ions could be switched to non-cubic sites by illumination with light of wavelength <5000 Å at 77 K. Cycling to room temperature or irradiation with light of wavelength >5500 Å converted the sites back to cubic symmetry.

These non-cubic sites have trigonal symmetry. If one assumes that the distorted site can be approximated by the simple addition of a term in $A_2^0\langle r^2\rangle$ to the cubic crystal field, the sign and magnitude of $(g_{\parallel}-g_{\perp})$ indicate that $A_2^0\langle r^2\rangle = +140\ \text{cm}^{-1}$. The optical spectrum of the trigonal centre is difficult to observe because it is obscured by the much more intense cubic spectrum. However, two new lines are observed in positions in quite good accord with the above model.

The nature of these centres is uncertain. One possibility is that of an F_i^- ion in a shallow trap at $(\frac{1}{2}\frac{1}{2}\frac{1}{2})$, originally suggested by Sabisky and Anderson, but it is difficult to devise a switching mechanism for this model. Anderson (private communication) has since suggested that the mechanism involved may be similar to that in the trivalent photochromic centres (see Chapter 7). An F centre at one of the *nn* F^- ion sites would form a paramagnetic system with an even number of electrons (as would $Tm^+(4f^{14})$ and a vacancy) which would not exhibit epr. Illumination might ionize cubic Tm^{2+} sites to form non-magnetic Tm^{3+} and free electrons which might then become trapped by the *F* centre forming a paired spin on the *nn* F^- ion site, which would produce a trigonal field with quite large positive $A_2^0\langle r^2\rangle$.

In support of the first model the g-values are similar to those of Yb^{3+} in Tr_4 trigonal centres in SrF_2 and BaF_2, which are thought to be due to F_i^- compensation at $(\frac{1}{2}\frac{1}{2}\frac{1}{2})$ interstitial sites, and $(g_{\parallel}-g_{\perp})$ is considerably smaller than that produced by O_s^{2-} substitution for a *nn* F^- ion in the $Tr(O_s^{2-})$ centre of Yb^{3+} (and the evidence for Gd^{3+} discussed below suggests that two electrons at a *nn* F^- ion vacancy may produce an even larger crystal field).

Another type of trigonal centre is thought to be produced by H^- substitution in a *nn* F^- ion site. Newman (1969) correlates the epr of this site in CaF_2 with the local mode of H^- in the trigonal centres mentioned in § 6.3.1. Marsh (1970) has shown that using a model in which the H^- ion is treated as a hard sphere, which was shown to be a fairly reasonable approximation in § 6.3.2, he could calculate the displacement of nearby atoms which minimize the energy. The resulting sign and magnitude of $A_2^0\langle r^2\rangle$ were calculated on a point-charge model and found to be in good agreement with the values deduced from $(g_{\parallel}-g_{\perp})$; they were found to vary from $-190\ \text{cm}^{-1}$ for CaF_2 to $-130\ \text{cm}^{-1}$ for BaF_2. The local modes of H^- ions have also been observed for Re^{2+}–H_s^- trigonal centres (Newman, 1969) with Yb, Eu, and Sm; and, after x- or

γ-irradiation to reduce the trivalent rare-earth ion, for Nd, Dy, Ho, and Er, but not for Ce, Pr, Gd or Tb (§ 2.5.5); but epr has been observed only for Tm^{2+}. Note that the ions for which this type of centre is not observed are those which form photochromic centres (§ 7.1).

The strangest of all Tm^{2+} non-cubic sites is described by Forrester and McLaughlan (1965) for crystals doped with Tm_2O_3 and irradiated with γ-rays. The spectrum has rhombic symmetry with large anisotropy. The ratio $A_{\alpha\alpha}/g_{\alpha}$ of the principal values of the hyperfine tensor and the g-tensor is different in the three principal directions α (569 MHz for [100], 467 MHz for [001] and inaccurately determined for $[1\bar{1}0]$, whereas all other known spectra of Tm^{2+} in cubic and axial sites have values of A/g very close to 325 MHz). Different values for this ratio in different principal directions is usually caused by J mixing. For Tm^{2+} with the next J manifold at 10 000 cm^{-1} this implies a very large crystal field, such as might be associated with a strongly bonded covalent complex. A possibility considered by Forrester and McLaughlan is that the resonance is due to Tm^{4+}, which is isoelectronic with Er^{3+}. Er^{3+} has not been observed at rhombic sites, but none of the axial sites has as low a mean g-value as 2·89; whereas one rhombic site for Yb^{3+}, isoelectronic with Tm^{2+}, does have a mean g-value of 3·06. However, there is no rhombic site of Yb^{3+} whose individual g-values resemble those of the Tm^{2+} site. The possibility that a $(Tm_2O_2)^{2+}$ complex replaces $(Ca_2F_2)^{2+}$ is rather discounted in a subsequent paper (McLaughlan *et al.*, 1966) as $(Yb_2O_2)^{2+}$ does not seem to occur in crystals grown in the same way with Yb_2O_3, but the possibility cannot be ruled out. It is suggested that γ-irradiation may remove an electron from one of the Tm^{3+} ions leaving a Tm^{2+} site of the correct symmetry. An endor experiment is clearly required to elucidate the atomic arrangement around this site.

We discuss another type of rhombic site for Tm^{2+} in § 6.6 which deals with pairs of interacting ions.

6.5.9. f^7

A number of non-cubic sites have been observed for Gd^{3+} where the ground state is $4f^7\ {}^8S_{\frac{7}{2}}$. The g-value is always close to 1·99 and the spin-Hamiltonian contains terms producing zero-field splittings which are due to high-order perturbations of the crystal field but are not simply related to the crystal-field parameters (see e.g. Wybourne, 1966). The sixth-order terms in the spin-Hamiltonian are very small and comparatively inaccurately determined, so we shall quote only second- and fourth-order parameters. Although these are not simply related to the crystal-field parameters their relative magnitudes for different sites do give an indication of the relative magnitudes of the crystal fields. Following Sierro (1963) we shall express the spin-Hamiltonian parameters in terms of coefficients C_n^m, where the spin-Hamiltonian, eqn (6.3), is written in terms of *spin* operators V_n^m which transform like spherical harmonics $Y_n^m(\theta\phi)$

$$\mathscr{H}_{\mathrm{cf}} = \sum_{mn} C_n^m V_n^m. \tag{6.3}$$

For cubic symmetry the value of C_4^0 is comparable for CaF_2, SrF_2, BaF_2, and CdF_2, and is about 0·1 cm^{-1}. Table 6.13 lists the parameters

TABLE 6.13

Spin-Hamiltonian parameters for Gd^{3+} *sites* (cm^{-1})

Tetragonal

Parameter	CaF_2(a)	SrF_2(b)	CdF_2(c)
C_2^0	1·929(6)	1·456(10)	0·054(3)
C_4^0	0·057(4)	0·049(8)	0·116(2)
C_4^4	0·045(2)	0·017(6)	0·067(2)

Trigonal

Type	Parameter	CaF_2(b)	SrF_2(b,e)	BaF_2(b,e)	CdF_2(d)
	C_2^0	0·582(1)	0·794	1·161	
I	C_4^0	0·035(1)			
	C_4^3	0·025(5)			
	C_2^0	2·160(2)	2·354(10)	2·21	0·9937(10)
II	C_4^0	−0·065(2)	−0·050(10)		−0·0713(10)
	C_4^3	0·065(15)	0·029(10)		0·0830(10)
	C_2^0		0·1784(10)	0·1767(10)	
III	C_4^0		0·1013(12)	0·0890(12)	
	C_4^3		0·1025(12)	0·0910(12)	

(a) Sierro and Lacroix (1960). (b) Sierro (1963). (c) Baker and Williams (1961). (d) Moret, Weber and Lacroix (1968). (e) Bevolo and Sook Lee (1972).

C_2^0 and C_4^m found for various sites; for comparison, $C = C_4^0$ in cubic symmetry.

Tetragonal sites with large axial terms C_2^0 are found for CaF_2 and SrF_2, presumably corresponding to $Tg(F_i^-)$ sites, but no tetragonal site is found for BaF_2, as predicted by Brown *et al.* (1969). In CdF_2 the axial field is very much smaller, presumably corresponding to a more distant charge-compensation mechanism. The identification of the site as

$Tg(F_i^-)$ is supported by the similarity of the parameters to those for $Tg(H_i^-)$ definitely identified by Jones *et al.* (1969) by the local mode of the hydride ion.

Several different types of trigonal centre are found as indicated by the order of magnitude of their C_n^m. They are produced by heat treatment in hydrolyzing conditions. Sierro (1963) suggests that Type II, with a strong crystal field, is produced by O^{2-} substitution for a *nn* F^- ion, like the $Tr(O_s^{2-})$ centre for Yb^{3+}. Type I, having significantly weaker crystal field, may be formed by OH^- replacing one of the *nn* F^- ions. Such a centre is not found for Yb^{3+}, although it was at one time suggested as a model for one of the trigonal Yb^{3+} centres. It is not clear whether this type I centre for Gd^{3+} is the same as $Tr(O_{4s}^{-2})$ for Yb^{3+}, as there is rather a small difference in the values of $A_2^0\langle r^2\rangle$ between $Tr(O_{4s}^{2-})$ and $Tr(O_s^{2-})$ for Yb^{3+}, in contrast to a large difference in the crystal fields for the two centres of Gd^{3+}. However, the lack of direct relationship between the spin-Hamiltonian and crystal-field parameters may make this comparison misleading. Type III was attributed by Sierro to a movement of the Gd^{3+} ion towards one of its *nn* F^- ion neighbours, as the trigonal field is weak. Such displacements do appear to occur when the site of the magnetic ion is large compared with the radius of the ion (see e.g. Sochava, Tolparov, and Kovalev, 1971; Richardson and Gruber, 1972), and this would account for its occurrence in BaF_2 and SrF_2 but not in CaF_2. Another possibility is that there is a charge-compensating F_i^- ion at $(\frac{1}{2}\frac{1}{2}\frac{1}{2})$ interstitial sites; Brown *et al.* (1969) suggest that this site is more stable than the $Tg(F_i^-)$ site in SrF_2 and BaF_2.

Bevolo and Sook Lee (1970*a*) have observed a fourth type of trigonal centre with $C_2^0 = 3{\cdot}75(1)$ cm^{-1} and $C_4^0 = -0{\cdot}030(4)$ cm^{-1}. This centre was produced by γ-irradiation of crystals containing $Tg(F_i^-)$ sites. The crystal field is even larger than for O^{2-} compensation, and may be due to a pair of electrons in an F^- ion vacancy (see § 7.3.2 for further discussion). Complete spin-Hamiltonian parameters for this centre are given by Yang *et al.* (1972), who also report the observation of a similar centre in SrF_2 (see also Bevolo and Sook Lee, 1970*b*).

Epr spectra of Tb^{4+}, which also has the configuration $4f^7$, have been reported by Kornienko and Rybaltovskii (1972). Crystals containing Tb^{3+} grown in a fluoridizing atmosphere were γ-irradiated, producing two rather similar trigonal Tb^{4+} spectra whose spin-Hamiltonian parameters are listed in Table 6.14. It is not clear why the C_4^0 and C_6^0 terms are comparable with C_2^0; Table 6.13 shows that C_4^n terms are

TABLE 6.14
Spin-Hamiltonian parameters for Tb^{4+} *ions in* CaF_2 (cm^{-1})

Type	C_2^0	C_4^0	C_6^0
I	0·492(9)	0·54(1)	0·53(1)
II	0·484(9)	0·57(1)	0·45(1)

usually considerably smaller than C_2^0, though not for type III centres, and for all of the previous examples $C_6^0 \approx 0{\cdot}001\ cm^{-1}$. Kornienko and Rybaltovskii suggest that when Tb^{3+} is oxidized to Tb^{4+} the decrease in ionic radius of about 15% makes the $Tg(F_i^-)$ site unstable and the $T(F_i^-)$ site the more stable (Brown *et al.*, 1969). It is suggested that one of the trigonal sites is associated with an F_i^- ion at $(\frac{1}{2}\frac{1}{2}\frac{1}{2})$, and that the other has a second F_i^- ion at $(-\frac{1}{2}-\frac{1}{2}-\frac{1}{2})$ so completing the charge compensation. The thermal properties are said to support this model, but there is no endor evidence. Also the difference between the crystal-field parameters for the two sites hardly seems large enough to be explained by a mechanism which one would expect roughly to double the crystal-field parameters.

6.5.10. *Non-cubic sites in* ThO_2 *and* CeO_2

Table 6.9 (p.382) lists a number of non-cubic sites in ThO_2 and CeO_2, mainly with trigonal symmetry. Here the trivalent rare-earth ion requires an additional positive charge for compensation. In § 6.3.1 we discussed endor evidence which showed that for some sites this is achieved by an F_s^- ion at a ligand O^{2-} site. For most of the sites the sign of $(g_{\parallel}-g_{\perp})$ is opposite to that for trigonal sites in alkaline earth fluorides.

Trigonal sites were reported for Gd^{3+} in ThO_2 but not measured. They were measured for Eu^{2+}, but here the zero-field splitting was so large that only the lowest doublet was observed with $g_{\parallel} = 1{\cdot}9678(5)$ and $g_{\perp} = 7{\cdot}876(2)$; that the resonance was due to Eu was confirmed by the hyperfine structure.

Non-cubic resonances have been reported for $Pu^{4+}(5f^3)$ in ThO_2. Here charge compensation is unnecessary, and it is suggested that the disparity in cation sizes allows the smaller substitutional Pu^{4+} ion to move to an off-centre position in the lattice (Richardson and Gruber, 1972).

6.5.11. *Non-cubic sites of* $3d^n$ *configurations*

Ions of the $3d$-group are not readily incorporated in alkaline-earth fluorides. Most measurements reported for non-cubic sites are for $3d^3$, either V^{2+} or Cr^{3+}, but Co^{2+} has been reported in an axial site in CaF_2 and two tetragonal sites have been observed for Fe^{2+} in CdF_2 (Zaripov, Kropotov, Livanova, and Stepanov, 1968*a*). The $3d^3$ centres observed all have quenched orbital angular momentum and long relaxation times, so that epr is observable at room temperature or 77 K. This is a characteristic of octahedral coordination, though Zaripov, Kropotov, Livanova, and Stepanov (1967) have tried to account for the quenching

TABLE 6.15
Spin-Hamiltonian parameters for $3d^3$ *ions* (b_2^0 *in* GHz)

		Cr^{3+}		
Crystal	$g_{\parallel}$	$g_{\perp}$	b^0	Ref.
CdF_2	1·966	1·97	54·0	a
PbF_2	1·970	1·969	37·05	b
CaF_2	1·961	1·97	59·4	c
		V^{2+}		
Crystal	$g_{\parallel}$	$g_{\perp}$	b_2^0	Ref.
CaF_2	1·935	1·943	42·31	c
CdF_2	1·943	1·955	43·44	a
SrF_2	1·927	1·944	47·62	d

[a] Zaripov, Kropotov, Livanova, and Stepanov (1968*a*). [b] Abdulsabirov, Zaitov, Zaripov, Livanova, and Stepanov (1970). [c] Zaripov, Kropotov, Livanova, and Stepanov (1967). [d] Zaripov, Kropotov, Livanova, and Stepanov (1968*b*).

of the orbital angular momentum in cubic coordination in terms of a very strong trigonal distortion. The spin-Hamiltonian parameters measured for $3d^3$ are given in Table 6.15.

Marshall and Nistor (1972) claim to have observed $Fe^{+}(3d^7)$ in tetragonal sites in ThO_2. This seems an unlikely valency for iron in this crystal; that iron is responsible for the epr is confirmed by ^{57}Fe hyperfine structure, and that the effective spin of the ion $S = \frac{3}{2}$ is inferred by a rather subtle argument. Only one transition is observed from a ground doublet, but large second-order Zeeman effects show that there is a low-lying state at about 3 cm^{-1}, suggesting that $S > \frac{1}{2}$. For a tetragonal site there are terms in the spin-Hamiltonian of type $C_4^4 V_4^4$ in eqn (6.3). For $S \geqslant 2$ such terms would lead to a variation as $\cos 4\phi$ in the g-value as the external field H was rotated in the plane perpendicular to the tetragonal axis. That no such variation is observed is taken as evidence that $S < 2$. However, if $S \geqslant 2$ this lack of variation merely sets an upper limit to the magnitude of C_4^4.

From a consideration of the stability of Fe^{+} at a Th^{4+} site it appears to us much more likely that the iron is in the Fe^{3+} state with $S = \frac{5}{2}$, though the epr data indicates a large crystal-field splitting (cf. Eu^{2+} in § 6.5.11). It seems that a strong enough case has not been made for Fe^{+} to rule out this possibility.

6.5.12. *Spin-lattice relaxation*

From time to time we have mentioned another way in which epr has been used to give information about the paramagnetic centres. If there are low-lying excited states, as there are for rare-earth ions, the spin-lattice relaxation rate contains a contribution from the Orbach process which varies exponentially with T. The complete relaxation rate is given by the following expression:

$$T_1^{-1} = AT + BT^9 + C\exp(-\epsilon/kT),$$

where A depends upon the fourth power of the microwave frequency, BT^7 becomes $B'T^9$ for ions with an even number of electrons, and ϵ is the energy of the first-excited state. At low temperatures the first term is dominant, and the other two terms become important at higher temperatures. Depending upon the magnitude of the parameters it is often quite difficult to detect the third term describing the Orbach mechanism. If ϵ is small enough, typically less than 100 cm^{-1} it is usually possible to detect the exponential dependence over a small temperature range, and hence to obtain a value of ϵ. If no exponential region is observed one can often put a lower limit on the value of ϵ. Table 6.16 summarizes the values of ϵ found in this way.

6.5.13. *Epr in a linear applied electric field*

If an electric field is applied to an ion which is in a site without inversion symmetry the epr line undergoes a shift which is linear in the electric field. As inversion in the cation site is one of the symmetry operations of the host crystal there will always be an equivalent site generated by such inversion for which the electric field effect has the opposite sign. Hence the epr line of the site shows a division into two components whose separation varies linearly with electric field E. As nearly all of the non-cubic sites discussed in this chapter lack inversion symmetry they are susceptible to study in an applied electric field.

TABLE 6.16

Energies of first-excited states derived from the Orbach relaxation rates.

Ion	Crystal	Symmetry	ϵ (cm^{-1})	Ref.
Nd^{3+}	CaF_2	Tg	74(5)	K64
	SrF_2	Tg	59	An71
Tb^{3+}	SrF_2	Tg_1	123	An68a
Er^{3+}	CaF_2	Tg	21(2)	Vo65
		Tr_2	50(2)	Vo65
		Tr_4	18(2)	Bo67
	SrF_2	Tr	85	A67c
	BaF_2	Tg	70	Z64
		Tr	75	A67c
	CeO_2	Tr	83	An68d
Yb^{3+}	CaF_2	Tg	>125	B64
		Tr_1	80(10)	Ka69
	SrF_2	Tr_4	>150	An68b
	BaF_2	Tr_4	>150	An68b
Ce^{3+}	CaF_2	Tg	>150	B64

For key to references see Table 6.9.

Kiel and Mims (1972) have observed such effects for Ce^{3+} ions in $Tg(F_i^-)$ sites in CaF_2, SrF_2, and BaF_2, which they describe by a spin-Hamiltonian:

$$\mathscr{H}_{\text{elec}} = E_x R_{15}(H_x S_z + H_z S_x) + E_y T_{15}(H_y S_y + H_z S_y) + \\ + E_z[T_{31}(H_x S_x + H_y S_y) + T_{33} H_z S_z].$$

The magnitudes of the electric-field parameters T_{nm} have been estimated theoretically by taking into account both the odd and even harmonics in the new crystal-field potentials which are induced at the Ce^{3+} ion when the electric field is applied to the sample. The basic crystal-field calculations were made using a point-charge, point-dipole model. The values obtained in this way were larger than the experimentally determined parameters, but can be brought into better agreement with the experimental values by allowing for screening effects and distortions in the lattice caused by the F^- ion (see §§ 6.3.2 and 6.3.3). An attempt was also made to interpret the changes in the parameters through the series of host lattices.

The fairly good agreement between theoretical and experimental values of the T_{nm} parameters indicates that the qualitative understanding of the electric-field effects is satisfactory. The remaining disagreements are due to the effects of shielding, distortion, and covalency, but the problem is so complex that one cannot hope to deduce a great deal from the details of the discrepancies. These electric-field parameters should be correctly predicted by a satisfactory theory of the $Tg(F_i^-)$ site, but it does not appear that they will be very helpful directly in constructing such a theory.

6.5.14. *Epr: a summary*

This discussion has shown that the main information available from epr studies is an identification of the paramagnetic ion occupying the site and of the symmetry of the site. Although epr gives evidence of a great number of sites of different symmetry, it is rare that one can deduce much more about the site without additional endor or optical evidence.

6.6. Interactions between magnetic ions

A large number of measurements have been made of energy transfer between paramagnetic ions in alkaline earth fluorides in connection with laser applications. The energy transfer is related to the interactions between the magnetic ions, but in such systems many different magnetic-ion sites are involved and many different pair configurations, and the effects measured are an average over all types of sites and pair configurations. Somewhat more specific information has been obtained in one recent experiment (Patel, Cavenett, and Davies, private communication) on CaF_2 doped with both Mn and Ce, in which it has been demonstrated that the ground state epr of Mn^{2+} may be observed by monitoring the change in luminescent intensity of either Mn^{2+} or Ce^{3+} ions. The spin dependence of the luminescence is claimed to be evidence that the exchange interaction is responsible for the energy transfer mechanism between the Ce sensitizer and the Mn activator: this is the first time that such a direct demonstration has been made. Usually it has not been possible to obtain more that very general information from measurement of energy transfer processes.

Investigations have been made however for both Nd^{3+} and Tm^{2+} of interactions between nearest-neighbour pairs in well-characterized sites using epr. In both cases the line joining the ions is along $\langle 110 \rangle$, called the z axis, and $\langle 001 \rangle$ is called the y axis; hence both the g-tensor and the interaction tensor have principal axes along x, y, and z. For Nd^{3+} it is probable that there is local charge compensation by two F_i^- ions arranged in the xz plane as shown in Fig. 6.11. The site of each ion has a large departure from cubic symmetry. In contrast, for Tm^{2+} pairs no charge compensation is required, so that there is little distortion at each Tm^{2+} site. The interaction may be written:

$$\begin{aligned}\mathscr{H} &= K_{xx}S_{1x}S_{2x}+K_{yy}S_{1y}S_{2y}+K_{zz}S_{1z}S_{2z} \\ &= J\mathbf{S}_1 \, . \, \mathbf{S}_2+A_{xx}S_{1x}S_{2x}+A_{yy}S_{1y}S_{2y}+A_{zz}S_{1z}S_{2z},\end{aligned}$$

where $A_{xx}+A_{yy}+A_{zz} = 0$. If both ions of the pair are identical, as they are for even isotopes of Nd^{3+}, the isotropic term couples the spins to form total spin states $\mathbf{S} = \mathbf{S}_1+\mathbf{S}_2 = 0, 1$ between which epr transitions are forbidden, so that epr allows the measurement of the parameters $A_{\alpha\alpha}$ only (see e.g. Baker, 1971). However, for Tm^{2+} the hyperfine interaction admixes the two spin states and allows transitions which enable one to measure both J and the components of $\mathbf{A}$. The parameters

TABLE 6.17

Spin-Hamiltonian parameters for pairs of interacting ions in nearest-neighbour positions.

	Tm^{2+}		Nd^{3+}	
Parameter	CaF_2	SrF_2	CaF_2	SrF_2
g_x			2·08(5)	2·26(2)
g_y	3·385(5)	3·522(5)	3·35(10)	3·62(2)
g_z	3·581(5)	3·383(5)	2·00(15)	1·95(2)
$A_{xx(\text{dip})}$	+0·0904	+0·0753	+0·031	+0·024
A'_{xx}	−0·0968(12)	−0·0746(12)	+0·035(15)	−0·005(1)
$A_{yy(\text{dip})}$	+0·0904	+0·0753	+0·044	+0·036
A'_{yy}	+0·0246(6)	+0·0192(6)	+0·118(15)	−0·008(1)
$A_{zz(\text{dip})}$	−0·1808	−0·1505	−0·075	−0·059
A'_{zz}	+0·0722(6)	+0·0533(6)	−0·143(11)	+0·012(1)
J	−0·0885(6)	−0·0321(6)	—	—

Interaction parameters are expressed in cm^{-1}. The measured values of $A_{\alpha\alpha}$ may be obtained by adding $A_{\alpha\alpha(\text{dip})}$, the dipolar contribution, to $A'_{\alpha\alpha}$, the non-dipolar contribution due to superexchange.

Tm^{2+} Baker and Marsh, 1971.

Nd^{3+} Kask and Kornienko, 1968.

measured for CaF_2 and SrF_2 are listed in Table 6.17. Preliminary measurements by Baker and Marsh (1971) for Tm^{2+} pairs in BaF_2 give $J = +0{\cdot}0134(10)\ cm^{-1}$ and $A_{yy} = +0{\cdot}0833\ cm^{-1}$.

The two sets of measurements for Nd^{3+} and Tm^{2+} show similar features. There are sizable non-dipolar contributions to the interaction, presumably due to superexchange, and the magnitude is similar for the two ions. Both sets show smaller interaction in SrF_2 than in CaF_2, corresponding to a decrease of exchange as the interionic distance increases. So far, no attempt has been made to calculate the magnitude of this superexchange from first principles, but these systems offer a good example for this type of calculation. Endor measurements of transferred hyperfine interaction have been made for Tm^{2+} ions in cubic sites and for Nd^{3+} ions in $Tg(F_i^-)$ sites. The pair sites discussed above correspond to the close juxtaposition of two such single-ion sites, and the wavefunctions of the paramagnetic ion may not be changed appreciably by the proximity of the neighbouring ion. If this is so one has good information about the overlap and interaction of the metal-ion and ligand-ion wavefunctions, which is essential for meaningful superexchange calculations.

6.7. Conclusion

This review has shown that there is a great deal of information available about non-cubic sites of magnetic ions. Some sites, particularly the tetragonal $Tg(F_i^-)$ site, are quite well characterized and reasonably well understood. When one examines almost any aspect of such a site in detail one comes across difficulties which are associated with three main problems: (a) distortions of the positions of the ions around the magnetic ion from their sites in the ideal undistorted lattice (b) the effects of overlap and covalent interaction with ligand and possibly more distant ions, and (c) the effect of electrostatic shielding by closed-shell electrons, and the effect of spin polarization of closed-shell electrons by exchange interaction. It is fairly clear that these effects have to be taken into account correctly before one can obtain a complete understanding of the properties of a magnetic ion in a particular site. Unfortunately the various parameters which can be measured depend in a rather complex manner on all of the effects, so that it is very difficult to make any logical deductions from the experimental data which greatly clarify the problem. It begins to look as though the most hopeful approach is from the theoretical side along the lines being pioneered by Newman (see the review in Newman, 1971), where one starts using the best available wavefunctions and the best available estimates of ionic positions. One would then attempt to calculate crystal fields taking full account of the effects listed above. The starting data would have to be refined by an iterative procedure in an attempt to predict correctly all of the parameters which are available to measurement. The great wealth of experimental information which is available, or is in principle obtainable, about these centres, particularly for rare-earth ions in alkaline earth halides, makes them a very good model system for a thorough theoretical and experimental investigation of the interaction of impurity ions with a lattice.

References

ABDULSABIROV, R. YU, ZAITOV, M. M., ZARIPOV, M. M., LIVANOVA, L. D., and STEPANOV, V. G. (1970). *Sov. Phys.—Sol. State* **11,** 2971–5.

AL'TSHULER, N. S., EREMIN, M. V., LUKS, R. K., and STOLOV, A. L. (1970). *Ibid* **11,** 2921–30.

AMSTER, R. L. and WIGGINS, C. S. (1969). *J. electrochem. Soc.*, **116,** 68–73.

ANTIPIN, A. A., KURKIN, I. N., LIVANOVA, L. D., POTVOROVA, L. Z., and SHEKUN, L. YA. (1965). *Sov. Phys.—Sol. State* **7,** 1271–2.

——, LIVANOVA, L. D., and SHEKUN, L. YA. (1968). *Ibid* **10,** 1025–9.

ASHBURNER, I. J., NEWMAN, R. C., and MCLAUGHLAN, S. D. (1968). *Phys. Lett.* **27A,** 212–4.

BAKER, J. M. (1971). *Rep. Prog. Phys.* **34,** 109–73.

—— and BLAKE, W. B. J., (1970). *Proc. R. Soc.* **A316,** 63–80.

—— and MARSH, D. (1971). *Phys. Lett.* **35A,** 415–6.

—— and WILLIAMS, F. B. (1961). *Proc. phys. Soc. Lond.* **78,** 1340–52.

——, HAYES, W., and JONES, D. A. (1959). *Ibid* **73,** 942–5.

——, DAVIES, E. R., and HURRELL, J. P. (1968). *Proc. R. Soc.* **A308,** 403–31.

——, BLAKE, W. B. J., and COPLAND, G. M. (1969). *Ibid* **309,** 119–39.

——, DAVIES, E. R., and REDDY, T. RS. (1969). *Phys. Lett.* **29A,** 118–9.

——, ——, —— (1972). *Contemp. Phys.* **13,** 45–59.

BEAUMONT, J. H., HARMER, A. L., and HAYES, W. (1972). *J. Phys. C. Solid State Phys.* **5,** 1475.

BEVOLO, A. J. and SOOK LEE. (1970*a*). *Phys. Rev. Lett.* **24,** 1276–8.

—— —— (1970*b*). *Proc. of the XVIth Colloque Ampere* (Bucharest) p. 933.

—— —— (1972). *Bull. Am. Phys. Soc.* **17,** 309 (DG8).

BLEANEY, B., LLEWELLYN, P. M., and JONES, D. A. (1956). *Proc. phys. Soc. Lond.* **69B,** 858–60.

BOCCARA, A. Thesis Paris, 1971.

BOKII, G. B., GAIGEROVA, L. S., GAIDUK, M. I., DUDNIK, O. F., and MURAV'EV, E. N., (1968). *Sov. Phys.—Doklady* **13,** 123–6.

BOWDEN, C. M., MEYER, H. C., MCDONALD, P. F., and STETTLER, J. D. (1969). *J. Phys. Chem. Solids* **30,** 1535–47.

BROWN, M. R., ROOTS, K. G., WILLIAMS, M. J., SHAND, W. A., GROTER, C., and KAY, H. F. (1969). *J. chem. Phys.* **50,** 891–9.

BYER, N. E., ENSIGN, T. C., and MULARIE, W. M. (1972). *Bull. Am. Phys. Soc.* **17,** 310 (DG15).

CATLOW, C. R. A. (1973). *J. Phys. C. Solid State Phys.*, **6,** 64–70.

DETRIO, J. A., YANNEY, P. P., FERRALLI, W. M., WARE, D. M., and DONLAN, V. L. (1970). *J. Chem. Phys.* **53,** 4372–7.

DVIR, M. and LOW, W. (1960). *Proc. Phys. Soc. Lond* **75,** 136–8.

ENSIGN, T. C., BYER, N. E., and MULARIE, W. M. (1972). *Bull. Am. Phys. Soc.* **17,** 310 (DG14).

ENSIGN, T. C. and BYER, N. E. (1972). *Phys. Rev.* **B6,** 3227–39: 1973, *Phys. Rev.* **B7,** 907–12.

EREMIN, M. V., LUKS, R. K., and STOLOV, A. L. (1971). *Sov. Phys.—Sol. State* **12,** 2820–5.

EVANS, R. C. (1966). *Crystal chemistry* (London: CUP).

EVANS, H. W. and MCLAUGHLAN, S. D. (1966). *Phys. Lett.* **23,** 638–9.

FINN, C. B. P., ORBACH, R., and WOLF, W. P. (1961). *Proc. phys. Soc. Lond.* **77,** 261–8.

FORRESTER, P. A., and MCLAUGHLAN, S. D. (1965). *Phys. Rev.* **138,** A1682–8.

GIL'FANOV, F. Z., DOBKINA, Zh. S., STOLOV, A. L., and LIVANOVA, L. D. (1966). *Sov. Phys.—Opt. Spectr.* **20**, 152–6.
——, LIVANOVA, L. D., ORLOV, M. S., and STOLOV, A. L. (1970). *Sov. Phys.—Sol. State* **11**, 1779–82.
——, ——, STOLOV, A. L., and KHODYREV, YU. P. (1967). *Sov. Phys. Opt. Spectr.* **23**, 231–4.
HARGREAVES, W. A. (1967). *Phys. Rev.* **156**, 331–42.
—— (1972). *Ibid* **B6**, 3417–22.
HUTCHINGS, M. T. (1964). In *Solid state physics*, eds. F. Seitz and D. Turnbull, (Academic Press: New York & London) Vol. 16, p. 227–73.
INOUE, M. (1963). *Phys. Rev. Lett.* **11**, 196–7.
IVOILOVA, E. Kh. and LEUSHIN, A. M. (1972). *Sov. Phys.—Sol. State* **13**, 1914–7.
JONES, G. D., PELED, S., ROSENWAKS, R., and YATSIV, S. (1969). *Phys. Rev.* **183**, 353–68.
KASK, N. E. and KORNIENKO, L. S. (1968). *Sov. Phys.—Sol. State* **9**, 1795–7.
——, ——, and FAKIR, M. (1964). *Ibid* **6**, 430–3.
——, ——, and RYBALTOVSKII, A. O. (1965). *Ibid* **7**, 532–3.
——, ——, —— (1966). *Ibid* **7**, 2614–9.
——, ——, and LARIONTSEV, E. G. (1967). *Ibid* **8**, 2058–62.
KIEL, A. and MIMS, W. B. (1972). *Phys. Rev.* **B6**, 34–9.
KINGSLEY, J. D. and PRENER, J. S. (1962). *Ibid* **126**, 458–65.
KIRO, D. and LOW, W. (1970). In *Magnetic resonance*, ed. C. K. Coogan *et al.* (Plenum Press: New York & London) p. 247–69.
KIRTON, J. and MCLAUGHLAN, S. D. (1967). *Phys. Rev.* **155**, 279–84.
—— and WHITE, A. M. (1969). *Ibid* **178**, 543–7.
KORNIENKO, L. S. and RYBALTOVSKII, A. O. (1972). *Sov. Phys.—Sol. State* **13**, 1609–15.
LEA, K. R., LEASK, M. J. M., and WOLF, W. P. (1962). *J. Phys. Chem. Solids* **23**, 1381–405.
LLEWELLYN, P. M. (1956). D. Phil. Thesis, Oxford University.
LUKS, R. K., SAITKULOV, I. G., and STOLOV, A. L. (1969). *Sov. Phys.—Sol. State* **11**, 210–2.
——, LIVANOVA, L. D., and STOLOV, A. L. (1970). *Ibid* **11**, 1810–4.
MCLAUGHLAN, S. D. (1967). *Phys. Rev.* **160**, 287–9.
——, FORRESTER, P. A., and FRAY, A. F. (1966). *Ibid* **146**, 344–9.
——, NEWMAN, R. C. (1965). *Phys. Lett.* **19**, 552–4.
MANTHEY, W. J. (1972). Ph.D. Thesis, Chicago.
MARSH, D. (1972). *J. Phys. C, Solid State Phys.* **5**, 863–70.
MARSHALL, S. A. and NISTOR, S. U. (1972). *Phys. Rev.* **B6**, 24–34.
MORET, J. M., WEBER, J., and LACROIX, R. (1968). *Helv. Phys. Acta* **41**, 243–50.
MICHOULIER, J. and WASIELA, A. (1971). *C.r. hebd. seanc. Acad. Sci., Paris* **271**, B, 1002–5.
MINER, G. K., GRAHAM, T. P., and JOHNSTON, G. T. (1972). *J. chem. Phys.* **57**, 1263–70.
NARA, H. and SCHLESINGER, M. (1971). *Phys. Rev.* **B3**, 58–64.
—— —— (1972). *J. Phys. C, Solid State Phys.* **5**, 606–14.
NEWMAN, D. J. (1971). *Adv. Phys.*, **20**, 197.
NEWMAN, R. C. (1969). *Adv. Phys.* **18**, 545–663.
O'HARE, M. (1971). *Phys. Rev.* **B3**, 3603–7.
—— (1972). *J. chem. Phys.*, **57**, 3838–43.
OSIKO, V. V. and SHCHERBAKOV, I. A. (1971). *Sov. Phys.—Sol. State* **13**, 820–4.

Rabbiner, N. (1967). *J. opt. Soc. Am.* **57,** 1376–80.
—— (1969). *Ibid* **59,** 588–91.
Ranon, U. and Hyde, J. S. (1966). *Phys. Rev.* **141,** 259–74.
—— and Low, W. (1963). *Ibid* **132,** 1609–11.
—— and Yariv, A. (1964). *Phys. Lett.* **9,** 17–19.
Rector, C. W., Pandey, B. C., and Moos, H. W. (1966). *J. chem. Phys.* **45,** 171–9.
Reddy, T. Rs., Davies, E. R., Baker, J. M., Chambers, D. N., Newman, R. C., and Ozbay, B. (1971). *Phys. Lett.* **36A,** 231–2.
Richardson, R. P. and Gruber, J. B. (1972). *J. chem. Phys.* **56,** 256–60.
Rohrig, R. and Schneider, J. (1969). *Phys. Lett.* **30A,** 371–2.
Sabisky, E. S. and Anderson, C. H. (1967). *Phys. Rev.* **159,** 234–8.
Schlesinger, M. and Kwan, C. T. (1971). *Ibid* **B3,** 2852–5.
—— and Nerenberg, M. (1969). *Ibid* **178,** 568–71.
Secemski, E., Kiro, D., Low, W., and Schipper, D. J. (1970). *Phys. Lett.* **31A,** 45–6.
Sierro, J. (1963). *Helv. Phys. Acta* **36,** 505–29.
—— and Lacroix, R. (1960). *C.r. hebd seanc Acad. Sci., Paris* **250,** 2686–7.
Smirnov, A. I. (1970). *Sov. Phys.—Sol. State* **12,** 590–3.
Smith, R. P. and Cole, T. (1970). *J. chem. Phys.* **52,** 1286–92.
Sochava, L. S., Tolparov, Yu N., and Kovalev, N. N. (1971). *Sov. Phys.—Sol. State* **13,** 1219–22.
Stacy, J. J., Edelstein, N., McLaughlin, R. D., and Conway, J. C. (1973). *J. chem. Phys.* **58,** 807–10.
Title, R. S., Sorokin, P. P., Stevenson, M. J., Pettit, G. D., Scardefield, J. E., and Lankard, J. R. (1962). *Phys. Rev.* **128,** 62–6.
Toledano, J. C. (1972). *J. chem. Phys.* **57,** 1046–50; 4468–72.
van Gorkom, G. G. P. (1970). *J. Phys. Chem. Solids* **31,** 905–12.
Voron'ko Yu K., Keminskii, A. A., and Osiko, V. V. (1966*a*). *Sov. Phys. JETP,* **23,** 10–15.
——, ——, —— (1966*b*). *Ibid* **22,** 295–300.
——, Mikaelyan, R. G., and Osiko, V. V. (1968). *Ibid* **26,** 318–20.
——, Osiko, V. V., and Shcherbakov, I. A. (1969). *Ibid* **29,** 86–90.
Weber, M. J. and Bierig, R. W. (1964). *Phys. Rev.* **134,** A1492–503.
Wetsel, G. C. and Donoho, P. L. (1965). *Ibid* **139,** A334–7.
Wybourne, B. G. (1966). *Ibid* **148,** 317–27.
Yang, C. C., Bevolo, A. J., and Sook Lee (1972). *Bull. Am. Phys. Soc.* **17,** 309 (DG7).
Yariv, A. (1962). *Phys. Rev.* **128,** 1588–92.
Zaripov, M. M., Kropotov, V. S., Livanova, L. D., and Stepanov, V. G. (1967). *Sov. Phys.—Sol. State* **9,** 155–8.
——, ——, ——, —— (1968*a*). *Ibid* **10,** 262–3.
——, ——, ——, —— (1968*b*). *Ibid* **9,** 2346.
Ždánský, K. and Edgar, A. (1971). *Phys. Rev.* **B3,** 2133–41.
Zonn, Z. N., Katyshev, A. N., Mitrofanov, Yu V., and Pol'skii, Yu E. (1969). *Sov. Phys.—Sol. State* **11,** 284–7.

7

OXIDATION-REDUCTION REACTIONS AND PHOTOCHROMISM

7.1. Introduction

We emphasized in Chapter 4 that coloration of CaF_2 at room temperature by X-rays is strongly affected by some impurities. Yttrium occurs commonly as an impurity in CaF_2 and SrF_2 and its effect on coloration properties caused confusion in the interpretation of early work on colour-centre phenomena (§ 4.1). CaF_2 containing a low concentration of impurities is strongly resistant to coloration by ionizing radiation at room temperature (§ 4.9). However, undoped CaF_2 obtained from most commercial sources displays a pronounced four-band optical spectrum after exposure to X-rays at room temperature (Smakula, 1950; Barile, 1952; Messner and Smakula, 1960; Scouler and Smakula, 1960). Scouler and Smakula showed that the intensity of the four-band spectrum could be enhanced by doping with yttrium. It was suggested by O'Connor and Chen (1963*a*,*b*) and Görlich, Karras, Ludke, Mothes, and Reimann (1963) that the bands were excitations of Y^{2+} created by reduction of Y^{3+} by irradiation and an attempt was made to interpret the spectrum on this basis by Theissing, Ewanisky, Kaplan, and Gross (1969). However, polarized luminescence measurements by Görlich, Karras, Kötitz, and Rauch (1968) showed that the four-band spectrum was due to a centre with trigonal symmetry. These authors described the corresponding four-band spectrum in SrF_2 and suggested that the bands were excitations of a complex consisting of Y^{3+} associated with a nearest *F* centre.

It was noticed by Staebler and Kiss (1967, 1969) that the optical absorption spectra of CaF_2 containing La, Ce, Gd, Tb, or Lu ions after additive coloration were very similar to the four-band spectrum of yttrium-doped crystals and were not strongly dependent on the particular impurity (Fig. 7.1). The spectrum produced by additive coloration of CaF_2:Y (Fig. 7.1(a)) is the same as the four-band spectrum produced by irradiation at room temperature. The centres giving rise to the spectra shown in Fig. 7.1 also have trigonal symmetry and we shall show in § 7.3 that a model consisting of a trivalent impurity ion associated with a nearest fluorine vacancy, with two electrons added (giving electrical neutrality), accounts for the spectra.

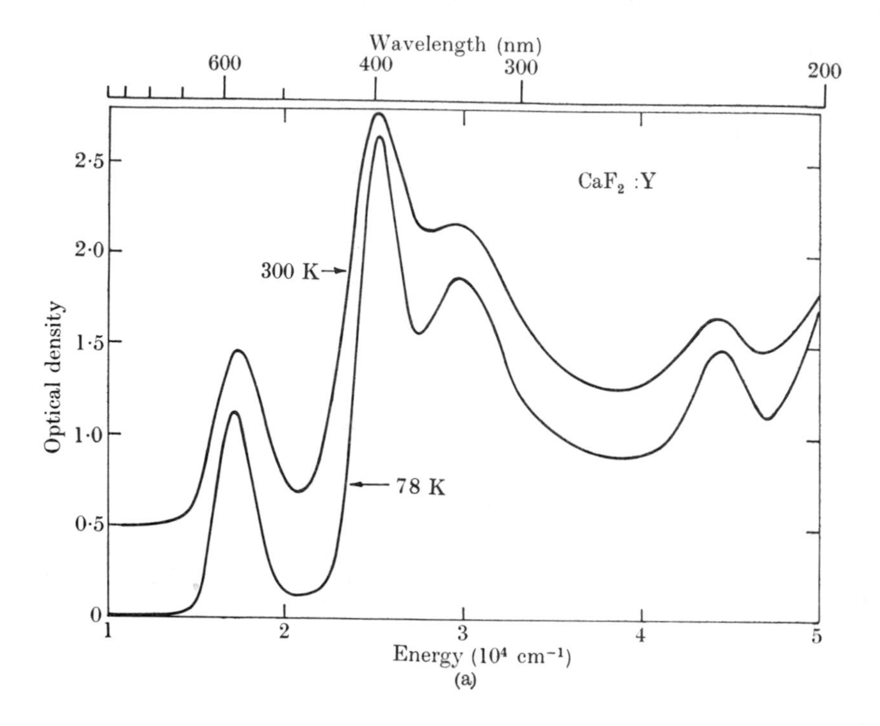
Wavelength (nm)
600
400
300
200
CaF$_2$:Y
300 K
78 K
Optical density
2·5
2·0
1·5
1·0
0·5
0
1
2
3
4
5
Energy (10^4 cm^{-1})

(a)

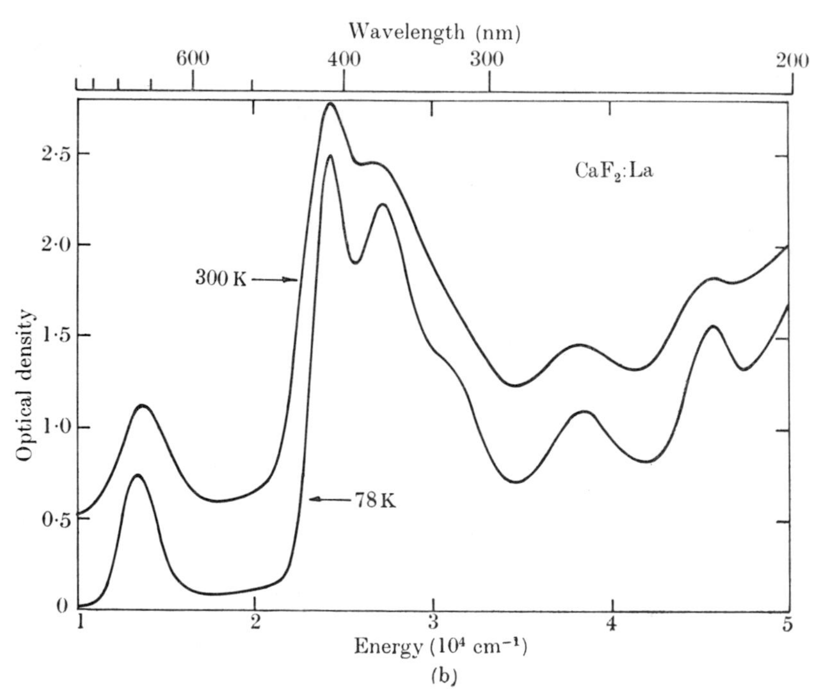
Wavelength (nm)
600
400
300
200
CaF$_2$:La
300 K
78 K
Optical density
2·5
2·0
1·5
1·0
0·5
0
1
2
3
4
5
Energy (10^4 cm^{-1})

(b)

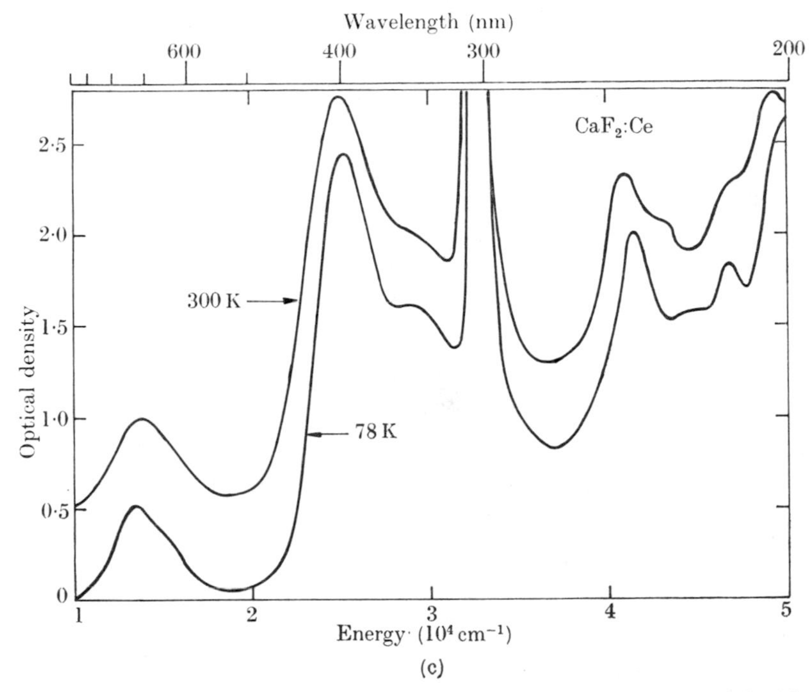

FIG. 7.1. Optical absorption spectrum of additively coloured CaF_2 doped with (a) Y, (b) La, and (c) Ce. The intense absorption band for Ce is due to the trivalent rare earth and is present before additive coloration. The other bands are due to a trigonal complex, referred to as the PC centre in the text. The absorption intensity corresponds to a PC-centre concentration of roughly 10^{17}/c.c. (after Staebler and Schnatterly, 1971).

If CaF_2 containing the above trigonal centres associated with La, Ce, Gd, or Tb is irradiated at room temperature with ultraviolet light ($\lambda < 400$ nm) the absorption of the trigonal centres decreases and a new absorption grows in the visible region (Fig. 7.2). This colour change is thermally unstable and the crystals return to the original state within days to weeks at room temperature. However, the return process can occur within seconds or less when the crystal is exposed to visible light. This switching process is reversible and the materials are known as photochromics. Materials of this type have important practical applications for information storage and display (Kiss, 1970; Faughnan, Staebler, and Kiss, 1971).

The photochromic behaviour of these crystals arises from ionization of the neutral trigonal centres (henceforth referred to as PC centres), forming PC^+ centres, and trapping of the ionized electron at another lattice-defect site. We shall show in § 7.3 that in some cases the trapping

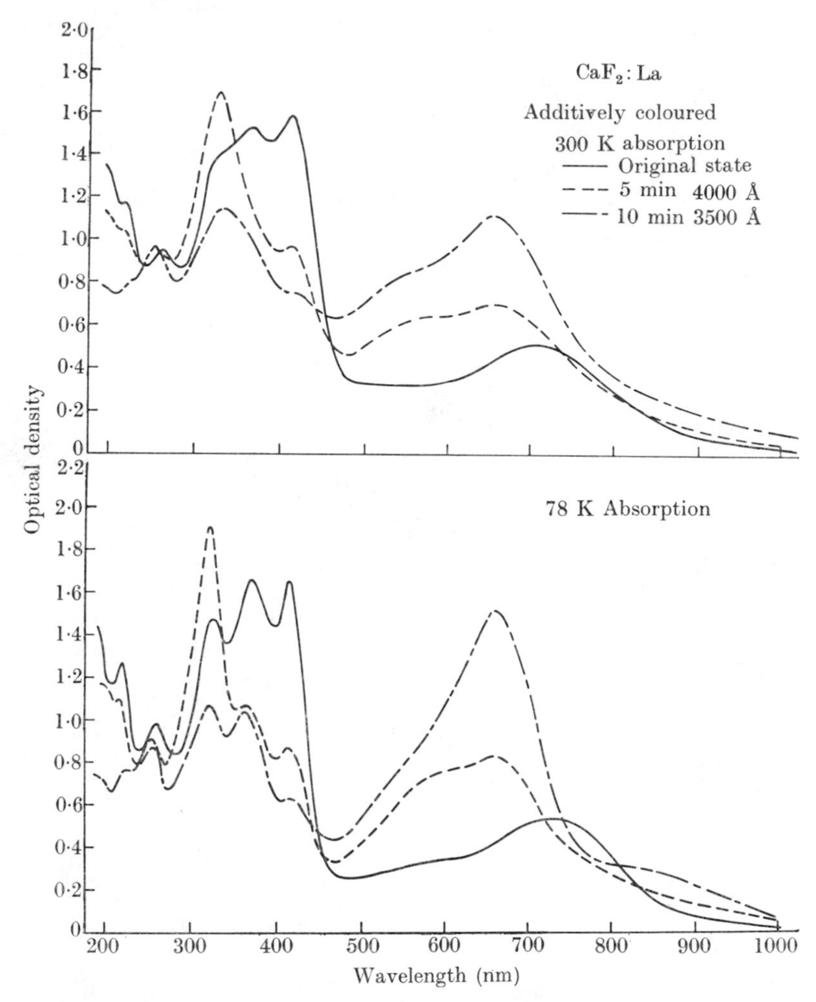

FIG. 7.2. Photochromic switching of CaF_2:La induced by two sequential exposures first at 400 nm and then at 350 nm (after Staebler, 1970).

site has been identified as RE^{3+} which is converted to RE^{2+}. The RE^{2+} ion gives rise to the absorption in the visible region in the switched state (Fig. 7.2). The return to the normal state proceeds by ionization of RE^{2+} to RE^{3+} and the conversion of PC^{+} to PC. The ionization processes do not appear to be thermally assisted; photochromic switching occurs at temperatures as low as $\sim$2·4 K with no observable reduction in efficiency. In the case of Y and Lu doped CaF_2 the electron trap is not stable at room temperature; these materials exhibit photochromic behaviour at 77 K but not at room temperature (§ 7.2).

Another photochromic model, localized charge transfer within the PC centre (Bernhardt, Görlich, and Kötitz, 1971), has been used to account for the dynamics of colour changes in CaF_2:La (Bernhardt, 1971). This model, however, is inconsistent with the detailed optical and epr results described in § 7.3.

Photochromic effects were observed in coloured fluorite nearly a half-century ago (Przibram, 1956). Such effects were first seen in synthetic CaF_2 by Welber (1965) using crystals doped with two species of rare-earth ions. In this case exposure to light at 4 K transfers electrons reversibly from one species to the other and the oxidation-reduction of the rare-earth ions causes a colour change. Arkhangel'skaya and Kiseleva (1970) demonstrated this process in double-doped SrF_2 at room temperature. It was shown by Taylor (1968) that CaF_2 doped with CeO_2 changed colour on exposure at room temperature to short-wavelength ultraviolet radiation. The irradiation produced two bands peaking at 378 and 577 nm. The bands could be destroyed by bleaching with light of longer wavelength ($\sim$380 nm) and the cycle could then be repeated.

Knowledge of the oxidation and reduction of rare-earth ions in CaF_2 by ionizing radiation is important for the understanding of photochromic CaF_2 and we shall deal with these effects in § 7.2 (see also Fong, 1967). We shall discuss the optical and epr spectra of photochromic CaF_2 in § 7.3.

7.2. Oxidation and reduction of rare-earth ions

The tripositive state is characteristic of rare-earth ions in compounds but other valence states occur, generally with $4f^0$, $4f^7$, or $4f^{14}$ configurations. These are achieved in the most stable non-tripositive species Ce^{4+}, Eu^{2+}, Tb^{4+}, and Yb^{2+}. In addition, Sm^{2+} and Tm^{2+} are close to stable configurations and also occur in compounds (Moeller, 1961). It has been emphasized by Moeller that the chemistry of yttrium is similar to that of some rare earths.

The rare earths and yttrium commonly dissolve in CaF_2 in the trivalent state and charge compensation under reducing conditions may be achieved by interstitial fluorine ions (Chapter 6). It was shown however that some RE^{3+} ions in CaF_2 could be reduced to RE^{2+} by X-irradiation (Hayes and Twidell, 1961), by additive coloration (Kiss and Yocom, 1964) and by electrolysis (Fong, 1964). All RE ions except Gd, Tb, and Lu have been identified in the divalent state after X- or γ-irradiation of CaF_2 at room temperature or lower (Chapter 5). With the additional exception of La and Ce, these ions are reduced to the divalent state by additive coloration, i.e. chemical reduction, and have the ground configuration $4f^n$. By contrast, the other RE ions are converted to PC centres by additive coloration (Phillips and Duncan, 1971) and are expected to have the divalent configuration $4f^n5d^1$ (McClure and Kiss, 1963). For Y, which also forms the PC centre, the divalent configuration is $4d^1$. It appears at first glance that the low *d* orbital may decide the formation of PC centres rather than RE^{2+}. In the case of Pr the $5d$ level is only 0·5 eV above the ground state and additive coloration of CaF_2:Pr gives both Pr^{2+} and PC centres with almost equal probability (Staebler, 1970). However, whether or not the *d* orbital plays a prominent role in the formation of PC centres has not been established. The ionic rearrangements accompanying the formation of PC centres by additive coloration are also not established. It seems worthwhile pointing out, however, that the PC centre is, in effect, a combination of RE^{3+}–F_i^- and an *M* centre.

Reduction by ionizing radiation depends on the symmetry of the rare-earth site. Hayes and Twidell (1961) showed, using epr methods, that irradiation of CaF_2:Tm at room temperature or at 77 K with 50 kV X-rays resulted in conversion of Tm^{3+} ions to Tm^{2+} ions in sites with cubic symmetry. They suggested that only those Tm^{3+} ions remote from charge-compensating defects trapped electrons, because of the repulsive coulomb effect of a negative charge compensator. They also observed the epr spectra of Ho^{2+} and La^{2+} ions produced in cubic sites in CaF_2 by X-irradiation at room temperature (Hayes and Twidell, 1963*a*,*b*). Fong (1964) investigated reduction of Dy^{3+} to Dy^{2+} in CaF_2 by γ-irradiation and demonstrated saturation of Dy^{2+} production. Sabisky (1965*a*,*b*) studied conversion of Ho^{3+} to Ho^{2+} in CaF_2 by γ-irradiation at room temperature and found that the fractional concentration of holmium converted to the divalent state decreases with increasing doping level; this fraction is only about 3% in a crystal containing a concentration of holmium of 0·38% molar. Sabisky (1965*a*,*b*) also showed that γ-irradiation of CaF_2:Ho at 77 K results in production of Ho^{2+} ions in cubic sites only. Reduction of Eu^{3+} to Eu^{2+} in CaF_2 by γ-irradiation at room temperature was investigated by Arkhangel'skaya and Kiseleva (1967); they discussed possible effects of fluorine ions on the equilibrium between Eu^{3+} and Eu^{2+} under irradiation. The possibility of reduction of Eu^{3+} to Eu^{2+} in fluorite mineral by natural radiation had been suggested earlier by Przibram (1956).

Arkhangel'skaya and Feofilov (1966) studied the relative ease of reduction of rare-earth ions to the divalent state in SrF_2 by double doping with rare earths. They noted that Ce, Gd, and Tb were not reduced to the divalent state by γ-irradiation at room temperature. Effects of double doping in CaF_2 were investigated by Kornienko and Rybaltovskii (1972) who found that γ-irradiation of CaF_2:(Tb, Sm) and CaF_2:(Tb:Yb) at room temperature resulted in oxidation of Tb^{3+} to Tb^{4+} with a corresponding reduction of Sm^{3+} and Yb^{3+} to the divalent state. The crystals showed two epr spectra of Tb^{4+}, each with trigonal symmetry (see § 6.5.9 for a discussion of these spectra). The fact that the Tb^{4+} symmetry is less than cubic is not surprising since it is possible that a negative charge compensator would attract holes to Tb^{3+}. The Tb^{4+} complexes are stable at room temperature only in the double-doped crystals because Sm^{3+} and Yb^{3+} provide deep electron traps not available in crystals doped with Tb only.

The stimulation of ionic motion by irradiation at room temperature was demonstrated by Kask and Kornienko (1968) who showed that Er^{3+} ions in cubic sites in CaF_2 are partly converted to tetragonal sites by γ-irradiation at room temperature but not at 77 K. Twidell (1970) found, using epr methods, that irradiation of CaF_2:Er at 273 K with a low-pressure mercury lamp with a quartz envelope partly converted tetragonal Er^{3+} into cubic Er^{3+}. It appears that both experiments involve radiation-induced motion of fluorine interstitials. Motion of fluorine interstitials was also invoked to account for effects of γ-irradiation of CaF_2:Nd at room temperature (Batygov and Osiko, 1972). Motion of charge-compensating H_i^- ions induced by ultraviolet light and X-rays at 77 K has been observed in CaF_2 containing RE^{3+}–H_i^- pairs (§ 2.5.5) by Ashburner and Newman (1972).

It was found by Staebler (1970) using optical methods that γ-irradiation of CaF_2:Ce at 78 K resulted in conversion of Ce^{3+} ions to Ce^{2+} ions in sites with cubic symmetry (see Chapter 5) but no PC centres were formed. Irradiation at room temperature produced Ce^{2+} in cubic sites and PC centres but not PC^+ centres. The Ce^{2+} centres are produced initially but quickly saturate and fall in intensity as the PC centres begin to grow (Fig. 7.3). The mechanisms by which irradiation converts RE^{3+} ions into PC complexes is not understood but it appears that the formation of RE^{2+} may be a first stage and that thermally-activated ionic motion is involved. If PC^+ centres are an intermediate stage in the formation of PC centres it seems that the capture cross sections of PC^+ centres for electrons must be very large.

PC centres are formed by room-temperature γ-irradiation for all of

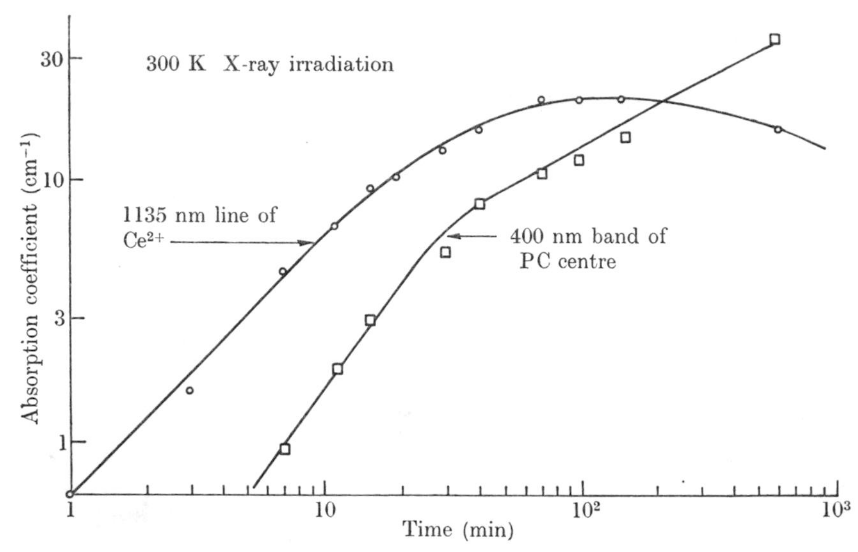

FIG. 7.3. Time dependence of X-ray coloration of CaF_2:La at room temperature (after Staebler, 1970).

the other photochromic ions, presumably by the same mechanism. Wagner and Mascarenhas (1971) have correlated the reduction in numbers of tetragonal Ce^{3+}–F_i^- complexes in CaF_2 produced by X-irradiation at room temperature with increase in PC centres and they find that approximately one PC centre is formed for each four Ce^{3+}–F_i^- complexes destroyed.

Both CaF_2:Y and CaF_2:Lu show photochromic behaviour at 78 K after additive coloration or γ-irradiation at room temperature. However, the ultraviolet-induced colour change bleaches on warming to room temperature. The absence of photochromic behaviour at room temperature in these materials is due to thermal instability of the electron trap. The absorption caused by ultraviolet excitation of PC centres in CaF_2:Y at 78 K is similar to that produced by X-irradiation of CaF_2:Y at 78 K (Fig. 7.4) (Scouler and Smakula, 1960; Görlich, Karras, Kötitz, and Ullman, 1967); the broad band at 550 nm and the long tail into the infrared appear to belong to the trap. Analogy with the photochromic rare earths suggests that the electron may be trapped as Y^{2+}. However, there is as yet no convincing epr data to support this suggestion (see § 7.3) and the structure of this trap is not established.

Reduction by ionizing radiation only partially converts RE^{3+} ions to RE^{2+} ions, and these are readily bleached thermally or optically because of a balancing concentration of hole centres in the crystal. On the other

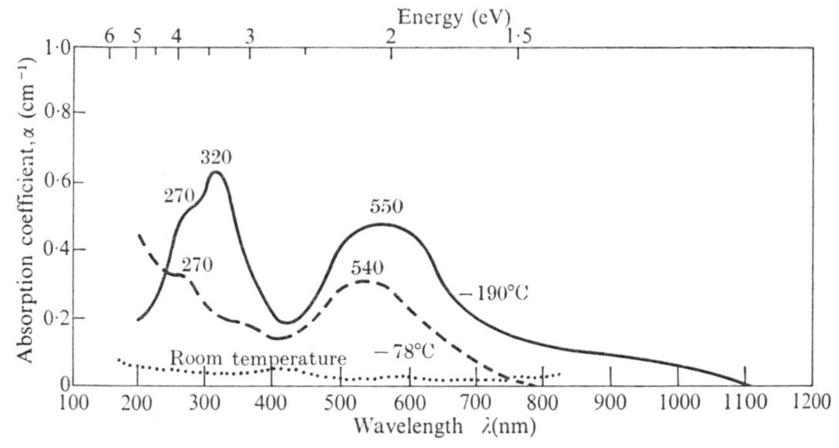

FIG. 7.4. Coloration of CaF_2:Y by 2·5 MeV electrons at 83 K and thermal bleaching produced by warming to 195 K and room temperature (after Scouler and Smakula, 1960). The absorption in the visible region is the spectrum of the trapped electron in photochromic CaF_2:Y.

hand, additive coloration can result in almost complete reduction and the conversion is stable. The photochromic materials behave similarly. Repeated switching of PC centres produced by γ-irradiation gradually bleaches all coloration in the crystal. Additively coloured crystals do not show this fatigue, and can be cycled many times.

Kiss and Staebler (1965) found that Dy^{2+} and Tm^{2+} ions produced in CaF_2 by γ-irradiation could be reoxidized to the trivalent state by thermally-induced liberation of holes or by optical excitation of RE^{2+} resulting in ionization of an electron into the conduction band. Capture of a hole by RE^{2+} or ionization of RE^{2+} produces RE^{3+} in an excited state and Kiss and Staebler (1965) found that the subsequent luminescence was characteristic of RE^{3+} ions in sites with cubic symmetry.

An extensive study of thermally-induced reoxidation of rare-earth ions in CaF_2 after X-irradiation at 77 K was made by Merz and Pershan (1967*a*). Warming the crystals gave rise to glow peaks which did not vary appreciably in temperature with rare-earth doping. With the particular crystals used by Merz and Pershan the glow peaks occurred at about 101, 113, 125, 150, 190, 244, 330, and 383 K. Spectral decomposition showed that, in some cases, the light emitted in the glow peaks below room temperature was due largely to RE^{3+} ions in cubic sites (Merz and Pershan, 1967*b*). The emission from cubic RE^{3+} was most clearly seen with Gd and was also evident for Pr and Dy. The temperatures at which the glow peaks occur are determined by hole traps of different stabilities. The peaks at 101, 113, and 125 K cover a range in which V_K centres move in the lattice (§ 4.6.3). H centres decay at about 170 K in CaF_2 and may have a range of stabilities in rare-earth doped CaF_2 as high at 190 K (§ 4.6.2).

The structures of higher-temperature hole centres giving rise to glow peaks are not known (§ 4.6.3). Merz and Pershan (1967*b*) found that the glow peaks above room temperature were due to RE^{3+} centres with symmetry less than cubic. They suggested that these centres were formed as a result of breakup of hole traps consisting of fluorine aggregates, with diffusion of interstitial fluorine atoms to a divalent rare earth, converting RE^{2+} into RE^{3+}–F_i^-. Studies similar to those made by Merz and Pershan were also carried out on CaF_2, SrF_2, and BaF_2 by Arkhangel'skaya (1964, 1965); on CaF_2 by Batsanov, Korobeinikova, Kazakov, and Kobets (1969), and on SrF_2 and BaF_2 by Luks and Stolov (1970).

Experiments of a similar type have also been carried out by Schlesinger and Whippey (1967, 1968, 1969, 1971) on CaF_2 doped with Gd, Ce, Ho, and Dy. These authors conclude that the low-temperature glow peaks occurring after irradiation at 77 K can be due to RE^{3+} ions in non-cubic sites. Their results, particularly on oxidized crystals, demonstrate a strong dependence on sample history. The reduction probability for ions in various sites depends on their relative concentrations and capture cross-sections. In crystals with a relatively low concentration of cubic sites, reduction of non-cubic sites appears to be possible.

In concluding this section we should like to point out that many investigations of thermoluminescence of undoped CaF_2 have been carried out (see, e.g. Przibram, 1956; Ratnam and Bose, 1966; Fleming, 1968, and Sunta, 1970). Such investigations are of interest in connection with geological dating (see, e.g. Kaufhold and Herr, 1968) and radiation dosimetry (see e.g. Brooke and Schayer, 1966; McCullough, Fullerton, and Cameron, 1972). It is possible that some of the glow peaks observed in these investigations are associated with rare-earth impurities.

7.3. Optical and paramagnetic resonance spectra

It appears from the experimental investigations described here and from theoretical work (Alig, 1971) that the two electrons associated with the RE^{3+} vacancy complex in the PC centre occupy perturbed orbitals of the F centre type in the ground state. The complex may be described in an approximate sense as an RE^{3+} ion with a $4f^n$ configuration associated with an F^- centre (an anion vacancy containing two electrons). The PC^+ centre may be regarded as an RE^{3+} ion associated with an F centre. Alternatively, the PC and PC^+ centres may be considered to be an RE^{2+} ion associated, respectively, with an F centre and an anion vacancy. Both the PC and PC^+ centres have trigonal (C_{3v}) symmetry.

A schematic representation of the energy levels of the PC^+ centre given by Alig is shown in Fig. 7.5. In drawing Fig. 7.5 it is assumed

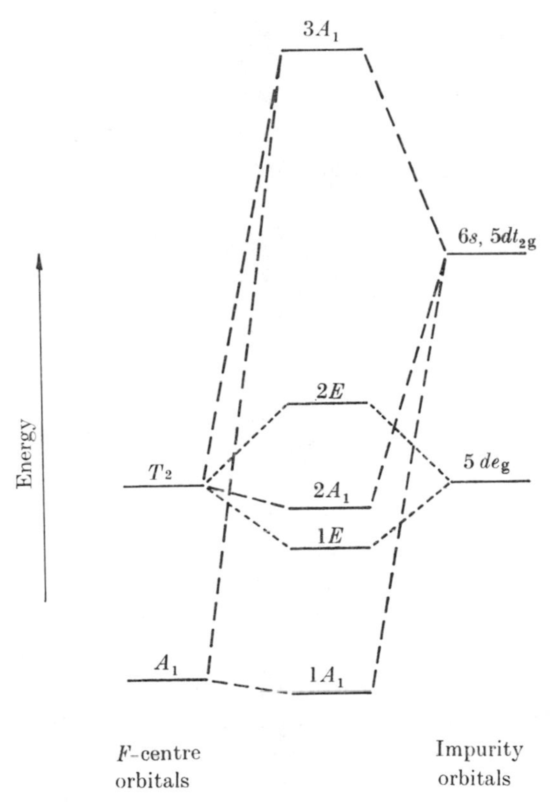

FIG. 7.5. Schematic diagram of the defect-impurity orbital energy levels of the ionized photochromic centre (C_{3v} symmetry). Energy levels of the isolated F centre (T_d symmetry) are shown on the left and crystal-field energy levels of the isolated rare-earth impurity ion (O_h symmetry) are shown on the right (after Alig, 1971).

that the $5d$ and $6s$ orbitals of the rare-earth ion overlap the F centre wavefunction but that the orbitals of the $4f^n$ configuration are so tightly bound to the rare-earth core that they do not contribute significantly to the bonding. In the PC centre the spin of the added electron pairs with the spin of the $1A_1$ electron of PC^+.

7.3.1. *Optical absorption spectra*

The most striking similarity about the various spectra of Fig. 7.1 is the intense band at about 400 nm. Irradiation in this band or at shorter wavelengths produces the photochromic effect at room temperature for La, Ce, Gd, and Tb and this is illustrated for CaF_2:La in Fig. 7.2. In the switched state a prominent absorption occurs at about 330 nm and this is an excitation of the PC^+ centre. Absorption by the trapped electron occurs in the visible region (500–700 nm) and it is increased relative to that of the PC^+ centre by further bleaching at wavelengths shorter than 400 nm (Fig. 7.2). This second step appears to be due to ionic processes (Staebler 1974).

The trapped electron has been positively identified as the RE^{2+} ion for La by epr (see § 7.3.2) and for Ce by optical absorption (see § 5.3.2) In all cases, the visible absorption of the trapped electron can also be obtained by exposing an uncoloured crystal to ionizing radiation at 78 K or lower, a treatment that usually produces the RE^{2+} spectrum in RE-doped CaF_2. A lack of electron traps in adequate numbers appears to limit the photochromic switching; 100% conversion of PC to PC^+ has not been achieved. In fact, heavily reduced crystals are not photochromic (Phillips and Duncan, 1971) because nearly all of the RE ions are converted to PC centres and very few RE^{3+} ions remain to act as electron traps.

To establish the origins of the bands in photochromic CaF_2 Staebler and Schnatterly (1971) investigated the linear optical dichroism produced in the bands by bleaching with linearly polarized light. These experiments are of a similar kind to those described for M and R centres in Chapter 4. In centres with trigonal symmetry absorption occurs for π-polarized light ($A \leftrightarrow A$ transitions) or for σ-polarized light ($A \leftrightarrow E$) transitions. Absorption of polarized light with the electric vector ϵ parallel to [111] by the PC and PC^+ centres gives rise to preferential reorientation of the centres; however, absorption of light with $\epsilon \parallel [100]$ does not produce reorientation and this confirms the trigonal structure of the centres. For the La, Ce, Gd, and Tb systems the PC centre is difficult to reorient directly and in these cases the centres were reoriented in the ionized state by exposure to linearly polarized light absorbed by the ~ 330 nm band of the PC^+ centre. Fig. 7.6*a* shows the dichroism produced in CaF_2:La treated in this way; in addition to the anisotropic absorption of the PC^+ centre there is a weak anisotropic contribution from the PC centre (see Fig. 7.6*b*) and an isotropic contribution from La^{2+}. These results show that the absorption

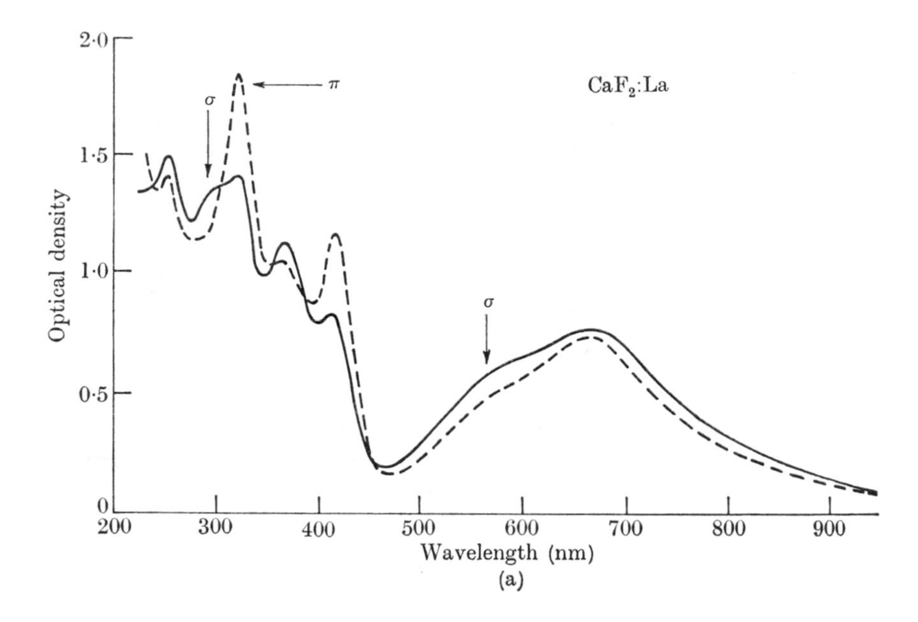

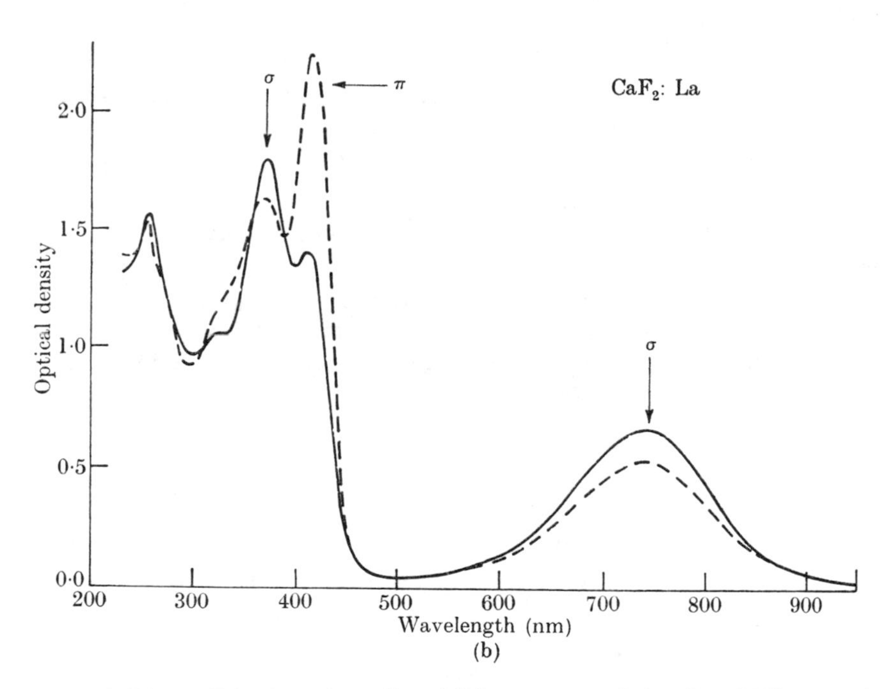

FIG. 7.6. Linear dichroism of (a) aligned PC^+ centres, and (b) aligned PC centres in CaF_2:La at 78 K. Arrows indicate the symmetry of the three most prominent bands in each case (after Staebler and Schnatterly, 1971).

band of the PC^+ centres are σ-, π-, and σ-polarized in order of increasing energy and this is consistent with Fig. 7.5.

The anisotropic absorption of aligned PC centres can now be studied by reversing the photochromic process. This is done by irradiating the crystals with visible light at low temperatures to ensure that no further reorientation takes place. The aligned PC^+ centres trap the photo-released electrons, resulting in aligned PC centres. The dichroism of PC centres in CaF_2:La is shown in Fig. 7.6*b*. Again the absorption bands are σ-, π-, and σ-polarized in order of increasing energy. The spectrum of the PC centre is similar to that of the PC^+ centre, but shifted somewhat to lower energy. Apparently, the added electron in the ground state does not greatly perturb the excited states of the complex.

For Y and Lu doped CaF_2 the PC centres reorient readily under excitation in the lowest energy band (at $\sim$600 nm; Fig. 7.1) with polarized light with $\epsilon \parallel [111]$. The optically-induced reorientation of PC and PC^+ centres is thermally activated and does not occur below about 200 K. The aligned PC centres may be thermally reoriented and the activation energy for this process in CaF_2:La is $1{\cdot}23 \pm 0{\cdot}05$ eV; the activation energy decreases with decreasing radius of the RE^{3+} ion involved in the PC centre (Staebler and Schnatterly, 1971). This result suggests that as the rare-earth ion becomes smaller the anion vacancy moves more easily about it. Similar results are obtained for F_A centres (an F centre associated with a monovalent impurity cation) in alkali halides (Lüty, 1968).

For those materials showing photochromic behaviour at room temperature extended exposure to ultraviolet light at room temperature results in conversion of PC^+ centres to RE^{2+} ions and mobile anion vacancies (Staebler, 1974). This is the second-stage coloration of Fig. 7.2 and is about two orders of magnitude less efficient than conversion of PC to PC^+.

Loh (1971) has extended absorption measurements of CaF_2 containing PC centres into the vacuum ultraviolet. All the photochromic impurities show absorption bands in the region 40 000–55 000 cm^{-1}. In addition, the La, Ce, and Lu doped crystals showed a very intense band at about 62 000 cm^{-1} which is presumably of the charge-transfer type. Loh has discussed possible origins of the bands.

7.3.2. *Paramagnetic resonance spectra*

The epr spectra of photochromic CaF_2 were investigated by Anderson and Sabisky (1971) and, unless otherwise stated, the results given in this section were obtained by them. Trivalent lanthanum has the ground configuration $4f^0$ and its ground state is not paramagnetic. The PC centre in CaF_2:La contains two additional paired electrons and is also not paramagnetic. However, conversion of the PC centre to PC^+ produces epr spectra of La^{2+} and the PC^+ centre. The epr spectrum of $La^{2+}(4f^05d^1)$ shows a characteristic Jahn-Teller effect (see § 5.3.4). Above 13 K the epr spectrum is isotropic but below this temperature a Jahn-Teller distortion gives a tetragonal spectrum which may be fitted to the spin Hamiltonian

$$\mathscr{H} = g_{\parallel}\beta H_z S_z + g_{\perp}\beta(H_x S_x + H_y S_y) + A_{\parallel} I_z S_z + A_{\perp}(I_x S_x + I_y S_y) \quad (7.1)$$

with $S = \frac{1}{2}$ and $I = \frac{7}{2}$ (Table 7.1). The epr spectrum of the PC^+ centre

TABLE 7.1

Parameters of the epr spectra of centres in photochromic CaF_2

Centre	Symmetry	$g_{\parallel}$	$g_{\perp}$	$A_{\parallel}$ (10^{-4} cm^{-1})	$A_{\perp}$ (10^{-4} cm^{-1})
La^{2+}†	tetragonal	$2{\cdot}00 \pm 0{\cdot}01$	$1{\cdot}904 \pm 0{\cdot}002$	37 ± 9	$62{\cdot}6 \pm 1$
PC^+(La)‡	trigonal	$1{\cdot}99 \pm 0{\cdot}01$	$1{\cdot}96 \pm 0{\cdot}01$	237 ± 20	197 ± 20
PC^+(Lu)‡	trigonal	$2{\cdot}0 \pm 0{\cdot}05$	—	300 ± 40	—
PC^+(Tb)‡	trigonal	$<0{\cdot}6$	$9{\cdot}076 \pm 0{\cdot}008$	—	1172 ± 7
PC(Ce)‡	trigonal	$1{\cdot}803 \pm 0{\cdot}004$	$1{\cdot}176 \pm 0{\cdot}002$	—	—

† Hayes and Twidell (1963*b*). ‡ Anderson and Sabisky (1971).

shows an eight-line hfs characteristic of the lanthanum nucleus; the spectrum has trigonal symmetry and may also be fitted to eqn (7.1). The large hfs constants and the long spin-lattice relaxation time of the PC^+ centre ($T_1 \sim 100$ sec at 1·3 K) suggest that the resonance is associated with an *s*-type electron rather than with the *d* electron of La^{2+}. The *g*-value of the PC^+ centre and its anisotropy was qualitatively explained by Alig; he found that the major contribution to the anisotropy comes from orthogonalizing the *F* centre wave function to the core states of the trivalent rare-earth ion.

The ion Lu^{3+} has the ground configuration $4f^{14}$ and is not paramagnetic. As in the case of CaF_2:La the PC centre in CaF_2:Lu is non-paramagnetic but the PC^+ centre shows an epr spectrum which fits eqn (7.1) (Table 7.1). ^{175}Lu is the only stable isotope and its nuclear spin of $\frac{7}{2}$ gives an eight-line hfs. The epr spectrum of the PC^+ centre in both CaF_2:La and CaF_2:Lu has a partly resolved fluorine hfs. No resonance due to Lu^{2+} has been found.

The Tb^{3+} ion has the ground configuration $4f^8$ and the ground configuration of the PC^+ centre may be written $4f^8$–e. Because of the coupling between the F centre electron and the f electrons the appearance of the epr spectrum of the PC^+ centre is quite different from that found with La and Lu. Tb has a nuclear spin of $\frac{3}{2}$ and the parameters of eqn (7.1) for the trigonal spectrum are given in Table 7.1. The epr lines do not show resolved fluorine hfs and the linewidths are similar to those of the Tb^{3+} resonance in CaF_2 (§ 6.5.5). The epr spectrum may be understood if we assume that the ${}^2S_{\frac{1}{2}}$ ground state of the F centre is coupled to the lowest spin-orbit component, ${}^7F_6$, of the $4f^8$ configuration of Tb^{3+} to form states with J-values of $\frac{13}{2}$ and $\frac{11}{2}$; the g and A-values are consistent with a $J = \frac{13}{2}$ ground state (Anderson and Sabisky, 1971). No resonances have been reported for the PC centre or for Tb^{2+}.

The ground configuration of Ce^{3+} is $4f^1$ and the PC centre has the ground configuration $4f^1$–e^2. The e^2 electrons are coupled strongly together and the epr spectrum (Table 7.1) of the PC centre resembles that of $Ce^{3+}(4f^1)$ in a trigonal environment (§ 6.5.1). The lowest spin-orbit manifold, ${}^2F_{\frac{5}{2}}$, of Ce^{3+} is split by the cubic field of CaF_2 into a doublet (Γ_7) and a quartet (Γ_8) with the latter lower. Taking $\langle 111 \rangle$ as the z axis, the quartet may be characterized by the wave functions $|\pm\frac{3}{2}\rangle$ and $\pm\sqrt{\frac{4}{9}}\,|\pm\frac{1}{2}\rangle \pm \sqrt{\frac{5}{9}}\,|\mp\frac{5}{2}\rangle$. A trigonal distortion splits the quartet into two doublets and the g-values of the PC centre are close to those expected for a small perturbation of the $(\pm\frac{1}{2}, \pm\frac{5}{2})$ doublet. Anderson and Sabisky (1971) did not observe a resonance for the PC^+ centre of cerium but point out that the ground state may be a singlet. Attempts to observe epr of Ce^{2+} were not successful and reasons for this are suggested.

Complex epr spectra were observed in photochromic CaF_2:Gd by Anderson and Sabisky (1971) but detailed studies have not yet been made. An epr spectrum of Gd with trigonal symmetry has been found by Bevolo and Lee (1970) after X-irradiation of CaF_2:Gd at room temperature. The crystal-field splitting of the ${}^8S_{\frac{7}{2}}$ state of Gd^{3+} is very large (§ 6.5.9) and it is possible that the spectrum may be due to the PC centre.

An epr spectrum was observed near $g = 2$ in the switched state of photochromic CaF_2:Y by Anderson and Sabisky (1971) but its complexity made useful measurements very difficult. O'Connor and Chen (1964) reported observation of the epr spectrum of Y^{2+} in CaF_2 but this has not been confirmed by other investigators.

Magneto-optical measurements were also carried out on the centres in the photochromic crystals by Staebler and Schnatterly (1971) and the results were consistent with the epr measurements of Anderson and Sabisky (1971).

In conclusion we emphasize that the stability of the colour change in the switched photochromic crystals depends on the depth of the traps available to the ionized electrons. In CaF_2:La the epr measurements show that the ionized electron is trapped by La^{3+} forming La^{2+}. In the case of CaF_2:Ce optical studies showed that the ionized electron converts Ce^{3+} to Ce^{2+}. At the present time no positive identification of the trapped electrons has been made for Tb, Gd, Lu, or Y doped crystals. In addition, the formation mechanism of the PC centre is not yet understood, and little is known of the mechanisms by which the centre is optically ionized. There is clearly a need for further exploration of these interesting systems.

References

ALIG, R. C. (1971). *Phys. Rev.* **B3,** 536.

ASHBURNER, I. J. and NEWMAN, R. C. (1972). *J. Phys. C, Solid State Phys.* **5,** L283.

ANDERSON, C. H. and SABISKY, E. S. (1971). *Phys. Rev.* **B3,** 527.

ARKHANGEL'SKAYA, V. A. (1964). *Opt. Spectrosc.* **16,** 343.

—— (1965). *Ibid* **18,** 46.

—— and FEOFILOV, P. P. (1966). *Ibid* **20,** 90.

—— and KISILEVA, M. N. (1968). *Sov. Phys.—Sol. State,* **9,** 2774 (Translated from *Fiz. Tver. Tela* **9,** 3523, 1967).

——, ——, (1970). *Opt. Spectrosc.* **29,** 149.

BARILE, J. (1952). *J. Chem. Phys.* **20,** 297.

BATSANOV, S. S., KOROBEINIKOVA, V. N., KAZAKOV, V. P., and KOBETS, L. I. (1971). *Opt. Spectrosc* **30,** 265.

BATYGOV, S. Kh. and OSIKO, V. V. (1972). *Sov. Phys.—Sol. State,* **13,** 1886.

BEVOLO, A. J. and LEE, S. (1970). *Phys. Rev. Lett.* **24,** 1276.

BERNHARDT, H. J. (1971). *Phys. Stat. Sol.* (a) **8,** 539.

——, GÖRLICH, P., and KÖTITZ, G. (1971). *Ibid* **6,** 479.

BROOKE, C. and SCHAYER, R. (1966). *Symposium on solid state and chemical radiation dosimetry* (Int. Atomic Energy Agency, Vienna), Paper No. SM78/20.

FAUGHNAN, B. W., STAEBLER, D. L., and KISS, Z. J. (1971). *Applied solid state science,* Vol. 2, ed. R. Wolfe, Academic Press, Inc., New York.

FLEMING, R. J. (1968). *Phys. Stat. Sol.* **27,** K57.

FONG, F. F. (1964). *J. chem. Phys.* **41,** 245, 2291.

—— (1967). *Progress in solid state chemistry,* **3,** 135 (Ed. H. Reiss) Pergamon Press.

GÖRLICH, P., KARRAS, H., KÖTITZ, G., and RAUCH, R. (1968). *Phys. Stat. Sol.* **27,** 109.

——, ——, ——, and ULLMAN, P. (1967). *Ibid* **23,** 313.

——, ——, LUDKE, W., MOTHES, H., and REIMANN, R. (1963). *Ibid* **3,** 478.

HAYES, W. and TWIDELL, J. W. (1961). *J. chem. Phys.* **35,** 1521.

—— —— (1963*a*). *Proc. phys. Soc. Lond.* **81,** 371.

—— —— (1963*b*). *Ibid* **82,** 330.

KASK, N. E. and KORNIENKO, L. S. (1968). *Sov. Phys.—Sol. State,* **9,** 1670.

KAUFHOLD, J. and HERR, W. (1968). *Thermoluminescence of geological materials*, ed. D. J. McDougall, Academic Press, Inc., London, New York, Ch. 3.7.
KISS, Z. J. (1970). *Physics Today*, **23,** 42.
—— and STAEBLER, D. L. (1965). *Phys. Rev. Lett.* **14,** 691.
—— and Yocom, P. N. (1964). *J. chem. Phys.* **41,** 1511.
KORNIENKO, L. S. and RYBALTOVSKII, A. O. (1972). *Sov. Phys.—Sol. State*, **13,** 1609 (Translated from *Fiz. Tver. Tela*, **13,** 1919, 1971).
LOH, E. (1971). *Phys. Rev.* **B4,** 2002.
LUKS, R. K. and STOLOV, A. L. (1970). *Opt. Spectrosc.* **29,** 170.
LÜTY, F. (1968). *Physics of color centers*, ed. W. B. Fowler, Academic Press, New York, Ch. 3.
McCULLOUGH, E. C., FULLERTON, G. D., and CAMERON, J. R. (1972). *J. appl. Phys.* **43,** 77.
McCLURE, D. S. and KISS, Z. J. (1963). *J. chem. Phys.* **39,** 3251.
MERZ, J. L. and PERSHAN, P. S. (1967*a*). *Phys. Rev.* **162,** 217.
—— —— (1967*b*). *Ibid* **162,** 235.
MESSNER, D. and SMAKULA, A. (1960). *Ibid* **120,** 1162.
MOELLER, T. (1961). The rare earths (ed. F. H. Spedding and A. H. Doane) John Wiley and Sons, New York and London, p. 9.
O'CONNOR, J. R. and CHEN, J. H. (1963*a*). *Phys. Rev.* **130,** 1790.
—— —— (1963*b*). *J. Chem. Phys. Solids*, **24,** 1382.
—— —— (1964). *Appl. Phys. Lett.* **5,** 100.
PHILLIPS, W. and DUNCAN, R. C. Jr. (1971). *Mettall. Trans.* **2,** 769.
PRZIBRAM, K. (1956). *Irradiation colours and luminescence*, Pergamon Press, London.
RATNAM, V. V. and BOSE, H. N. (1966). *Phys. Stat. Sol.* **15,** 309.
SABISKY, E. (1965*a*). *J. appl. Phys.* **36,** 802, 1788.
—— (1965*b*). Ph.D. Thesis, University of Pennsylvania.
SCHLESINGER, M. and WHIPPEY, P. W. (1967). *Phys. Rev.* **162,** 286.
—— —— (1968). *Ibid* **171,** 361.
—— —— (1969). *Ibid* **177,** 563.
—— —— (1971). *Ibid* **B3,** 2852.
SCOULER, W. J. and SMAKULA, A. (1960). *Ibid* **120,** 1154.
SMAKULA, A. (1950). *Ibid* **77,** 408.
STAEBLER, D. L. (1970). Thesis, Princeton University.
—— (1974). (to be published).
—— and KISS, Z. J. (1967). *Bull. Am. Phys. Soc.* **12,** 670.
—— —— (1969). *Appl. Phys. Lett.* **14,** 93.
—— and SCHNATTERLY, S. E. (1971). *Phys. Rev.* **B3,** 516.
SUNTA, C. M. (1970). *J. Phys. C*, **3,** 1978.
TAYLOR, M. J. (1968). *Phys. Lett.* **27A,** 32.
THEISSING, H. H., EWANISKY, T. F., KAPLAN, R. J., and GROSS, D. W. (1969). *J. chem. Phys.* **50,** 2657.
TWIDELL, J. W. (1970). *J. Phys. Chem. Sol.* **31,** 299.
WELBER, B. (1965). *J. chem. Phys.* **42,** 4262.
WAGNER, J. and MASCARENHAS, S. (1971). *Phys. Rev. Lett.* **27,** 1514.

AUTHOR INDEX

Page numbers in italic signify a full reference

SUBJECT INDEX